D0203770

Dynamic properties of forest ecosystems

THE INTERNATIONAL BIOLOGICAL PROGRAMME

The International Biological Programme was established by the International Council of Scientific Unions in 1964 as a counterpart of the International Geophysical Year. The subject of the IBP was defined as 'The Biological Basis of Productivity and Human Welfare', and the reason for its establishment was recognition that the rapidly increasing human population called for a better understanding of the environment as a basis for the rational management of natural resources. This could be achieved only on the basis of scientific knowledge, which in many fields of biology and in many parts of the world was felt to be inadequate. At the same time it was recognized that human activities were creating rapid and comprehensive changes in the environment. Thus, in terms of human welfare, the reason for the IBP lay in its promotion of basic knowledge relevant to the needs of man.

The IBP provided the first occasion on which biologists throughout the world were challenged to work together for a common cause. It involved an integrated and concerted examination of a wide range of problems. The Programme was coordinated through a series of seven sections representing the major subject areas of research. Four of these sections were concerned with the study of biological productivity on land, in freshwater, and in the seas, together with the processes of photosynthesis and nitrogen fixation. Three sections were concerned with adaptability of human populations, conservation of ecosystems and the use of biological resources.

After a decade of work, the Programme terminated in June 1974 and this series of volumes brings together, in the form of syntheses, the results of national and international activities.

INTERNATIONAL BIOLOGICAL PROGRAMME 23

Dynamic properties of forest ecosystems

Edited by

D. E. REICHLE

Environmental Sciences Division, Oak Ridge National Laboratory
Oak Ridge, Tennessee, US.

CAMBRIDGE UNIVERSITY PRESS

CAMBRIDGE

LONDON   NEW YORK   NEW ROCHELLE

MELBOURNE   SYDNEY

Published by the Press Syndicate of the University of Cambridge
The Pitt Building, Trumpington Street, Cambridge CB2 1RP
32 East 57th Street, New York, NY 10022, USA
296 Beaconsfield Parade, Middle Park, Melbourne 3206, Australia

© Cambridge University Press 1981

First published 1981

Photoset and printed in Malta by Interprint Limited

Library of Congress Cataloging in Publication Data

Main entry under title:

Dynamic properties of forest ecosystems.

(International Biological Programme; 23)

Includes bibliographical references and index.

1. Forest ecology. I. Reichle, David E.
II. Series.
WK938.F6D83   574.5′264   78-72093

ISBN 0 521 22508 6

This volume is dedicated to the scientists of the IBP

Contents

Table des Matières

Содержание

Contenido

Contributors and collaborators

Contributors

P. Benecke, Institut für Bodenkunde und Waldernährung, Der Universität Göttingen, Göttingen, Fed. Rep. of Germany

Robert L. Burgess, Environmental Sciences Division, Oak Ridge National Laboratory, Oak Ridge, TN 37830, USA

Dale W. Cole, Forest Resources Ar-10, University of Washington Seattle, WA 98195, USA

Donald L. DeAngelis, Environmental Sciences Division, Oak Ridge National Laboratory, Oak Ridge, TN 37830, USA

Nelson T. Edwards, Environmental Sciences Division, Oak Ridge National Laboratory, Oak Ridge, TN 37830, USA

Alan R. Ek, School of Forestry, University of Minnesota, St. Paul, MN 55101, USA

André Galoux, Station de Recherches des Eaux et Forêts, Ministère de l'Agriculture, B-1990 Groenendaal-Hoeilaart, Belgium

Robert H. Gardner, Environmental Sciences Division, Oak Ridge National Laboratory, Oak Ridge, TN 37830, USA

G. Gietl, Bayerische Forstliche Versuchs und Forschungsanstalt, D-8000 Munich, Fed. Rep. of Germany

Robert A. Goldstein, Electric Power Research Institute, 3412 Hillview Avenue, Palo Alto, CA 94304, USA

H. Hager, Institut für Forstliche Standortsforschungsanstalt, Universität für Bodenkultur, A-1190 Vienna, Austria

W. Frank Harris, Environmental Sciences Division, Oak Ridge National Laboratory, Oak Ridge, TN 37830, USA

W. Carter Johnson, Department of Biology, Virginia Polytechnic Institute and State University, Blacksburg, VA 24060, USA

C. Kayser, Niedersächsische Landes-verwaltungsamt, Hannover, Fed. Rep. of Germany

P. K. Khanna, Institut für Bodenkunde und Waldernährung Der Universität Göttingen, Göttingen, Fed. Rep. of Germany

O. Kiese, Institut für Geographie und Länderkunde Westfällischer, Wilhelms-Universität, Münster, Fed. Rep. of Germany

K. R. Knoerr, School of Forestry, Duke University, Durham, NC 27706, USA

Orie L. Loucks, The Institute of Ecology Holcomb Institute, Butler University Indianapolis, IW 46208, USA

J. B. Mankin, Computer Sciences Division, Oak Ridge National Laboratory, Oak Ridge, TN 37830, USA

R. Mayer, Institut für Bodenkunde und Waldernährung Der Universität Göttingen, Göttingen, Fed. Rep. of Germany

Samuel B. McLaughlin, Environmental Sciences Division, Oak Ridge National Laboratory, Oak Ridge, TN 37830, USA

Robert A. Monserud, Intermountain Forest and Range Experiment Station, US Forest Service, Moscow, ID 83843, USA

Contributors and collaborators

C. E. Murphy, Environmental Sciences Section, Savannah River Operations Office,

E. I. DuPont de Nemours Co, Aiken, S. Carolina, 29801, USA

Robert V. O'Neill, Environmental Sciences Division, Oak Ridge National Laboratory, Oak Ridge, TN 37830, USA

Maurice Rapp, C.N.R.S., Centre d'Etudes Phytosociologiques et Ecologiques L. Emberger, 34033 Montpellier, France

David E. Reichle, Environmental Sciences Division, Oak Ridge National Laboratory, Oak Ridge, TN 37830, USA

James J. Rogers, Rocky Mtn. Forest & Range Experiment Station, Forestry Sciences Lab., ASU Campus, Tempe, AZ 83281, USA

G. Schnock, Faculty of Sciences, Free University of Brussels, Brussels, B-1160 Belgium

Herman H. Shugart, Environmental Sciences Division, Oak Ridge National Laboratory, Oak Ridge, TN 37830, USA

T. R. Sinclair, U.S.D.A., SWC – Microclimate Investigations, Bradfield Hall, Cornell University, Ithaca, NY 14850, USA

Philip Sollins, Forest Sciences Laboratory, 3200 Jefferson Way, Oregon State University, Corvallis, OR 97331, USA

Wayne T. Swank, Coweeta Hydrologic Laboratory, US Forest Service, Franklin, NC 28734, USA

G. L. Swartzman, College of Forest Resources, University of Washington, Seattle, WA 98195, USA

Professor Bernhard Ulrich, Institut für Bodenkunde und Waldernährung der Universität Göttingen, Göttingen, Fed. Rep. of Germany

Richard H. Waring, School of Forestry, Oregon State University, Corvallis, OR 97331, USA

Collaborators

V. A. Abrashko, Komarov Botanical Institute, Academy of Sciences of the USSR, Leningrad 197022, USSR

V. A. Alexeev, Komarov Botanical Institute, Academy of Sciences of the USSR, Leningrad 197022, USSR

J. P. E. Anderson, Institut für Bodenbiologie, D-33 Braunschweig-Volkenrode, Bundesallee 50, Fed. Rep. of Germany

Folke Andersson, Swedish Coniferous Forest Project, Agricultural College, S – 750 07 Uppsala 7, Sweden

Takashi Ando, Shikoku Branch, Government Forest Experiment Station, 915 TEI Asakura, Kochi 780, Japan

Peter Attiwill, Botany School, University of Melbourne, Parkville, Victoria 3052, Australia

Desh Bandhu, Department of Botany, Swami Shraddhanand College, University of Delhi, Alipur-Delhi 110036, India

G. L. Baskerville, Forestry Branch Dep., Fisheries and Forestry, Fredericton, New Brunswick, Canada

O. N. Bauer, Zoological Institute, University Embarken 1, Leningrad 184, USSR

A. Baumgartner, Lehrstuhl für Bioklimatologie und angewandte Meteorologie, der Universität München, Fed. Rep. of Germany

Bernier, Universitet Laval, Quebec, Canada

C. Bindiu, Institutul de Cercetari, Proiectari si Documentare Silvica, Bucuresti 505, Pipera 46, Sector 2, Rumania

Vladimir Biskupsky, VULH-VS

Bratislava, Drienova 5, CS-829
74 Bratislava, Czechoslovakia

J. A. Bullock, Department of Zoology,
University of Leicester, Leicester
Le1 7RH, England

Elia Colczyk, Zaklad Ochrony
Przyrody Pan U1. Lubicz 46,
31-512 Krakow, Poland

J. B. Cragg, Faculty of Environmental
Design, Univ. of Calgary, Calgary,
Alberta T2N 1N4, Canada

Eilif Dahl, Botanical Institute,
Norwegian Agric. College,
Vollebekk, Norway

S. Denaeyer-De Smet, Laboratoire
de Botanique Systematique et
d'Ecologie, Universite Libre de
Bruxelles B-1050 Bruxelles 5,
Belgium

K. G. Djhalilov, Institute of Botany,
Academy of Science, Azerb. SSR,
Baku, USSR

K. H. Domsch, Institut für
Bodenbiologie, D-33 Braunschweig-
Volkenrode, Bundesallee 50, Fed.
Rep. of Germany

N. Donita, Institutul de Cercetari,
Proiectari si Documentare Silvica,
Bucuresti 505, Pipera 46, Sector 2,
Rumania

Gina Douglas, IBP Publications
Office, The Linnean Society,
Burlington House, Piccadilly,
London W1V OLQ, England

P. Duvigneaud, Laboratoire de
Botanique, Systematique et
d'Ecologie, Universite Libre de
Bruxelles, B-1050 Bruxelles 5,
Belgium

Robert Edmonds, University of
Washington, Seattle, WA 98105 USA

Heinz Ellenberg, Botanical Institute,
D-34 Gottingen, Untere Karspule
2, Fed. Rep. of Germany

A. A. Esmekanova, Institute of
Botany, Academy of Science,
Azerb. SSR, Baku, USSR

Janusz B. Falinski, Stacja
Geobotaniczna U.W., Bialowieza
Woj., Bialystok, Poland

A. G. Floyd, Forestry Commission
of New South Wales, Coff's
Harbour, Australia

Q. Foruqi, Department of Botany,
Banaras Hindu University,
Varanasi-5, India

Jerry F. Franklin, Pacific Northwest
Forest Experiment Station,
Oregon State University, Corvallis,
OR 97331, USA

Thomas Frey, Institute of Zoology
and Botany Estonian Academy of
Sciences, Tartu 202400, Anne
34-23, Estonian USSR

S. P. Gessel, School of Forestry,
University of Washington, Seattle,
WA 98105, USA

V. Giacomini, Institute Botan. Univ.
Citta Universitaria, Roma, Italy

Elia Golczyk, Zaklad Ochrony
Przyrody Pan, UL Lubicz 46,
31-512 Krakow, Poland

Alan G. Gordon, Ontario Ministry
of Natural Resources, Forest
Research Branch, Sault Ste. Marie,
Ontario, Canada

T. K. Goryshina, Leningrad State
University, Leningrad, USSR

Charles C. Grier, Forest Research
Laboratory, Oregon State University,
Corvallis, OR 97331, USA

Paavo Havas, Botanical Institute,
University of Oulu, Torikatu 15,
Oulu 10, Finland

Hans Heller, Lehrstuhl fur
Geobotanik, D-34 Gottingen,
Untere Karspule 2, Fed. Rep. of
Germany

Gray S. Henderson, School of Forestry,
University of Missouri Columbia,
MO 65201, USA

Joan Hett, Forest Resources AR-10,
University of Washington, Seattle,
WA 98105, USA

T. Hosokawa, Dep. Biology, Faculty of Sci., Kyushu University, Fukuoka, Japan

Hakan Hytteborn, Institute of Ecological Botany, University of Uppsala, Box 559, S-751 22 Uppsala 1, Sweden

Krystyn Izdebski, Instytut Biologii UMCS, Zaklad Ekologii i Ochrony Przyrody, Lublin, ul. Akademicka 19, Poland

Paul Jakucs, Botanical Institute, L. Kossuth University, H-4010 Debrecen, Hungary

Jenik, NA Piskach 89, 160 00 Prague 6, Czechoslovakia

J. R. Jorgensen, Forestry Sciences Laboratory, Research Triangle Park, NC 27709, USA

P. Kallio, Dept. Botany, University of Turku, Turku 2, Finland

V. G. Karpov, Geobotany Department, Komarov Botanical Institute, Leningrad, USSR

N. I. Kazimirov, Karelian Branch of The Russian S.S.R. Academy of Sciences, Petrozovodsk, USSR

Heribert Kerner, Institut fur Forstsamenkunde und Pflanzenzuchtung, D-8 Munchen 40, Amalienstrasse 52, Fed. Rep. of Germany

Russell S. Kinerson, Dept. of Botany, University of New Hampshire, Durham, NH 03824, USA

Yuzo Kitazawa, Department of Biology, Tokyo Metropolitan University, Fukazawa Setagaya-Ku, Tokyo-158, Japan

H. Klinge, Max Planck Institut für Limnologie, Abteilung Tropenoikologie, D 232 Ploen, Postfach 165, Fed. Rep. of Germany

Thim Komkris, XX Appl. Sc. Research Corp. Thailand, Bankhen, Bangkok, Thailand

Ferdinand Kubicek, Slovac Academy of Sciences, Botanical Institute, Dubravska 26, CS-809 00 Bratislava, Czechoslovakia

Jan Kvet, Institute of Botany, Czechoslovakian Academy of Science Dukelska 145, 37982 Trevon, Czechoslovakia

G. J. Lawson, Dept. of Biology, University of Wisconsin, Madison, WI 53706, USA

Matti Leikola, Forest Res. Inst., Unioninkatu 40A, SF-00170 Helsinki 17, Finland

G. Lemee, Universite de Paris-Sud, 91506-Orsay, Paris, France

Salvatore Leonardi, Instituto di Botanica, Universita di Catania, Via Antonino Longo 19, I-95125 Catania, Italy

Helmut Lieth, Dept. of Botany, University of North Carolina, Chapel Hill, NC 27514, USA

L. Lindgren, Lunds Universitat, Avd. for Ecologisk Botanik, Helgonavaegen 5, S-223 62 Lund, Sweden

Adam Lomnicki, Nature Conservation Research Centre, ul. Lubiez 46, Pl-31-512 Krakøw, Poland

Paul Lossaint, Centre Nat. de la Recherche Scientifique, Route de Mende – F-34 Montpellier, France

Ricardo Luti, Facultad de Ciencias Ex. Fis, y Naturales, Av. Velez Sarsfield 299, Cordoba, Argentina

Francois Malaisse, Service de Botanique et Climatologie, Université Nationale Du Zaire, B.P. 1825, Lubumbashi, Zaire

E. Medwecka-Kornas, Nature Conservation Research Center, Polish Academy of Sciences, ul. Lubiez 46, PL-31-512 Krakow, Poland

O. G. Merzoev, Institute of Botany, Academy of Sciences, Azerb. SSR, Baku, USSR

R. Misra, Department of Botany, Banaras Hindu University, Varanasi-221005, India

V. Mocanu, Institutul de Cercetari, Proiectari si Documentare Silvica, Bucuresti 505, Pipera 46, Sector 2, Rumania

A. A. Molchanov, Forest Laboratory, Academy of Sciences of the USSR, Uspenskoye, Moscow District, Odintsov Region, USSR

Y. I. Molotovsky, Institute of Botany, Academy of Science, Tadjikistan SSR, Dushanbe, USSR

Carl D. Monk. Dept. of Botany, University of Georgia, Athens, GA 30601, USA

M. Monsi, Botany Department, Tokyo Metropolitan University, Fukasawa Setagaya, Tokyo, Japan

Harold Mooney, Dept. of Biological Sciences, Stanford University, Palo Alto, CA 94302, USA

R. M. Morozova, Karelian Branch of the Russian S.S.R., Academy of Sciences, Petrozovodsk, USSR

Bengt Nihlgard, Lunds Universitat, Avd. for Ecologisk Botanik, Helgonavaegen 5, S-223 62 Lund, Sweden

H. Ogawa, Biology Department, Osaka City University, Sugimoto-Cho, Sumiyoshi-Ku, Osaka, Japan

Hans Persson, Institute of Ecological Botany, University of Uppsala, Box 559, S-751 22, Uppsala 1, Sweden

Henning Petersen, Jordbundsbiologisk Institut, Strandkaer, Femmoller, DK-8400 Ebeltoft, Denmark

E. Poli, Ist. Botanica Univ. Via A. Longo 19, Catania, Italy

I. Popescu-Zeletin, Institutul de Cercetari si Amenajari Silvice, Bucuresti 505, Pipera 46, Sector 2, Rumania

Juan Puigdefabregas, Centro Pirenaico de Biologia Experimental, AP. 64 Jaca (Huesca), Egipciacas, 15 – Barcelona – 1, Spain

C. W. Ralston, School of Forestry, Duke University, Durham, NC 27706, USA

S. S. Ramam, Dept. of Botany, Banaras Hindu University, Varanasi, Uttar Pradesh, India

L. Reintam, Soil Science and Agronomy Chemistry, Estonian Agric. Academy, Tartu, Estonia, USSR

W. A. Rodrigues, Inst. Nac. Amazonico Centro Pesquis. Florest., Manaus Amazonas, Brazil

Sanga Sabhasri, Faculty of Forestry, Katsetsart University, Bangkok, Thailand

I. S. Safarov, Institute of Botany, Academy of Science, Azerb. SSR, Baku, USSR

J. E. Satchell, The Nature Conservancy, Merlewood Research Station, Grange-Over-Sands Lancashire, England

T. Satoo, Department of Forestry, University of Tokyo, Tokyo 113, Japan

D. Satyanarayana, Department of Botany, Banaras Hindu University, Varanasi, Uttar Pradesh, India

V. K. Sharma, Department of Botany, Banaras Hindu University, Varanasi, Uttar Pradesh, India

T. Shidei, Department of Forestry, Kyoto University, Kyoto, Japan

T. Simon, Department of Plant Systematics, Eotros University, Museum Krt. 4(a), Budapest VIII

R. P. Singh, Department of Botany,

Contributors and collaborators

Banaras Hindu University, Varanasi, Uttar Pradesh, India

R. L. Specht, Botany Department, University of Queensland, St. Lucia 4067, Queensland, Australia

C. O. Tamm, Department of Forest Ecology, College of Forestry, S-104 05 Stockholm 50, Sweden

Nicola Tarsia, Centro zi Sperimentadione, Agricola e Forestale, P. O. Box 9079, Roma, Italy

H. M. Thamdrup, Naturhistorisk Museum, University of Aarhus, DK-8000 Aarhus C, Denmark

Tadeus Traczyk, Institute of Ecology, Djiekanow Lesny K. Warszawy, 05-150 Limnianki, Poland

A. I. Utkin, Laboratory of Forest Science, Odintzovo District, Moscow 130043, USSR

K. Van Cleve, Forest Soils Laboratory, University of Alaska, Fairbanks, AK 99701, USA

J. van der Drift, Institute for Biological Field Research, Kemperbergerweg 11, Arnhem, Netherlands

X. Vyskot, Dept. Silviculture, Brno Univ. of Agriculture, Brno, Czechoslovakia

L. J. Webb, CSIRO Rainforest Ecol. Unit Div., Brisbane, Australia

C. G. Wells, Forestry Sciences Laboratory, Research Triangle Park, NC 27709, USA

Robert H. Whittaker, Ecology and Systematics, Cornell University, Ithaca, NY 14850, USA

G. M. Woodwell, The Ecosystems Center, Marine Biol. Laboratory, Woods Hole, MA 02543, USA

Preface

The evolution of IBP has been described in the first volume of this series of 'synthesis' volumes. Whilst the record is complete insofar as it concerns IBP as an entity, it makes few references to discussions and draft plans already in being for international collaboration prior to the emergence of IBP. It is, therefore, worth recalling that proposals for an international study of forest production had been discussed in the late 1950s by Dr H. Ellenberg and Professor J. D. Ovington. It was their hope that a limited number of parameters related to primary production in woodlands could be investigated at various sites in Europe. If this attempt at a standardized approach to woodland production were to prove successful, then it was their intention to extend the program to other parts of the world. Their proposal was welcomed by members of the Commission on Ecology of the International Union for the Conservation of Nature and Natural Resources (IUCN) who realized that the emerging science of production ecology could assist the Commission in its task of advising on methods for the conservation of natural ecosystems. But for the birth of IBP, there can be little doubt that the Ellenberg–Ovington plan would have been implemented by IUCN.

When, in 1960, plans for an IBP surfaced at a meeting of the International Union of Biological Sciences at Neuchâtel the investigation of forest productivity was seen as an essential component of such a program. Furthermore, it was decided that the study of primary production should not be confined to woodlands and forests but should be extended to cover other ecosystems. During the period 1960–3, the proposal was worked out in greater detail by a Committee chaired by Dr H. Ellenberg. It was proposed that the productivity of terrestrial ecosystems should be studied from three major points of view – those of ecology, physiology and conservation – and Dr H. Ellenberg was appointed Chairman of the subcommittee responsible for the ecological approach. It was this committee which later emerged as IBP–Terrestrial Productivity – abbreviated IBP(PT). Up to that point, attention had been concentrated on the measurement of primary production. It was decided to include the role of consumers and a meeting devoted to secondary production was held in Paris in 1963. The outcome of this preparatory phase was a paper prepared by the two 'originals', Dr Ellenberg and Dr Ovington, with me to give an animal biologist's point of view. This working paper was considered during the First General Assembly of IBP (Paris, 1964).

Prior to the Second General Assembly (Paris, 1966) the PT committee reviewed all woodland and forest projects submitted by national committees. Dr Ellenberg's influence was such that the first fully operational IBP project was at Virelles, Belgium. There a site selected by Dr P. Duvigneaud in consultation with Dr Ellenberg had become operational in 1964. Perhaps I should add that the Virelles project has another claim to fame. It never received any money from IBP sources throughout the whole period of IBP yet its contribution to the development of the Woodland Theme was of considerable importance.

It was in developing the Woodlands Biome program that Dr Ellenberg and Dr Duvigneaud stressed time and time again the need to have *minimum* and *maximum* programs. National committees favoured the 'big science approach (very often with inadequate resources) and the minimum program, which would have extended the range of sites beyond that which actually materialized, failed to gain support. As the list of projects in this synthesis volume shows, there were relatively few projects in the developing world. Gaps, however, were not confined to such areas. In spite of a very extensive proposal submitted to the PT committee in 1964, no major IBP investigation materialized in the coniferous forest zone of Canada.

It is worth recalling the aims of the Woodlands Theme as they were described in IBP News No. 13 (1969): 'The IBP program is concerned with the total production of trees, shrubs and ground vegetation in woodlands, and the flow of primary production to the major consumers, both of the grazing and detritus food chains. Productivity data obtained within IBP will enable comparisons to be made between mixed and single species stands, and between even and uneven aged communities. Studies on the production and energy flow of widely distributed genera such as *Pinus* and *Eucalyptus* will provide information on the variation of performance, structure and functioning of living communities under different environmental and management conditions, thus allowing us to better manage, on a long term, sustained yield basis, our important woodland resources.'

It will be seen that these aims make no mention of the use of systems dynamics techniques although, by the time that they were formulated, a systems approach, under the influence of the USA program, was becoming a way of life in IBP(PT). The Woodlands Biome vied with the Grasslands Biome to establish systems ecology on a sound theoretical basis. The Woodlands Workshop held at Oak Ridge in 1972 demonstrated the great strides which had been made in the functional analysis of ecosystems during IBP. There, non-linear seasonal simulation models were produced for deciduous, boreal coniferous, tropical deciduous, and broad-leaved evergreen forest from some thirty sets of IBP data. The same Workshop also showed that scientists from many parts of the world were prepared to make their results freely accessible to others in the program. This willingness to share unpublished results, to have them criticized and to have them incorporated in the anonymity of global models, has been one of the great achievements of IBP.

The overall aim of IBP – The Biological Basis of Productivity and Human Welfare – demanded that the program should contribute towards the development of rational procedures for managing the biosphere. The Woodlands Biome studies have helped to refine and make more predictive management plans not only for woodland areas but for total landscapes. It is evident from documentation already produced for the Program on Man and the Biosphere, especially the reports dealing with different types of land use, management practices and the impact of human activities on ecosystems, that it must build upon the solid achievements of the Woodlands Biome studies.

J. B. Cragg
Killam Memorial Professor,
Faculty of Environmental Design,
University of Calgary, Calgary, Canada

Foreword

The International Biological Programme was a unique experience for environmental biologists – its inception, participation, execution and accomplishments. The IBP was originally conceived to be a scientific program to learn about the Earth's productive capacity for the welfare of mankind. From these early agronomic orientations of yield, participating scientists emphasized the need for understanding the ecological processes governing the productivity of the diversity of world ecosystems. Scientists participated in the IBP sometimes because the IBP had received their national support through the International Union of Biological Sciences, more often because they had funded research projects, and always because they had a deep commitment to the scientific issues and environmental objectives of the IBP.

I first became involved with the IBP in the summer of 1966 while attending an international symposium on the *Productivity of Terrestrial Ecosystems* (Petrusewicz, 1967) at Jabłonna, Poland, that was organized by Professor Kazimer Petrusewicz. His leadership and friendship have been an inspiration to many of us through the years. At this meeting, Professor Paul Duvigneaud and I were elected to co-chair the Woodlands Working Group of the Terrestrial Productivity Section. Professor Francois Boulière served as convenor of the PT Section and Dr Malcolm Hadley as his staff assistant. Without the leadership and dogged persistence of these two, much would not have been accomplished.

Implementation of an international woodlands program began with a workshop in 1968 in Tennessee, USA, published as *Analysis of Temperate Forest Ecosystems* (Reichle, 1970). This was followed by *Productivity of Forest Ecosystems of the World* (Duvigneaud, 1971), a UNESCO symposium held in 1969 in Brussels, Belgium. In 1971 a workshop in Sweden (Rosswall, 1971) refined conceptual approaches and analytical measurements for forest studies. Three workshops were held to facilitate exchange and analysis of data being produced at the nearly 120 international forest research sites. The data files of the first workshop in 1972 in Tennessee, USA, were published as *Modeling Forest Ecosystems* (Reichle, O'Neill & Olson, 1972). From the next workshop in 1974 organized by Professors Ellenberg and Ulrich in Göttingen, Germany, an updated data file was prepared and published as *Data Analysis and Data Synthesis of Forest Ecosystems* (Ulrich, Mayer & Heller, 1974). A common computer format was established, and computer storage and analysis of data was assumed by the Oak Ridge National Laboratory, USA. In 1975 one last workshop was held in Jädraas, Sweden, to update the data files. The results of this sequence of efforts is the data base upon which this synthesis volume of the woodlands IBP group is based.

Years of research effort, countless miles of travel and numerous meetings by innumerable scientists are the foundation for this volume. A dedicated IBP staff in London provided continuing impetus, as well as coordination with CUP. In 1974 in Göttingen, participants in the Woodlands Group developed the outline for the synthesis effort. Chapter editors were selected to coordinate the contributions of colleagues and to begin the analysis and interpretation of data provided by the 117 research sites in 22 different countries that collaborated in this effort.

This volume, thus, represents the integrated product of all collaborating scientists. This was a unique venture and a courageous effort. The success of the endeavor depended upon the willingness to share and exchange unpublished data. It required different 'schools' of ecology to meld and complement. It necessitated breaking both scientific and language barriers. It helped to usher in a new generation of ecologists and a new kind of ecology. It provided the conceptual and empirical data base for a new wave of international environmental programs. Most importantly, it led to the development of an international community of ecologists whose mutual respect and scientific collaboration continue long after the official end of the IBP.

D. E. Reichle
Oak Ridge

References

Duvigneaud, P. (ed.) (1971). *Productivity of forest ecosystems.* UNESCO, Paris.

Petrusewicz, K. (ed.) (1967). *Secondary productivity of terrestrial ecosystems (principles and methods).* (2 vols.) Panstwowe Wydawnictwo Naukowe, Warsaw.

Reichle, D. E. (ed.) (1970). *Analysis of temperate forest ecosystems.* Springer Verlag, Berlin–Heidelberg–New York.

Reichle, D. E., O'Neill, R. V. & Olson, J. S. (1972). *Modeling forest ecosystems.* EDFB–IBP–73–7. Oak Ridge National Laboratory, Oak Ridge, Tennessee, USA.

Rosswall, T. (ed.) (1971). *Systems analysis in northern coniferous forests – IBP workshop.* Bull. 14 Ecol. Res. Com., Swedish Natural Research Council.

Ulrich, B., Mayer, R. & Heller, H. (1974). *Data analyses and data synthesis of forest ecosystems.* Göttingen Bodenkundliche Berichte 30, Göttingen.

1. Physiognomy and phytosociology of the international woodlands research sites*

R. L. BURGESS

Contents

The advent of the International Biological Programme led to a large number of cooperative ecosystem studies that would not have flourished in another milieu. Ecologists throughout the world, eventually from more than 60 nations, agreed to investigate ecosystem structure and function in both native and managed systems in ways that they hoped would permit refined comparison and, eventually, global synthesis. The juxtaposition of the various land masses comprising the circumpolar arctic Tundra led to early comparisons and international cooperation (Wielgolaski & Rosswall, 1971; Bliss & Wielgolaski, 1973; Wielgolaski, 1975ab), although the rigors of the arctic environment necessarily limited the number of study sites. Grasslands and deserts received extensive treatment (Orians & Solbrig, 1977; Simpson, 1977; Mabry, Hunziker & Difeo, 1977; Innis, 1978), as did Mediterranean ecosystems, (di Castri & Mooney, 1973; Mooney, 1977). A combination of low populations of scientists and location in an array of developing nations again limited the number of participating grassland and desert sites.

The vast temperate and boreal regions of the earth, heavily concentrated in the northern hemisphere, have long harbored the population centers of the world. In these regions, too, the great universities and centers of research have developed. Consequently, the northern forests of the globe

*Research supported by the Eastern Deciduous Forest Biome, US–IBP, funded by the National Science Foundation under Interagency Agreement AG–199, DEB76–00761 with the US Department of Energy under contract W–7405–eng–26 with Union Carbide Corporation. Contribution No. 320 from the Eastern Deciduous Forest Biome, US–IBP. Publication No. 1228, Environmental Sciences Division, ORNL.

1

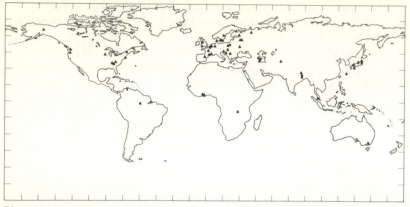

Fig. 1.1. The distribution of research sites in the International Woodlands Data Set (IBP). Each symbol may indicate more than one field site. Projection is Smithsonian Standard Earth II.

have been well-studied relative to other vegetation types, and the impetus of the IBP built an ecosystem edifice on an already appreciable foundation of knowledge (Reichle, 1970; Reichle, Franklin & Goodall, 1975). Scientists from 23 countries have now contributed to the IBP Woodlands Data Set, with information from 117 study sites (Fig. 1.1). These data, available in computerized form at Oak Ridge National Laboratory, now permit a suite of quantitative analyses, many of which appear elsewhere in this volume.

The purpose of this chapter, however, is to describe and qualitatively compare the forests studied as a result of the IBP. While interaction among scientists has continually expanded internationally, many ecologists have never had the opportunity to appreciate the great Douglas fir forests of the Pacific northwest in the United States, the magnificent beech forests of Europe, or the impressive Japanese *Cryptomeria* communities. Thus, it is important that an introduction to the breadth of the forest systems represented in the IBP be undertaken. While it is not possible to treat each site in depth, broad classes of forests can be characterized, and both the physiognomy and the phytosociology can be discussed solely on the basis of the Woodlands Data Set (see Chapter 11).

Classification system

In attempting to analyze and compare a large array of forest ecosystems, a degree of stratification is a prerequisite. In order to provide a working classification that would serve the 117 sites of the Woodlands Data Set, a variety of existing classification schemes were examined, including those of Braun-Blanquet (1932), Fosberg (1967), UNESCO (1973) and Ellenberg & Mueller-Dombois (1975). While each of these had useful features, none

Table 1.1. *Site number, name, forest type, and overstory composition of ecosystems in the International Woodlands Data Set. Forests have been delineated by climate (*Tr*, tropical; *M*, Mediterranean; *Te*, temperate; *Bo*, Boreal), life-form (*BL*, broad-leaved; *NL*, needle-leaved), behavior (*D*, deciduous; *E*, evergreen), and status (*natural is assumed; plantation indicated by /*P*). (*Adapted from DeAngelis et al., 1978)*

Site no.	Name	Forest type	Principal overstory components
1	Mt Disappointment, Australia	MBLE	*Eucalyptus obliqua*
2	Virelles, Belgium	TeBLD	*Quercus robur, Carpinus betulus, Fagus sylvatica, Acer campestris, Prunus avium*
3	Manaus, Brazil	TrBLE	Leguminosae, Sapotaceae, Lauraceae, Palmae, Annonaceae, Lecythidaceae, Rubiaceae, Burseraceae
4–7	Ontario site region 5, Canada	BoNLE	*Picea rubens, P. mariana, P. glauca, Pinus strobus, Abies balsamea, Tsuga canadensis, Thuja occidentalis*
8	Bab, Czechoslovakia	TeBLD	*Carpinus betulus, Acer campestris, Quercus cerris, Quercus petraea, Cornus mas, Crataegus oxyacantha, Ulmus carpinifolia*
9	Hestehaven, Denmark	TeBLD	*Fagus sylvatica*
10	Oulu, Finland	BoNLE	*Picea excelsa*
11	Fontainebleau, France	TeBLD	*Fagus sylvatica*
12	Madeleine, France	MBLD	*Quercus ilex*
13	Rouquet, France	MBLD	*Quercus ilex*
14	Sikfokut, Hungary	TeBLD	*Quercus petraea, Quercus cerris*
15–16	Chakia Forest, Varanasi, India	TrBLD	*Shorea robusta, Buchanania lanzan*
17	Chakia, India	TrBLD	*Shorea robusta, Buchanania lanzan, Anogeissus latifolia, Terminalia tomentosa*
18–23 and 30–35	Sal Plantation Gorakhpur Forest Division, India	TrBLD/P	*Shorea robusta*
24–29	Teak Plantation Gorakhpur Forest Division, India	TrBLD/P	*Tectona grandis*
36	Banco (plateau), Ivory Coast	TrBLE	*Turraeanthus africana, Dacryodes klaineana, Strombosia glaucescens, Berlinia confusa, Coula edulis, Chrysophyllum sp., Combretodendron africanum, Allanblackia floribunda*

(*Table* 1.1 *continued*)

Site no.	Name	Forest type	Principal overstory components
37	Yapo, Ivory Coast	TrBLE	*Dacryodes klaineana, Allanblackia floribunda, Coula edulis, Strombosia glaucescens, Scottelia chevalieri, Scytopetalum thieghemii, Piptadeniastrum africanum, Tarrietia utilis*
38	JPTF–66–Koiwai, Japan	TeNLD/P	*Larix leptolepis*
39	Ashu, Kyoto, Japan	TeBLD	*Fagus crenata, Carpinus laxiflora, Quercus mongolica*
40	Okinawa, Japan	TeBLE	*Castanopsis cuspidata*
41	JPTF–Okita, Okita, Japan	TeNLE	*Pinus densiflora*
42	Shigayama, Japan	TeNLE	*Tsuga diversifolia, Abies mariesii, Betula ermani*
43	JPTF–70 Yusuhara Kubotaniyama, Japan	TeNLE	*Tsuga sieboldii, Chamaecyparis obtusa, Pinus densiflora*
44	JPTF–71 Yusuhara Takatoriyama, Japan	TeNLE	*Abies firma*
45	Pasoh, West Malaysia	TrBLE	Dipterocarpaceae, Fagaceae, Burseraceae, Leguminosae, Euphorbiaceae, Myrtaceae
46	Meerdink, Netherlands	TeBLD	*Quercus petraea, Fagus sylvatica, Sorbus aucuparia, Frangula alnus*
47	Geobotanical Station, Bialowieza, Poland	TeBLD	*Carpinus betulus, Tilia cordata, Picea excelsa, Acer platanoides, Quercus robur, Ulmus campestris, Fraxinus excelsior*
48	Ispina, Niepolomice near Krakow, Poland	TeBLD	*Quercus robur, Carpinus betulus, Tilia cordata*
49	Kampinos National Park, Poland	TeNLE	*Pinus sylvestris, Quercus robur, Betula verrucosa*
50	Babadag, Site 1, Rumania	TeBLD	*Quercus pubescens*
51	Babadag, Site 2, Rumania	TeBLD	*Quercus pedunculiflora, Acer tataricum*
52–53	Sinaia, Site 1, Rumania	TeBLD	*Fagus sylvatica, Abies alba*
54	San Juan, Spain	TeNLE	*Pinus sylvestris*
55	Andersby Angsbackar III, Sweden	TeBLD	*Quercus robur, Betula pubescens, Betula verrucosa*
56	Kongalund Beech Site, Sweden	TeBLD	*Fagus sylvatica*
57	Kongalund Spruce Site, Sweden	BoNLE/P	*Picea abies*
58	Langarod, Sweden	TeBLD	*Fagus sylvatica*
59	Linnebjer, Sweden	TeBLD	*Quercus robur, Tilia cordata, Sorbus aucuparia, Corylus avellana*
60	Oved, Sweden	TeBLD	*Fagus sylvatica*
61	Koinas, Arkangelsk Region, USSR	BoNLE	*Picea abies*

(Table 1.1 continued)

Site no.	Name	Forest type	Principal overstory components
62–63	Caucasus Birch, Azerbaijan, USSR	TeBLD	*Betula pendula*
64–67	Tallish Ironwood, Azerbaijan, USSR	TeBLD	*Parrotia persica*
68–69	Tallish Oak, Azerbaijan, USSR	TeBLD	*Quercus castaneifolia, Zelkova carpinifolia, Parrotia persica*
70–71	Les Na Vorskl, Plot 7, Belgorod Region, USSR	TeBLD	*Quercus robur, Tilia cordata, Acer platanoides, Ulmus scabra*
72	Central Forest Reserve, USSR	BoNLE	*Picea abies*
73–89	Southern Karelian Spruce, Karelia, USSR	BoNLE	*Picea abies*
90	Tigrovaya Floodplain, Tadjikistan, USSR	TeBLD	*Populus prunosa, Elaeagnus angustifolia*
91	Meathop Wood, United Kingdom	TeBLD	*Quercus petraea, Quercus robur, Betula pendula, Betula pubescens, Fraxinus excelsior, Corylus avellana*
92–94	Black Spruce, Alaska, USA	BoNLE	*Picea mariana*
95	Hubbard Brook, New Hampshire, USA	TeBLD	*Acer saccharum, Betula lutea, Fagus grandifolia*
96	Brookhaven, New York, USA	TeBLD	*Quercus alba, Quercus coccinea, Pinus rigida*
97	Watershed 1, Coweeta, North Carolina, USA	TeNLE/P	*Pinus strobus*
98	Watershed 18, Coweeta, North Carolina, USA	TeBLD	*Acer rubrum, Quercus prinus*
99	Duke Forest, North Carolina, USA	TeNLE/P	*Pinus taeda*
100	Saxapahaw, North Carolina, USA	TeNLE/P	*Pinus taeda*
101	Andrews Experimental Forest, Watershed 10, Oregon, USA	TeNLE	*Pseudotsuga menziesii, Tsuga heterophylla, Castanopsis chrysophylla, Thuja plicata, Pinus lambertiana, Acer macrophyllum*
102	Liriodendron Site, Oak Ridge, Tennessee, USA	TeBLD	*Liriodendron tulipifera, Quercus spp., Carya tomentosa, Pinus echinata*
103–106	Walker Branch, Oak Ridge, Tennessee, USA	TeBLD	*Quercus alba, Quercus prinus, Quercus velutina, Pinus echinata, Liriodendron tulipifera*
107	Thompson Research Center, Seattle, Washington, USA	TeNLE/P	*Pseudotsuga menziesii, Tsuga heterophylla*
108	Noe Woods (Lake Wingra), Wisconsin, USA	TeBLD	*Quercus alba, Quercus velutina, Prunus serotina*
109	Nakoma Urban Forest, Wisconsin, USA	TeBLD	*Quercus alba, Quercus velutina*

(*Table* 1.1 *continued*)

Site no.	Name	Forest type	Principal overstory components
110	Fa. Eglharting, Abt. 27A, Federal Republic of Germany	BoNLE/P	*Picea abies*
111–113	Solling Project, Federal Republic of Germany	TeBLD	*Fagus sylvatica*
114–116	Solling Project, Federal Republic of Germany	BoNLE	*Picea abies*
117	Lubumbashi, Zaire	TrBLD	*Marquesia macroura*

included environmental (climatic) zonation or management status. The Holdridge system (Holdridge, 1964) was also consulted, and a degree of gross bioclimatic categorization was incorporated in the final classification.

A hierarchical scheme was used, ultimately consisting of four major units – climate, life-form, behavior, and status. The last three are dichotomous, while four divisions of the climatic regime were necessary to encompass all woodland sites. Twenty-six research sites were classed as *Tropical*, three as *Mediterranean*, 55 as *Temperate*, and 33 as *Boreal*. With respect to life-form, 72 sites were dominated by broad-leaved species, while 45 were needle-leaved. Behavior was more evenly divided; 65 stands are deciduous, and 52 evergreen. Finally, 89 sites consist of natural forest, while 28 are managed plantations. There is a reasonable representation of all categories with the exception of Mediterranean woodlands, where a paucity of sites relates, at least in part, to the small number of contributing nations with Mediterranean climate.

All 117 sites, categorized according to the derived classification, are listed in Table 1.1. This array correlates with the Woodlands Data Set (DeAngelis, Gardner & Shugart, Chapter 11), and the nomenclature is used in the other chapters as well.

Mediterranean woodlands

Three sites are available that are representative of woodlands in Mediterranean climates, one in Australia and two in southern France. Characterized by warm, dry summers and cool moist winters (Emberger *et al.*, 1963), the vegetation of these regions is unique, consisting for the most part of sclerophyllous trees and shrubs with an understory of herbs and grasses (di Castri & Mooney, 1973) (Fig. 1.2).

In Australia, the IBP site is located in the Mt Disappointment State Forest in southern Victoria. It is dominated exclusively by messmate or messmate stringy bark (*Eucalyptus obliqua*) (Rule, 1967) that forms a continuous overstory approaching 30 m in height. The stand is somewhat over 50 years old, formed on a Krasnozem (a red, friable, porous soil) overlying metamorphosed mudstones, sandstones, and shales of Silurian age. The soils are well drained and acid, with pH ranging from about 5.2 to 5.9. Mean annual precipitation is 962 mm. The region is frost-free, with a mean annual temperature of 11°C.

The understory is relatively sparse, consisting primarily of shrubs (*Xanthorrhoea australis, Adenanthos terminalis, Acacia* spp., *Hakea* spp.) from 0.5 to 2 m tall, perennial grasses (*Themeda australis, Stipa pubescens*), and an admixture of herbs (*Lepidosperma canescens, Lomandra dura, Kennedia prostrata*). Recently published documents do not yield much additional information on species composition, although Adamson & Osborn (1924), Lawrence (1939), and Specht (1972–3) are useful references. The site has a basal area of 58.7 m²/ha, at densities ranging from 705 to 946 trees per hectare. The leaf area index is 4.1, and above-ground standing crop is 31.2 kg/m².

The two sites in southern France straddle the city of Montpellier, home of one of the world's great schools of phytosociology. The site at Madeleine lies at 10 m, while Rouquet is at 180 m above mean sea level. Both sites are dominated by holly-leaf oak (*Quercus ilex*), reaching heights of 11 m at Rouquet and 15 m at Madeleine. Rouquet has been coppiced, and the present stand age is about 150 yr. Based on height, Madeleine appears older, although supporting data on age are not available. Additional height may be a function of density, as Madeleine has a density of only 527 trees per hectare, while Rouquet displays 1440 trees per hectare. Herbaceous understory is poorly developed at Rouquet, while that at Madeleine is 90%

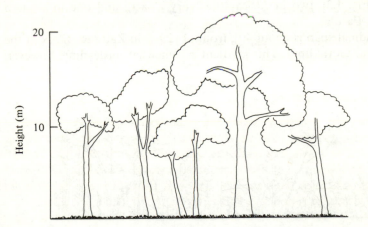

Fig. 1.2. Profile diagram of Medterranean Woodland, southeastern Australia. Overstory of *Eucalyptus obliqua*.

Hedera helix (English ivy) (Lossaint, 1973). Other species typically present are *Ruscus aculeatus*, *Rosa sempervirens*, *Vinca major*, and *Asparagus acutifolius*. All of these, like the overstory, are both sclerophyllous and evergreen.

These sites are also frost-free, with mean annual temperatures of 14.1 and 13.4°C. Precipitation is 754 mm at Madeleine, but rises to 987 mm at the higher elevation at Rouquet. Both sites are on dolomitic limestone, with slightly basic, rendzina-type soils. Drainage is good, with the soils containing 8 to 9% organic matter.

In disturbed areas, *Quercus ilex* is replaced by *Q. coccifera*, a shrubby, sclerophyllous scrub oak that dominates the 'garrigue' of the Mediterranean coast (Reed, 1954). Grazing pressure then creates, through time, a grassy pasture composed predominantly of *Brachypodium ramosum*.

The Mediterranean woodlands are not, of course, extensive on a global scale, as most of the native communities are dominated by shrubs. The small representation of the sites does, however, give a reasonable characterization of the type, and a range of both environmental characteristics and phytosociological parameters.

Tropical forests

The tropical forests represented in the IBP Woodlands Data Set are represented by 26 sites, 18 of which are managed plantations at two locations in northeastern India. The remaining eight sites are evenly divided between true rain forest and broad-leaved deciduous forests on the subtropical fringe. The three great regions of tropical rain forest are represented (Aubert de la Rüe, Bourlière & Harroy, 1957), by a neotropical site in Brazil (Manaus) (Fig. 1.3), two in Africa (the Ivory Coast), and one in western Malaysia at Pasoh.

The latitudinal span is about 40°, from 11° 29′ S in Zaire to 27° N at the Indian sites at Gorakhpur. This gradient is somewhat misleading, however,

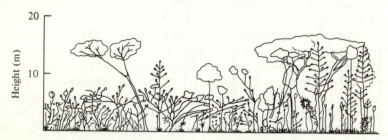

Fig. 1.3. Profile diagram of tropical deciduous forest near Manaus, Brazil. Modified from Huetz de Lemps, 1970.

as all but Lubumbashi, Zaire, lie in the northern hemisphere. Among the rain-forest sites, annual precipitation ranges from 1739 mm at Yapo in the Ivory Coast to 2095 mm nearby at Banco. All four sites are lowland, ranging to only 100 m above mean sea level at Pasoh. Mean annual temperature is 26–7°C.

Four deciduous sites, three at Chakia in India and one in Zaire, illustrate a dry season dormancy. Consequently, the growing seasons are 270 days (nine months) at Chakia, but only 118 days (four months) at Lubumbashi. The latter lies at an elevation of 1208 m, while Chakia is only 350 m above mean sea level. Precipitation at Chakia is 844 mm, with a mean annual temperature of 30°C. Comparable values for Lubumbashi are 1273 mm and 20.3°C, respectively. Thus, the site in Zaire receives half again as much rain as Chakia, in roughly half the length of time. The effects of the monsoons that sweep north off the Bay of Bengal are evident particularly when compared to the decidedly continental climate in the interior Congo basin.

The 18 Indian broad-leaved plantations lie at 81 m, with a mean annual temperature of 27.5°C and a rainfall of 1158 mm. These plantations, six each of teak, sal, and mixed teak and sal, differ only in age. No data are available on soils or underlying geology.

Soils in the eight natural forests of this series (the four rain-forest and four deciduous sites) are pale yellow to reddish latosols, formed mostly on ancient igneous or metamorphic rock. Drainage is medium to rather poor. The rain-forest soils are strongly acid, with pH ranging from 3.1 to 5.1, while those of the deciduous sites are less so, with values from 5.0 to 6.8.

Tropical rain forest

Tropical rain forest, the epitome of forest everywhere, is a highly productive and diverse formation — one that has been both an inspiration and an enigma to plant geographers and ecologists since the days of Alexander von Humboldt. The forest is well stratified, often with five well-defined layers, including emergent, canopy, sub-canopy, shrub, and herbaceous synusiae (Richards, 1952; Fig. 1.4). Among the tree species, dominance, as it is commonly recognized in temperate forests, has little relevance; the most abundant species may be present at densities of no more than two or three per hectare. Therefore, it is often the practice to discuss dominance in terms of plant families, where several genera may collectively assume a position of importance in the community (Meggers, Ayensu & Duckworth, 1973). Among the four IBP sites, this practice has been followed, although specific information is available for major species at the two sites in the Ivory Coast. The dominants come either from tropical families (Palmae, Olacaceae) or from large, cosmopolitan ones (Leguminosae, Rosaceae), as is apparent in Table 1.2. Note that species of Burseraceae and Leguminosae

9

Fig. 1.4. Profile diagram of tropical rain forest showing layers (synusiae) of woody components. Modified from Richards (1952).

are dominant at all four sites. The former is a tropical family of 16 genera and approximately 500 species, all trees or shrubs. Major genera are *Protium*, *Commiphora*, *Boswellia*, *Bursera*, *Canarium*, and *Santiria*. Conversely, the legumes are the third largest family of flowering plants, with about 600 genera and over 12 000 species. There is a heavy concentration of legumes in the tropical regions, particularly in the Amazon Basin of South America.

The forest at Manaus is mature, 38 m tall, with a standing crop of 40.6 kg/m^2. Basal area is 30.7 m^2/ha, but the stocking density is questionable. Tree density probably approaches 1000 per hectare. The mature forests at Banco and Yapo are physiognomically similar, with basal areas of 30 and 31 m^2/ha, and standing crops of 51 and 45 kg/m^2, respectively. Reported densities are 265 trees/ha at Banco and 427 trees/ha at Yapo.

The Ivory Coast sites share the dominants *Allenblackia floribunda* (Guttiferae), *Coula edulis* and *Strombosia glaucescens* (Olacaceae), and *Dacyrodes klaineana* (Burseraceae). In addition, *Turreanthus africana*, the leguminous *Berlinia confusa*, *Combretodendron africanum*, and *Chrysophyllum* sp. are present at Banco. Yapo includes *Scottelia chevalieri*, *Scytopetalum thieghemii*, *Tarrietia utilis*, and the legume, *Piptadeniastrum africanum*. Neither subcanopy nor understory species lists are available for either site, but Aubréville (1938) (reproduced in Richards 1952, *p*. 51) gives a list of 74 species (both over- and understories) in the forest at Massa Me, Ivory Coast. His list includes *Piptadeniastrum*, *Combretodendron*, *Strombosia*, *Scottellia*, *Allanblackia*, *Coula*, etc., so species composition is similar to the forests at Banco and Yapo. In addition to reproduction of the dominant tree species, the understory includes representatives of *Cola* (two species), *Diospyros* (two species), *Baphia* (two species), *Albizzia* (legumes), *Hannoa*, *Myrianthus*, and *Napoleona*.

The site at Pasoh, Malaysia (Fig. 1.5), is dominated by dipterocarps

10

Table 1.2. *Dominant plant families in the overstory of four tropical rain-forest sites represented in the Woodlands Data Set. Presence indicated by × and absence by —*

Family	Manaus (Brazil)	Blanco (Ivory Coast)	Yapo (Ivory Coast)	Pasoh (Malaysia)
Annonaceae	×	—	—	—
Barringtoniaceae	—	×	—	—
Burseraceae	×	×	×	×
Dipterocarpaceae	—	—	—	×
Euphorbiaceae	—	—	—	×
Fagaceae	—	—	—	×
Flacourtiaceae	—	—	×	—
Guttiferae	—	×	×	—
Lauraceae	×	—	—	—
Lecythidaceae	×	—	—	—
Leguminosae	×	×	×	×
Meliaceae	—	×	—	—
Moraceae	×	—	—	—
Myrtaceae	—	—	—	×
Olacaceae	—	×	×	—
Palmae	×	—	—	—
Rosaceae	×	—	—	—
Rubiaceae	×	—	—	—
Sapotaceae	×	×	—	—
Scytopetalaceae	—	—	×	—
Sterculiaceae	—	—	×	—
Violaceae	×	—	—	—

(probably *Dipterocarpus, Dryanobalanus,* and *Hopea*) and *Trigonobalanus* in the Fagaceae. The stand is mature, 35–40 m tall, has a basal area of 27 m^2/ha, a density of 591 trees/ha, and a standing crop (aboveground) of 35.4 kg/m^2. No specific information is available on understory structure and composition, although Whitmore (1975) and Walter (1964) give examples that are remarkably similar, in most respects, to the rain-forest site in Malaysia.

Tropical deciduous forest

Tropical deciduous forest, widespread in all three major regions, is represented by four sites, one in Zaire and three in India (Fig. 1.6). As noted above, the site at Lubumbashi is the only southern hemisphere representative. The Indian sites at Chakia, near Varanasi, lie at 25°N.

Marquesia macroura (Dipterocarpaceae) dominates the site at Lubumbashi, reaching a height of 18 mm at an age of 120 yr. The site is representative of the Miombo woodland (Knapp, 1973), which stretches south

Fig. 1.5. Tropical rain forest, Kuala Lumpur, Malaysia. Photo by D. E. Reichle.

of the equator from Kenya and Tanzania as far as northern Namibia (Walter, 1973). At a density of 446 trees/hectare, the stand has a basal area of $22m^2$/ha and a standing crop of about 17 kg/m^2.

At Chakia, the three sites differ primarily in age, ranging from 38 yr (Sharma) to 120 yr (Singh). The site studied by Bandhu is intermediate at 60 yr of age. Despite a 9-month growing season, the forests are not tall, ranging only to about 18 m. All three sites are dominated by *Shorea robusta* (sal) which, with teak (*Tectona grandis*), constitutes the most commercially valuable forest species on the subcontinent. At the sites, *Buchanania lanzan* (Anacardiaceae) and *Terminalia tomentosa* (Combretaceae) are common associates. Other principal tree species include *Anogeissus latifolia* (Combretaceae), *Madhuca indica* (Sapotaceae), *and Diospyros melanoxylon* (Ebenaceae), one of the Indian species of ebony. Substory is dominated by *Carissa opaca*, a large shrub, and *Zizyphus xylophyrus*, a small thorn. Champion & Seth (1968) give extensive tables of species composition for what they term the 'dry sal-bearing forest', along with abundant meteorological data from the nearby station at Varanasi.

Tropical deciduous plantations

Data are available on 18 stands of managed forest plantation in the

Fig. 1.6. Tropical broad-leaved deciduous forest, central India. *Buchanania lanzan* and *Anogeissus latifolia* in young scrub forest. Photo by D. E. Reichle.

Gorakhpur Forest Division of Uttar Pradesh. These are in three groups of six stands each; sal (*Shorea robusta*) plantations, teak (*Tectona grandis*) plantations, and mixed sal and teak. The mixed sites represent a gradient of 5, 8, 14, 26, 30, and 40 yr of age, matched by the teak plantations. The sal forests are 10, 16, 22, 28, 35, and 38 yr old. The three series illustrate a linear increase in height with age, and a negative exponential decrease in stand density. Biomass (standing crop, above-ground) ranges from a low of 3.8 kg/m^2 in the 10-yr-old sal plantation to a high of 115 kg/m^2 in the 40-yr-old mixed site. Leaf area index is more complex, but ranges from a low of 5.76 in the 22-yr-old sal plantation to 30.6, again, in the oldest mixed site. Productivity, both above- and below-ground, correlates also with stand age and species composition. The experiment demonstrates, in this case, the value of species mixtures compared with monocultures.

Temperate forests

Fifty-five sites in 15 countries represent the temperate forests of the IBP. The vast majority (42 sites) are natural stands of broad-leaved deciduous forest (Fig. 1.7). The other 13 sites form two small groups, needle-leaved

Fig. 1.7. Temperate forest complex, Great Smoky Mountains, Tennessee, USA. Mixed conifer (*Picea rubens, Abies fraseri, Tsuga canadensis*) in the foreground, mixed deciduous (*Acer, Fagus, Quercus, Liriodendron, Tilia, Aesculus*) on the distant slopes. Photo by D. E. Reichle.

14

evergreen forests and needle-leaved evergreen plantations, with seven and four sites, respectively. Finally, the Japanese contribution to the IBP provides one site each of temperate broad-leaved evergreen forest and needle-leaved deciduous plantation.

The United States is represented by 15 IBP sites, the USSR by 11, and Japan by seven. Sweden contains five sites, Poland, Rumania, and West Germany three each, while single sites are in Belgium, Czechoslovakia, Denmark, France, Hungary, the Netherlands, Spain, and the United Kingdom. Thus, 15 IBP sites are North American, seven are Asian (all in Japan), and 33 are European (including those in the Soviet Union).

Latitudinally, the sites range from 26° 47′ N in Okinawa (the broad-leaved evergreen forest) to 60° N at Andersby Angsbackar in Sweden. Elevations run from 22 m below mean sea level at Tallish Oak 2 near Baku on the Caspian coast, to 2000 m at the Caucasus Birch sites nearby.

The soils of the temperate forests are extremely variable, including podzols, podzolic soils (both yellow and gray-brown), brown forest soils, rendzinas, and humic gleys. They are formed on glacial deposits of outwash and morainic till, on alluvium, and as residual soils over igneous, sedimentary, and metamorphic lithology. Granites and basalts, gneiss and schist, limestone, sandstone, shale, and volcanic materials are all present in the array of temperate woodland sites. Most soils are acid, some strongly so, with pH ranging to 3.5. Babadag 1, in Rumania, reports a range from 6.5 to 7.5. The only basic soil in the group is an alluvial, meadow type, toogai soil on a floodplain at Tigrovaya Balka in the USSR, where pH runs 7.4 to 8.4.

In the following sections, each of the five temperate forest classes is treated individually. The four small groups of sites form a logical base from which to lead into a discussion of the main body of temperature deciduous forest sites.

Temperate broad-leaved evergreen forest

Only one site, in Okinawa, in representative of this type. The stand is 55-yr-old, dominated by the Japanese chinkapin, *Castanopsis cuspidata* var. *sieboldii*. The canopy is at approximately 12 m. The understory is composed of *Schefflera octophylla*, *Neolitsea sericea*, *Distylium racemosum*, and *Styrax japonica*, medium shrubs up to 2 m in height. The field layer is dominated by species of *Cyathea*, a tree fern characteristic of the moist tropics and subtropics, but ranging into warm temperature environments in the western Pacific.

The site has a density of 2900 trees/ha with a basal area of 47.9 m^2/ha, and an aboveground standing crop of a little over 19 kg/m^2. The region is frost-free, and the heavy rainfall (2630 mm) provides for the evergreen character of this temperate forest type.

15

Temperate needle-leaved deciduous plantation

Again, a single site in Japan provides a representative of a plantation of deciduous conifers. The 39-yr-old stand, almost 20 m tall, is dominated by an Asiatic larch, *Larix leptolepis* (Fig. 1.8). Shrubs and small trees in the understory include a mulberry, *Morus bombycis*, a plum, *Prunus grayana*, *Viburnum opulus* var. *galvescens*, and *Stachyurus praecox*. The field layer is predominantly *Calyptranthe petiolu* var. *ovalifolia*, a hydrangea. The site has a basal area of 37 m²/ha at a density of 1155 trees/ha. Standing crop (above-ground) is about 17 kg/m².

Temperate needle-leaved evergreen forests

Seven sites are available in the Woodlands Data Set that represent temperate conifer forests (Figs. 1.9, 1.10), four in Japan, one in Poland, one in Spain, and one in the United States. The Japanese sites, two hemlock, one fir, and a pine forest, range in elevation from 300 to 1790 m above sea level. The pine forest is young, only 20 yr old, but the others are mature forests of 145, 290, and 443 yr of age. Soils are brown forest types on sandstones and shales of Cretaceous age, or, in the subalpine hemlock site (Shigayama), a podzolic soil formed on Pleistocene volcanic lava. All are strongly acid, with pH values ranging from 3.6 to 5.4.

Yusuhara Kubotaniyama is dominated by a hemlock (*Tsuga sieboldii*), a cypress (*Chamaecyparis obtusa*) whose range is confined to Japan and Taiwan (Numata, 1974), and *Pinus densiflora*. Woody understory is composed of *Cleyera japonica*, *Eurya japonica*, *Illicium religiosum*, *Clethra barbinervis*, and a couple of small oaks, *Quercus acuta* and *Q. salicina*. The field layer is virtually devoid of plants. The site exhibits a density of 1948

Fig. 1.8. Profile diagram of temperate needle-leaved deciduous forest, central Honshu, Japan. Overstory of *Larix leptolepis*. From a photograph in Numata (1974).

16

Fig. 1.9. Temperate needle-leaved evergreen forest, Wyoming, USA. Overstory of *Pseudotsuga menziesii*, *Pinus flexilis*, and *Picea engelmannii*. Photo by D. E. Reichle.

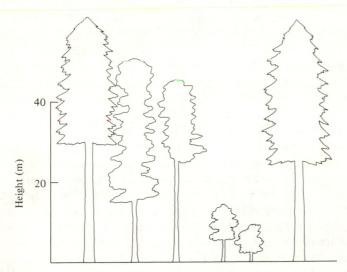

Fig. 1.10. Profile diagram of temperate needle-leaved evergreen forest, Washington, USA. Overstory of *Picea sitchensis* and *Tsuga heterophylla*. Modified from Knapp (1965).

17

trees/ha, yielding a basal area of 88 m^2/ha and a standing crop (aboveground) of 56.7 kg/ha. The area has a mean annual temperature of 13.6°C and an annual precipitation of 2748 mm.

Yusuhara Takatoriyama, 300 m lower in elevation, is dominated by a fir, *Abies firma*. At 145 yr, the stand reaches over 26 m in height. Understory species include *Actinodaphne lancifolia, Castanopsis cuspidata,* and *Illicium religiosum*. Again, there is almost nothing in the field layer. The site has a basal area of 81.7 m^2/ha, a density of 2077 trees/ha and an aboveground standing crop of about 50 kg/m^2.

The site at Okita, further north on Honshu, lies at 300 m above sea level. It is dominated by *Pinus densiflora*, with saplings of *Quercus serrata* in the understory. The field layer is mostly composed of the sedge, *Carex lanceolata*, and the common, cosmopolitan bracken fern, *Pteridium aquilinum* var. *latiusculum*. The stand is only 20 yr old, with a present height of a little over 10 m. Density is high at 6600 trees/ha. Basal area is 32 m^2/ha, but aboveground standing crop is only about 9 kg/m^2. The area has a mean annual precipitation of 1467 mm. The growing season is 180 days.

Shigayama, at 1790 m elevation, has a similar climate – 180-day growing season, 11.7°C annual temperature, and 1455 mm of rain. The stand is a mixture of hemlock (*Tsuga diversifolia*), fir (*Abies mariesii*), and a birch (*Betula ermani*). No information on the understory is available. At a height of 18 m and a density of 1200 trees/ha, the site exhibits a standing crop of 19 kg/ha and a basal area of 53.1 m^2/ha.

The Polish site, in Kampinos National Park near Warsaw, is developed on a well-drained podzol soil over glacial till (Fig. 1.11). Mean annual temperature is about 10°C, with precipitation of 548 mm. The growing season approximates 207 days. The stand is a mixture of Scots pine (*Pinus sylvestris*), oak (*Quercus robur*), and the European birch (*Betula verrucosa*). Characteristic shrubs are a blueberry (*Vaccinium myrtillus*), the common heather (*Calluna vulgaris*), and a dogwood (*Cornus sanguinea*). The field layer is predominantly composed of an anemone (*Anemone nemorosa*), a sorrel (*Oxalis acetosella*), the mayflower (*Maianthemum bifolium*), and the circumpolar reindeer lichen (*Cladonia rangiferina*). Density is 1020 trees/ha, which, at a height of 25 m, yields a standing crop of 26.5 kg/m^2.

San Juan, Spain, is also dominated by Scots pine (*P. sylvestris*), with small admixtures of European beech (*Fagus sylvatica*) and blueberry (*Vaccinium myrtillus*). The holly, *Ilex aquifolium*, dominates the substory. The field layer is composed of *Pseudoscleropodium purum* and the grass, *Brachypodium pinnatum*. The site lies at an elevation of 1230 m, with a mean annual temperature of 8°C and 802 mm of precipitation. Stand height is 15 m at an age of 140 yr. Basal area equals 52.3 m^2/ha and aboveground standing crop is 20.4 kg/m^2 at a stocking density of 1916 trees/ha.

Finally, Watershed No. 10 on the H. J. Andrews Experimental Forest in

18

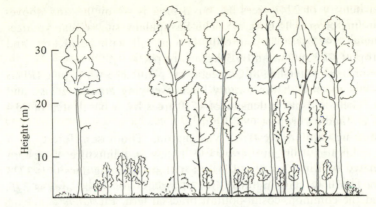

Fig. 1.11. Profile diagram of temperate needle-leaved evergreen forest. Overstory of *Pinus sylvestris*, central Poland. Modified from Szafer (1966).

Oregon, USA, ranges in elevation from 430 to 670 m above mean sea level. With a growing season of 150 days, the area has a mean annual temperature of 8.5°C and an annual precipitation of 2250 mm. The site is old-growth Douglas fir (*Pseudotsuga menziesii*) forest, reaching to 60 m in height at an age of 450 yr. Associates include western hemlock (*Tsuga heterophylla*), western red cedar (*Thuja plicata*), *Pinus lambertiana*, and *Castanopsis chrysophylla*. The area, recently designated as a Biosphere Reserve in the United Nations Man and the Biosphere (MAB) program, exhibits a density of only 290 trees/ha, but reaches a basal area of 62.7 m²/ha and an above-ground standing crop of over 77 kg/m².

Temperate needle-leaved evergreen plantations

These four plantations in the Woodlands Data Set are all in the United States, three in North Carolina and one in the state of Washington. Two sites, Saxapahaw and the Duke Forest, are between 140 and 150 m above sea level on the Appalachian Piedmont. Watershed 1 at the Coweeta Hydrologic Laboratory ranges between 706 and 988 m in the mountains of extreme western North Carolina. Growing season on the Piedmont is 231 days, dropping to 150 days in the mountains. The annual temperature of 15.6°C at Duke and Saxapahaw drops to 13.6°C at Coweeta, but precipitation increases from 1150 mm to 1628 mm. The Thompson Research Center site near Seattle has a growing season of 214 days. At an altitude of 210 m, the mean annual temperature is 9.8°C, and annual precipitation is 1360 mm, about half coming during the growing season.

Both Piedmont sites are of Loblolly pine (*Pinus taeda*), a widespread and commercially important species. Duke Forest is 16 yr old, the trees reaching

15 m. At a density of 2243 trees/ha, basal area is 49 m²/ha and above-ground standing crop is 15.6 kg/m². The Saxapahaw site, a year younger and less dense (1470 trees/ha), is only 11.6 m tall with basal area and standing crop of 41.3 m²/ha and 9.3 kg/m², respectively.

The experimental watershed at Coweeta was planted to white pine (*Pinus strobus*) 15 yr ago on a shallow, stony loam overlying granite, gneiss, and mica schist. The stand has a density of 1760 trees/ha, a basal area of 23.4 m²/ha, and a standing crop of 69.6 kg/m².

The remaining site in western Washington, Thompson forest, is a Douglas fir plantation currently 46 yr old. It has an admixture of western hemlock (*Tsuga heterophylla*). Stand height is 18 m, stocking density is 2223 trees/ha, aboveground standing crop is 17.4 kg/m², and basal area is 37.7 m²/ha.

Temperate broad-leaved deciduous forests

Natural stands of north temperate deciduous forest constitute the largest single block of sites in the IBP Woodlands Data Set (Fig. 1.12). Forty-two stands in 14 countries present a broad range of related forest types over a rather large geographical area. Eleven sites lie in the Soviet Union, ten in the United States, five in Sweden, three each in Rumania and West Germany, two in Poland, and single sites in Belgium, Czechoslovakia, Denmark, England, France, Hungary, the Netherlands, and Japan. The 42 sites range in elevation from below sea level to 2000 m, and as far north as 60°N.

There is an almost universal dominance by members of the Fagaceae, with all but five sites reporting oak (*Quercus*), beech (*Fagus*), or both, in the overstory. *Quercus* is a large genus, and dominant species include *Q. alba*, *Q.*

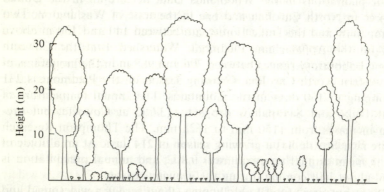

Fig. 1.12. Profile diagram of temperate broad-leaved deciduous forest, Michigan, USA. Overstory of *Acer saccharum* and *Fagus grandifolia*. Modified from Knapp (1965).

20

coccinea, Q. prinus, Q. velutina, and *Q. rubra* in North America (Knapp, 1965), *Q. robur, Q. cerris, Q. petraea, Q. pubescens*, and *Q. pedunculiflora* in Europe, *Q. mongolica* var. *grosserrata* in Japan, and *Q. castaneifolia*, and *Q. robur* in the trans-Caspian region (Berg, 1950). Similarly, *Fagus crenata* in Japan, *F. sylvatica* in Europe, and *F. grandifolia* in North America are very common. The other five stands (four in the Caucasus) are dominated by *Betula glandulosa, Parrotia persica* (Persian ironwood), and, in the United States, tulip poplar (*Liriodendron tulipifera*).

Carpinus (hornbeam) is a common associate, including *C. laxiflora* and *C. tschonoskii* in Japan, *C. caucasica* near Baku, and *C. betulus* in Europe. In North America, *C. caroliniana* is present in most of the deciduous forest region, but is not a major constituent of the US sites. The genus *Acer* (maple) is also quite widespread, often as an understory component. Species include *A. japonicum* and *A. sieboldiana* (Japan), *A. platanoides* in the Vorskla River sites, USSR, and in Poland, *A. campestris* and *A. tataricum* in Europe, and *A. saccharum, A. rubrum*, and *A. spicatum* in the eastern United States (Braun, 1950).

The site at Ashu, Japan also includes *Magnolia salicifolia* and *Prunus grayana* in the substory, with *Rhus ambigua* and *Schizophragma hydrangeoides* (shrubs), and *Viola vaginata, Oxalis griffithii*, and *Sasa* sp. (a small bamboo) in the field layer.

The four sites near Baku, dominated by *Parrotia*, all support species of *Rosa* and the evergreen shrub, *Ruscus*, in the field layer. Tallish Oak sites, also near Baku but in the montane foothills of the Caucasus, also contain *Zelkova carpinifolia* (Ulmaceae), *Alnus barbata* (an alder), and the widespread European ash, *Fraxinus excelsior*. Two stands, Caucasus Birch 1 and 2, are dominated by *Betula glandulosa*, again with roses (*Rosa* sp.) in the undergrowth. Tigrovaya Balka contains *Populus pruinosa* (a poplar) and Russian olive (*Elaeagnus angustifolia*) as primary species in the floodplain forest, along with *Tamarix ramosissima* and a group of tall grasses (*Phragmites, Imperata*, and *Erianthus*) in the understory. The two Vorskla River sites, 7 and 8, add a linden (*Tilia cordata*) and an elm (*Ulmus scabra*) to the list of dominants, with an apple (*Malus sylvestris*) and a crab (*Crataegus currisepala*), *two species of Euonymus*, and *Thelyeriania sanguinea* comprising most of the substory. The field or herbaceous layer includes a sedge (*Carex pilosa*), an umbellifer (*Aegopodium podagraria*), *Scilla sibirica, Stellaria holostea, Gagea lutea* (a lily), *Anemone ranunculoides*, and *Corydalis halleri*.

Virelles, Belgium contains *Prunus avium* in the canopy, and *Corylus avellana* (European hazel, a small tree) in the substory. The field layer is composed of ivy (*Hedera helix*), a mint (*Lamium galeobdolon*) that is widely distributed in European forests, and species of *Scilla, Narcissus, Cardamine*, and *Mercurialis*. Bab, in Czechoslovakia, includes *Cornus mas, Crataegus*

21

oxyacantha, Ulmus campestris, and *Rosa canina* as principal species in addition to the dominant *Quercus, Acer* and *Carpinus.* Extensive analysis of forest dynamics as well as species composition is available in Biskupsky (1975). Hestehaven, Denmark, combines an understory of *Anemone nemorosa, Melica uniflora, Asperula odorata,* and *Carex sylvatica,* with its overstory of beech (*Fagus sylvatica*). This is similar to the forest of Fontainebleau, France, that adds a fescue (*Festuca heterophylla*) and the evergreen shrub (*Ruscus aculeatus*) (Reed, 1954). Meerdink, the Netherlands, is dominated by *Quercus petraea, Fagus sylvatica, Sorbus aucuparia* (mountain ash), and a buckthorn, *Frangula alnus.* Bracken fern, English ivy, and a blackberry (*Rubus spidigerus*) constitute the understory.

A group of six sites in Hungary, Rumania, and Poland share some similarities as well as differences. Sikfokut, Hungary, an oak forest of *Quercus petraea* and *Q. cerris,* possesses a substory of two maples, two dogwoods, a crab, a *Euonymus,* and a privet (*Ligustrum*). The field layer has sedges (*Carex montana* and *C. michelii*), grasses (*Poa nemoralis, Melica uniflora,* and *Dactylis polygama*), and legumes (*Lathyrus niger* and *L. vernus*) along with *Fragaria vesca* (strawberry), *Chrysanthemum coryombosum,* and the bedstraw, *Galium schultesii.* Bialowieza, in northeastern

Fig. 1.13. Temperate deciduous forest of *Quercus robur, Carpinus betulus,* and *Tilia cordata,* Bialowieza, Poland. European Wisent, *Bison bonasus,* part of a reestablished herd, in the foreground. Photo by D. E. Reichle.

Poland, and Ispina, near Krakow, are fine examples of European oak-hornbeam forest. Bialowieza has some spruce (*Picea excelsa*) and Norway maple (*Acer platanoides*) that add to the dominant *Quercus, Carpinus*, and *Tilia cordata* (Fig. 1.13). Szafer (1966) and Aulak (1970) give extensive information on the herbaceous strata of these forest types.

Two sites at Babadag (1 and 2) and Sinaia 1 in Rumania represent contrasting types. Sinaia is 65% beech and 35% fir (*Abies alba*), with *Pulmonaria rubra, Dentaria glandulosa, Symphytum cordatum, Asperula odorata*, and *Oxalis acetosella* in the herb layer. The two at Babadag are sub-Mediterranean oak forests, with *Quercus pubescens*, the smoke tree, *Cotinus coggygria, Galium dasypodium*, and peony (*Paeonia peregrina*) as principal associated species. Babadag 2 is dominated by *Quercus pedunculiflora* and *Acer tataricum*, with *Brachypodium pinnatum* and *Centaurea stenolepis* (a thistle) in the understory (Biskupsky, 1975).

The five Swedish sites include two oak forests (Andersby Angsbackar and Linnebjer) and three beech stands (Kongalund, Langarod, and Oved). At Linnebjer, *Tilia, Sorbus*, and *Corylus* share the canopy with *Quercus robur*, while at Andersby Angsbackar, two species of birch (*Betula pubescens* and *B. verrucosa*) accompany the oak. *Corylus*, a currant (*Ribes alpinum*), and a honeysuckle (*Lonicera xylosteum*) are prominent in the understory. In the beech forests at Kongalund, Langarod, and Oved, *Fagus sylvatica* is the single dominant tree species. *Anemone nemorosa, A. ranunculoides, Oxalis acetosella, Allium ursinum, Mercurialis perennis, Stellaria nemorum, Lamium galeobdolon*, and *Deschampsia flexuosa* appear in the field layers.

The intensively managed beech forests of northern Germany (Fig. 1.14) are represented by three sites, B1, B3, and B4, that are part of the total Solling Project (Ellenberg, 1971). In England, Meathop Wood is dominated by two oaks (*Quercus robur* and *Q. petraea*), two birches (*Betula pendula* and *B. pubescens*), *Corylus avellana*, and *Fraxinus excelsior* (Edlin, 1958).

The North American sites include one in New Hampshire (Hubbard Brook), one in New York (Brookhaven on Long Island), one in North Carolina (Coweeta 18), five in east Tennessee on the Oak Ridge Reservation, and two at Madison, Wisconsin. Hubbard Brook is a 110-yr-old stand of climax deciduous forest dominated by maple (*Acer saccharum*), yellow birch (*Betula lutea*), and beech (*Fagus grandifolia*). The substory contains mountain or vine maple (*Acer spicatum*) and *Viburnum alnifolium*, while the field layer supports two ferns, *Dryopteris spinulosa* and *Dennstaedtia punctilobula*. The oak–pine forest at Brookhaven is dominated by a canopy of white and scarlet oak (*Quercus alba* and *Q. coccinea*) and pitch pine (*Pinus rigida*) with some black oak (*Q. velutina*). The ericaceous understory consists of shrubby blueberries (*Vaccinium vacillans*) and huckleberries (*Gaylussacia baccata*), and *Gaultheria procumbens*. Coweeta Watershed 18, a basic oak–hickory forest, has *Quercus prinus* and *Acer rubrum* in the overstory with a dense

Fig. 1.14. Winter aspect of temperate deciduous forest, Solling, Federal Republic of Germany, with experimental apparatus for stemflow and nutrient cycling studies. Overstory of *Fagus sylvatica*. Photo by D. E. Reichle.

layer of large shrubs, *Rhododendron maximum* and *Kalmia latifolia*, both evergreen ericads, in the understory.

The five sites at Oak Ridge represent one mesic forest of *Liriodendron tulipifera* (tulip poplar; Fig. 1.15), and four less mesic sites of the oak–hickory complex. Oaks (*Quercus alba, Q. rubra, Q. velutina, Q. prinus*), and hickories (*Carya tomentosa, C. ovata*, and others) are important components (Fig. 1.16). Associates, forming a distinct understory tree stratum include redbud (*Cercis canadensis*), dogwood (*Cornus florida*), sourwood (*Oxydendron arboreum*), black gum (*Nyssa sylvatica*) and shortleaf pine (*Pinus echinata*).

Finally, two sites in south central Wisconsin represent interesting contrasts. One site, Noe Woods, is a natural forest in the University of Wisconsin Arboretum. It is dominated by white and black oaks (*Quercus alba* and *Q. velutina*) and black cherry (*Prunus serotina*). The understory contains a number of shrubs and vines (e.g., *Corylus americana, Parthenocissus quinquefolia*) and numerous herbs (*Circaea quadrisulcata, Phryma leptostachya, Galium* spp.). A short distance away lies the Nakoma Urban Forest, actually a residential area of the city of Madison that was literally carved out of the pre-existing forest. The overstory composition is

Fig. 1.15 Winter aspect of temperate deciduous forest, Tennessee, USA, with towers for meteorological and primary production studies. Overstory of *Liriodendron tulipifera*, *Quercus prinus*, and *Pinus echinata*. Oak Ridge National Laboratory photo.

Fig. 1.16 Summer aspect of temperate deciduous forest, Tennessee, USA. Overstory of *Liriodendron tulipifera*, *Quercus prinus*, and *Pinus echinata*. Oak Ridge National Laboratory photo.

similar, but the substory has been replaced with ornamental honeysuckles (*Lonicera* spp.) and lilacs (*Syringa vulgaris*), and the planted lawns produce a field layer of almost pure Kentucky bluegrass (*Poa pratensis*).

Boreal forests

Thirty-three sites of the Woodlands Data Set are classified as boreal forest. These include 28 natural forests (Fig. 1.17) and five plantations, one in Sweden and four in West Germany. The majority of sites (19) are in the USSR, with four in Canada, one in Finland, one in Rumania, and three in Alaska (US). All are dominated by spruce (*Picea* spp. (Fig. 1.18)), although fir (*Abies* spp.) is also common. The sites range in latitude from about 45° N in Canada to 66° N at Oulu, Finland. Most are lowland sites (< 500 m) with the highest reported elevation at Sinaia 2, Rumania, 1010 m above mean sea level.

Boreal needle-leaved evergreen plantations

The five plantations are all of *Picea abies*. All are growing in acid (pH 3.3–6.5, mostly 3.5–4.0) brown earths overlying sandstone, shale, or stony moraine. All are average to well-drained. Kongalund, Sweden, has a

Fig. 1.17. Boreal needle-leaved evergreen forest, Switzerland. Overstory of *Picea abies*. Photo by D. E. Reichle.

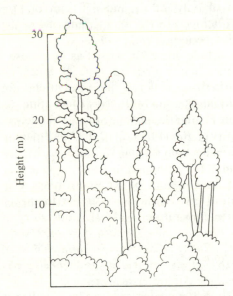

Fig. 1.18. Profile diagram of boreal needle-leaved evergreen forest, central Sweden. Overstory of *Picea abies*, *Pinus sylvestris*, and *Betula verrucosa*. From a photograph by Lars Bergström.

27

growing season of 230 days, a mean annual temperature of 7°C, and 800 mm of precipitation. By contrast, the three Solling sites (F1, F2, and F3) have a 132-day season, a mean temperature of 5.9°C, and 1063 mm of precipitation. The remaining site, Faeglharting, has a growing season of 160–70 days, 7°C mean annual temperature, and 875 mm of rainfall.

The stands range in age from 34 yr (Solling F3) to 115 yr at F2. Kongalund is highly productive; at 60 yr the trees are 25 m tall, and show a basal area of 55.6 m²/ha and 30.8 kg/m² aboveground standing crop at a density of 880 trees/ha. Sparse raspberries (*Rubus idaeus*) and sheep sorrel (*Oxalis acetosella*) are present in the ground layer. The Solling sites show variable parameters with little correlation with age or height. Ellenberg (1971) gives additional details on structure and composition.

Boreal needle-leaved evergreen forests

Data on 17 boreal sites in the USSR were provided by N. I. Kazimirov and R. M. Morozova. Eleven of these are classed, in the nomenclature of Braun-Blanquet (1932) as 'Myrtillosum', i.e., with an understory of the blueberry, *Vaccinium myrtillus*. These form an age series from 22 yr (Site 8) through 138 yr (Site 17). Heights range from 2.6 m to 22.8 m, basal area from 10.6 m²/ha to 40.5 (Site 16) at 126 yr of age. Leaf area index changes little, ranging from 1.8 at age 22 yr, peaking at 3.8 (age 82 yr), then dropping to 2.4 in the oldest stands. The series is exposed to a growing season of 150 days, a mean annual temperature of only 2.2°C, and 650 mm of precipitation. However, growing season temperature is 11.9°C, and 380 of the 650 mm comes as rainfall during this period.

The other six sites of this group include one each classed as Oxalidoso–Mytrillosum (*Oxalis acetosella* and *Vaccinium myrtillus*) (Site 4), Uliginio–Herbosum (*Vaccinium uliginosum* with herbaceous flora) (Site 7), Polytrichosum (species of the moss genus *Polytrichum*, usually *P. commune*) (Site 6), Oxalidosum (Site 5), Vacciniosum (Site 2), and Cladinosum Saxatile (the lichen, *Cladonia* and the small, shrubby, raspberry, *Rubus saxatilis*) (Site 1). All of these sites range between 37 and 45 yr of age, between 4.2 and 12.2 m in height, and between 13.3 and 25.4 m²/ha in basal area.

The Central Forest Reserve, with its 11-yr-old spruce forest, has a 128-day growing season. Temperature and precipitation are similar to the 17 sites above, although growing season temperature is 14.4°C. Understory in the taiga site is composed of *Vaccinium myrtillus*, *Linnaea borealis*, *Oxalis*, *Maianthemum*, *Trientalis europaea*, *Rubus saxatilis*, *Stellaria holostea*, *Luzula pilosa*, and two ferns, *Dryopteris austriaca* and *Thelypteris phegopteris*.

The Koinas site, near Archangelsk, is the coldest site in the Woodlands Data Set, with a mean annual temperature of −1.2°C, and a growing season only 76 days long. Mean annual precipitation approximates 500 mm. Under the spruce overstory are the common shrubby juniper (*Juniperus communis*), two blueberries or bilberries (*Vaccinium myrtillus* and *V. vitis-idaea*), and a ground layer of the boreal moss, *Hylocomium proliferum*.

The site at Oulu, Finland is similar to Koinas. The spruce is *Picea excelsa*, the moss is *H. splendens* (joined by *Pleurozium schreberi*), but the blueberries are identical. The stand is 260 yr of age, 16.2 m high, and is formed on a strong podzol overlying dolomitic quartzite. At a stocking density of 550 trees/ha, the forest has a basal area of 22.5 m²/ha and a standing crop (aboveground) of 10.2 kg/m².

Sinaia 2, in Rumania, lies at elevations between 960 and 1010 m above mean sea level. *Abies alba* is the dominant tree, with an understory of *Oxalis acetosella* and the moss, *Pleurozium schreberi*. At a far lower latitude than the above sites (45° 23′ N), at 110 yr the trees are 36 m tall. Basal area is 76.6 m²/ha, and standing crop is 47 kg/m². Density is 485 trees/ha. This is accomplished during a growing season of 175 days, a mean annual temperature of 5.1°C, and 1025 mm of precipitation.

The Alaskan sites are all dominated by black spruce, *Picea mariana*, the characteristic species of the North American taiga. Substory principal components are the shrubs *Ledum groenlandicum*, *Vaccinium uliginosum*, *V. vitis-idaea*, *Rosa acicularis*, alder (*Alnus crispa*), and willows (*Salix* spp.). *Geocaulon lividum*, *Ptilium crista-castrensis*, and *Equisetum* spp. are joined in the field layer by a number of mosses (*Pleurozium schreberi*, *Hylocomium splendens*, *Polytrichum perinum*, *Dicranum* spp., and *Sphagnum girgenschnii*) and lichens (*Cladonia rangiferina*, *Cladonia* spp., *Peltigera apthosa*, *P. scabrosa*, and *Nephroma arcticium*).

One stand, 130 yr old, reaches a height of 13.7 m, with a basal area of 34.7 m²/ha and 23.1 kg/m² aboveground standing crop. The other two sites, by contrast, at 51 and 55 yr of age, are only 2.9 and 3.1 m tall, respectively. At extremely high densities, however (27 335 and 14 820 trees/ha, respectively), the stands reach basal areas of 35.8 and 41.9 m²/ha and aboveground standing crops of 11.1 and 16.4 kg/m².

Finally, four sites in southern Ontario, Canada, conclude the boreal forest series. These sites (1, 2, 3, and 4) form a gradient in soil moisture from dry through fresh, moist, and wet. All but the last are on ferro-humic podzols, the wet site formed on a hydric humisol. All are very acid, pH ranging from 3.7 to 4.5. The stands (all but Site 3, moist) lie between 465 and 503 m above sea level. Site 3 is at 80 m elevation. Regional growing season is 167 days in length, with a mean annual temperature of 4.0°C, and 1242 mm of precipitation.

Site 1 has a diverse overstory composed of red spruce (*Picea rubens*), black spruce (*P. mariana*), and white spruce (*P. glauca*), white pine (*Pinus strobus*), balsam fir (*Abies balsamea*), hemlock (*Tsuga canadensis*), white

Table 1.3. *Summary of biological and environmental parameters for sites included in the Woodlands Data Set. Unless otherwise indicated, values are means and standard errors based on the number of stands included*

	Forest type				
	Mediterranean broad-leaved evergreen	Tropical broad-leaved evergreen	Tropical broad-leaved deciduous	Tropical broad-leaved deciduous plantation	Temperate broad-leaved evergreen
No. of sites	3	4	2	14	1
Stand age (range, yr)	51–150	Mature	60–120	5–40	55
Basal area (m²/ha)	41.3±8.7	29.8±0.8	26.3±4.3	29.4±3.6	47.9
Height (m)	17.1±4.3	37.8±0.3	13.8±2.2	14.3±1.0	12.0
Leaf area index	4.3±0.2	—	—	9.2±0.8	6.0
Leaf biomass (g/m²)	694±6	—	—	834±112	770
Branch and bole biomass (g/m²)	27825±1625	37126±1749	16494±2324	14366±2539	18558
Branch and bole increment (g/m²/yr)	—	—	747±63	1009±140	983
Aboveground standing crop (g/m²)	28753±1759	43266±3111	17200±2404	15200±2581	19328.5
Net primary production (g/m²/yr)	748±104	1549±44	1304±194	1631±184	1368
Total litterfall (g)	370±14	—	639±39	—	—
Total leaffall (g)	217±28	654±94	496±70	639±82	385
Belowground standing crop (g/m²)	—	—	2908±360	3459±545	—
Soil top organic matter (g/m²)	1482±342	—	—	719±89	809
Average annual temperature (°C)	12.9±0.9	26.5±0.4	—	27.5	21.5
Average annual precipitation (mm)	908±77	1851±82	1058±214	1158	2630
Length of growing season (days)	365	365	194±76	—	365
Growing season temperature (°C)	12.9±0.9	26.5±0.4	—	—	21.5
Growing season precipitation (mm)	908±77	1851±82	—	—	2630

	Temperate deciduous needle-leaved plantation	Temperate needle-leaved evergreen	Temperate needle-leaved evergreen plantation	Temperate broad-leaved deciduous	Boreal needle-leaved evergreen	Boreal needle-leaved evergreen plantation
No. of sites	1	5	5	19	9	5
Stand age (range, yr)	39	85–290	15–36	30–200	51–130	34–115
Basal area (m²/ha)	37.3	68.8±9	34.5±5.3	23.7±1.1	32.8±1.7	46.2±4.5
Height (m)	19.4	21.0±2	15.0±1.4	20.8±1.4	17.2±17	25.8±2.4
Leaf area index	6.7	8.8±1	—	5.2±0.3	7.6±0.4	—
Leaf biomass (g/m²)	359	932±213	647±90	350±29	964±90	1371±281
Branch and bole biomass (g/m²)	16080	22496±5490	11249±1647	16249±2562	12443±1876	23081±3083
Branch and bole increment (g/m²/yr)	580	382±74	743±144	359±36	135±29	699±108
Aboveground standing crop (g/m²)	16938	21437±4000	11918±1732	17352±2235	13917±1869	24452±2930
Net primary production (g/m²/yr)	939	1159±236	1249±157	918±74	516±60	1128±177
Total litterfall (g)	—	408±29	578±106	528±58	349±60	—
Total leaffall (g)	359	201±18	348±66	342±13	230±42	344±32
Belowground standing crop (g/m²)	3794	—	3116±319	3799±451	3810±551	6005±917
Soil top organic matter (g/m²)	1390	—	—	757±169	3776±798	—
Average annual temperature (°C)	16.2	6.1±1.9	13.6±1.1	9.9±0.8	0.25±1.56	6.6±0.4
Average annual precipitation (mm)	1806	935±270	1338±89	917±115	514±89	913±78
Length of growing season (days)	150	196±8	201±16	198±14	106±20	186±29
Growing season temperature (°C)	—	11.6±0.04	18.4±1.2	15.0±1.0	13.0±0.6	13.4±0.8
Growing season precipitation (mm)	—	639±142	690±31	499±81	264±60	452±54

cedar (*Thuja occidentalis*), white or paper birch (*Betula papyrifera*), and red maple (*Acer rubrum*). The substory contains young individuals of the overstory species, plus *Amelanchier arborea*. The field layer exhibits *Vaccinium angustifolium*, bracken fern (*Pteridium aquilinum*), and the reindeer lichen, *Cladonia rangiferina*. Stand age is 84 yr, and the canopy is at a height of 15 m. Basal area is 32.3 m²/ha, standing crop is 12.1 kg/m², and the stand density is 3311 trees/ha.

Site 2 has a canopy of red spruce, balsam fir, white cedar, and red maple. Additions to the substory and field layer include beech (*Fagus grandifolia*), *Vaccinium brittonii, V. myrtilloides, Gaultheria hispidula*, and *Coptis groenlandica*. The stand is 246 yr old and reaches a height of 18.3 m. Density is 3870 trees/ha, basal area is 46.8 m²/ha, and aboveground standing crop is 26.4 kg/m².

Site 3 again has red spruce, hemlock, balsam fir, white cedar, and yellow birch in the overstory, along with beech. Four species of maples, *Acer pennsylvanicum, A. saccharum, A. spicatum*, and *A. rubrum*, and beaked hazel (*Corylus cornuta*) share the understory, while the field layer includes *Oxalis montana, Aralia nudicaulis*, the fern, *Dryopteris intermedia*, and the shining clubmoss, *Lycopodium lucidulum*. The stand is 212 yr old, 25.6 m high, and has a basal area of 59.4 m²/ha. The density is 3065 stems/ha and aboveground standing crop totals a little over 8 kg/m².

The wet stand, Site 4, is restricted to red and black spruce in the canopy layer. Balsam fir, along with red spruce, appears in the substory. The field layer includes a sedge, *Carex trisperma*, and two species of *Sphagnum, S. capillaceum* and *S. palustre* that reach positions of importance. This site harbors a forest 135 yr old. The trees are 17.4 m tall, basal area is about 28 m²/ha, and above-ground standing crop equals 12 kg/m² at a density of 8031 stems/ha.

Summary

In order to derive comparative values for the 11 major forest types, the Woodlands Data Set was analyzed to provide means and standard errors for 18 biological and environmental parameters (Table 1.3). In several instances, (Tropical broad-leaved deciduous plantation, Temperate broad-leaved deciduous, and Boreal needle-leaved evergreen), a subset of the total number of sites in each category was used to ensure a more uniform analysis of all parameters. Sites with many missing values for the selected suite of parameters were eliminated from consideration in the preparation of Table 1.3. DeAngelis *et al.* (Chapter 11, this volume), give all available data for each site incorporated in the Woodlands Data Set.

The biological parameters chosen for inclusion in the table are those for which there is the greatest degree of confidence in methods of measurement.

In spite of early attempts to standardize methodologies (the IBP Handbook series published by Blackwell, for example), forest researchers at the IBP sites used a variety of techniques for measurement of various ecosystem characters. The 13 biological parameters in Table 1.3, however, are sufficiently well understood so that the values expressed are ecologically meaningful. Perusal of Table 1.3 results in a generalized knowledge of both the magnitudes and the variability in the forest ecosystems studied during the International Biological Programme. The data, collectively, represent a major source of information on the world's forest resource from which additional analyses and syntheses will continue to spring.

In the pages that follow in the volume, a series of chapters synthesizes much of the quantitative data that were generated in these 117 forest and woodland sites during the International Biological Programme. Analyses and comparisons of a broad array of ecosystem parameters treat the scant information presented here in considerably more detail. It is hoped, however, that the qualitative, descriptive overview of this chapter will serve as an introduction to what follows, and will lead to a greater ultimate level of understanding on the part of the readership.

References

Adamson, R. S. & Osborn, T. G. B. (1924). The ecology of the Eucalyptus forests of the Mount Lofty ranges (Adelaide District), South Australia, *Trans. Roy. Soc. S. Austral.* **48**, 87–141.

Aubert de la Rüe, E., Bourlière, R. & Harroy, J. P. (1957). *The tropics.* Alfred A. Knopf, New York.

Aubréville, A. (1938). La forêt coloniale: les forêts de l'Afrique occidentale française. *Ann. Acad. Sci. Colon., Paris,* **9**, 1–245.

Aulak, W. (1970). Studies on herb layer production in the *Circaeo-Alnetum* Oberd. 1953 Association. *Ekol. Polska,* **18**, 411–27.

Berg, L. S. (1950). *Natural regions of the USSR.* The Macmillan Co., New York.

Biskupsky, V. (ed.). (1975). *Research Project Bab. IBP Progress Report II.* Veda Publ. House, Slovak Acad. Sci., Bratislava, Czechoslovakia.

Bliss, L. C. & Wielgolaski, F. E. (1973). *Primary production and production processes, Tundra Biome. Proceedings of the Conference, Dublin, Ireland, April 1973.* Tundra Biome Steering Committee, Univ. Alberta, Edmonton, Alberta, Canada.

Braun, E. L. (1950). *Deciduous forests of eastern North America.* Blakiston Co., Philadelphia.

Braun-Blanquet, J. (1932). *Plant sociology. The study of plant communities.* (Transl. G. D. Fuller & H. S. Conard). McGraw-Hill Book Co., New York.

Champion, H. G. & Seth, S. K. (1968). *The forest types of India.* Government of India Press, Delhi.

DeAngelis, D. L., Gardner, R. H. & O'Neill, R. V. (1978). *Productivity in temperate woodlands. CRC Handbook on Nutrition and Food.* Chemical Rubber Co. Press, Cleveland.

di Castri, F. & Mooney, H. A. (ed.). (1973). *Mediterranean type ecosystems. Origin*

33

and structure. Ecological studies 7. Springer-Verlag, New York-Heidelberg-Berlin.

Edlin, H. L. (1958). *England's forests.* Faber & Faber, Ltd. London.

Ellenberg, H. (1971). *Integrated experimental ecology. Methods and results of ecosystem research in the German Solling Project.* Springer-Verlag, New York.

Ellenberg, H. & Mueller-Dombois, D. (1975). Tentative physiognomic-ecological classification of plant formations of the earth. Appendix B. pp. 466–79. In *Aims and methods of vegetation ecology,* ed. D. Mueller-Dombois & H. Ellenberg. John Wiley & Sons, New York.

Emberger, L., Gaussen, H., Kassas, M. & de Phillipis, M. (1963). *Bio-climatic map of the Mediterranean Zone. Arid Zone Research XXI.* UNESCO-FAO, Paris.

Fosberg, F. R. (1967). A classification of vegetation for general purposes. Appendix 1, pp. 73–120. In *Guide to the checksheet for IBP areas,* ed. G. F. Peterken. Blackwell Scientific Publications, Oxford.

Holdridge, L. R. (1964). *Life zone ecology.* Center for Tropical Studies, San Jose, Costa Rica.

Heutz de Lemps, A. (1970). *La végétation de la terre.* Masson et Cie, Paris.

Innis, G. (ed.). (1978). *Grassland simulation model. Ecological studies 26.* Springer-Verlag, New York-Heidelberg-Berlin.

Knapp, R. (1965). *Die Vegetation von Nord- und Mittelamerika und der Hawaii-Inseln.* Gustav Fischer Verlag, Stuttgart.

Knapp, R. (1973). *Die Vegetation von Afrika.* Gustav Fischer Verlag, Stuttgart.

Lawrence, A. O. (1939). Notes on *Eucalyptus obliqua* (messmate) regrowth forests in central Victoria. *Austral. For.* **4,** 4–10.

Lossaint, P. (1973). Soil-vegetation relationships in Mediterranean ecosystems of southern France, pp. 199–210. In *Mediterranean type ecosystems. Origin and structure. Ecological studies* 7, ed. F. di Castri & H. A. Mooney. Springer-Verlag, New York-Heidelberg – Berlin.

Mabry, T. J., Hunziker, J. H. & Difeo, D. R. Jr. (ed.). (1977). *Creosote bush. Biology and chemistry of* Larrea *in New World deserts.* Dowden, Hutchinson & Ross, Stroudsburg, Pennsylvania.

Meggers, B. J., Ayensu, E. S. & Duckworth, W. D. (1973). *Tropical forest ecosystems in Africa and South America: a comparative review.* Smithsonian Institution Press, Washington, D. C.

Mooney, H. A. (ed.). (1977). *Convergent evolution in Chile and California. Mediterranean climate ecosystems.* Dowden, Hutchinson & Ross, Stroudsburg, Pennsylvania.

Numata, M. (1974). *The flora and vegetation of Japan.* Elsevier Publ. Co., Amsterdam.

Orians, G. H., & Solbrig, O. T. (ed.). (1977). *Convergent evolution in warm deserts. An examination of strategies and patterns in deserts of Argentina and the United States.* Dowden, Hutchinson & Ross, Stroudsburg, Pennsylvania.

Reed. J. L. (1954). *Forests of France.* Faber & Faber, Ltd. London.

Reichle, D. E. (1970). *Analysis of temperate forest ecosystems.* Springer-Verlag, New York.

Reichle, D. E., Franklin, J. F. & Goodall, D. W. (ed.). (1975). *Productivity of world ecosystems.* National Academy of Sciences, Washington, D. C.

Richards, P. W. (1952). *The tropical rain forest.* Cambridge University Press, Cambridge.

Rule, A. (1967). *Forests of Australia.* Angus & Robertson, Ltd., Sydney.

Simpson, B. B. (ed.). (1977). *Mesquite. Its biology in two desert scrub ecosystems.* Dowden, Hutchinson & Ross, Stroudsburg, Pennsylvania.
Specht, R. L. (1972). *The vegetation of South Australia.* Govt. Printer, Adelaide.
Specht, R. L. (1973). Structure and functional response of ecosystems in the Mediterranean climate of Australia, pp. 113–20. In *Mediterranean type ecosystems. Origin and structure. Ecological Studies 7*, ed. F. di Castri & H. A. Mooney, Springer-Verlag, New York-Heidelberg-Berlin.
Szafer, W. (ed.). (1966). *The vegetation of Poland.* Pergamon Press, Oxford.
Tseplyaev, V. P. (1965). The forests of the USSR. *Israel Program for Scientific Translations, Jerusalem.*
UNESCO. (1973). *International classification and mapping of vegetation.* UNESCO, Paris.
Walter, H. (1964). *Die Vegetation der Erde. Band I. Die tropischen und subtropischen Zonen.* Gustav Fisher Verlag, Stuttgart.
Walter, H. (1968). *Die Vegetation der Erde. Band II. Die gemassigten und arktischen Zonen.* Gustav Fisher Verlag, Stuttgart.
Walter, H. (1973). *Vegetation of the earth in relation to climate and the eco-physiological conditions.* Springer-Verlag, New York.
Whitmore, T. C. (1975). *Tropical rain forests of the Far East.* Clarendon Press, Oxford.
Wielgolaski, F. E. & Rosswall, T. (1972). *Tundra Biome proceedings, IV international meeting on the biological productivity of tundra, Leningrad, USSR, October 1971.* Tundra Biome Steering Committee, Wenner-Gren Center, Stockholm, Sweden.
Wielgolaski, F. E. (1975a). *Fennoscandian tundra ecosystems. Part 1. Plants and microorganisms.* Springer-Verlag, New York.
Wielgolaski, F. E. (1975b). *Fennoscandian tundra ecosystems. Part 2. Animals and systems analysis.* Springer-Verlag, New York.

2. Growth, aging and succession

O. L. LOUCKS, A. R. EK, W. C. JOHNSON & R. A. MONSERUD

Contents

In this volume, ecosystems are understood as units of biota and environment in which organisms and the non-living requirements for life interact as a system to produce exchanges of materials and the continuance of mixed populations of reproducing species. All aspects of forest ecosystems are viewed, therefore, as potentially dynamic, particularly their development over long periods of time. This development should be examined as an ecosystem process, a function of the interactions between organisms and their environment over time. However, in addition to understanding development of interactions over time, it is necessary to recognize the dynamic nature of the age distribution of individuals making up species in long-lived forest ecosystems. Continual gradual change in the age or size classes of species making up the dominants in these systems, and understanding the pattern and the influences which bring the change about, thus, is an essential part of the analysis of woodland ecosystems.

This chapter, therefore, consists of three sections, designed to describe current thought on the growth, aging and succession of the forest ecosystem. An initial section by Alan R. Ek and Robert A. Monserud describes recent innovations in analysis of the patterns of tree development in forest and subsequent characterization of these patterns in the form of mathematical models. The emphasis is on the methodology for characterizing forest stand dynamics in a form directly applicable to evaluating alternative forest manipulations as management practices.

The following section, by W. Carter Johnson, takes a larger view of stand development processes and describes regional successional models as an aid both in understanding succession as a process, and more importantly, to provide an understanding of human influences in altering the forest composition of large areas of the continents.

The third section, by O. L. Loucks, discusses ecosystem diversity in relation to regions, their environment, productivity and especially age from disturbance (i.e., secondary successional time). This paper builds on research over the past decade concerning the processes governing community diversity and attempts to view this aspect of forest ecosystems – and its associated ideas, stability and productivity – as time-dependent functions that can greatly influence carbon and nutrient processing.

The chapter editor regrets the unfortunate over-emphasis of North American literature in this section. Strenuous efforts were made to obtain several review papers on the subject for Europe and Asia. The material appears to be available for these areas, but could not be arranged without personal contact. Happily, the principles described in the following papers seem likely to have wide application.

Methodology for modeling forest stand dynamics*

The development of individual trees and stands involves complex processes and observed changes which vary greatly with genotype, stage of development, and environmental factors. Still, there are numerous patterns that virtually all trees follow in their development. Analysis of these patterns and their characterization in the form of mathematical models form the objectives of this section. While most of these patterns might be traced to rather basic biochemical processes within individual trees, the level of abstraction in the models emphasized here extends only to physically measureable exterior tree characteristics. More complex models have been developed (e.g., Wilson & Howard, 1968; *see* Murphy, Hesketh & Strain, 1972 for additional references), but as yet they have not been effectively extended to the analysis of aggregate stand development. The emphasis here is on a methodology for modeling forest stand dynamics that is directly applicable to the evaluation of alternative forest treatment practices.

* Research supported by the Eastern Deciduous Forest Biome, US-IBP, funded by the National Science Foundation under Interagency agreement AG-199, BMS69–01147 A09 with the US. Department of Energy–Oak Ridge National Laboratory. Contribution No. 340 from the Eastern Deciduous Forest Biome, US–IBP.

The nature and components of change in forest stands

Tree growth

Virtually all exterior tree dimensions develop in a sigmoid manner with respect to time. This relationship might be expressed by the integral form of the Richards function (Richards, 1959) as:

$$Y = \beta_1 (1 - e^{-\beta_2 t})^{\beta_3}, \tag{1}$$

where the βs are constants, t is time and Y is some tree characteristic, such as total height. In this expression β_1 is the upper asymptote, β_2 a rate parameter and β_3 is a rate and shape parameter. Aside from the fact that growth may be seasonally distinct (i.e., a difference model might be more appropriate), this model is often sufficient to characterize a wide range of species or tree characteristics. Only the βs need to be changed in most cases to reflect changes in the definition of the dependent variable. (*See* Grosenbaugh, 1965 for a variety of sigmoid models.)

Given one tree characteristic, such as total height, other dimensions (e.g., stem diameter, crown width, branch length, etc.) are often related in an allometric fashion. Thus the power function:

$$Y_2 = \beta_1 Y_1^{\beta_2}, \tag{2}$$

where Y_1 and Y_2 are two different tree characteristics, may provide an accurate description of these relationships. Likewise, more complex descriptions of tree shapes have been developed involving more than one independent variable. The length of branch free bole (Y_3), for example, has been expressed by Ek (1974a) as:

$$Y_3 = Y_1 e^{-\beta_1 Y_1^{\beta_2}(Y_2 + \beta_3)^{\beta_4}}, \tag{3}$$

where Y_1 and Y_2 are tree height and diameter, respectively.

Returning to the basic sigmoid curve (eqn 1.1), it should also be emphasized that the position of a tree on its own or expected curve of development is often indicative of its sensitivity or ability to respond to changes in its environment (Hari, Leikola & Rasanen, 1970). While much is known of the general patterns of a tree's responses to its environment (*see* Fowells, 1965), only recently are we learning about the mechanisms involved in these responses (Kozlowski, 1971a,b).

Stand development patterns

A great deal has been synthesized to date on patterns of stand development (*see* Meyer *et al.*, 1961; Turnbull, 1963; Assmann, 1970). For even-aged stands there is a gradual competition-induced reduction through time of stems per unit area is an approximately exponential decay pattern. Thus stand density in terms of stems per unit area tends to converge asymptotically with time. Furthermore, it is well understood from yield table construction that good sites tend to carry fewer trees at a given age than poorer sites, perhaps due to greater crown and size class differentiation on the former areas. Intraspecific root grafting has also been shown to be a common occurrence, developing at early stand ages (Eis, 1972). The consequence here is that trees may respond more as a unit than as individuals.

Size-class distributions, whether in terms of tree diameters or heights, also show a competition-induced trend from positive skewness to symmetrical shapes, often approaching normality with time. Changes in species composition of stands likewise depend upon competition and especially on the differential ability of the species involved to reproduce and develop to maturity under a given set of conditions. Turnbull (1963) has postulated likely patterns of species composition and stand growth, but it is difficult to generalize here due to the complex and variable nature of competitive stresses induced over long time horizons. Spatial patterns are also known to vary between species and over time due largely to specific requirements for reproduction and later competition (Payandeh, 1974). A common trend is from clustered patterns at early ages to near random and uniform spacings at later points in time. The aggregate yield or development of a stand for most products (e.g., biomass, board feet), like the individual stem, assumes an approximately sigmoid pattern.

The uneven-aged forest stand is similar to the even-aged condition, but more complex in that several to many age classes may exist in the form of various size classes and/or species groups in the stand. The individual size or age classes involved develop in a manner similar to that described above, but competition is evident both within and between these classes. In the classic but rare case of an all-aged stand, reproduction and mortality among the various size classes may effectively stabilize the aggregate size and age-class distributions over time (Meyer, 1952).

The evolution of forest stand growth modeling

There is a long history of growth and yield research at the stand level. Indeed, perhaps more data have been gathered on growth, yield, and stand development than on any other aspect of forestry research (Bella, 1970). From the standpoint of modeling the responses of stands to various treatments, five distinct approaches are evident: traditional yield table methodology, differential or difference equations, stochastic processes, distributional methods, and individual tree simulation models.

Traditional yield tables

The earliest approach involved extrapolation from conventional normal, empirical, or variable density yield tables (Husch, Miller & Beers, 1972). Such tables for even-aged stands are usually in the form of yield tabulations for various site qualities for a range of stand ages. A normal yield table is constructed from plot data on stands of assumed similar stocking, i.e., where the site is judged to be fully utilized (normal or fully stocked). An empirical yield table differs by the use of plots of average rather than full stocking. Variable density yield tables provide yields for a range of densities for each age and site combination. For uneven-aged stands, yields to a rotation age are replaced by periodic yields over some specified time interval. In North America most yield tables were developed in the period 1920–50, following methodology developed earlier in Europe. Since this methodology was intended to describe the undisturbed development of natural stands, the resulting tables are of limited usefulness for comparing silvicultural alternatives. A detailed worldwide review of such approaches is provided by Vuokila (1966) and Curtis (1972).

Differential-difference equation models

Since the 1960s, differential or difference equations have been used to describe the rates of change of various components of stand development. Buckman (1962) expressed the growth rate of even-aged stands as a function of age, site, and stocking in a differential expression. Numerical integration over time then provided yield predictions. Clutter (1963), with a similar approach, developed rate models which could be integrated to provide closed form expressions for yield predictions. Moser & Hall (1969) applied these techniques to uneven-aged hardwood stands by deriving time-dependent nonlinear yield functions from the integration of rate equations which did not have time or age as an independent variable. Moser (1972) also quantified separately the three basic growth components: ingrowth, mortality, and survivor growth. Moser (1974) then developed systems of nonlinear differential equations for predicting the development of the growth components by size or diameter classes. In an earlier study, Leary (1968) developed a system of ten simultaneous difference equations for survivor growth by height classes. Leary (1970) later developed a differential system of equations for diameter class development. Leary (1970) and Leary & Skog (1972) further discussed the theoretical aspects of such simultaneous systems and presented an analytical framework for system parameter estimation and solution.

An alternate approach was taken by Ek (1974b) in dealing with uneven-aged hardwood stands. He described ingrowth, mortality, and survivor

41

growth, in terms of the number of stems per two inch diameter class, with nonlinear difference models. The system is recursive, while those of Moser and Leary are simultaneous. An advantage of this approach is that it readily facilitates the application of mathematical programming methodology to develop optimal tree-size–class-distributions (Adams & Ek, 1974). This follows from the fact that, in a recursive system, the distribution at any point in time is a function only of initial conditions.

The work of Turnbull (1963) and Pienaar (1965) also deserves note here for their extension of the Lotka–Volterra theory (Lotka, 1924) to include forest population dynamics for unmanaged even-aged stands of mixed and pure species composition. In particular, they found the Chapman-Richards function sufficiently flexible to serve as a simple general model for forest stand growth.

Stochastic processes

The Markov process is the approach used most frequently in stochastically based descriptions of stand development. Rudra (1968) and Burner & Moser (1973) used Markov processes to describe the natural transition of stems through diameter classes over time for even- and uneven-aged stands, respectively. Stephens & Waggoner (1970) assumed a stationary Markov model to describe 40-yr changes in species composition of mixed hardwood stands in Connecticut. Peden, Williams & Frayer (1973) also described a Markov model for stand table projection for which they obtained closed form expressions for expectations, variances, and covariances of tree counts.

Suzuki & Umemura (1974) also used a Markov model to predict future diameter distributions, but time was considered continuous rather than discrete. By postulating an exponential model for the mean and variance of growth, they were able to obtain solutions to the general Kolmogorov equations. By further incorporation of an absorbing barrier in the boundary conditions to account for mortality due to suppression, they obtained positively skewed diameter distributions that were actually a sum of two normal distributions with equal variance but different means. Dress (1968) presented a general frame-work for a stochastic individual tree model that is similar to the approach taken by Suzuki & Umemura.

Leak (1969) employed a variation of the simple death process in projecting the development of an even-aged northern hardwood stand. Rather than estimating survival probabilities directly as a function of time, Leak expressed chances of mortality as a function of mean stand diameter. It was further assumed that trees within a size class died independently and with equal probabilities. Leak (1970a) later extended this model to predict future numbers of trees by diameter classes.

The utility of pure stochastic models appears to be more limited than that of deterministic approaches, due largely to the substantial data bases required to develop meaningful transition probabilities. However, they may have advantages for some growth-related optimization research involving mathematical programming (Adams, 1975).

Distributional approaches

As was noted earlier (e.g., *see* Turnbull, 1963), size-class distributions for even-aged stands develop in a rather regular manner over time. Using this knowledge, Clutter & Bennett (1965) characterized size-class distribution in old field slash pine plantations with the beta probability density function. They then estimated distribution parameters as functions of age, site quality, and initial stand density. Specification of these three values thus provided for indirect projection of a range of stand characteristics, depending on the size classes considered.

The beta density was also used to predict diameter distributions in natural yellow-poplar stands (McGee & Della-Bianca, 1967) and in loblolly pine plantations (Burkhart & Strub, 1974). Zöhrer (1972) discussed techniques for estimating the parameters of the beta density function and the applicability of this model to various types of diameter distributions.

The utility of the beta density function arises from its ability to assume a wide variety of shapes. This flexibility is also a property of the Weibull density function; however, the Weibull function is mathematically more tractable than the beta. Consequently, the Weibull function has become increasingly popular for characterizing stand diameter distributions. Bailey (1973) and Clutter & Allison (1974) have used the Weibull function to predict the development of *Pinus radiata* stands in New Zealand. Burkhart and Strub (1974) also compared this model with the beta density function and obtained generally better fits with the Weibull model for loblolly pine plantations. A number of useful techniques for estimating the Weibull parameters are discussed by Bailey & Dell (1973).

As with traditional yield tables, however, few silvicultural alternatives can be examined with such distributional approaches, since only the undisturbed development of stands is described. Irregular size-class distributions resulting from management or other disturbances do not lend themselves to characterization by the above rather simple distribution models.

Individual tree simulation models

The most recent approach – individual tree simulation models – differs substantially from the preceding methodology. Here the stand is described by *individual trees* rather than *aggregate stand* characteristics such as basal

43

area or number of trees. These individual trees can then be combined to form stands.

Individual tree models of forest stands provide many capabilities not available to models based on stand aggregates (Hatch, 1971). Aggregate models do not easily lend themselves to all-aged or mixed species stand analyses or the consideration of tree spatial patterns. As long as a tree is represented in the model as an individual entity with a unique location, however, virtually any stand condition and treatment regime can be examined. Hatch also points out that individual tree models can easily be adapted to the study of management alternatives quite removed from those dealing with wood or fiber production (e.g., watershed management).

Distance-independent individual tree models

As noted by Munro (1974), individual tree models may or may not consider inter-tree distances (i.e., locations) as essential attributes. Models of the latter type, called *distance-independent* by Munro, have been constructed by Goulding (1972), Stage (1973) and Lemmon & Schumacher (1962). In this type of model, trees are usually grown individually or by size classes according to some function of aggregate stand characteristics. While these models lack the sensitivity to spatial patterns of the distance-dependent models described below, they are computationally more efficient.

The above models have all been concerned with stands of a single species. Botkin, Janak & Wallis (1970a,b; 1972) developed a mixed species model of northern hardwood stands on the Hubbard Brook Experimental Forest. While this model does not consider individual tree locations, it does present an interesting approach to tree growth potential based on atmospheric energy inputs to photosynthesizing leaf area.

Distance-dependent individual tree models

With distance-dependent models, the interdependence of individual stems is frequently assessed by a competition index (CI), which is usually a function of a subject tree's size and location relative to the size and location of its competitors. Actual growth is then a function of potential growth and this measures competition. The heart of an individual tree simulation model is thus its CI and the manner in which growth is affected by this index (*see* Gerrard, 1966 for a review of various competition indices). Fig. 2.1 presents a simplified flow chart for this type of model.

Previous work with distance-dependent models has dealt almost exclusively with the problem of modeling single species even-aged stands. Models have been fitted to the following species: Douglas fir (Newnham, 1964; Arney, 1972; Lin, 1974) western hemlock (Lin, 1970), white spruce

44

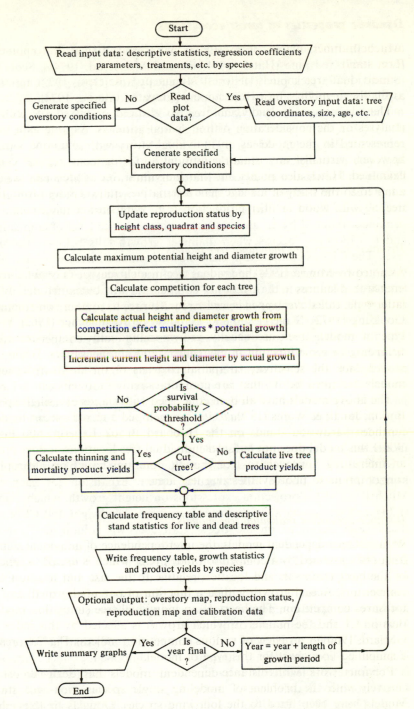

Fig. 2.1. Simplified flow chart for a distance dependent individual tree simulation model.

(Mitchell, 1967), black spruce (Ek & Monserud, 1974*a*), lodgepole pine (Lee, 1967), red pine (Hatch, 1971; Ek and Monserud, 1974*a*), Scots pine (Strand, 1972), jack pine (Hegyi, 1974), eucalyptus (Opie, 1972), trembling aspen (Bella, 1970), and hybrid poplar (Ek & Monserud, 1974*a*).

The earliest work on individual tree based stand models was Newnham's (1964) Douglas-fir simulator. Although operationally crude compared to recent simulators (c.f. Arney, 1972; Ek & Monserud, 1974*b*) Newnham's approach nevertheless became the foundation for the construction of more flexible and sophisticated models. Major assumptions of his model were: (i) a tree free from competition has the diameter growth rate of an open-grown tree of equal diameter, (ii) a tree subject to competition has its diameter increment reduced by an amount proportional to the level of competition, and (iii) mortality occurs when diameter growth falls below a threshold level. The CI was computed in the following manner: trees were first given open-grown crowns; next, the arc length of the subject tree's crown circumference that fell inside the crowns of competitors was computed; finally, the ratio of the tree's overlapped crown circumference to total crown circumference became the CI of the subject tree. It was further assumed that trees – which were restricted to a uniform square spacing – only competed within a distance of eight times the initial spacing. This time saving assumption proved untenable for close material spacing (e.g., 3.3 ft by 3.3 ft), as severe biases were introduced once potential open-grown crown radii extended past this arbitrary limit. Monserud & Ek (1974) later reviewed various approaches used to reduce this bias (termed plot edge bias) in distance-dependent individual tree models. Lee (1967) streamlined Newnham's model, while adapting it to lodgepole pine.

Mitchell (1967) adopted a somewhat different approach by simulating the horizontal projection of the irregular crown expansion for each tree. Mitchell based his projection expressions on height growth, which is more stable and predictable than diameter increment. The height and CI of each tree further depended on the tree's relative crown volume (i.e., ratio of actual to open-grown crown volume) and on average dominant height. As Bella (1970) pointed out, this competition index is an integrated expression for the competition the tree has undergone in the past, but it may fail to describe a tree's current competitive status, since two trees of equal size may not have the same competitive status, owing to differences in the sizes and locations of their respective competitors.

Mitchell added an element to his simulation model that has since been included by most modelers: random error components. Since variability is a characteristic feature of individual organisms – whether due to genetic or environmental differences – some unaccounted variation will always remain (Bella, 1970). In addition, such random variability also reflects inadequacies in the modeler's knowledge of the underlying developmental processes.

Lin (1970, 1974) simplified Newnham's approach by considering as a competitor only the tree in each quadrant that made the largest angle of incidence with the subject tree. The contribution these four trees made to the subject tree's CI was thus a function of the diameter and distance of the competitor from the subject tree. This index was used as an independent variable in the prediction of diameter in much the same manner as CI was used by Newnham (1964). Lin's model also provided for more realistic spacial patterns, since any real coordinates were allowed.

Bella (1970) refined Newnham's CI by considering the area of overlap of competing (open-grown) crowns rather than their perimeter overlap. Furthermore, each competitor's contribution to a subject tree's CI was exponentially weighted by the ratio of the diameters of the subject and competing trees. This weighting implied that, given two competitors with equal areas of overlap with the subject tree, more intense competition would come from the larger competitor. As with Newnham's, Lee's, and Lin's model, Bella's growth functions were based on diameter.

Hatch (1971) developed a growth potential index rather than a competition index for red pine. This index was based on the estimated crown surface area exposed to direct sunlight. In this case crown shapes were assumed to be conical. The analytical barriers to refinements along this line are large, especially if the asymmetrical crowns of broad-leaved trees are to be considered. The primary difference between this model and other simulators lies in Hatch's extensive use of stochastic processes to describe individual stem growth. It combines several aspects of Newnham's (1964) deterministic model with the theoretical framework for a stochastic individual tree model given by Dress (1968).

Arney (1972, 1974) developed an even more complex system in which individual whorls of branches compete. Diameters at successive internodes along the bole and height are annually incremented, with the open-grown potential cambial increments reduced according to the amount of competition each whorl is subjected to. The result is that the growth of each tree is registered along the entire length of the bole as a new sheath of wood. The size of the stem at a given age is thus the summation of the cambial growth sheaths at all previous ages. This growth procedure is thus physiological in emphasis, and attempts to describe the growth patterns noted by Duff & Nolan (1953, 7, 8) and Larson (1963). Hegyi (1974) adapted Arney's Douglas-fir simulator to jack pine, a multinodal species.

The sheath approach developed by Arney appears well suited to the simulation of species that maintain an excurrent branching habit (i.e., conifers), for the growth sheath is added to a well defined main bole. The approach is perhaps most appropriate for even-aged stand simulation. Uneven-aged stands could also be simulated, but this might require the input of a large amount of stem analysis data in order to specify the stand condition at the initiation of the simulation.

47

An alternate approach – which could properly be termed functional rather than physiological – was taken by Ek & Monserud (1974*a,b*) in their development of the simulator FOREST. This model has the capacity to handle uneven-aged mixed species stands and their associated reproduction. Mixed stand growth is facilitated by species-specific parameter sets. The facility for growing even- or uneven-aged stands is a byproduct of the overstory growth algorithms and the relatively small amount of input data required to specify any initial stand condition. The competition index used is a refinement of Bella's weighted area of overlap method, in which the weight is given by the ratio of the size (i.e., height × crown width) of the competitor to the subject tree. This CI is further adjusted by a 0–1 multiplier reflecting shade tolerance of the subject tree. Mortality is probabilistic and based on a tree's competitive status. A step-by-step formulation of the reproduction process is controlled by the subroutine REPRO. Reproduction processes explicitly considered include seed production, dispersal and germination, root sucker and basal sprout production, and the number of seedlings, root suckers and basal sprouts surviving, dying and reaching overstory status by age classes. These processes are considered individually on subplots or quadrats within the main plot; thus reproduction growth and survival are conditioned by the overstory size and spatial pattern.

In the functional approach of Ek & Monserud, height and diameter (at breast height) growth are determined by highly nonlinear functions of height and CI (*see* Ek & Monserud, 1974*b*, pp. 5–6). Crown length is then a function of height and diameter. The growth of trees with deliquescent as well as excurrent crown form are treated with equal ease, for there is no assumption of strong apical dominance. This model could also provide detailed output on stem form by using taper equations such as those described by Demaerschalk (1972). Stem volume could then be obtained by integrating the taper equation.

Ek & Monserud (1974*a*) further formalized calibration procedures by substituting nonlinear least squares for trial-and-error procedures. This procedure also provided an estimate of the magnitude of the individual tree variability associated with the particular model in question (e.g., height growth). Thus the standard error of the regression fit can be used as an estimate of the standard deviation of the error distribution for that model.

In Arney's model, stem diameter and mortality are considered on a whorl basis. Stem mortality then occurs as a function of aggregate whorl mortality. Alternatively, Ek and Monserud increment diameters only at breast height, and mortality is considered only for the stem as a unit. In spite of these differences, however, the basic relationships among competition and height and diameter growth and, mortality are quite similar. These general relationships are shown in Fig. 2.2.

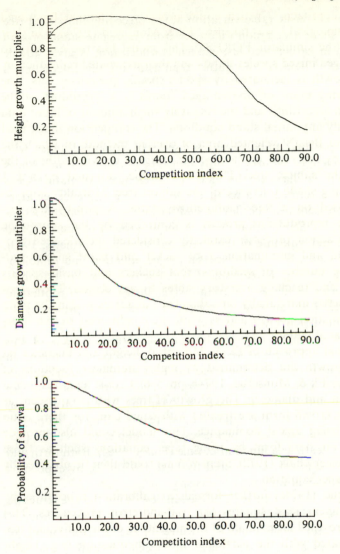

Fig. 2.2. Basic growth- and mortality-competition relationships assumed in individual tree simulation models. In the usual case it is assumed that actual growth = (potential growth) × (a competition-dependent multiplier).

Fertilization and/or drainage provide an example of the treatment analysis capabilities of this approach. Assuming the relationships in Fig. 2.2 invariant, the effects of site alternation on the entire stand's development may be estimated by simple changes in the potential height growth curve. Such treatments may also be studied by direct manipulation of height growth as described by Hegyi (1974).

Dynamic properties of forest ecosystems

Although there is considerable variation in the preceding individual tree models, they all share a common difference from aggregate stand growth models discussed in earlier sections: explicit consideration is given to the interdependence of individual trees rather than assuming this interaction is equal or nonexistent.

Reproduction models

The reproductive process can be viewed as a series of sequential stages, with successful reproduction occurring only after completion of all stages (Kozlowski, 1971b). To illustrate this view, the development of mature seeds and fruits may be described as a result of the following sequence: enlargement of the inflorescence in the flower bud; flowering; pollination; fertilization, growth, and differentiation of the embryo; growth of the fruit and seed to maturity; and, ripening of the fruits and strobili (Matthews, 1963). Major events subsequent to seed dispersal are germination and establishment of the young seedling (Kozlowski, 1971a). Factors critical to establishment are temperature, water and nutrient availability, ability of rootlets to penetrate the seed bed, amount of overhead shade, and proximity to competing vegetation (Kozlowski, 1971a).

Stochastic processes have been used to model various aspects of this reproductive process. The sequential nature of the processes resulting in seedling establishment provided the basis for Leak's (1968) birch regeneration model, which is an absorbing Markov chain with 29 states. (An application of this model was discussed by Marquis, 1969). Most reproduction models have ignored the sequential nature of seedling development, however. Simple birth and death processes, for example, predict the probability of birth or death as a function of time (Bailey, 1964). Both birth and death models were explicitly considered by Leak (1970b, 2), by expressing these rates as polynomial functions of density.

Life-table analyses have recently been used in describing the age structure of plant populations (Hawksworth, 1965; Waters, 1969; Hett & Loucks, 1968, 71; Hett, 1971a). A life-table is a listing of the distribution of all individuals in a population by age class. Age-specific mortality and survival rates can then be calculated. Implicitly assumed is that age-specific mortality and the input into the youngest age-class are both relatively constant over time. Hett (1971) and Hett & Loucks (1971b) studied the age distribution of sugar maple seedlings, and found that an increase in the size of the seed crop and seed viability (which Curtis (1959) observed were positively correlated) was offset by an increasing mortality rate among germinating seeds. This supported the assumption of constant ingrowth into the youngest age class. They further noted that ingrowth fluctuations after five years were not detectable. They also found that a power function provided a better fit to the survivorship curve than did an exponential decay function. This implied that seedlings are less susceptible to mortality after the critical establishment phase. In an earlier study of balsam fir, red maple, and white-pine seedlings, Hett & Loucks (1968) had also concluded that fluctuations in seed crop were not detectable past the five year age class.

Rather than treating reproduction processes directly, the differential equation growth models of Moser (1972, 4) and the difference equation models of Ek (1974*b*) utilized equations for the number of stems growing into the smallest measured size class. Ingrowth into the smallest measured diameter class during a growth period was expressed as a function of overstory stand basal area and total number of trees.

An alternate ingrowth model that is currently under investigation by the authors is similar to Ek's (1974*b*) stand table projection model, but based on reproduction height classes rather than overstory diameter classes. This model utilizes the competition index used for overstory trees by Ek & Monserud (1974*a*) as a predictor variable. Thus the model is distance-dependent, since reproduction growth and survival is expressed as a function of the overstory spatial pattern.

As previously described, Ek & Monserud (1974*a,b*) proposed a general sequential computer model for the reproductive process that more closely resembles Leak's (1968) absorbing Markov chain model than the other one-step reproduction models. The detail considered by such sequential models affords examination of a greater number of silvicultural alternatives. However, precise parameter estimation of detailed sequential models depends on more extensive data sets than do the single stage models.

Summary

The evolution of stand growth modeling can be characterized in a number of ways, but two apparent trends stand out. The first is the trend toward greater detail in stand description, i.e., the consideration of stand growth components (ingrowth, mortality, and survivor growth). This is followed by further breakdowns by size classes and species groups. More recently, individual tree models have led to consideration of several individual stem characteristics. A second trend tied to the more complex modeling noted above is the increasing sophistication of parameter estimation procedures. These have developed from graphical approaches through linear regression to the current interest in a number of nonlinear techniques. The incorporation of greater detail in model form and the improvement of parameter estimation techniques is made possible by coincident refinements in the digital computer.

It is evident that these trends will continue. However, the ability of the scientist to postulate more complex models may sometimes exceed para-meter estimation capabilities, data bases, or methodology for solving, analyzing, or validating the models. Such limitations are thus likely to alter or direct future trends in model development. It is clear to us, however, that available methodology is more than adequate to answer a wide range of questions concerning stand dynamics and treatment.

Regional succession modeling in deciduous and coniferous forests of North America *

Succession is a fundamental principle in ecology. Since the discovery of its universality, succession research in most countries has centered around the documentation of species replacement patterns within a diverse array of ecosystems, the exploration of underlying causes of such changes, and speculation about the composition of regional vegetation in the absence of disturbance agents. Recently, attention has turned toward two new aspects of succession research: analysis of changes in ecosystem properties (biomass, diversity, productivity) that accompany compositional shifts and the de-velopment of mathematical (often predictive) models of succession. Al-though specific objectives and model configurations vary considerably, the overall goal of most succession modeling projects has been to develop a system of equations which, when solved, characterize desired attributes of single, often generalized stands of vegetation or a number of concrete stands that are representative of a regional forest. These models then have direct utility in the study of natural succession processes, but perhaps most important (as will be shown later in this paper) they can serve to identify man's peculiar role in altering the development of vegetation stands or the overall vegetational composition of regions. Ultimately, models of this type can serve as tools to aid in the rational management of regional ecosystems.

Two of the major differences among succession or forest 'simulator' models relate to the degree to which environmental parameters drive the model and the spatial scale of application. Thus, in some succession models (e.g., Botkin *et al.* 1972) reproduction within the stand may be controlled by the quantity of light reaching the forest floor (a function of leaf area index) or the growth of trees may be regulated by species-dependent competition factors (see Ek, this chapter). In essence, these models predict certain process rates from explicit changes in environmental parameters. Other approaches do not deal explicitly with environmental parameters, but

* Research supported by the Eastern Deciduous Forest Biome, US–IBP, funded by the National Science Foundation under Interagency agreement AG–199, BMS69–01147 A09 with the US Dept. of Energy-Oak Ridge National Laboratory. The Oak Ridge National Laboratory is operated by the Union Carbide Corporation for the US DOE. Contributions No. 341 from the Eastern Deciduous Forest Biome, US–IBP. Publication No. 1521, Environmental Sciences Division, ORNL.

52

rather base the directions and rates of change on time-series data. For example, Leak (1970*b*) simulated population density in several forest types and density classes by deriving species' birth and death rates from permanent plot records. Olson & Cristofolini (1966) and Waggoner & Stephens (1970) used remeasurement data to predict future forest composition on undisturbed plots. Still another approach (Hett, 1971*a*) used serially-dated aerial photographs to determine parameters) for a landscape level succession model. The time-series approach will yield good estimates of future succession events as long as important environmental conditions (e.g., climate) remain relatively constant. With changing environmental conditions, however, the environmentally-based models are most appropriate in determining the impacts of such changes on successional processes. The general availability of permanent plot data for a number of small experimental forests (Bartlett Forest, New Hampshire; several state forests in Connecticut) and intensively monitored watersheds (Hubbard Brook, New Hampshire) has resulted in a number of models at this spatial scale. The increasing availability of US Forest Service Continuous Forest Inventory (CFI) data has stimulated their use in determining parameters for regional successional models. Although CFI data are collected primarily for forest inventory purposes, their usefulness for ecological research has been well established (Livingston, 1969; Webb, 1974).

Succession models have been applied to several spatial scales including individual forest plots (0.01 − 1 ha), small watersheds (1–10 km^2) and large regions (100–1000 km^2). The basic data in most cases, however, are from small permanent plots.

During the US–IBP, most research in forest succession was directed towards the development of models applicable to large land areas. Models representing two levels of complexity were constructed for a coniferous forest watershed in Oregon and mixed deciduous–coniferous landscapes in Michigan and Georgia. These models are analogous to many ecosystem models where systems of linear differential equations describe the flow of some material substance between compartments. In this particular application, compartment flow models of entire landscapes were developed to transform land area from one forest or nonforest category to another.

The Oregon and Michigan models simulate compositional changes in the absence of natural or man-induced perturbations. Hence, although these models refer specifically to existing (current) vegetation conditions, they essentially simulate how the areal extent of a number of forest types would be expected to change if autogenic succession operated alone. These results serve as the background against which the impact of perturbations can be assessed. The most recent stage of development stresses the effect of various perturbations in Georgia and focuses more closely on how man's use of the landscape influences regional vegetation structure and composition.

53

Natural succession models

Topology

Natural succession models were constructed for each of these forested regions. The Michigan model (Shugart, Crow & Hett, 1973), which illustrated a general approach to modeling succession over large regions, consisted of three sub-models defined by soil moisture classes (Fig. 2.3). Fifteen cover-states or compartments were recognized, and each compartment was subdivided into three sub-modules representing size classes (seedling–sapling, poletimber, and sawtimber). The linkages among compartments correspond to the classical patterns of forest succession in the region (Curtis, 1959).

Coniferous cover-states occur in a variety of positions along the successional gradient (Fig. 2.3). On hydric sites, however, the pioneer and transitional communities are coniferous while a predominantly deciduous type is terminal. On the xeric sites, intolerant oak is replaced by mixed oak. The mesic portion of the model is most complex with aspen, pin cherry and jack pine designated as pioneer cover-states. The more tolerant red and white pine types replace jack pine, and the abundant aspen type is rapidly replaced by northern hardwoods on upland sites and fir–spruce in the lowlands. Hemlock, sugar maple and fir–spruce are the terminal types in the mesic portion of the model.

The rates of area transfer among size classes are based on the time required for stands to grow from one size class to another. This rate is calculated as a function of the mean growth rate of dominant species. Rate constants between cover-states, however, are based on the persistence of the sawtimber size class of the preceding cover-state. Coefficients were derived from Curtis (1959), Fowells (1965), and Goff (1967).

The Oregon model (Edmonds, 1974) was developed for the H. J. Andrews watershed located in the Cascade Mountains. Because the major environmental gradient is altitudinal (temperature and moisture), the submodels were defined by elevation classes (Fig. 2.4). This differs from the sub-module organization devised for the Michigan model.

The model topology was derived from Franklin & Dyrness (1969, 1973). Four elevation zones are recognized, and each has a number of cover-states, some of which were further sub-divided into four size classes. An additional size class was needed for cover-states in the lower three zones because of the occurrence of Douglas fir, a pioneer species with a life span of over 650 yr. As in the Michigan model, transfer rates among size classes were based on diametric growth rates. Rates among cover-states were based on Franklin & Dyrness (1969, 73). Because a portion of the watershed has recently been clearcut, a clearcut cover-state was also identified.

All cover-states are dominated by coniferous species, although a de-

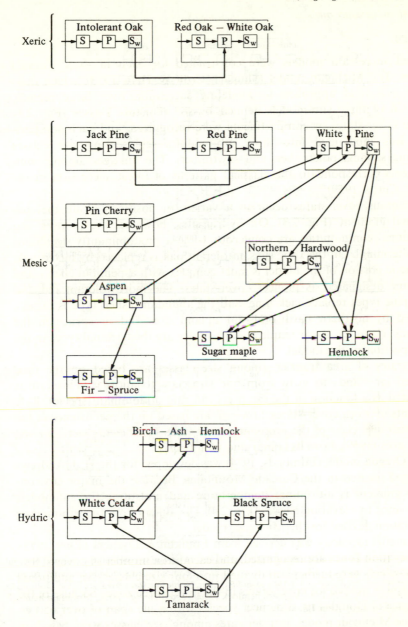

Fig. 2.3. Topology of the Michigan succession model. The blocks (compartments) indicate forest types identified by dominant tree species. The three blocks within each compartment indicate size classes (1, seedlings and saplings; 2, poletimber; 3, sawtimber) within forest types. Arrows represent transfers of land area during succession. After Shugart *et al.*, 1973.

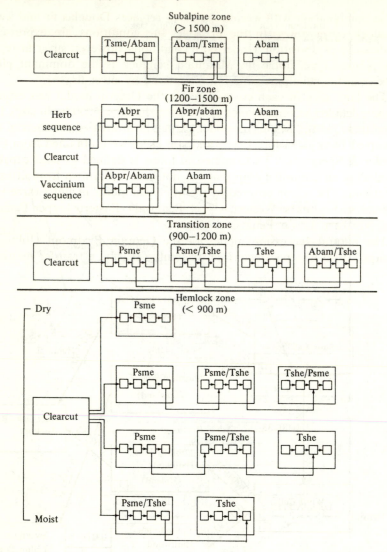

Fig. 2.4. Topology of the Oregon Cascades model (Andrews Experimental Forest). Natural succession patterns are indicated by arrows. Clear-cut compartments include land area recently cut-over. Abam, Pacific silver fir (*Abies amabilis*); Abpr, Noble fir (*Abies procera*); Psme, Douglas fir (*Pseudotsuga menziesii*); Tshe, Western hemlock (*Tsuga heterophylla*); Tsme, Mountain hemlock (*Tsuga mertensiana*). After Edmonds, 1974.

ciduous species (*Acer macrophyllum*) is important. In the hemlock zone, which is further sub-divided by soil moisture classes, Douglas fir is the pioneer species on most sites following natural disturbance or clear-cut. On the driest sites it may perpetuate itself, but under mesic conditions it is replaced by hemlock, the terminal type. In the transition zone, pacific silver fir in combination with western hemlock replaces Douglas fir and western hemlock cover-states. In the fir zone, site conditions are expressed by understory composition. Under both conditions, however, silver fir replaces the pioneer type, noble fir. Mountain hemlock is the dominant pioneer species in the subalpine zone but is replaced by silver fir.

The land area in each cover-state for the Oregon model was estimated from a detailed reconnaissance of the watershed (*ca.* 6000 ha in size). Initial conditions for the Michigan model, expressed in area of commercial forest occupied by a given forest type and size class, were obtained from Chase, Pfeifer & Spencer (1970). Commercial forest is defined as land capable of producing an economic crop that has not been withdrawn from utilization by statute. The region covered by the Michigan model includes three Forest Service survey units: Western Upper Peninsula, Eastern Upper Peninsula, and Northern Lower Peninsula.

The natural succession model for the Georgia Piedmont (Johnson & Sharpe, 1976) was parameterized differently. Briefly, a transfer matrix was

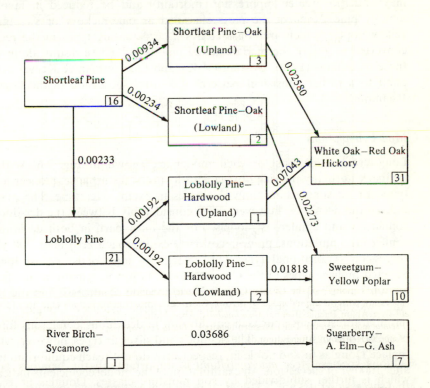

Fig. 2.5. Topology of the Georgia Piedmont natural succession model. Coefficients (*100) are percent of land area in donor compartment transferred to recipient compartment per year. Numbers in lower right insert of each compartment are percent of undisturbed plots in 1972. After Johnson & Sharpe, 1976.

constructed by analyzing compositional shifts on US Forest Service permanent plots relatively undisturbed by man during an 11-yr period (1961–72). The off-diagonal elements in the transfer matrix determined the succession patterns. Coefficients used in the differential equations were calculated using Newton's method (Mankin & Brooks, 1971).

The model topology and coefficients are shown in Fig. 2.5. The two major pine types are replaced by their upland or lowland pine–oak counterparts. The higher transfer rate from shortleaf pine to the upland rather than the lowland shortleaf pine–oak type reflects the general occurrence of shortleaf pine in association with oaks on upland, drier sites. Area from loblolly pine flowed equally to each pine–hardwood type. Both the mesic pine–hardwood types are replaced by sweetgum–yellow poplar, which occurs on mesic upland or well-drained lowland sites. The upland shortleaf pine–oak and loblolly pine–hardwood types are succeeded by oak–hickory. The minor flux from shortleaf pine to loblolly pine indicates that under certain conditions (mesic sites) where these species occur together, shortleaf pine may undergo greater suppression mortality and be reduced in favor of loblolly pine. None of the plots classified as oak–hickory or sweetgum–yellow poplar, which are considered self-perpetuating types in the region, converted to any of the pine or mixed pine types. These results along with those of Billings (1938), Oosting (1942), and Quarterman & Keever (1962) point to a rather consistent pattern of natural succession in the Georgia Piedmont.

Model simulations

Long-term simulations for each model are shown in Fig. 2.6. Without feedback normally provided by various forms of disturbance, pioneer forest types for all simulations show decreases in extent over time. The magnitude of the decreases, however, differ considerably between the deciduous–coniferous and coniferous models. For the most part, in both deciduous–coniferous simulations, pioneer cover-states approach zero after 200 yr of succession. Transitional types (e.g., Northern hardwoods) persist longer. Selected pioneer types in the Oregon model, however, remain in significant amounts after 500 yr of succession. Here, the rate of succession in the H. J. Andrews watershed is slowed dramatically by the persistence of Douglas fir. The terminal forest types approach equilibrium in deciduous–coniferous forests over 200–400 yr succession. The hemlock and silver fir types, however, which replace Douglas fir and noble fir, respectively, do not reach equilibrium until after 1000 yr of succession. An examination of the model coefficients indicates that succession rates are 2–3 times more rapid in eastern deciduous forests than in forests of the Oregon Cascades.

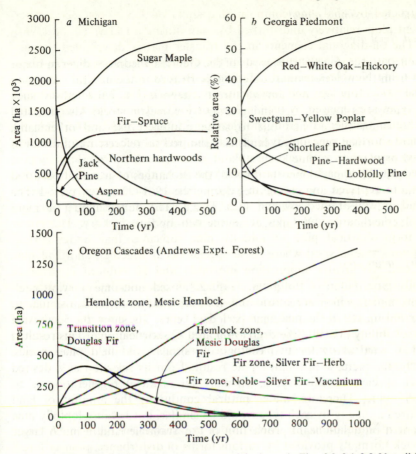

Fig. 2.6. Simulation of the natural succession models shown in Figs. 2.3, 2.4, 2.5. Not all forest types are shown. x-axes have identical scales. y-axes are actual or relative areal extent. (Shugart, Crow & Hett, 1973; Johnson & Sharpe, 1976).

In summary, the succession models presented thus far simulate the exchanges of land area that would be expected to take place in three forested regions of North America in the absence of significant natural or man-induced disturbances. The models were constructed using two different types of data generally available for estimating parameters: open literature and detailed records from permanent plots. Although the simulations shown would not be expected to reflect actual dynamics in urbanized landscapes or in regions where natural disturbances are substantial, they can serve as a background against which the impact of a host of man's disturbances can be assessed. The next section is an example of this technique.

59

Dynamic properties of forest ecosystems

Landscape dynamics model

Model topology

The current dynamics of the forests of the Georgia Piedmont differ in major ways from the unidirectional succession portrayed in the natural succession model. The flow of area among forest categories is redirected by such disturbances as logging, fire and grazing by domestic stock. Also, forested land is often cleared and put to some nonforest use and other land, primarily former agricultural land, is permitted to reforest naturally or is seeded or planted to pine.

The following model incorporates all the exchanges of area that occurred among forest types and size classes during the 1961–72 survey period. The generalized topology is shown in Fig. 2.7. The actual model is more complex, with some 300 linkages among sub-modules.

Model simulation

The initial simulation (solid line, Fig. 2.8) used non-time varying coefficients, and therefore is realistic only if future conditions remain similar to those during the remeasurement period. The results show the increasing area of loblolly pine and the decreasing area of shortleaf pine. This result is striking, particularly for two ecologically similar, old field pine species. Clearly, the increase in loblolly pine results from its selection as a desired timber species. In this region, virtually all artificial regeneration is to loblolly pine. Therefore, under natural conditions the amount of land colonized annually by each type would be similar, but because loblolly pine is planted on a significant proportion of the available land, a much larger

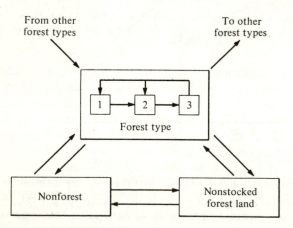

Fig. 2.7. Conceptual model of the flow of land area among landscape categories. Subcompartments within forest types are equivalent to stocking classes. After Johnson & Sharpe, 1976.

60

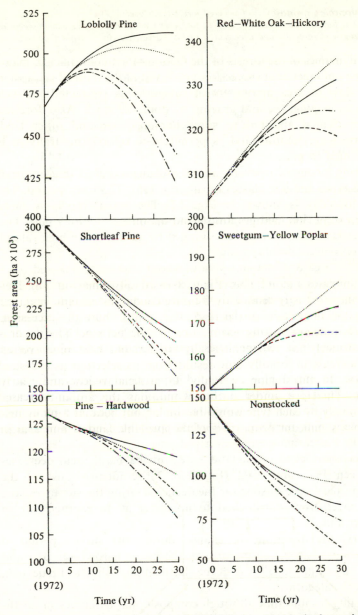

Fig. 2.8. Simulation of change in area of selected forest categories in the northern Piedmont of Georgia for a 30-yr period based on current rates of change, and on hypothetical changes in the rates of timber harvesting and land abandonment. The vertical scale was drawn differently for each graph in order to focus on differences among the scenarios rather than to stress a comparison among types. ——, Current dynamics; ---, decreasing reversions; ····, increasing harvest; —·—·—, decreasing reversions and increasing harvest. After Johnson & Sharpe, 1976.

Table 2.1. *Relative change in area (%) of selected forest types over a 30-yr period as simulated by the models for natural succession, current forest dynamics and a management scenario involving decreased land reversion and accelerated timber harvest for the Georgia Piedmont region*

Forest type	Natural succession	Current dynamics	Management scenario
Shortleaf Pine	−34	−33	−46
Loblolly Pine	−7	+10	−10
Pine–Hardwoods	−6	−6	−15
White Oak–Red Oak–Hickory	+16	+9	+6
Sweetgum–Yellow Poplar	+23	+17	+16
Nonstocked	—	−43	−50

percentage of nonstocked and nonforest land converts to loblolly pine. The decline in shortleaf pine is further encouraged by an unusually high rate of diversion to residential development.

The simulated area of sweetgum–yellow poplar and white oak–red oak–hickory also increases, but at a slower rate than loblolly pine (Fig. 2.8). This increase in area is strongly related to the selective harvest of pine in mixed stands; only a small portion of the increase is due to natural succession. Without more intensive cutting (conversion to the nonstocked category) and stand treatment (e.g., seeding), these mixed stands are being converted to hardwoods.

The area of the mixed pine–hardwood types remains low and nearly constant throughout the period. Loblolly pine–hardwood area increases slightly in response to the increasing area of loblolly pine and its contribution to loblolly pine–hardwood. Shortleaf pine–oak decreases in area in response to the decreasing area of shortleaf pine.

The area in nonstocked forest land (<2.3 m^2/ha basal area) decreases sharply between 1972 and 2002 (Fig. 2.8). More intensive use of the landscape is primarily responsible for the decline. During the survey period, area was removed from nonstocked to nonforest at an average rate of almost 2% per annum.

In summary, the directions of change during the survey period, if maintained, point to an increasing area of the major hardwood types and loblolly pine, largely at the expense of the shortleaf pine, shortleaf pine–oak, and nonstocked categories.

It is interesting to note that although compositional dynamics in north-central Georgia are heavily dominated by the hand of man, the simulated area of most of the major forest types follows the expected trends of natural succession (Table 2.1). For example, the decreases in area of shortleaf pine and pine–oak as simulated by both models (Figs. 2.6, 2.8) are nearly identical. Obviously, the perturbation agents are quite different in each case, but the combined effect of reduced reforestation on agricultural land and high rates of land-use change mimics the expected depletion curve for shortleaf pine if man-induced perturbations over the same period were eliminated. Also, natural mortality of pine in mixed stands directs that area to hardwood types, while the selective removal of pine during harvest in mixed stands accomplishes a similar result. Without diversions to other land-uses, the percent increase in the oak–hickory and sweetgum–yellow poplar types would be nearly identical to the percent increase due to natural succession alone. Of the major forest types, only loblolly pine exhibits opposing trends (Table 2.1). This reversal of the natural succession trend has only been possible through the expenditure of considerable amounts of energy (i.e., initial clearing for agriculture, seeding, planting) directed toward its perpetuation as a major forest type.

Land-use and forest management scenarios

In this section, three scenarios are hypothesized and their respective impacts on regional forest composition simulated. These were not implemented to make specific projections, but rather to demonstrate the use of time-varying coefficients in assessing certain long-term impacts of alternative land-use and forest management strategies. When used in this mode, succession models of this type have application to resource management problems.

Description of scenarios

In the first scenario, the rate of old field abandonment and reforestation is decreased linearly to zero over the 30-yr period. Both declining rates of old field abandonment and continued high rates of urbanization are considered likely for Georgia, especially adjacent to expanding metropolitan areas like Atlanta.

Second, the area of forest land harvested is increased 1% per year over the 1961–72 estimate. In this scenario all the increase in harvesting is assumed to occur on land maintained as forest after harvesting.

The third scenario combines the two above. At the end of the 30-yr simulation period land reversions to commercial forest are zero and the harvest rate of land in the forest sector is increased by 30%.

Impact of scenarios on selected forest types

The impact of the three scenarios (interrupted lines) in contrast to the constant coefficient model (solid line) is shown for the major forest types in Fig. 2.8. In general, decreases in reversion rates and/or increases in harvesting will cause a decline in the area of the pioneer pine and pine–hardwood types. The trend toward greater area of loblolly pine over the 30-yr period is slowed and then reversed by the new dynamics introduced with the three scenarios. Reduction in the rate of old field abandonment has an especially strong impact on the area in shortleaf and loblolly pine since additions of area to these forest types other than by harvesting (and often planting or seeding) are primarily by old field reforestation. An increase in harvesting also contributes toward a decrease in the extent of pine. The combined impact of the third scenario is an 18% reduction for loblolly pine and 20% for shortleaf pine.

The impact on the successionally mature forest types, white oak-red oak-hickory and sweetgum–yellow poplar, contrasts with the trends in the pine and pine–hardwood types. Increasing the harvest rates leads to slight increases in area due primarily to a more rapid conversion of mixed stands to hardwood through the selective removal of pine. Reducing land reversion affects these forest types as it does the pines. The combined impact is a small net reduction in the extent of oak–hickory over the simulation period. In the sweetgum–yellow poplar type, the combined impact follows the trend in the constant coefficient model.

Because a substantial portion of harvested land was heavily cut and directed to the nonstocked category during the survey period, an increase in harvesting slows the decline of nonstocked area (Fig. 2.8). The reduction of reverted land considered separately or together with harvesting, leads to more rapid rates of decline.

The third scenario, which is more likely than either of the other two, provides an interesting comparison with the simulation results of both the natural succession and current dynamics models (Table 2.1). Scenario three would result in a more rapid decrease in the shortleaf pine and pine–hardwood types and would reverse the trend toward increasing loblolly pine shown by the current dynamics simulation. The percent decrease for loblolly pine is similar to that simulated by the natural succession model, although the natural succession rate is conservative. Under the conditions in scenario three, the area in pine could be increased to levels shown by the constant coefficient similation only by substantially increasing management intensity in the region (e.g., seeding and planting after harvest, prescribed burning to control hardwood reproduction). Without such an increase, decreased reversions, increased harvesting, and constant diversions to non-forest uses will combine to reverse the long-term trend of increasing loblolly pine area in Georgia's northern Piedmont.

Conclusions

Our understanding of forest dynamics at the regional scale has often been limited by our inability to keep track of and integrate the effects of all the contributing factors. Mathematical modeling techniques provide the means to bring together, in quantitative form, a large number of these factors and their associated rates. As demonstrated, a series of models can be constructed from several types of data and used to portray desired aspects of regional forest dynamics. Natural succession models can conveniently be used to compare the replacement patterns and rates of forest succession in geographically distant regions. These analyses contribute to our understanding of the dynamic behavior of natural ecosystems. The development of regional models which utilize permanent plot data and include the impacts of man have particular relevance to modern ecological problems. Furthermore, the ultimate utility of such models is enhanced substantially by the creation and simulation of alternative land-use scenarios. Here, our best estimates of man's future impacts can easily be incorporated. Perhaps an optimal use of such techniques is to utilize the entire series, which permits a comparison of natural conditions with those that presently exist or those that are foreseen.

In the development of succession models during the US–IBP, we have tried to devise a model structure that was compatible with existing and forthcoming regional data bases such as those compiled by the US Forest Service. With the increasing availability of remeasurement data for many regions, models such as those described here can be easily and inexpensively constructed. We hope that this technique will prove to be useful in the analysis of man's impact on regional biological resources.

Woodland ecosystem composition in relation to environment, productivity and time*

Much recent research on woodland ecosystems has tried to assess the mechanisms by which one species does better in one place than in another and an entire community of species comes to function as an ecosystem. In previous work I have examined forest composition in relation to regional climate (Loucks, 1962a), environmental gradients (Loucks, 1962b), productivity (Loucks, 1970), and time (Loucks, 1970; Peet & Loucks, 1977). Research on entire watershed systems was undertaken as part of the US contribution to the International Biological Programme, and at the Eastern Deciduous Forest Biome (EDFB) sites, as elsewhere, climatic, edaphic and age-dependent factors greatly influenced the ecosystem processes under study. The objective of this paper is to outline the viewpoints taken by researchers in the EDFB, and at the Lake Wingra Watershed study site in

*Research sponsored by the Eastern Deciduous Forest Biome, US–IBP, funded by the National Science Foundation under Interagency Agreement AG–199, BMS76–00761 with the US Department of Energy–Oak Ridge National Laboratory. Eastern Deciduous Forest Biome Contribution Number 308.

Wisconsin (Noe Woods), to integrate across the many influences governing the diversity and dynamics of woodland ecosystems.

Whittaker (1956) and Curtis (1959) have demonstrated the usefulness of various approaches to the study of gradients in species composition. Greig-Smith (1964) has suggested that the relative influence of environmental factors in these gradients can be estimated through the direct use of linear multiple regression. Unfortunately, the response of a species to simple, direct influences on its population may be highly nonlinear. In fact, a species often has been shown to have one optimum for one or more environmental factors, when free of competition, and very different optima for these factors when studied in natural habitats. In addition, the observed response relationships do not operate independently of each other, because the effects of varying levels of second, third or other influences (climate, soil, biota or time) tend to mask some of the effects associated with the first. The following sections describe an approach to differentiating each of these influences on forest diversity.

Climatic and biogeographic elements in forest diversity

Analysis of forest ecosystems covering large continental areas necessarily requires a broad scale in differentiating controls over species composition and related ecosystem processes. Almost world-wide these influences have been incorporated into climatic or biogeographic regions which define local areas of comparative uniformity in potential species composition and climate. One of the classic examples comes from Europe, that of Rubner & Reinhold (1953), whose forest classification, the Natural Forest Pattern of Europe, is subtitled *A Basis for a European Silviculture*. The same attributes were used in *A forest classification for the maritime provinces* (Loucks, 1962a) to distinguish geographical units, each characterized by specific relationships between the vegetation, climate and soil. The criteria used were chosen to describe the gradations in species composition and diversity from one geographic area to another, providing at the same time the arbitrary scales along which regional boundaries can be located.

A basic work within which the Eastern Deciduous Forest Biome study was organized is the classification of the deciduous forests of North America by Braun (1950). Braun treats regional diversity by defining 'Forest Region' to denote a natural entity whose boundaries are determined in part by 'the limits of the more or less continuous ranges of characteristic species. Physiographic and climatic limits were used where information on the vegetation was insufficient for locating a regional boundary. Specific regions are singled out only when specific criteria are applied and the area is found to be relatively homogeneous in terms of the criteria.

For the areal units of physiography, climate and biotic diversity, the descriptive term obtained by applying the prefix 'eco-' to the term 'region' has been proposed (Loucks, 1962*a*) and adopted now in the US National Wetlands Classification (Cowardin *et al.*, 1977). 'Ecoregion' was used originally as the geographic unit within which ecological relationships between species and their environments are essentially similar, and within which vegetation manipulations may be expected to have similar results. So many facets of the vegetation are considered in different places that individual ecoregions will not necessarily be different in all aspects of the vegetation. Most important is the recognition of forest ecosystem classification as a multiple-level phenomenon, requiring ecosystem types to be an integral part of larger units.

Local edaphic control of ecosystem diversity

Within a region, and in the absence of disturbance-related processes, ecosystem diversity can be governed largely by local environmental influences. The general nature of the response by species to the effects of variations in water, nutrients, temperature, and light has been characterized in several ways as reviewed by Forsythe & Loucks (1972) and shown in Fig. 2.9. The dome-shaped response form in Fig. 2.9*a* occurs when a wide

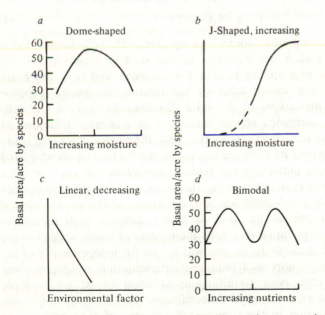

Fig. 2.9. Representative forms of the response of a species measure, basal area per acre, when plotted against environmental gradients.

enough range in the variable is present in the study universe to produce a low species abundance due to 'too little' as well as 'too much' of the environmental influence. If a more limited range of habitats is considered, giving a narrower range at one end or the other of the gradient, responses such as those shown in Fig. 2.9*b* and *c* may be observed. Curves of the type represented by Figs. 2.9*a* and *b* characterize the results of Curtis & McIntosh (1951), Whittaker (1956), Loucks (1962*b*), Bakuzis (1961), and Waring and Major (1964). More anomalous relations such as the bimodal distribution in Fig. 2.9*d* also can be demonstrated and are viewed as the results of interactions with other species or habitat factors (Bakuzis, 1961), or the presence of undifferentiated ecotypes (Waring & Major, 1964).

For most of these studies, sigmoid distributions of species at the ends of gradients, and dome-shaped distributions in the middle are an accepted paradigm. Simple regression methods have been used to investigate the relationships between environment and species abundance assuming a linear model for the sigmoid or gradient-end distributions, but more complex analyses including data transformation is required for the bell-shaped distributions in the center. A few studies have been undertaken to explore the extent to which the species richness along these environmental gradients, independent of perturbations, can be expressed in a multivariate model of environmental controls.

Forsythe & Loucks (1972) used a parabola model of the form

$$Y = a + bX + cX^2, \tag{4}$$

to investigate environmental control of species with bell-shaped distributions. In this model, X is a substrate quality such as moisture, Y is the relative importance of a species in that environment, and a, b and c are fitting coefficients. The model achieves linearization of species response curves within the limitations of the solid distributions that fill the bell-shaped form. The combined effects of several environmental factors on species composition (i.e., diversity) in the New Brunswick, Canada, study area was then examined by multiple regression methods. The standardized regression coefficients indicating the relative importance of each of eight environmental measures in determining the abundance of six forest species are shown in Table 2.2. Values are low for many of the environmental measures, but each species has one or more relatively high regression coefficients. Regression equations for abundance of each species were developed, but each multiple regression was solved for independently of the others. This is a step short of a balanced determination of species composition (or diversity) using simultaneous solution of all six multiple regressions from the environmental determinants.

The residual variation in the observed basal area of each species not accounted for by the multiple regression model is shown in Table 2.3. Most

Table 2.2. *Standardized regression coefficients[a] from multiple regression analysis (values show the relative importance of each factor in influencing the basal area of that species)*

Species	Log depth	Water-holding capacity	Runoff	Silt and clay	Depth of solum	$A_1 - A_2$	Aspect	Minimum temperature
Acer saccharum Marsh.	0.07	−0.03	0.08	0.00	−0.04	−0.03	−0.04	0.86
Abies balsamea L. Mill	−0.06	−0.05	0.03	0.02	0.11	0.01	−0.00	−0.77
Betula lutea Michx.f.	−0.07	−0.26	−0.16	−0.01	−0.09	−0.11	−0.07	−0.46
Picea mariana Mill.	−0.11	−0.02	−0.11	−0.02	−0.08	0.58	0.58	0.21
Picea glauca Moench	−0.25	−0.30	−0.04	−0.01	0.02	0.47	0.15	−0.39
Thuja occidentalis L.	0.54	0.31	0.11	−0.03	0.24	−0.26	−0.08	0.07

[a] All the species coefficients are from method 2 except for *Picea glauca*, which has less residual variation by the linear analysis of method 1. The sign of the entry must be multiplied by the sign of the coefficient in Table 2.3 in order to tell the general direction of influence of the factor.

Table 2.3. *Residual variation (expressed as percentage)[a] for estimates of relative basal area using multiple regression from habitat factors. Values show the success of the regression model in using the response relations to account for the actual variation in stand species importance*

Species	Residual variation
Acer saccharum Marsh.	20
Abies balsamea L. Mill.	35
Betula lutea Michx. f.	56
Picea mariana Mill.	22
Picea glauca Moench	35
Thuja occidentalis L.	44

[a] A figure of 62% or less is significant at the 0.001 level by the F test.

are low in comparison to other studies of this type, largely because the study area had been free of fire or human disturbance for at least 200 yr. The relatively high value for *Betula lutea* (Michx.) can be viewed as indicating a species whose present importance in the ecosystem has been more influenced by past disturbances (wind storms in this region) rather than the present edaphic control of the system.

Primary production and ecosystem diversity

The summary of ecosystem diversity and its supposed determinants, Fig. 2.10 (from Loucks, 1970) represents an interpretation based on literature in the past 20 yr. Authors have referred in general terms to the characteristic relationships between diversity, succession and primary production shown here as trend lines. Whittaker (1965) and Monk (1967) have suggested increase in diversity with the successional sequence [Fig. 2.10a]; others such as Margalef (1957, 68) and Odum (1963) have indicated that this curve probably levels off and may drop in the later stages of a successional sequence. Whittaker (1965) and Pielou (1966) have suggested that under certain conditions diversity may decrease with succession, as shown by the dotted line. MacArthur (1955) and the discussion by Connell & Orias (1964) of diversity under very stable environments suggest that a high stability is associated with a high level of productivity and diversity, with no clear indication of which is dependent on the other. Fig. 2.10c expresses the consensus that with higher levels of annual primary production there will be higher levels of community diversity (Margalef, 1968), although Odum (1963) and Whittaker (1965) have questioned this relationship. Their position is that there may be no consistent relationship, as shown by the horizontal dashed line. The axiomatic results of the three previous figures,

70

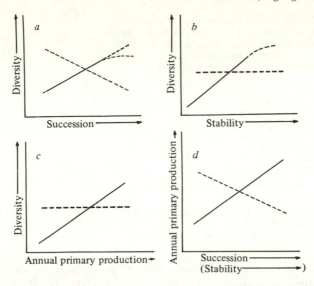

Fig. 2.10. A graphic summary of the relationships reported between diversity of species and other properties of biotic communities.

summarized in Fig. 2.10*d*, show that with more advanced succession and an increased stability, one might expect higher levels of production by the primary producers.

The production of dry-matter in forests in southern Wisconsin has been estimated by common forest mensurational techniques. The total basal area of living stems in square feet/acre has been used to estimate total plant biomass. A plot of basal area of stand against age of stand shows that the youngest stands support lowest basal area, the middle age range is highest, and the oldest stands support a low basal area. This is a well-documented phenomenon in forest biomass responses in which the high basal area developed by the initial growth of colonizing species is reached at some age determined by the longevity of the pioneer stand itself. Tables of forest yield for hardwood species also indicate that the annual increment of wood in the forest can be estimated very closely by assuming a 2% per annum increase in basal area. The total annual increase in biomass can be estimated reasonably well by multiplying the annual increase in basal area by the mean height of each stand. The increase in dry weight has been estimated by the product of volume and the mean specific gravity for the dominant species in each stand.

Fig. 2.11 summarizes the estimates of production of dry matter exclusive of tree foliage and herbs for the 30 forest sites distributed across the 10 segments of the southern Wisconsin upland continuum. The range in values among the three stands in each decimal segment is shown. The bars in

71

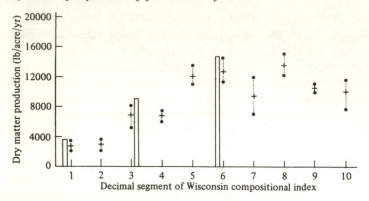

Fig. 2.11. Primary production fixed annually by the tree vegetation (exclusive of foliage), plotted by decimal segment along the southern Wisconsin forest gradient, to show the impact of the severe environment at the low end of the gradient.

segments 1, 3, and 6 are direct estimates of production of dry-matter obtained by dimension-analysis methods in three stands of the same compositional index as those adjacent to each bar. One of these is the Noe Woods stand studied as part of the Lake Wingra Basin work in the US–IBP.

The combined results show the profound influence of the environmental extremes at the low end of the compositional index. The six higher segments represent the peak levels of productivity, but there is some indication of a drop in primary production among the highest values of the gradient.

The results of an analysis for the 30 stands in Wisconsin are illustrated in Fig.2.12. They are presented as average diversity for three stands in each decimal segment of the Wisconsin upland compositional gradient (Curtis &

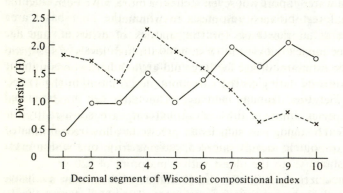

Fig. 2.12. The inverse relationship between diversity of species of the tree (O—O) and seedling (×---×) strata across the range in environments and composition represented in the southern Wisconsin compositional gradient.

McIntosh, 1951). The diversity of the overstory trees is low in the droughty, extreme environments of the low compositional index forests, ranging upward with small random fluctuations to a plateau level across the four right-hand segments of the gradient. On the other hand, seedling diversity begins relatively high, and after some fluctuations which are due in part to environmental differences, declines to a very low level across the three highest segments of the gradient. The first three decimal segments are all characterized by coarse-textured soils with bedrock frequently close to the surface. The fourth segment is not greatly different in forest composition, but is quite different in texture of substrate, depth of soil, and total water storage available to seedling and sapling layers. The result of the more favorable environment is, as one might expect, a greatly increased diversity of both seedling and tree layers, both reflecting in part the high light conditions of these relatively pioneer communities as well as the higher supply of moisture.

While the inverse relationship between the diversity of the understory and the overstory is interesting, and of considerable consequence to diversity relations across the landscape in southern Wisconsin, the great differences in environment across the compositional gradient mean that these results mix the two important elements influencing diversity: (1) the restrictions imposed by a severe environment, and (2) the variations that may be imposed as a direct result of productivity or community development over time within a restricted range of environment. Total 'available water capacity' has been estimated for each of the 30 stands to depths of both 36 and 60 inches. The mesic environments represented by the four highest decimal segments of the gradient are all characterized by an available water capacity of seven or more inches of water in the surface three feet of soil. Therefore, all stands, regardless of position along the compositional index, with an available water capacity of seven inches or more, have been used for the following analyses of diversity in mesic environments.

To establish still another base for the analysis of diversity, age determinations were carried out on the 18 stands with high levels of substrate water supply. The estimates of age for large old-growth hardwoods utilized both historical records dating to the surveyors' descriptions in the 1830s, and correlation between diameter and age. The latter are good for the shade-tolerant species whose widths of annual ring are consistent over extended periods. The aging was sufficiently precise to allow recognition of five age-classes among the stands sampled, ranging from one whose midpoint is just over 100 yr, to the oldest with a midpoint of 220 yr.

The annual production results for the 18 stands of maximum available water capacity are shown as a function of age, Fig. 2.13. A wave-form response is observed with the peak being reached at just under 200 yr of age. This response is well known for old-growth forests, but in these stands

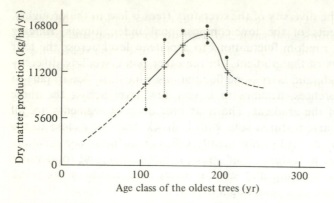

Fig. 2.13. Primary production fixed annually by the tree stems, plotted by age class of the oldest trees in each stand.

the original overstory species are no longer present and the reduction in growth rates is associated with a change in species composition. The change is a gradual increase in dominance by the few shade-tolerant species present, particularly *Acer saccharum*, in both the overstory and the understory layers. As long as a significant component of the pioneer species remains in the overstory canopy, primarily *Quercus* spp., the annual production of dry-matter can be maintained at a high level. As the basal area of these species decreases and is replaced by *Acer*, production decreases. Comparison with the results in Fig. 2.12 (and Fig. 2.16) shows the greatly reduced levels of diversity in the seedling, and to a lesser extent, the overstory layer, that is associated with the stands yielding the highest production as well as those dominated by *A. saccharum*.

Quantifying gradients in forest composition over time

Differentiating edaphic from time-related influences

However, there remains yet another difficulty in the interpretation of compositional relationships among forest ecosystems. Most forest studies have been limited by the absence of any specific measure of species structure within the system itself, and of the extent to which species composition varies with the changes in ecosystem structure over time. Thus, it is apparent that effective methods are essential for differentiating edaphic control of diversity from species patterns that are a response to time-dependent processes.

One attempt to differentiate edaphic from successional processes in the control of species diversity is the work of Peet & Loucks (1977). To determine what factors were important in controlling stand composition, this study examined all environmental data for each site, plotted the data in various scatter diagrams, and subjected them to correlation and principal components analysis. The two most important factor-complexes to emerge were recognized as a moisture-nutrient gradient and a stand dynamics or species replacement gradient. The analysis showed, further, that the sand content of the A_1 horizon provided an adequate physical measure of the edaphic component, the moisture-nutrient status of the system. Using weighted-average, species-occurrence values along the A_1 sand content gradient, these authors found a compositional gradient in which the most xeric species (*Quercus velutina* and *Quercus macrocarpa*) occupied the drought-susceptible environments, and *Acer saccharum* and *Ostrya virginiana* occupied the mesic environments.

Peet & Loucks (1977) then attempted to examine species responses associated with the invasion of tolerant species in the understory and the non-replacement of overstory elements. One measure used was based on the success of species in the second smallest of eight size-strata recognized in each forest. This measure, an index, was shown to be closely related to differences in forest composition associated with stand structure, but it proved a better indicator of regional trends in compositional change than of successional patterns within individual stands. Another measure of compositional dynamics was developed, based on the changing importance of each species across four uniform size strata in each stand. The shade-tolerant species are most frequent in the lower strata, while the intolerant species established following the most recent fire have their greatest importance in the larger size classes. The average percent change in the relative density of each species across the four strata was calculated. Each change was then weighted by the quantity of the species in the larger of the two strata relative to the total occurrence in all stands. This produced a weighted-average rate of change for each species (*see* Peet & Loucks, 1977).

This type of synthetic measure of stand dynamics has utility only in forest regions undergoing compositional change, but the research by Peet & Loucks shows it is entirely independent of edaphic or regional diversity influences. Also, to date, this gradient relates only to post-fire recovery and does not predict patterns following human disturbances.

The combination of the moisture-nutrients gradient and a successional dynamics gradient as orthogonal axes produces a two-dimensional representation which is a conceptually simple characterization of forest ecosystems. Fig. 2.14 illustrates the location of stands representing each of four strata in the 30 stands used. Variation can be followed for both gradients and the complete characterization is indicated by a simple ordered pair (*a,b*) with the first number signifying moisture-nutrients gradient position, and the second, the successional status position. Environmentally mesic systems occur along the left side of the two-dimensional representation and show

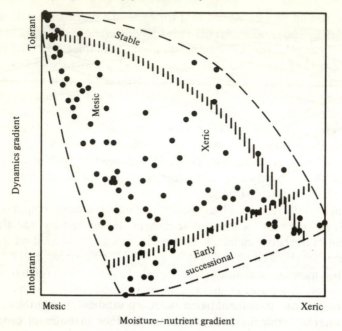

Fig. 2.14. A gradient representation of southern Wisconsin forests; each dot represents the location of one of four size strata from each of the 30 sampled stands. The abscissa is a moisture-nutrients gradient derived from soil sand content; the ordinate represents the species dynamics gradient related to stand successional status.

considerable variation in successional position. In contrast, the xeric stands on the right side do not exhibit much successional latitude, implying that the species replacement potential is less on the environmentally more severe sites.

With respect to developmental patterns it is evident that the shade-intolerant species are found across the bottom of the distribution in Fig. 2.14. These species were dominant in the long-established ecosystems present at the time of settlement during the last century (Curtis, 1959), before cessation of wildfires. They were found on both mesic and xeric environments. In contrast, stands undergoing little change in species composition, are found along the top of the distribution in Fig. 2.14. Some relatively xeric sites are undergoing relatively little change; these sites support open communities dominated by either *Quercus macrocarpa* or *Quercus velutina*. The environment here appears too severe for full canopy development of other understory species, and sufficient light penetrates for continued regeneration of the shade-intolerant species.

These results have led me (Loucks, 1978) to propose the bivariate representation of forest ecosystem composition shown in Fig. 2.15. The principle is that a compositional gradient attributable to stand structure (or

76

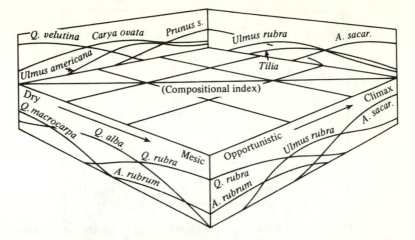

Fig. 2.15. Implied species gradients along a continuous two-dimensional gradient. The diagonal of the environmental and species replacement components represents the one-dimensional continuum of Curtis & McIntosh (1951).

species replacement) can be differentiated from the gradient attributable to edaphic control. The concept is represented here as the orthogonal distribution of communities associated with environmental control, on the one hand, or species change over time on the other, as illustrated in Fig. 2.15. The edaphic gradient ranges from dry to mesic, and can be examined as differences in soil texture, intercorrelated with differences in nutrient status and/or exposure. Independent of these measures, I postulate that stand structure can be used as a measure of apparent species replacement and that in forest ecosystems it takes place in a variety of environments. In our study area, a series of opportunistic species are adapted to the different habitats available, and in each habitat type there can be a progression toward a community that is a mix of the most shade-tolerant species capable of surviving there.

In Fig. 2.15 we see an implied species continuum along each of four gradients as shown along the end of each factor-complex. These continua recognize that the diagonal of the environmental and species replacement components represents the one-dimensional continuum apparent in the original work of Curtis & McIntosh (1951). These species distributions could now be examined as part of a regional ecosystem pattern in diversity related to aridity, productivity, and species replacement over time.

Recurrent patterns of diversity change over time

In the paper on the evolution of diversity and stability (Loucks, 1970), I examined the responses of species in the Great Lakes region as they respond over time to recurring perturbation. On theoretical grounds alone,

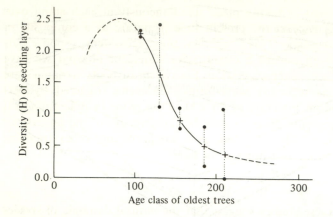

Fig. 2.16. Diversity of species in the seedling layer, plotted by age class of the oldest overstory trees.

one expects seedling diversity to be high immediately after disturbances, followed in time by a loss of species both in the sapling and tree layers as they develop, but particularly in the seedling layer beneath the saplings or immature trees. From evidence in the several papers that have looked at diversity in forests over time, and our knowledge of processes in the deciduous forest, we are able to say that the response for seedling diversity is a curve which drops to its lowest level shortly after the initial establishment of a young forest, then rises to a peak in forests of 100 to 150 years of age, and then falls off (Fig. 2.16).

The low diversity in the oldest age-classes sampled in the study results from domination of the seedling layer by one or a few very shade-tolerant species, *Acer saccharum*, in the western Great Lakes region of the United States. When the overstory forest reaches roughly 100 yr of age, the seedling layer is composed of scattered individuals of the overstory species which survive briefly as seedlings, but which usually do not enter the sapling layer. In addition, however, there are large numbers of individuals of the more shade-tolerant species entering the community at this point in time (Monk, 1967). In another 100–200 yr, there is no longer a canopy with any appreciable number of the pioneer species and therefore few seedlings of the species dependent on the original perturbation.

Given that perturbations such as fire, tornadoes, and hurricanes have been a part of the environment of North American forest ecosystems for as long as these systems have been evolving, a modern paradigm of ecosystem processes must consider the recurring post-disturbance transient as a natural property of the dynamic system.

The wave-form during the early years after establishment of a stand can be regarded as a gently rising sigmoid curve. The subsequent cresting and decline can be viewed as taking place simultaneously in each stand over time, or as taking place at the present time in nature as a composite of stands across a landscape. Taken together, these ecosystem response patterns represent a step toward a quantitative statement of what we have always known as secondary succession. Ecosystem diversity dynamics are viewed therefore as recurring wave-form phenomena triggered by perturbations at random intervals of 20–200 yr or more in different forest ecosystems. The response property may be the diversity of the understory layer, the productivity of the system as a whole, or any other transient phenomena responding as a function of time.

Conclusion

The results from these studies suggest that the natural dynamic of species composition in woodland ecosystems can best be thought of as a long-term stationary process involving random perturbations and recurring recovery transients. The concept is consistent with the known selection mechanisms operating during the evolution of these ecosystems and allows recognition of the mechanisms by which species are selected to carry out specialized functions in the ecosystem. Although geographic and environmental influences are often thought of as the principal isolating agents in speciation, and therefore of the control of diversity in ecosystems, the results presented here strongly suggest that separation over time in recurring transients is also an important isolating mechanism. For the deciduous forests studied during the US–IBP, the role of natural disturbances in the evolution and maintenance of ecosystem diversity appears to have been greatly underestimated.

Thus, we can conclude, further, that a relatively simple theory is available for the control of species diversity in forest ecosystems, utilizing readily available measures of regions, environments, productivity, and time. Most of the measures can be derived directly from stand analyses. The theory permits the differentiation of woodland ecosystem types and processes in a dynamic as well as a regional and edaphic context.

References

Adams, D. M. (1975). Derivation of optimal management guides: A survey of analytical approaches, pp. 1–9. In *Forest Modeling and Inventory*, ed. A. R. Ek, J. W. Balsiger, & L. C. Promnitz. School of Natural Resources, College of Agricultural and Life Sciences. University of Wisconsin Press, Madison.

Adams, D. M. & Ek, A. R. (1974). Optimizing the management of uneven-aged forest stands. *Canad. J. forest Res.* 4; 274–87.

Arney, J. D. (1972). Computer simulation of Douglas-fir tree and stand growth. Ph.D. Thesis. Oregon State University, Corvallis. (University Microfilms 72–70, 742.)

Arney, J. D. (1974). An Individual Tree Model for Stand Simulation in Douglas-fir, pp. 38–46. *In Growth models for tree and stand simulation*, ed. J. Fries. Royal College of Forestry, Research Notes No. 30, Stockholm.

Assman, E. (1970). *The Principles of Forest Yield Study*. Pergamon Press, New York.

Bailey, N. T. J. (1964). *The Elements of Stochastic Processes*. John Wiley & Sons, Inc., N.Y.

Bailey, R. L. (1973). Development of unthinned stands of *Pinus radiata* in New Zealand. Ph.D. Thesis, University Georgia, (Diss. Abstr. **33**, 4061–B.)

Bailey, R. L. Dell, T. R. (1973). Quantifying diameter distributions with the Weibull function. *Forest Sci.* **19**, 97–104.

Bakuzis, E. (1961). *Synecological coordinates and investigation of forest ecosystems*. Paper presented at 13th Congress of the International Union of Forest Research Organizations, Vienna.

Bella, I. E. (1970). Simulation of growth, yield and management of aspen. Ph.D. Thesis, University of British Columbia, Vancouver. (Nat. Lib. of Canada, Ottawa, Diss. Abstr. **31**, 6148–B.)

Billings, W. D. (1938). The structure and development of old field short-leaf pine stands and certain associated physical properties of the soil. *Ecol. Monogr.* **8**, 437–99.

Botkin, D. B., Janak, J. F. & Wallis, J. B. (1970a). *A simulator for northeastern forest growth; a contribution of the Hubbard Brook Ecosystem and IBM Research.* IBM Res. Rep. RC 3140. IBM Thomas J. Watson Research Center. Yorktown Heights, N.Y.

Botkin, D. B., Janak, J. F. & Wallis, J. R. (1970b). *The rationale, limitations and assumptions of a northeast forest simulator.* IBM Res. Rep. RC 3188. IBM Thomas J. Watson Research Center. Yorktown Heights, N.Y.

Botkin, D. B., Janak, J. F. & Wallis, J. R. (1972). Some ecological consequences of a computer model of forest growth. *J. Ecol.* **60**, 849–72.

Braun, E. L. (1950). *Deciduous Forests of Eastern North America*. Blakiston Co., Philadelphia.

Bruner, H. D. & Moser, J. W. (1973). A Markov chain approach to the prediction of diameter distributions in uneven-aged forest stands. *Canad. J. forest Res.* 3, 409–17.

Buckman, R. E. (1962). *Growth and yield of red pine in Minnesota*. USDA Tech. Bull. 1272.

Burkhart, H. E. & Strub, M. R. (1974). A model for simulation of planted loblolly pine stands, pp. 128–35. In *Growth models for tree and stand simulation*, ed. J. Fries. Royal College of Forestry, Research Notes No. 30, Stockholm.

Chase, C. D., Pfeifer, R. E. & Spencer, J. S. Jr. (1970). *The growing timber resource of Michigan*, 1966. USDA Forest Service Resource Bulletin NC–9. North Central For. Expt. Sta., St. Paul, MN.

Clutter, J. L. (1963). Compatible growth and yield models for loblolly pine. *Forest Sci.* **9**, 354–71.

Clutter, J. L. & Bennett, F. A. (1965). *Diameter distributions in old field slash pine plantations*. Georgia Forest Research Council Rept. No. 13.

Clutter, J. L. & Allison, B. J. (1974). A growth and yield model for *Pinus rediata* in New Zealand, pp. 136–66. In *Growth models for tree and stand simulation* ed. J. Fries. Royal College of Forestry, Research Notes No. 30. Stockholm.

Connell, J. H. & Orias, E. (1964). The ecological regulation of species diversity. *Amer. Naturalist*, **98**, 299–414.

Cowardin, L. M., Carter, V., Golet, F. C. & LaRoe, E. T. (1977). *Classification of wetlands and aquatic habitats of the United States.* US Fish and Wildlife Service Monograph. 100p.

Curtis, J. T. (1959). *The vegetation of Wisconsin: an ordination of plant communities.* University of Wisconsin Press, Madison.

Curtis, J. T. & McIntosh, R. P. (1951). An upland forest continuum in the prairie-forest border region of Wisconsin. *Ecology*, **32**, 476–96.

Curtis, R. O. (1972). Yield tables past and present. *J. Forest.* **70**, 28–32.

Demaerschalk, J. P. (1972). Converting volume equations to compatible taper equations. *Forest Sci.* **18**; 241–5.

Dress, P. E. (1968). Stochastic models for the simulation of even-aged forest stands. Paper presented at 1968 Soc. Amer. Forest. Annual Meeting, Philadelphia, Pa.

Duff, G. H. & Nolan, N. J. (1953). Growth and morphogenesis in the Canadian forest species. I. The controls of cambial and apical activity in *Pinus resinosa* Ait. *Canad. J. Bot.* **31**, 471–513.

Duff, G. H. & Nolan, N. J. (1957). Growth and morphogenesis in the Canadian forest species. II. Specific increments and their relation to the quantity and activity of growth in *Pinus resinosa* Ait. *Canad. J. Bot.* **35**, 527–72.

Duff, G. H. & Nolan, N. J. (1958). Growth and morphogenesis in the Canadian forest species. III. The time scale of morphogenesis at the stem apex of *Pinus resinosa* Ait. *Canad. J. Bot.* **36**, 687–710.

Edmonds, R. L. (ed.). (1974). *An initial synthesis of results in the Coniferous Forest Biome, 1970–73.* Bulletin No. 7, Coniferous Forest Biome, Ecosystem Analysis Studies, US–IBP.

Eis, S. (1972). Root grafts and their silvicultural implications. *Canad. J. Forest Res.* **2**, 111–20.

Ek, A. R. (1974*a*). *Dimensional relationships of forest and open-grown stems in Wisconsin.* University of Wisconsin Forestry Research Note 181.

Ek, A. R. (1974*b*). Nonlinear models for stand table projection in northern hardwood stands. *Canad. J. forest Res.* **4**, 23–7.

Ek, A. R. & Monserud, R. A. (1974*a*). Trials with program FOREST: Growth and reproduction simulation for mixed species forest stands, pp. 56–73. In *Growth models for tree and stand simulation*, ed. J. Fries. Royal College of Forestry, Research Notes No. 30, Stockholm.

Ek, A. R. & Monserud, R. A. (1974*b*). *FOREST: A computer model for simulating the growth and reproduction of mixed species forest stands.* University of Wisconsin, College of Agricultural and Life Sciences Research Report R2635.

Forsythe, W. L. & Loucks, O. L. (1972). A transformation for species response to habitat factors. *Ecology*, **53**, 1112–9.

Fowells, H. A. (ed.). (1965). *Silvics of forest trees of the United States.* USDA Handbook 271.

Franklin, J. F. & Dyrness, C. T. (1969). *Vegetation of Oregon and Washington. USDA Forest Services Research Paper PNW*–80.

Franklin, J. F. & Dyrness, C. T. (1973). *Natural vegetation of Oregon and Washington* USDA Forest Service General Technical Report. PNW–8.

Gerrard, D. J. (1966). *Competition quotient: a new measure of the competition affecting individual forest trees.* Michigan State Agricultural Experimental Station Research Bulletin 20.

Goff, F. G. (1967). Upland vegetation. In *Soil resources and forest ecology of Menominee County, Wisconsin.* Bulletin 85, University of Wisconsin Geology and Natural History Survey.

Goulding, C. J. (1972). Simulation techniques for a stochastic model of the growth of Douglas-fir. Ph.D. Thesis, University of British Columbia, Vancouver. (Nat. Lib. of Canada, Ottawa, *Diss. Abstr.* **33**, 5599–B.)

Grieg-Smith, P. (1964). *Quantitative Plant Ecology.* Butterworths, London.

Grosenbaugh, L. R. (1965). Generalization and reparameterization of some sigmoid and other nonlinear functions. *Biometrics,* **21**, 708–714.

Hari, P., Leikola, M. & Rasanen, P. (1970). A dynamic model of the daily height increment of plants. *Ann. Bot. Fenn.* **7**, 375–8.

Hatch, C. R. (1971). Simulation of an even-aged red pine stand in northern Minnesota. Ph.D. Thesis. Univ. of Minnesota, St. Paul. (University Microfilms 72–14, 314.)

Hawksworth, F. G. (1965). Life tables for two species of dwarf mistletoe. *Forest Sci.* **11**, 142–51.

Hegyi, F. (1974). A simulation model for managing jack pine stands. pp. 74–90. In *Growth models for tree and stand simulation,* ed. J. Fries. Royal College of Forestry, Research Notes No. 30, Stockholm.

Hett, J. M. (1971*a*). *Land-use changes in east Tennessee and a simulation model which describes these changes for three counties.* ORNL–IBP–71–8. Oak Ridge National Laboratory, Oak Ridge, TN.

Hett, J. M. (1971*b*). A dynamic analysis of age in sugar maple seedlings. *Ecology,* **52**, 1071–4.

Hett, J. M. & Loucks, O. L. (1968). Application of life table analysis to tree seedlings in Quetico Provincal Park, Ontario. *Forest Chron.* **442**, 29–32.

Hett, J. M. & Loucks, O. L. (1971). Sugar maple (*Acer saccharum* Marsh.) Seedling mortality. *J. Ecol.* **59**, 507–20.

Husch, B., Miller, C. I. & Beers, T. W. (1972). *Forest Mensuration,* 2nd Edn Ronald Press, N.Y.

Johnson, W. C. & Sharpe, D. M. (1976). Forest dynamics in the northern Georgia Piedmont. *Forest Sci.* **22**, 307–22.

Kozlowski, T. T. (1971*a*). *Growth and Development of Trees.* Vol. I. Academic Press, N.Y.

Kozlowski, T. T. (1971*b*). *Growth and Development of Trees.* Vol. II. Academic Press, N.Y.

Larson, P. R. (1963). *Stem form development of forest trees.* Forest Science Monograph 5.

Leak, W. B. (1968). Birch regeneration: a stochastic model. USDA Forest Service, N.E. Forest Experimental Station Research Note NE–85.

Leak, W. B. (1969). Stocking of northern hardwood regeneration based on experimental dropout rate. *Forest Chron.* **45**, 344–7.

Leak, W. B. (1970*a*). Sapling stand development: a compound exponential process. *Forest Sci.* **16**, 177–80.

Leak, W. B. (1970*b*). Successional change in northern hardwoods predicted by birth and death simulation. *Ecology,* **51**, 794–801.

Leak, W. B. (1972). Competitive exclusion in forest trees. *Nature, Lond.* **236**, 461–3.

Leary, R. A. (1968). A multi-dimensional model of even-aged forest growth. Ph.D. Thesis, Purdue University. (University Microfilms 69–7472.)

Leary, R. A. (1970). *System identification principles in studies of forest dynamics.* USDA Forest Service, North Central Forest Experimental Station Research Paper NC–45.

Leary, R. A. & Skog, K. E. (1972). A computational strategy for system identification in ecology. *Ecology,* **53**, 969–73.

Lee, Y. (1967). Stand models for lodgepole pine and limits to their application. Ph.D. Thesis. University of British Columbia, Vancouver. (Nat. Lib. of Canada, Ottawa, Diss. Abstr. **29**, 829–B.)

Lemmon, P. E. & Schumacher, F. X. (1962). Stocking density around ponderosa pine trees. *Forest Sci.* **8**, 397–402.

Lin, J. Y. (1970). Growing space index and stand simulation of young western hemlock in Oregon. D. F. Thesis, Duke University, Durham, North Carolina. (University Microfilms 70–20, 276.)

Lin, J. Y. (1974). Stand growth simulation models for Douglas-fir and western hemlock in the northwestern United States, pp. 102–18. In *Growth models for tree and stand simulation*, ed. J. Fries. Royal College of Forestry, Research Notes No. 30, Stockholm.

Livingston, D. A. (1969). Communities of the past, pp. 83–104. In *Essays in plant geography and ecology.* ed. K.N.H. Greenidge. Nova Scotia Museum, Halifax.

Lotka, A. J. (1942). *Elements of mathematical biology.* Reprinted by Dover, N.Y., 1956.

Loucks, O. L. (1962*a*). A Forest Classification for the Maritime Provinces. *Proc. Nova Scot. Inst. Sci.* **25**, 85–167.

Loucks, O. L. (1962*b*). Ordinating forest communities by means of environmental scalars and phytosociological indices. *Ecol. Monogr.* **32**, 137–66.

Loucks, O. L. (1970). Evolution of diversity, efficiency and community stability. *Amer. Zool.* **10**, 17–25.

Loucks, O. L. (1978). Comparison of edaphic and successional control of species composition in southern Wisconsin forests. In *Ecosystems of Dark Coniferous Forests of the Temperate Zone*, ed. T. A. Frey. Academy of Sciences of the Estonian SSR, Tartu.

MacArthur, R. H. (1955). Fluctuations of animal populations and a measure of community stability. *Ecology*, **36**, 533–6.

Mankin, J. B. & Brooks, A. A. (1971). Numerical methods for ecosystem analysis. ORNL–IBP 71–1. Oak Ridge National Laboratory, Oak Ridge, TN.

Margalef, D. R. (1957). La teoria de la informaeion en ecologia. *Men. Real Acad. Cienc. Art. Barcelona*, **32**, 373–449. (Trans. Soc. Gen. Syst. Res. **3**, 36–71.)

Margalef, D. R. (1968). *Perspective in Ecological Theory.* University of Chicago Press.

Marquis, D. A. (1969). Silvical requirements for natural birch regeneration, pp. 40–9. In *Birch Symposium Proceedings, August*, 1969. USDA Forest Service, N.E. Forest Experimental Station, Upper Darby, Pa.

Matthews, J. D. (1963). Factors affecting the production of seed by forest trees. *Forest Abstr.* **24**, i–xiii.

McGee, C. E. & Della-Bianca, L. (1967). *Diameter Distributions in Natural Yellow-Poplar Stands.* USDA Forest Service, Southeast forest Experimental Station Research Paper SE–25.

Meyer, H. A. (1952). Structure, growth and drain in balanced uneven-aged forests. *J. Forest.* **50**, 85–92.

Meyer, H. A., Recknagel, A. B., Stevenson, D. D. & Bartoo, R. A. (1961). *Forest Management*, 2nd Edn. Ronald Press, N.Y.

Mitchell, K. J. (1967). Simulation of the growth of even-aged stands of white spruce. Ph.D. Thesis. Yale University, New Haven, Conn. (Univ. Microfilms 68–5188).

Monk, C. D. (1967). Tree species diversity in the eastern deciduous forest with particular reference to north central Florida. *Amer. Naturalist*, **101**, 173–87.

83

Monserud, R. A. & Ek, A. R. (1974). Plot edge bias in forest growth simulation models. *Canad. J. Forest Res.* **4**, 419–23.

Moser, J. W. (1972). Dynamics of an uneven-aged forest stand. *Forest Sci.* **18**, 184–91.

Moser, J. W. (1974). A system of equations for the components of forest growth, pp. 260–88. *Growth models for tree and stand simulation*, ed. J. Fries. Royal College of Forestry, Stockholm.

Moser, J. W. & Hall, O. F. (1969). Deriving growth and yield functions for uneven-aged forest stands. *Forest Sci.* **15**, 183–8.

Munro, D. D. (1974). Forest growth models – a prognosis, pp. 7–21. In *Growth models for tree and stand simulation*, ed. J. Fries. Royal College of Forestry, Research Notes No. 30, Stockholm.

Murphy, C. E. Hesketh, J. D. & Strain, B. R. (ed.) (1972). *Modeling the growth of trees.* National Technical Information Service, US Dept. of Commerce, Springfield, Virginia.

Newnham, R. M. (1964). The development of a stand model for Douglas-fir. Ph.D. Thesis, University of British Columbia, Vancouver. (University Microfilms 64–4535.)

Odum, E. P. (1963). *Ecology.* Holt, Rinehart & Winston, New York.

Olson, J. S. & Cristofolini, G. (1966). Model simulation of Oak Ridge vegetation succession, pp. 106–7. In *Health Physics Division Annual Progress Report*, Oak Ridge National Laboratory, Oak Ridge, TN.

Oosting, H. J. (1942). An ecological analysis of the plant communities of Piedmont, North Carolina. *Amer. Midl. Nat.* **28**, 1–126.

Opie, J. L. (1972). Standism–a general model for simulating the growth of even-aged stands. IUFRO 3rd Conference Advisory Group of Forest Statisticians, Jouy-en-Josas, 7–11 Sept. 1970. Institut National de la Recherche Agronomique Publ. 72–3, 217–40.

Payandeh, B. (1974). Spatial pattern of trees in the major forest types of northern Ontario. *Canad. J. forest Res.* **4**, 8–14.

Peden, L. M., Williams, J. S. & Frayer, W. E. (1973). A Markov model for stand projection. *Forest Sci.* **19**, 303–14.

Peet, R. K. & Loucks, O. L. (1977). A gradient analysis of southern Wisconsin forests. *Ecology*, **58**, 485–99.

Pielou, E. C. (1966). Species-diversity and pattern-diversity in the study of ecological succession. *J. Theor. Biol.* **10**, 370–83.

Pienaar, L. V. (1965). Quantitative theory of forest growth. Ph.D. Thesis. University of Washington, Seattle. (University Microfilms 65–11, 485.)

Quarterman, E. & Keever, C. (1962). Southern mixed hardwood forest: climax in the southeastern coastal plain. *Ecol. Monogr.* **32**, 167–85.

Richards, F. J. (1959). A flexible growth function for empirical use. *J. Exp. Bot.* **10**, 290–300.

Rubner, K. & Reinhold, F. (1953). *Das Naturliche Waldbild Europas.* Paul Pary & Co.

Rudra, A. B. (1968). A stochastic model for the prediction of diameter distribution of even-aged forest stands. *Opsearch*, **5**, 59–73.

Shugart, H. H., Crow, T. R. & Hett, J. M. (1973). Forest succession models: a rationale and methodology for modeling forest succession over large regions. *Forest Sci.* **19**, 203–12.

Stage, A. R. (1973). Prognosis model for stand development. USDA Forest Service Intermountain Forest and Range Experimental Station Research Paper INT–137.

Strand, L. (1972). A model for stand growth. IUFRO 3rd Conference Advisory Group of Forest Statisticians, Jouy-en-Josas, 7–11 Sept. 1970. Institut National de la Recherche Agronomique Publ. 72–3, 207–16.

Stephens, G. R. & Waggoner, P. E. (1970). *The forests anticipated from* 40 *years of natural transitions in mixed hardwoods.* Conn. Agr. Exp. Sta., New Haven, Conn. Bull. No. 707.

Suzuki, T. & Umemura, T. (1974). Forest transition as a stochastic process, pp. 358–79. In *Growth models for tree and stand simulation*, ed. J. Fries. Royal College of Forestry, Research Notes No. 30, Stockholm.

Turnbull, K. J. (1963). Population dynamics in mixed forest stands. Ph.D. Thesis, Univ. of Washington, Seattle. (Univ. Microfilms 64–437.)

Vuokila, Y. (1966). Functions for variable density yield tables of pine based on temporary sample plots. *Commun. Inst. For. Fenn.* **60**, 1–86.

Waggoner, P. E. & Stephens, G. R. (1970). Transition probabilities for a forest. *Nature, Lond.* **225**, 1160–1.

Waring, R. H. & Major, J. (1964). Some vegetational aspects of the California coast redwood region in relation to gradients of moisture, light, nutrients, and temperature. *Ecol. Monogr.* **34**, 167–215.

Waters, W. E. (1969). The life table approach to analysis of insect impact. *J. Forest.* **67**, 300–4.

Webb, T. (1974). Corresponding patterns of pollen and vegetation in lower Michigan: a comparison of quantitative data. *Ecology*, **55**, 17–28.

Whittaker, R. H. (1956). Vegetation of the Great Smokey Mountains. *Ecol. Monogr.* **26**, 1–80.

Whittaker, R. H. (1960). Vegetation of the Siskiyou Mountains, Oregon and California. *Ecol. Monogr.* **30**, 279–338.

Whittaker, R. H. (1965). Dominance and diversity in land plant communities. *Science, Wash.* **147**, 250–60.

Wilson, B. F. & Howard, R. A. (1968). A computer model for cambial activity. *Forest Sci.* **14**, 77–90.

Zöhrer, F. (1972). The beta-distribution for best fit of stem diameter distributions. IUFRO 3rd Conference Advisory Group of Forest Statisticians. Jouy-en-Josas, 7–11 Sept. 1970. Institut National de la Recherche Agronomique Publ. 72–3, 91–106.

3. Radiation, heat, water and carbon dioxide balances

A. GALOUX, P. BENECKE, G. GIETL, H. HAGER, C. KAYSER, O. KIESE, K. R. KNOERR,* C. E. MURPHY,* G. SCHNOCK & T. R. SINCLAIR*

Contents

*Research supported by the Eastern Deciduous Forest Biome, US–IBP, funded by the National Science Foundation under Interagency Agreement AG–199, DEB76–00761 with the US Department of Energy under contract W–7405–eng–26 with Union Carbide Corporation. Contribution No. 343 from the Eastern Deciduous Forest Biome, US-IBP.

Radiation flux through the atmosphere–biosphere system

A forest is a biospheric system made up of biomass surrounded by the scattered components of its metabolic processes. It may be regarded as a biochemical continuum or as a vast population of relatively closely related molecules located in a permanent energy radiation field. This system is traversed by a flux of energy, photons, emitted by a source, the sun, which serves to organize the system, to maintain it in a stationary state, and which is subsequently dispersed in space, the sink in a degraded form, infrared radiation.

In the non-equilibrium or steady-state systems the energy flow engenders cycles of matter, which represent an increase in organization that is accompanied by a local diminution of entropy. The absorption of photons, in particular, permits the formation in the system of chemical compounds with a high level of potential energy and a high electronic temperature compared with thermodynamic temperature. These compounds change by yielding up their potential energy, and new combinations with a low chemical potential leave the system. Thus, the same atoms pass successively through states of rearrangement of unequal chemical potential, which is the condition of biogeochemical cycles.

Similarly, absorption and transfer of radiant energy and heat transfer by conduction, convection, and conversion occur in biomass, the soil mass, water, and air masses; energy gradients of internal movement (kinetic temperature, change of phase) and of external movement (kinetic energy) are created. These gradients give rise to atmospheric cycles (turbulent convection, advection) and the water cycle (evapotranspiration, precipitation). If we take a sufficient period of time with a zero storage term an approximate steady state is established. The whole earth is a stationary closed system with an inflow of solar short-wave radiation balanced by an outflow of infrared to outer space, and an almost perfect conservation of matter inside the system.

Radiation and earth

From space, different types of radiation are received on the earth; these are cosmic, sun wind, Röntgen and ultraviolet, visible, infrared, high frequency radiations. A perpendicular surface in the vicinity of the earth intercepts a short-wave radiation energy of 2 cal cm^{-2} min^{-1} or 4.2×10^{15} kWh day^{-1}. As well as electromagnetic radiant energy emitted by the sun, the earth receives about 1000 tons of space matter every day in the form of meteorites. When radiation encounters matter, electronic motion results in absorption or emission of quanta with the highest energy level, X-ray, ultraviolet, visible radiations, vibrational motion with near infrared and infrared, and rotational motion with small energy quanta, infrared and microwaves. So radiation flux and sun wind first encounter a magnetic field, then pass through a radiation shell, the Van Allen belt, before they reach the exosphere where Röntgen and UV radiations are dissociating and ionizing some gases as hydrogen (H). Further, in the ionosphere the same short-wave radiation dissociates oxygen into ozone. In the mesosphere, the formation of ozone and the recombination of oxygen by the action of ultraviolet radiation is one of the most important processes of the atmosphere in regard to life.

In the stratosphere and troposphere, the absorption of some bands in the spectrum of the short-wave sun radiation by some components of the atmosphere, and principally the absorption of infrared radiation emitted by the earth surface, the vegetation cover, maintains the kinetic temperature under certain limits. A further consequence is the process of evaporation at the surface and condensation in the clouds with vertical movements of masses that lead to electrical discharges, that is, high frequency waves.

It is here at the base of the atmosphere that life can appear, the biosphere. The atmosphere protects organisms from highly energetic, very short waves such as cosmic, X- and UV-rays and it is the source of important downward and upward infrared radiation fluxes. This balances the energy deficit of organisms. This thermal process, the greenhouse effect, increases the earth's temperature by 35°C from −21°C to 14.3°C, the average temperature of the surfaces of the earth. On the way through the atmosphere, solar radiation is attenuated and modified in spectral quality by some absorption and scattering processes of the molecules, Rayleigh scattering, inversely proportional to the fourth power of the wavelength. The absorption of solar radiation is a selective process in the spectral range and is related to the presence of some components such as water vapor, carbon dioxide, ozone, dusts. The evolution of short-wave and long-wave radiation on the way from space to the soil absorbing surface is given in Fig. 3.1, which expresses the qualitative and relative quantitative changes.

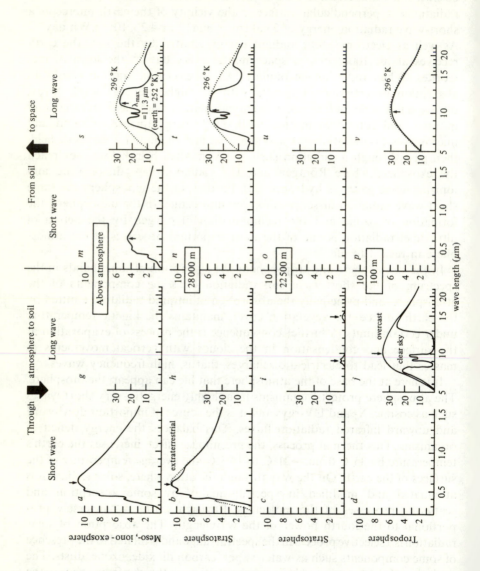

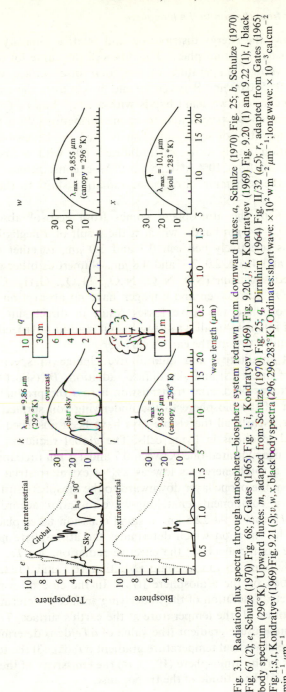

Fig. 3.1. Radiation flux spectra through atmosphere–biosphere system redrawn from downward fluxes: *a*, Schulze (1970) Fig. 25; *b*, Schulze (1970) Fig. 67 (2); *e*, Schulze (1970) Fig. 68; *f*, Gates (1965) Fig. 1; *i*, Kondratyev (1969) Fig. 9.20; *j*, *k*, Kondratyev (1969) Fig. 9.20 (1) and 9.22 (1); *l*, black body spectrum (296°K). Upward fluxes: *m*, adapted from Schulze (1970) Fig. 25; *q*, Dirmhirn (1964) Fig. II/32 (*a*,5); *r*, adapted from Gates (1965) Fig.1; *s*,*t*, Kondratyev (1969) Fig. 9.21 (5); *v*, *w*,*x*, black body spectra (296,296,283°K). Ordinates: short wave: $\times 10^2\,\mathrm{w\,m^{-2}\,\mu m^{-1}}$; long wave: $\times 10^{-3}\,\mathrm{cal\,cm^{-2}\,min^{-1}\,\mu m^{-1}}$.

91

Radiation flux from the sun to the biosphere

Fig. 3.1*b* shows the energy distribution and relative intensity in the solar spectrum outside the atmosphere. This fits well the curve for a black body at 5900°K, average temperature of the solar disc surface with spectral maximum at 0.55–0.60 μm. Between 20 and 30 km from the earth's surface, absorption occurs on two main bands with a small loss ($\pm 6\%$), and gives rise to dissociation of oxygen and to ozone formation. With the Rayleigh scattering, a quantity of solar radiation is used in the visible range and as sky radiation partly downward diffused (*D*) with a spectral maximum at 0.45 μm (loss 10% at 0.55 μm). As a consequence, the spectral maximum of the remaining direct solar radiation shifts to the near infrared.

A high loss in the atmosphere comes from aerosol absorption and scattering in 0.3–1.4 μm bands, which is the result of a negligible displace-which absorbs selectively between 0.7 and 3.0 μm, together with carbon dioxide mass at 1.4, 1.6, 2.0, 2.7, and 4.8 μm. Numerous other gases absorb in the infrared spectrum (NO, N_2O, N_2O_4, N_2O_5, C_3H_8, C_2H_6, C_2H_4, CH_4, H_2S, CO), but oxygen and nitrogen have no absorption band in the infrared. All of these successive losses result in direct sun radiation, a diminution of intensity in all wavelengths but also in polarized and diffuse sky radiation (also toward the earth's surface).

The largest qualitative and quantitative change in spectral intensity distribution of direct (*S*) and sky diffuse radiation (*D*) (short-wave global radiation, *G*) takes place after transmission across the production, chlorophyllous layer of the biosphere to the underlying soil surface, nonchlorophyllous layer. Nearly all the visible rays have been absorbed and used in photochemical reactions in living cells. The global radiation spectrum is almost a pure near-infrared one between 0.7 and 1.2 μm (maximum at 0.75–0.9 μm), with a total energy as low as 2% of extraterrestrial direct solar radiation. From the ozone layer downward to the biosphere surface, absorbing atmospheric components emit long-wave radiation according to their temperature ($I = \varepsilon \sigma T^4$). At 22.5 km, a small emission takes place only at 9 and 15 μm. The equation which determines the atmosphere net radiation contains three components: (1) the effective radiation, $A - Te$, where Te is terrestrial radiation, (2) the upgoing radiation (*U*) and (3) the solar radiation absorbed by the atmosphere (G_{abs}). In the case of normal atmosphere, the effective radiation of the underlying surface is determined by five physical quantities: (1) the temperature at the earth's surface, $Te = \varepsilon \sigma T^4$, (2) the vertical temperature gradient (the value of dTe/dz is determined first of all by the value of vertical temperature gradient dT/dz), (3) the total content of water vapor in the atmosphere (W_{atm}), (4) the emissivity of the underlying surface (ε) and (5) the altitude of the tropopause.

Calculations and observations show that the G_{abs} value is much smaller than the other components. The atmospheric net radiation is, therefore, determined mainly by the thermal radiative influx $(A - Te) < U$ and, consequently, the atmospheric net radiation is negative. This is due to the fact that the atmosphere absorbs only the earth thermal radiation, and considerably less of the solar radiation, while its emission is directed towards the earth's surface and towards space. At 0.1 km, atmospheric downward radiation, A, with clear sky, ranges from 3 to 60 μm with three maxima at 7.5, 9.5 and 14 μm. With fully overcast sky with water droplets, the spectral curve approaches very close to a black body spectrum with a maximum at about 10.5 μm. Downward oriented long-wave radiation flux with this spectrum is received by the chlorophyllous layer together with global radiation. At the soil surface under the chlorophyllous layer, long-wave radiation flux has about the same spectrum, but the atmosphere as an emitting source is now negligible and related to the size of canopy openings. The biomasses of the earth's surface with their own surface temperatures which are higher than air temperature during the daytime are a new source of infrared radiation downward and upward with an emissivity (ε) close to 1.0, in fact, about 0.97.

Radiation flux from the biosphere to space

Fig. 3.1 also shows the short-wave and long-wave upward radiation spectrum at the same levels as the downward spectrum. At the soil surface under the chlorophyllous layer, the spectrum of reflected short-wave radiation is almost deprived of visible rays (1/40 of the value of the global radiation outside the biosphere) and represents quantitatively a very small radiant flow. On the biosphere's chlorophyllous surfaces, reflected global $(D + S)$ radiation flow shows a spectral distribution of light shifted to the near infrared, related to the selectively absorbing property of chlorophyll in the visible bands. The reflected portion varies from 5 to 25% of extraterrestrial solar radiation flow, according to optical properties of the light or dark-green chlorophyll surfaces and to thickness, structure, or mass of the foliage. The reflected short-wave solar radiation (aS) spectrum outside the atmosphere has about the same distribution as the extraterrestrial solar spectrum (S). Reflection and scattering by dust and upper cloud surfaces do not change the spectrum, but Rayleigh scattering is larger for short wave lengths. This gives the albedo of the earth a shift to blue as a spectrum of a black body at 8000°K.

As to the upward long-wave radiation spectrum, the soil surface under the chlorophyllous layer with a low temperature gives rise to a blackbody spectrum with λ_{max} of 10.1 μm. Just above the upper surface of vegetation, the spectrum is that of a black body with emissivity 0.97 and leaf temperature higher than air temperature during day-time. At 0.1 km, spectrum distribution is that of a black body radiation (earth surfaces and 100 m-thick humid air) with maximum intensivity near 10 μm, according to vegetation surfaces and air layer temperatures. At the ozone level, the spectrum curve has its maximum intensity at 11.5 μm, but the bands 5 to 8, 9.5, 13–17 μm are weakly represented. The upward long wave radiation flux (U) is most important in relation to the steady state of the atmospheric–biospheric system, the surface of which the average temperature at $z=0$ is $+14.3°$C. As incoming radiation balances upward radiation, $(1-0.34)\pi R^2 S = 4\pi R^2 \varepsilon\sigma T^4$ where S is 2 cal cm^{-2} min^{-1}, R is the radius of earth, σ is $8.26\cdot10^{-11}$ cal cm^{-2}°K^{-1}, σT^4 is the Stefan Boltzmann law with T 252°K ($-21°$C). This is the effective radiative average temperature of the earth. So from the biosphere to extra-atmospheric level, one can observe the shift of λ_{max} from about 10 to 11.3 μm.

Net radiation balances

If we look at the general changes of short and long-wave spectra inside the atmosphere and the biosphere, we can observe a decrease of solar radiation short-wave flux and an increase of long-wave flux from upper atmosphere to soil surface. The maximum intensity of solar radiation spectrum shifts from 0.55 μm towards near the infrared region (0.7–0.94 μm) or reduced to a pure near-infra-red spectrum at the soil surface (0.7–1.2 μm). Conversely, maximum spectral intensity for long-waves shifts to smaller wave-lengths on approaching the vegetation surface.

The albedo values of solar radiation increase from soil surfaces to space. The upward long-wave radiation spectrum indicates a diminution of total intensity and a shift to smaller waves from space to biosphere. As a result, the peaks of the downward and upward radiation flux spectrum diverge from each other from soil surfaces toward space. The significance is that the temperatures of waves, $h\nu=kT$ (where k is the Boltzmann constant, ν the wave frequency, h, Planck's constant), are diverging from biosphere to space (the radiant energy degradation process).

Variation of net radiation balance and its components from biosphere to space

If one makes up the total balance (net radiation balance, $Q=D+S(=G)-aG+A-U$; U is upward radiation) at any level of the atmosphere–biospheric system, some clarification is necessary. Fig. 3.2 allows us to form an idea of possible variations of radiative fluxes and net radiation balance with altitude in clear and cloudy skies. Direct solar radiation and global radiation fluxes decrease approaching the biosphere owing to air, dusts, and

94

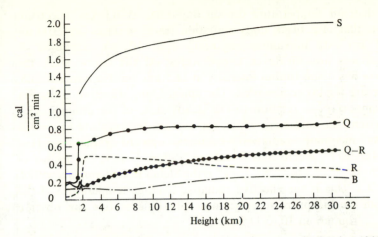

Fig. 3.2. Vertical profiles of net radiation and its components, 23 October, 1964, on horizontal surface except that *S* was received on a surface perpendicular to the sun's rays. *S*, solar radiation; *Q*, global radiation on a horizontal surface; *Q − R*, global radiation − reflected global radiation; *R*, reflected global radiation; *B*, radiation balance. After Kondratyev, 1969.

water density. At the level of clouds, reflection may reach up to 80%, then decrease regularly with altitude to 30%. Consequently, net radiation is low at the soil surface of the biosphere, increases rapidly up to chlorophyllous surfaces and then increases constantly over long altitude variations up to a high level. States close to radiative equilibrium above the troposphere are noticeable. Decrease of $(G − aG)$ from space to the biosphere is compensated by an increase of $(A − U)$.

Total heat balance of an atmosphere–biosphere system

On the other hand, Table 3.1 gives a total balance for an atmospheric–biospheric system (growth period) at latitude 50°N (Virelles IBP site). The beginning and the end of the period have been fixed in the spring and in the fall by taking the two days where the temperatures of the system, including the soil, biomass air, and air in the low atmosphere, were the most similar. At the level of the ecosystem, the balance results from measurements; at the extra-atmospheric level, the balance results from estimations of extra-atmospheric sun radiation according to the tables of Linacre (1969). As the net production of an ecosystem (NEP) is positive, the radiation net balance at the extra-atmospheric level is not zero. We can compare this balance with some yearly balances given by Möller (1957, in Schulze, 1970) for latitude 50°N, Schulze (1970) for Hamburg, and London for the northern hemisphere (one year). One can see that these balances are not too divergent from one another. Measured balances (%) at the earth's surface for Virelles (growing period) and Hamburg (year) are close.

95

Table 3.1. *Radiation balance in atmosphere–biosphere systems (cal cm^{-2} min^{-1} with percentage in brackets)*

	Northern hemisphere, London 1957 (in Kondratyev, 1969) Yearly balance	Latitude 50° N (Möller 1957–69 in Schulze, 1970) Yearly balance	Hamburg latitude 53° N (Schulze, 1970) Yearly balance	Virelles latitude 50° N longitude 4° 20′ E (Galoux) 25.5.67–25.10.67
Above atmosphere				
S_o	+0.500 (+100)	+0.395 (+100)	+0.382 (+100)	+0.541 (+100)
aS_o	−0.176 (−35.2)	−0.155 (−39.2)		
U	−0.324 (−64.8)	−0.300 (−75.9)		
Balance	0.00 (+0.00)	−0.060 (−15.2)		
Atmosphere				
S_o, G abs.	+0.087 (+17.4)	+0.070 (+17.7)		
$(A - Te)$ abs.	+0.090 (+18.0)	−0.240 (−60.7)		
Total absorbed	+0.177 (+35.4)			
U_{atm}	−0.324 (−64.8)			
Balance	−0.147 (−29.4)	−0.170 (−43.0)		
Ecosystem				
G_{abs}	+0.237 (+47.4)	+0.175 (+44.3)		+0.212 (+39.2)
$(A - Te)$ upward radiated	−0.90 (−18.0)	−0.065 (−16.4)		−0.094 (−17.4)
Balance	+0.147 (+29.4)	+0.110 (+27.9)	+0.078 (+20.3)	+0.118 (+21.8)

96

Entropy flow and entropy production

Let us consider the atmosphere–biosphere system as consisting of two open systems which exchange with each other energy, in the forms of both heat and work, and matter through their boundaries, which are themselves included in a closed system exchanging energy but not matter with the exterior; in fact almost no matter is exchanged. The first system or open phase (Phase I) is composed of living organisms limited within their porous external membranes. The second one (Phase II) is the ambient exterior environment with its vast population of molecules which are metabolic products of living organisms. The boundaries of the closed system are the outer limit of the atmosphere through which practically only radiant energy is exchanged with space.

In a closed system, $dE = d_eE$ and $d_i\,E = 0$, the energy supplied by the exterior during the time span dt is equal to the sum of the heat and the mechanical work performed at the boundaries of the system: $dE = dQ - pdV$, ($d_eE =$ exchange of energy inside and outside the system). Moreover, the system may be considered in the stationary state; then positive entropy production (dS) has to be compensated for by a negative flow of entropy and

$$\frac{dS}{dt} = \frac{d_eS}{dt} + \frac{d_iS}{dt} = 0,$$

with

$$\frac{d_iS}{dt} > 0 \text{ and } \frac{d_eS}{dt} < 0. \tag{3.1}$$

The total heat received by the system is zero (steady state):

$$dQ = d_e{}^IQ + d_e{}^{II}Q = 0,$$

and

$$\frac{d_eS}{dt} = \frac{d_eQ}{dt}\left(\frac{1}{T^I} - \frac{1}{T^{II}}\right) < 0. \tag{3.2}$$

The entropy flow (incoming and outgoing radiant energy) is negative, because energy received at high temperature T^I is returned to space at a lower temperature T^{II} (Fig. 3.1).

This last equation may be applied at the upper atmosphere surface. Radiant energy $d_e{}^IQ$ leaving the surface of the sun with temperature $5900°K$ carries an entropy $d_e{}^IQ/T^I$. If $d_e{}^IQ$ is absorbed by the system, an equivalent amount of energy is reradiated into space at temperature T^{II}. Then the system maintains a steady state (volume, pressure, temperature of atmosphere and biosphere in the earth gravity field). If we know $d_e{}^IQ$ and $d_e{}^{II}Q$ in each $\Delta\lambda$ band, we can calculate the entropy flow; the entropy production corresponds to the work that has been done within the entire system and includes biological production in the biosphere together with work in the atmosphere, maintenance of volume, pressure and temperature of gases, convection, kinetic and electrical work. The same may apply to any atmospheric level in the steady state, that is the mesosphere, troposphere, biosphere.

This is the significance of the wider spacing of spectral maximum for incoming and outgoing radiant fluxes from soil to the outer atmosphere:

$$T^{I} - T^{II}_{biosphere} < T^{I} - T^{II}_{upper\ atmosphere} \qquad (3.3)$$

The corresponding work performed at biosphere level is much smaller than the work done in the atmosphere, but it seems to be the most exceptional process of energy conversion in the universe. In the photochemical reaction of living pigments, some of the light energy is trapped in the free energy increase which is needed to form energy-rich chemical compounds in chloroplasts. These energy currencies lead to the reductive fixation of carbon dioxide into a carbohydrate in photosynthesis. This is the basic process of biological productivity and of all the derived kinds of work performed by living organisms (Galoux, 1974).

Radiation, instruments and use

Radiation is one of the most important site factors, since it is the driving force of all climatological features and all energy and material exchange processes in ecosystems. Standard equipment to study exchange processes in a complex community such as a forest takes time and money for implementation. The staff require very intensive and non-interrupted work periods. Difficulties are numerous and unexpected. The choice of site has to take into account a large number of prerequisites. Table 3.2 presents instrumentation, measurements and limitations derived from several IBP projects in forest sites.

Heat and vapor fluxes, applicability and limitations of methods and instruments

There are several methods for the determination of heat and vapor fluxes. Apart from the eddy correlation method, most of them are variations of the Sverdrup method or the aeroydnamic method, which are described below.

The aerodynamic exchange

The turbulent exchange of air properties may be described by the exchange coefficient A. Under adiabatic conditions the exchange coefficient A_a is , according to Prandtl:

$$A_a = \rho_a l^2 \frac{\delta u}{\delta z}. \qquad (3.4)$$

In this equation ρ_a is the density of the air, $\frac{\delta u}{\delta z}$ is the vertical gradient of the horizontal wind speed and l is the mixing length; l can be calculated by the formula

$$l = K(z + z_o), \qquad (3.5)$$

where K is von Karman's constant (0.4), z is the height and z_o is the roughness length. The value of z_o is computed from the logarithmic wind profile equation:

$$u = \frac{u^+}{K} \ln \frac{z - d + z_o}{z_o}, \qquad (3.6)$$

where u^+ is the friction velocity, calculated from the measured wind speeds, and $-d + z_o = D$ is according to Lettau the 'zero point displacement parameter'. Using the Richardson number R_i and the adiabatic corrections ϕl. for sensible heat and ϕv for latent heat, the exchange coefficient is also applicable for non-adiabatic stratifications. Also, in this case the fluxes of latent heat V and sensible heat L are determined by: ·

$$V = A_a \cdot \phi v(R_i) \cdot r(\delta q/\delta z), \text{ and} \qquad (3.7)$$
$$L = A_a \cdot \phi l(R_i) \cdot c_p(\delta T/\delta z + \gamma a), \qquad (3.8)$$

where the adiabatic temperature gradient γa is very small compared with $\delta T/\delta z$ and, therefore, negligible.

Using the method of aerodynamic exchange, it is possible to calculate the exchange of latent heat and sensible heat over a forest with the following conditions. It is necessary to determine a distinct wind profile, and the convective exchange may not be too large compared with the turbulent diffusion.

The Sverdrup method

The basic idea of this method is the energy balance equation

$$Q + B + L + V = 0, \qquad (3.9)$$

where Q is the net radiation, B is the heat flux into the soil, L is the sensible heat flux and V is the latent heat flux. If the energy balance equation is applied to a forest, the terms P and K have to be included in the formula, where P is the sensible heat flux into the plant material and K is the chemical fixed energy. In this case the formula of the energy balance

99

Table 3.2. *Radiation, instruments, and use*

Object	Radiation flux	Dimension	Instrument	Limitation	Point of measurement (height above soil)	Service	Time interval of measurement
1. Ecological and microclimatological research	Incoming global radiation (0.3–3μ on horizontal surface)	cal cm^{-2} day^{-1}	Robitsch Aktinograph (mechanical)	No obstacle on horizon	Without obstacle in the horizon >5° in a distance of 100 m; above the canopy on tower; in open area 1.20 m above ground level	Horizontal position, calibrate twice/year with standard thermopile. Change recording paper each 4 or 7 days, clean glass protection, change drying material	Continuous recording, daily values in the period (yr)
2. Ecological and microclimatological research; total energy balance	Global radiation (0.3–3μ) reflected global radiation (0.3–3μ) net radiation balance (0.3–100μ) on horizontal surface)	cal cm^{-2} h^{-1} cal cm^{-2} h^{-1}	Pyranometer, thermopiles, solarimeters Net radiometer	Electrical power	Above the canopy on tower 2 m above canopy and 3 m wide from the tower	Horizontal position, once or twice each year with standard thermopile, clean glass protection, change drying material, change polyethylene protection 2 or 3 times each year	Continuous or max. 12-s recording hourly and daily values during the period (yr)
3. Ecological and microclimatological research; total energy balance;	Global radiation (0.3–3μ) reflected global radiation (0.3–3μ) net radiation balance (0.3–100μ) on horizontal surface	cal cm^{-2} h^{-1} cal cm^{-2} h^{-1} cal cm^{-2} h^{-1}	Pyranometer, thermopiles, solarimeters Net radiometer	Electrical power	Without obstacle in the horizon >5 in a distance of 100 m;	Horizontal position, calibrate once or twice each year with standard thermopile, clean	Continuous or max. 12-s recording hourly and daily values during

Purpose	Quantity measured	Units	Instrument	Power	Location	Maintenance	Period
photosynthesis	Light (0.3–0.7μ)	lux h^{-1} or ratio lux—calories of global radiation	Luxmeter		above the canopy on tower; in open area 1.20m above ground level	glass protection, change drying material, change polyethylene protection 2 or 3 times each year.	the period (yr)
4. Ecological and microclimatological research; total energy balance; photosynthesis; energy balances within canopy and energy fluxes (different levels)	Global radiation (0.3–3μ) Reflected global radiation (0.3–3μ) Net radiation balance (0.3–100μ) but with separated measurements of incoming and outcoming long waves radiation	cal cm^{-2} h^{-1} cal cm^{-2} h^{-1} cal cm^{-2} h^{-1}	Pyranometer thermopiles solarimeters Net radiometer	Electrical power	Above the canopy on tower	Horizontal position, calibrate once or twice each year with standard thermopile, clean glass protection, change drying material, charge polythylene protection 2 or 3 times each year	Continuous or max. 12-s recording hourly and daily values during the period (yr)
	light (0.3–0.7μ) global radiation (0.3–3μ) reflected global radiation (0.3–3μ) net radiation balance (0.3–100μ on horizontal surface)	lux h^{-1} ratio lux – calories of global radiation	Luxmeter Pyranometer thermopiles solarimeters Net radiometer		Several heights within canopy in relation to vegetation layers 1 to 2 m above herb layer and soil surface (several or few mobile sensors)	Horizontal position, calibrate once or twice each year with standard thermopile, clean glass protection, change drying material change polyehtylene protection 2 or 3 times each year	Continuous recording or recording periods related to weather types and phenological phases

101

equation is extended in the following way:

$$Q + B + L + V + P + K = 0. \tag{3.9a}$$

As a reference level for the energy balance of a forest we assume usually a plane at the average upper height of the stand. All fluxes which are directed towards the reference plane are calculated positive, all others, which are averted, are defined as negative. It is possible to evaluate the single terms of the energy balance equation. Q can be measured directly; B can be computed from temperature profiles and from the soil density and the specific heat of the soil profile; P can be evaluated from the tree temperature and from the volume of the stand, which is usually expressed as the height of an homogenous wood layer and the specific heat of green wood. For an approximation it is possible to substitute, instead of tree temperature, air temperature of the air within the stand. K, the chemical fixed energy, is in comparison to the other terms of the energy balance equation very small and is therefore in most cases neglected. But it is possible to determine K over the increment of the forest stand as the energy which is necessary for the net production. L and V can be calculated through the use of the turbulent exchange coefficient in the following way:

$$V = A_a \cdot \phi v(R_i) \cdot r(\delta q / \delta z), \tag{3.10}$$

$$L = A_a \cdot \phi l(R_i) \cdot c_p(\delta T / \delta z + \gamma a). \tag{3.11}$$

Under the assumption that the exchange coefficient for latent heat is equal to the exchange coefficient for sensible heat, one can omit the turbulent exchange coefficient by use of the so-called Bowen ratio, L/V. From the two previous equations, the following formula results:

$$L/V = c_p(\delta T / \delta z) / r(\delta q / \delta z). \tag{3.12}$$

This equation was used by Sverdrup to solve the energy balance equation either for L or for V:

$$L = (Q + B + P)/[1 + (r/c_p)(\delta q/\delta z)/(\delta T/\delta z)], \text{ and} \tag{3.13}$$

$$V = (Q + B + P)/[1 + (c_p/r)(\delta T/\delta z)/(\delta q/\delta z)], \tag{3.14}$$

where r is the heat of vaporization of water, c_p the specific heat of the air under constant pressure, $\delta q/\delta z$ the specific humidity gradient, and $\delta T/\delta z$ is the temperature gradient.

Due to the dependence on gradient measurement, the Sverdrup method is not applicable for strongly turbulent exchange, when the gradients approach zero. Because of this, the method is applicable for high energy exchange but not for nearly indifferent stratification with an adiabatic exchange coefficient. According to Berz, the limitation for the application of the Sverdrup method over grassland, including the accuracy of the instruments, is $Q + B \geqq 50$ mcal cm^{-2} min^{-1}. Under these limitations only half of the time, but more than 90% of the energy exchanges, are represented.

The eddy correlation method

Any properties of the air, for example sensible heat and latent heat, are moved vertically by the eddies. In this case, the vertical transport of a parameter at a fixed point is the product of the fluctuation of this parameter times the fluctuations of the vertical wind speed, averaged over time. Accordingly,

$$V = \rho_a \overline{q'\, w'}, \text{ and} \qquad (3.15)$$

$$L = \rho_a c_p \overline{T'\, w'}, \qquad (3.16)$$

where ρ_a is the density of air, c_p is the specific heat of the air, w' is the instantaneous deviation of the vertical wind speed from its mean value and q' and T' are the instantaneous deviations of latent heat and sensible heat of the mean values. However, by this method the flux of any parameter is measured only at a certain point, and the values are not representative for the whole of the stand, because the surface of a forest is too heterogeneous. Therefore, it is advisable in using this method to measure in an array considering the horizontal heterogeneity of the surface.

Instrumentation

Micrometeorological instruments are described in detail by Monteith (1972) in the IBP handbook No. 22 In using the methods described above one should pay attention to the following factors. For profile methods (Sverdrup method and the method of aerodynamic exchange), the instruments should be calibrated not only to absolute values but to each other. In this way the systematic error can be minimized. In order to achieve a high degree of accuracy one should use values averaged from recordings of very small time intervals. An example for meteorological instrumentation of measuring and calculating the heat and vapor exchanges by a combined profile method (Sverdrup method and the method of aerodynamic exchange) is shown in the section on carbon dioxide (p. 106). This instrumentation has been used at the IBP Eberger forest site in a 30-m high spruce forest. The Evapotron (Dyer & Maher, 1965) and the Fluxatron (Dyer & Alii, 1957) were developed especially for the eddy correlation method, to measure the latent heat flux. When using other instruments one should make certain that all the eddies are registered. Generally sensors with a time resolution less than a second are sufficient. A factor to be remembered in using this method is the large volume of data which accumulates due to the short time resolution.

Carbon dioxide: methods, instrumentation, and limitations

In order to estimate and calculate carbon dioxide fluxes and flux balances of forest stands, one can apply two distinctly different monitoring techniques, with specific advantages and drawbacks. There exists no ideal solution to the question of which method works best and what techniques should be employed under general circumstances. The researcher has to decide for each specific case the best applicable method to meet accuracy and resolution needs for his study. The different methods and appropriate instrumentation recommendations, with the assumptions and limitations of each method, are described in this section.

Profile methods

Common to the profile methods is the measurement of the different meteorological parameters at different heights above and within a forest stand (i.e. radiation, net radiation, temperature, humidity, windspeed, carbon dioxide concentrations, or other atmospheric qualities) and the computation from these data through different approaches of the fluxes of concern. The shortcoming of the profile methods in general is the additive behavior of the random error of measurement when gradients are determined.

The aerodynamic method

The turbulent exchange coefficient can be derived from the so-called logarithmic windprofile under neutral thermal conditions. At a given height (z_r) it is only dependent upon the ambient windspeed and upon a surface-and-windspeed-dependent proportionality factor, (f):

$$f = \frac{u^*}{u} = \frac{k}{\ln \dfrac{z_r}{z_0}}, \qquad (3.17)$$

where z_r is the reduced height, k is von Karman's constant (0.4), z_0 the roughness parameter, and u^* the mixing velocity. The turbulent exchange coefficient for a given height above the stand (z_r) is then:

$$A_a = k \, z_r \, f \, \rho_a \, u. \qquad (3.18)$$

Carbon dioxide fluxes are then found as the product of the CO_2 gradient over the respective height interval times the turbulent exchange coefficient. Thermal stability or instability corrections can be applied to the turbulent exchange coefficient through the use of Richardson numbers.

One of the basic assumptions for this computation method for carbon dioxide-flux balances is that the turbulent exchange of momentum is

104

assumed to be equal to the turbulent exchange of carbon dioxide. Short-comings of the turbulent exchange method are that it is only applicable for exchange and carbon dioxide-flux computation under neutral thermal (adiabatic) conditions, and that the turbulence theory has only validity for a certain height interval above the stand. Computation of carbon dioxide fluxes within the stand and underneath the canopy by the turbulent exchange method is not possible, but there is the possibility of (empirically) calibrated extrapolation into the stand space.

Erroneous results can be obtained from turbulent exchange methods, despite the Richardson number correction, when there are low windspeeds and high radiative energy inputs.

Sverdrup method

In this case the exchange coefficient for given heights is calculated by the use of the Bowen ratio (L/V) and by the reversal of the energy balance equation and solution of this equation for the turbulent diffusion coefficient (eqn 3.19, and 3.20):

$$S_z - B = K_z \left(c_p \rho_a \frac{\delta\theta}{\delta z} + r_w \frac{\delta a}{\delta z} \right), \tag{3.19}$$

$$K_z = \frac{S_z - B}{c_p \rho_a \dfrac{\delta\theta}{\delta z} + r_w \dfrac{\delta a}{\delta z}}, \tag{3.20}$$

where S_z is the net radiation for a height z, B is the soil heat flux, K_z the turbulent diffusion coefficient for height z, C_p is the specific heat of air, r_w the heat of evaporation, ρ_a the mean density of air, $\delta\theta/\delta z$ the temperature gradient, and $\delta a/\delta z$ the absolute humidity gradient. The carbon dioxide flux is determined in an analogous way to the aerodynamic method after evaluation of the turbulent exchange coefficient.

The basic assumption for the Sverdrup method is that:

$$K_{CO_2} = K_V = K_L. \tag{3.21}$$

The turbulent diffusion for carbon dioxide is equal to the turbulent diffusion of sensible and latent heat. Advantages of the Sverdrup method are that it works throughout the canopy space where the turbulence theory is not valid. A severe shortcoming of this method is that it does not work or gives highly erroneous values in the case when the net radiation approaches zero or when the gradients of sensible and latent heat become very small (i.e. when there is high turbulent mixing, which tends to decrease gradients). In many cases best results are obtained when the aerodynamic and the Sverdrup method are combined.

105

Instrumentation for the profile methods

Different micrometeorological elements have to be monitored in and above a homogeneous forest stand to apply one or both of the two different profile methods. For the aerodynamic method only profiles of the horizontal windspeed above the zero plane have to be measured with light-weight cup anemometers. If a stability correction with Ri-numbers should be employed, a temperature profile (which could be taken with Pt-100 or thermistor sensors) should be measured. If the Sverdrup method is used, more elaborate instrumentation is necessary. Profiles of net radiation air temperature and humidity have to be measured, which should be done with either psychrometers, or thermistors and dew cells. To determine the soil heat flux (*B*) one can either employ heat flux plates, or ground thermometers and moisture probes.

An example of a workable instrumentation set-up can be seen in Fig. 3.3. This set-up was successfully used for the IBP-site 'Ebersberger-Forst' for both profile methods. A similar set-up, which was also designed by Prof A. Baumgartner, was employed on the other German IBP-site in the Solling.

For both methods an adequate method for sampling carbon dioxide gradients would be to take carbon dioxide probes from corresponding

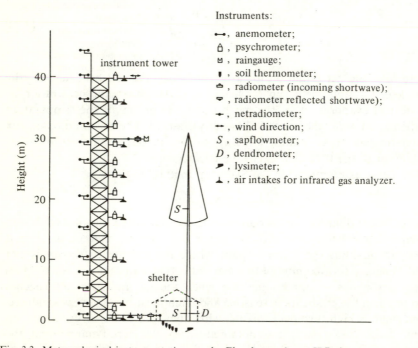

Fig. 3.3. Meteorological instrumentation on the Ebersberger forest IBP site.

heights of measurement, and employ either wet chemical or infrared carbon dioxide analysis. Since the wet chemical analysis (alkali absorbent for carbon dioxide-containing air probes and titration) is very tedious, it is only used for long term carbon dioxide averages. Continuous monitoring of carbon dioxide concentrations at different heights is only possible by infrared gas analysis and an intake, pump and valve switch system, which draws air continuously from different heights and switches the probe stream onto the gas analyser. Special care has to be taken to eliminate the water vapor error, either by employing bandfilters or a gas cooler to freeze out water vapor, or by use of a desiccation tower.

In Fig. 3.4 the carbon dioxide concentration and gradient monitoring system of the IBP 'Ebersberger-Forst' is shown in more detail. This carbon

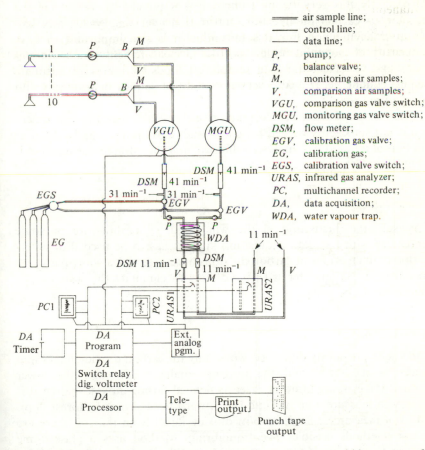

———	air sample line;
———	control line;
———	data line;
P,	pump;
B,	balance valve;
M,	monitoring air samples;
V,	comparison air samples;
VGU,	comparison gas valve switch;
MGU,	monitoring gas valve switch;
DSM,	flow meter;
EGV,	calibration gas valve;
EG,	calibration gas;
EGS,	calibration valve switch;
URAS,	infrared gas analyzer;
PC,	multichannel recorder;
DA,	data acquisition;
WDA,	water vapour trap.

Fig. 3.4. Scheme of the carbon dioxide air sampling and data acquisition system of the Ebersberger forest IBP site, Munich, Germany.

dioxide monitoring system was employed together with the sensor configuration as shown in Fig. 3.1 and yielded good results for long time periods.

The eddy correlation method

The vertical flux (or eddy transport), F, of any atmospheric parameter is equal to the fluctuation of the parameter times the fluctuation of the vertical windspeed:

$$F_{CO_2} = \overline{C'_{CO_2} W'},\tag{3.22}$$

where F_{CO_2} is the instantaneous flux of carbon dioxide C'_{CO_2} the fluctuation of carbon dioxide concentration, and $W' = w - \bar{w} =$ fluctuation of the vertical windspeed. By this very strong technique it is possible to determine the instantaneous carbon dioxide flux. Integration over time yields flux balances at a given point. For this technique, it is also important to have homogeneity of the stand (unless horizontal measurement arrays are employed), and to have quite long sampling periods to guarantee adequate statistical confidence. It is also very important to employ instruments with adequate time resolution.

Shortcomings of the eddy correlation method are that the instrumentation is not sufficiently rugged for use in and above forest stands, the time resolution of infrared gas analysis is limited, and the infrared gas analysis is susceptible to aerosols.

Instrumentation

To measure the fluctuation of the windspeed, either hot-wire or sonic anemometry with fast data acquisition systems has to be used. For determination of fluctuations of carbon dioxide concentrations, a short path, free beam infrared gas analyzer with bandfilters to eliminate the water-vapor influence would be adequate.

Soil respiration

The methods previously described work only for carbon dioxide fluxes and balances of the stand and the upper boundary. But since the lower boundary of a growing forest is a clearly defined carbon dioxide source, it is necessary to measure and evaluate the upward fluxes from the forest floor. For *in situ* measurements of carbon dioxide fluxes from the soil there are only two methods feasible. The Equilibrium Method uses a glass dome placed on the soil, with a carbon dioxide equilibrium established between soil air and air in the dome. Samples are drawn periodically and analyzed.

108

In the Flow–Uptake method, carbon dioxide-free or air with known carbon dioxide content is drawn into a glass dome placed with the open side on the soil. Air samples are drawn from the outlet port into a gas analyser. The difference between inlet and outlet carbon dioxide concentration represents the carbon dioxide flux per unit soil area.

Considerations for data acquisition

Due to the fact that manpower for data collection has become more expensive and more difficult to obtain, a fully automatic data acquisition system with an output of computer processable records seems the most feasible solution to the data collection problem. This solution is in the long run cheaper and more accurate than the processing of analog records from the different sensing instruments. The fully automatic data acquisition system can be designed either for automatic data logging or for data processing with on-line plausibility testing routines. The latter is the best method, because it allows the researcher to have a tight control of the monitoring process and to detect instrumental errors in the shortest time period. When selecting a computer system, the main emphasis should be placed on the availability of good and fast service for hardware and software, the versatility of the computer system under consideration, and compatibility with other computer systems. A data record on magnetic tape or disc should have preference to a record on punch tape or other paper records.

Water exchange processes, limitations in use of instruments and methods

Atmospheric water

Rain in the open

Measuring the rainfall in the open at forest sites requires either a clearing or a tower extending to the upper region of the canopy. The clearing should be close to the experimental sites and of sufficient size, i.e. the diameter of the clearing should be at least five times as large as the height of the surrounding trees. It is advisable to place the rain gauge into a soil pit in order to have the opening of the gauge at about the level of the surrounding soil surface. Measurement of rainfall in the open in the upper region of the canopy has the advantage of better comparability with the measurement beneath the canopy (*see* the next section) and thus provides more reliable data for determining the interception as the difference between the rainfall above and below the canopy. The gauge should be positioned as deep as possible, but in such a way that no branch or top of a tree enters a funnel

109

shaped region extending from the opening of the gauge with an angle of about 30⁰ upwards (relative to a horizontal plane). Recording gauges are available as well as rain-collecting devices, which need to be checked from time to time, depending on the rainfall amount. In order to meet freezing temperatures, a heating system may be installed. Energy supply can be provided either by electricity or propane gas; the former is more manageable but requires a power line. Snow will be melted but not at the same time-rate at which it falls. Heavy snowfall or very low temperatures may exceed the heating capacity.

Snow

Snow is principally recorded by pail-shaped collectors, which are simply put on the ground under the canopy as well as in the nearby clearing. Furthermore, weekly snow samples can be taken randomly for weighing by means of a cylindrical pipe pushed vertically through the snow to the ground. Finally, snowfall can be recorded by the heatable recording rain gauges (see above).

Canopy drip and throughfall

Because of the irregular distribution, canopy drip and throughfall are optimally recorded by 16-m long troughs with 2.5 m² open catch. Three such troughs are used in parallel, one or two of them being outfitted with recording equipment. Additionally, 25–30 small rain gauges should be used, distributed systematically 5 m apart. Both methods compare fairly well, although even the relatively big troughs show an unexpected standard deviation, which may run as high as 20% of the mean value.

Soil water

Water storage, movement and seepage in the soil are generally calculated from tensiometer readings. The tensiometers are placed around typical trees in order to take into account the distance from the trunks as a parameter of the distribution of soil moisture. Measurements should be taken at five to six different levels to a depth of 1.30 m. Because of the heterogenity of the distribution of the soil moisture, up to 50 parallel tensiometers should be used at each depth, thus requiring 200 to 300 tensiometers per site. Readings must be taken two or three times a week, or a recording tensiometer using pressure transducers may be constructed.

Tensiometers have a limited application. At suctions higher than about 800 to 1000 mbar, tensiometers do not work and other methods must be employed. Possibilities are thermocouple psychrometers, aquapots, gypsum blocks, etc. The use of tensiometers is limited in soils with high stone (rock) content as well as in drier soils. Water content measurements should be made simultaneously to tensiometer readings mainly to determine if hysteresis phenomena are involved. One may use a neutron moisture probe for this purpose. A high precision method is the use of a double tube gamma-ray meter, if used stationary. Continuous recording on punch tapes is advisable. In any case one will find it necessary to take soil samples for determining the water content by weighing and drying. This has to be done for calibration and for obtaining a reference water content (to which the change of water content as measured with the gamma-ray meter can be calculated). For calculation of water content, water movement either up- or downward at different depths, the water characteristics (water suction against water content) and the permeability characteristic (permeability against water content) must be known. The same is true if one intends to simulate the changes of moisture content over a certain time period.

Modeling the physical environment

Introduction to environmental models

A model is an abstraction of the real world in the sense that it does not try to reproduce an exact replica of the thing being studied but only those features of reality that are of interest. A mathematical model does this by having a set of mathematical expressions which have the same quantitative relationship as certain factors of the system being modeled. Models dealing with the physical environment of the forest attempt to describe the physical processes in the atmosphere, soil, and forest biomass as they are related to the physiological processes of the forest stand.

The model developed here (Fig. 3.5) demonstrates the way in which these processes are related. The boxes represent parts of the ecosystem, while each arrow designated a transfer of energy or mass. Certain simplifying assumptions are incorporated in this model. All fluxes of energy and mass are assumed to move vertically. This condition is approximated in forest stands which are large compared with the length of their edges and are reasonably homogeneous in composition. It is also necessary that the soil be permeable enough for almost all water to infiltrate, and loss from drainage be small or of known quantity.

A model of the energy, water, and carbon dioxide exchange in the forest

The atmosphere and the plant. The exchange of energy and matter between the plant and the atmosphere is mediated by the process of diffusion. In the atmosphere away from the plant surfaces, the diffusion is turbulent in

111

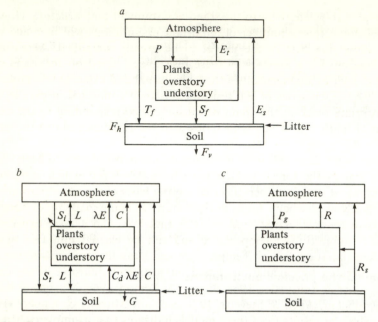

Fig. 3.5. *a*, diagrammatic representation of the water balance, *b*, energy balance and, *c* carbon dioxide exchange in a forest.

nature, i.e. energy and matter are transfered by eddies in the atmosphere. Very near the surfaces the diffusion must take place through a laminar boundary layer where molecular motion is the means of transfer. Diffusion of water vapor and carbon dioxide into the leaf through the stoma is also through molecular diffusion.

One way of expressing the relationship of the fluxes along the diffusion path from the free atmosphere to the leaf or into the leaf is to use a resistance analogy. In this analogy the path is conceived of as a group of series and parallel resistances corresponding to various parts of the diffusion pathway. The basic set of resistances in the pathways consist of a resistance for diffusion above the canopy from the bulk air mass (r_b), a resistance for transfer in the stand space (r_{st}), a resistance for diffusion across the laminar boundary layer (r_{bl}), a resistance for diffusion through the stoma of the leaves (r_s) for water vapor and carbon dioxide, and two internal resistances through the cell wall (r_{cw}) and the protoplasm (r_p) for carbon dioxide. The first two of the resistances are for turbulent diffusion; the rest are for molecular diffusion in air, except the cell wall and protoplasmic resistances in which the processes are molecular diffusion through the cell materials.

In order to determine realistic values of these resistances, it is necessary

112

to be able to relate them to the stand environment and structure. However, before doing this, a number of important assumptions will be made which will allow this part of the model to be greatly simplified. A reasonable arrangement for a network of these resistances, based on the stand structure, is a series arrangement of the leaf internal and external resistances (r_{bl}, r_s, r_{cw}, r_p). Each series of leaf resistances is then connected in parallel to its appropriate place in the chain of resistances corresponding to the canopy diffusion path (r_{st}), which is in series with the resistance above the canopy (r_b).

Since is not desirable from a computational point of view to keep track of every leaf in the forest, models have been developed which assigned the leaves into a fewer number of classes on the basis of the leaf environment. A commonly used scheme looks at the canopy as being made up of horizontal layers of finite thickness, and assumes that leaves in a layer experience a similar humidity and temperature and can be classed into two or more incident radiation classes. Temperature and humidity gradients in the stand are generated by calculations based on turbulent diffusion in the canopy (Waggoner & Reifsnyder, 1968; Stewart & Lemon, 1969; Murphy & Knoerr, 1970; Sinclair, Allen & Stewart, 1971, Sinclair, Murphy & Knoerr (1975).

Analysis of results from these models indicates that a simpler scheme can be used to simulate exchanges from the entire canopy (Monteith, 1965; Stewart & Thom, 1973; Sinclair, Murphy & Knoerr, 1975). In these models the entire canopy is treated as though it was one large, average leaf for energy and water transfer. Because of the non-linear response of photosynthesis to light, it is necessary to stratify the leaves into sunlit and shade classes for carbon dioxide-flux modeling.

The following equations can be used to calculate the resistances for carbon dioxide transport. To obtain the appropriate resistances for heat and water-vapor transport, the molecular diffusivities must be taken into account by multiplying by the appropriate ratio of the molecular diffusivities. The turbulent resistances are the same for all of the fluxes. The cell-wall resistance does not exist for heat and water-vapor transport.

$$r_{cw} = \frac{Wt}{k_c A}, \tag{3.23}$$

$$r_{sCO_2} = r_{sm} + \frac{a}{b+L} + f(T) + g(\psi_l), \tag{3.24}$$

$$r_{blCO_2} = c \, D^n \, u^{-m}, \tag{3.25}$$

$$r_{st} = Bh/K = Bh/\{K_h \exp[c(z-h)]\}, \text{ and} \tag{3.26}$$

$$r_b = \frac{1}{u_* k} \left[\ln \frac{z_r - d}{h - d} - (\psi_{zr} - \psi_h) \right], \tag{3.27}$$

113

where k_c is the diffusivity of carbon dioxide in water, A the ratio of internal cell surface area to leaf surface area, W the thickness of cell walls, t the tortuousity factor, r_{sm} the minimum stomatal resistance, L the light intensity incident on a leaf, $f(T)$ a function of temperature, sometimes linear, $g(\psi_l)$ a function of water potential of the leaf, often hyperbolic, D the aerodynamic dimension of leaf, u the wind speed, h the height of canopy above ground, K the average turbulent diffusivity for carbon dioxide over distance Bh, u_* the friction velocity ($uk/[\ln z - d/z_0 - \psi_m]$), k, von Karman's constant ($\cong 0.4$), z_r the reference height for atmospheric input parameters, d the aerodynamic displacement height, ψ_x the aerodynamic profile correction for bouyancy at height x, c the empirical extinction coefficient for wind speed in the canopy, B the fraction of height corresponding to effective diffusion path in the canopy, often approximated as $d + z_0$, and z_0 the aerodynamic roughness length.

The total resistances to transfer are then

$$r_h = r_b + r_{st} + r_{bl} A, \qquad (3.28)$$

and

$$r_w = r_b + r_{st} + \cfrac{1}{A_s/(r_{bl} + r_b)_s + A_{sh}/(r_{bl} + r_s)_{sh}}, \qquad (3.29)$$

where one resistance is used as an average for the whole stand for energy and water transfer. Because the internal sink for carbon dioxide is very dependent on light intensity, the sun leaves and shade leaves must be dealt with separately for the carbon dioxide transfer. In this case all the resistance from the air layer above the stand through the cytoplasmic resistance is linked in parallel for shade leaves and then again for sunlit leaves, giving a value of net photosynthesis for each group of leaves. This value is then multiplied by the leaf area in each class to give the net photosynthesis for the stand.

The preceding discussion provides for the diffusive aspects of heat, water-vapor, and carbon-dioxide flux between a forest and the atmosphere. What remains to be developed is a means of specifying the source strength for these fluxes at their point of origin. The source strength can be expressed as the value of the state variable at the origin. For the heat, water-vapor, and carbon dioxide fluxes the state variables are temperature (T), water-vapor concentration (specific humidity, q, and carbon dioxide concentration (CO_2).

The equilibrium leaf temperature is affected by all the fluxes of energy which are exchanged between the atmosphere and the leaf. These are the sensible heat exchange (C), latent heat exchange (λE), absorbed solar (αS) and terrestrial radiation (L), and reradiation (R). The defining equations are:

114

$$C = \frac{\rho c_p}{r_h}(T_l - T_a), \tag{3.30}$$

$$\lambda E = \frac{p}{r_w}(q_l - q_a), \tag{3.31}$$

$$R_n = \alpha S + L - R, \tag{3.32}$$

where the subscript l indicates the values at the leaf surface and the subscript a is for the values in the atmosphere near the leaf. R_n is the residual of the radiation balance, the net radiation.

Using the above relationships with the assumption that the leaf specific humidity is the saturation value at leaf temperature and can be adequately expressed as

$$q_l = q_s(T_a) + \frac{\delta q_s}{\delta T}(T_l - T_a), \tag{3.33}$$

and defining

$$\Delta = \frac{\delta q_s}{\delta T}, \tag{3.34}$$

then the evaporation can be solved for as

$$E = \frac{\dfrac{\Delta(R_n - G)}{c_p} + \dfrac{\rho(q_s(T_a) - q_a)}{r_h}}{\dfrac{\lambda\Delta}{c_p} + r_w/r_h}. \tag{3.35}$$

The value of the heat flux to the soil (G) can often be neglected. This is the form of the combination or Penman equation (Penman, 1948) first worked out by Monteith (1965).

The equation (eqn 3.34) was first proposed by Penman for monthly water budgets. However, it should do best when calculated for shorter periods over which the parameters are more nearly constant. We have made tests which seem to indicate the half-hour values summed over days with sinusoidal radiation inputs give approximately the same value of evaporation as one calculation using the average conditions for the day. Thus, the time period in these cases is a matter of resolution, not accuracy. The sensible heat flux can then be found as

$$C = R_n - E - G. \tag{3.36}$$

The use of this equation allows us to calculate the energy exchange and the water-vapor exchange as proportional to the latent-heat term in the energy balance.

To find the source strengths for carbon dioxide in the leaves, a mathematical relationship must be found for gross photosynthesis (P_g) and respiration rates as they vary with light (I), temperature (T), and carbon dioxide concentration ($[CO_2]$) in the leaf. A number of models exist for doing this which are able to reproduce known curves; we have chosen to use a model developed by Sinclair (1972). The relationships are:

$$R = R_x \exp\left[9000 \ln (Q_{10}) \left(\frac{1}{T_x} - \frac{1}{T} \right) \right], \text{ and} \tag{3.37}$$

$$P_g = \frac{k_1 k_2 k_3 A_o N_o [CO_2] I}{(k_1 [CO_2] + 1)(k_2 I + 1)}. \tag{3.38}$$

The parameters k_1, k_2, k_3, r_p, Q_{10}, A_o, N_o, and R_x must be determined from experimental data. This requires a considerable experimental effort since there are likely to be seasonal changes in these parameters (Higginbotham, 1974).

It is important to notice that in the above formulations (eqns 3.37, 3.38) the evaporation is not limited by the effect of the plant water stress on the leaf specific humidity. Although the status of the plant water stress is a function of the water potential, the water potential in the leaf almost never approaches a value that would lower the specific humidity by 5% and normally does not lower it by more than 2%. The control of evaporation by the plant, when water stress increases, is through the closing of the leaf stoma which increases the stomatal resistance. This means that the relationship between stomatal resistance and leaf water stress is critical in the dynamic calculation of the stand water balance. Leaf water potential is one variable that is closely related to the water stress of the leaf and it has been shown to be a good predictor of the effect of water stress as related to stomatal resistance (*see* summaries in Slatyer, 1967; Kramer, 1969).

To model energy, water and carbon dioxide exchange for a forest it is necessary to be able to model the water status of the foliage. Since this is related to the water content (water potential) of the soil, a complete model must predict the dynamics of the soil–water–plant system. This can be done by a consideration of the soil–water–plant relationships and the calculation of a running soil water balance.

Plant–water–soil relationships. The liquid movement of water through the plant is modeled by using the electrical analog scheme first suggested by van den Honert (1948). In this model the flux of water through the plant is defined as being proportional to the difference in water potential between the soil solution next to the roots and the water potential at the leaf mesophyll cells where evaporation is taking place, and inversely proportional to a resistance to liquid flow. The plant resistance can be thought

116

of as the sum of a number of resistances in series representing the root, stem, branches and leaves.

The values of these resistances were calculated from evaluation of the structure of the roots, stem, branches, and leaves of loblolly pine (*Pinus taeda* L.). The relationships used were taken from Harlow (1931), Cowan & Milthorpe (1965), Briggs (1967), and Koch (1972). The values found for this species are listed in Table 3.3.

The root resistance probably varies with the temperature and soil moisture. However, Kramer & Kozlowski (1960) showed evidence that the root resistance of pines does not vary greatly with the temperature and Andrews & Neuman (1969) indicated that the pararhizal interaction between the roots and the soil should not have a large effect on stands having as great a root mass as shown in these plantations (Kinerson, 1975). Therefore, an average root resistance value was considered adequate for the calculations. The total plant resistance r_{tp} as given by summing the above is 8.5 day/bar-cm.

The response of the stomata to soil water potential and evaporation rate can be modeled by using the following two equations iteratively with the evaporation equation given previously (eqn 3.13) to obtain leaf water potential, ψ:

$$\psi = \psi_s - \frac{E}{r_{tp}}, \tag{3.39}$$

$$r_s = \frac{r_{sm} + \dfrac{a}{b+L}}{1 + c\psi\left(r_m + \dfrac{a}{b+L}\right)}, \tag{3.40}$$

where a, b, c are empirical constants and r_{sm} is the minimum stomatal resistance for the specific form of eqn 3.24 used in the simulations.

The soil-water potential (ψ_s) is a complicated function of the soil-water content, i.e. the water characteristic curve of the soil. In order to model the soil-water dynamics, it is necessary to calculate a water content in the rooting zone from a running soil-water balance, and use the water characteristic curve to get the soil-water potential. If the soil is stratified into layers of different water characteristic curves, it may be necessary to simulate the water balance in more than one layer. In the loblolly pine

Table 3.3. *Resistance to liquid water movement in loblolly pine (day bar-cm^{-1})*

Roots	Stem	Branches	Needles
1.0	1.7	1.4	4.4

plantation where we have been working, the soil has two distinct layers. We have chosen to keep a water balance for each layer. During some periods of the year a perched water table is present above the interface between the two layers. The model also keeps track of this water depth.

The water balance can be expressed by:

$$P - O - E_t - F_v - F_h - \Delta W = 0, \tag{3.41}$$

where the change in soil water content (ΔW) is a result of the difference in inputs and outputs to the soil layer. The precipitation (P) is an input to the model and O is the surface runoff. The evapotranspiration (E_t) includes the transpiration which has been dealt with above and the evaporation of precipitation intercepted by the canopy and the litter. The vertical flux of water (F_v) is the exchange by drainage or upward diffusion from the other soil layers. There is also some horizontal drainage (F_h) along the interface between layers. The horizontal drainage is at a maximum when a perched water table is present.

The drainage terms are calculated by using exact solutions to the Darcy flow equation which approximate the conditions found in the field. Black, Gardner & Thurtell (1969) have demonstrated that drainage from wet soil columns can be approximated by:

$$F_v = C_1 \exp[C_2(W - C_3)], \tag{3.42}$$

where the constants C_1, C_2, and C_3 must be determined empirically. This equation is shown to be a solution to the Darcy equation for conditions when gravity forces dominate the flow.

If gravity flow is determined by eqn 3.42, the diffusional term of the Darcy equation can be dealt with separately. A solution to the diffusional term for two semi-infinite layers at constant initial moisture contents is

$$\psi_1 = \frac{\psi_{01} - \psi_{02}}{1 + \sigma}\left[1 + \sigma \exp\left(\frac{(|z|)}{2\sqrt{K_1 t}}\right)\right] z < z_0,$$

$$\psi_2 = \frac{\psi_{01} - \psi_{02}}{1 + \sigma}\left[1 - \exp\left(\frac{z}{2\sqrt{K_2 t}}\right)\right] z > z_0, \tag{3.43}$$

$$\sigma = \frac{K_2\sqrt{k_1}}{K_1\sqrt{k_2}}$$

where z is depth, ψ is water potential, K is diffusivity, and k is conductivity. The soil-water flux across the interface can be calculated by using the characteristic curve and finding the change in water content over the period of time desired. This equation is a crude approximation of what goes on under natural conditions but seems to work well for the conditions and soils under which the simulations were done.

118

The interception term (I_w) is calculated from an empirical relationship determined for a stand of loblolly pines of approximately the same density as the study site stand:

$$I_w = 0.305(1 - \exp(-0.815\ P)).\tag{3.44}$$

Intercepted rainfall is assumed to evaporate completely before transpiration begins.

It is also assumed that infiltration capacity at the surface is high and all precipitation after interception enters the soil until the upper layer is saturated. Rainfall reaching the soil surface after the soil has become saturated is assigned to surface runoff.

With the above calculations an estimate of the soil moisture and soil potential can be made. This value is used to determine the equilibrium leaf-water potential. The stomatal resistance term of the evaporation and photosynthesis equations is calculated with this value of water potential. This allows a simulation of the seasonal time series of energy, water and carbon dioxide exchange in the stand.

Simulation results

A simulation of the loblolly pine stand at Saxapahaw, North Carolina (Triangle Site, EDFB, US–IBP) was run for the climatic conditions of 1972. Fig. 3.6 shows the daily input values of precipitation, solar radiation, and temperature. Fig. 3.7 demonstrates the simulated values of evaporation, soil moisture and net photosynthesis. It is obvious that for this year the soil moisture conditions were never limiting. Under these conditions the evaporation and the carbon dioxide exchange are most greatly influenced by the solar radiation input during a given day.

Direct measurements of the simulated values exist for only a few days

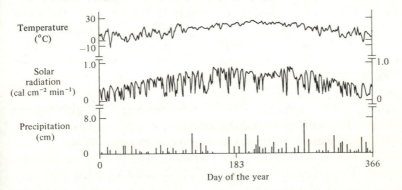

Fig. 3.6. Time series of selected input parameters used in the simulation of the stand physical environment.

119

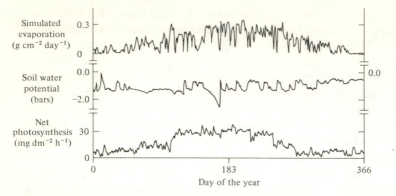

Fig. 3.7. Time series of simulated evaporation, soil water potential and daily average net photosynthesis.

during the year. They agree quite well with the simulated values but this is certainly not a definitive validation of the model. One addition check we were able to make on the water balance was to compare the seasonal evaporation to the average for the eastern Piedmont of the United States. The values agree very well and allow some additional confidence in the simulation.

Radiation exchange and transfer in forest ecosystems

Short wave radiation flux (global radiation)

Variation of incoming global radiation flux density

It has been observed that the optical properties of receiving surfaces may have an influence on the intensity of incident global radiation. Global radiation at the level of the biospheric systems also depends on the reflection coefficient of the surfaces as a portion of the reflected solar energy is rescattered back to the earth (Deirmerdjian & Sekera, 1954; Dirmhirn, 1964). This also appears in the global radiation values on sites close to a sea border. Lütske (1966) measured (from April to September) 0.054 less global radiation than above an adjacent meadow. Zuffa (1975) made the same observation at the Bab IBP Site (Slovakia): above forest 100%; March 104.5% on free area value; 99.5% in April; 99.4% in May; 101.4% in June; 101.5% in July; 104.1% in August, 107.9% in September; 108.4% in October. This illustrates the necessity of having adequate equipment just above the experimental ecosystem, if one is to make precise energy balances.

Reflection on upper surfaces

Numerous values for albedo have been established above vegetation surfaces (Krinoy, 1953; Penndorf, 1956; Shachori, Birbebak & Birbebak,

120

1964; Stanhill & Michaeli, 1965; Van Miegroet, 1965; Leonard, 1968; Jarvis, 1976; Susmel, Viola & Bassato, 1976). Instantaneous measurements of albedo range from 0.04 to 0.15 for coniferous forests with fine needles and from 0.12 to 0.20 for deciduous forests. Krinov (1953) gives the following albedo ranking for different forests: fir < pine < birch < aspen. Most authors have recognized that albedo in the visible range is smaller than for global radiation. Continuous measurements of albedo have been made only for intensive studies of radiant energy in forest ecosystems. According to variations of albedo with many factors over the forest, study of radiant energy over a vegetative period requires either frequent sampling or continuous measurement. Variation of the albedo coefficient with sun height is greatest up to about unity. Variations with such situations as clear sky, overcast day and mist have been observed by continuous measurements. Grulois (1968) reported that oak forests during the leafless period have an albedo of 0.112 for clear sky and 0.166 for overcast days; during the foliated period, this changes from 0.155 for clear sky to 0.195 for cloudy days (Table 3.4). A similar trend has been observed in Scotland for Sitka spruce: albedo was higher during misty days, but the reverse situation was not observed (Dirmhirn, 1964, for beech forests; Kiese, 1972, for the IBP Site at the Solling; Landsberg, Jarvis & Slater, 1973).

Variations in albedo with growth phases of the forest are important in the deciduous biome. Albedo is small (0.105–0.116) during the leafless period and increases very quickly up to 0.20 by foliation in spring. For example, Roussel (1953) has reported an increase of 0.04 in comparison with leafy period average albedo. In the fall during the yellowing of foliage, albedo rises to 0.22 and remains relatively high when the brown foliage falls to the ground (Grulois, 1968) (Fig. 3.8). Muhenberg (in Budyko, 1974) gives

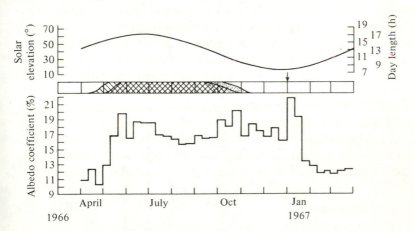

Fig. 3.8. Maximum sun height, day length and albedo coefficient (continuous measurements). Oak forest at Virelles, Belgium. After Grulois, 1968. The arrow indicates snowfall.

121

Table 3.4. *Mean values of coefficients of albedo, interception and transmission of short-wave radiation according to phenophases and cloud cover*

Forest type and author	Cloud cover	Foliated phase Radiation			Leafless phase Radiation		
		Reflected	Intercepted	Transmitted	Reflected	Intercepted	Transmitted
Oak forest, Virelles (Grulois, 1968)		*Long-term continuous measurements*					
	Clear sky (10 days)	0.155	0.789	0.056	0.112	0.688	0.200
	Overcast (10 days)	0.194	0.724	0.082	0.165	0.444	0.390
	Rainy (10 days)	0.195			0.167		
Beech forest, Sofling (Kiese, 1972)	Clear sky (June–Sept)	0.145	0.786	0.069		0.097 (May)	
	Overcast (June–Sept)	0.125	0.780	0.095			
		Periodic and instantaneous values					
Pinus taeda, N. Carolina (Gay & Knoerr, 1975)	Spring	0.107					
	Autumn	0.128					
Scots pine and beech (Lützke, 1966)		0.132				0.06	
Scots pine, Great Britain (Stewart, 1971)		0.109					

Scots pine, Great Britain (Rutter, 1968)	0.100–0.150
Picea sitchensis, Picea glauca Great Britain (Jarvis, 1976)	0.150
After Muhenberg (in Budyko 1974)	
Stable snow in high latitude (above 60°)	0.80
Forest with stable snow cover	0.45
Steppe and forest between snow melting and mean daily temperature, 10°C	0.13
Tundra, steppe, deciduous forest from mean temperature 10°C to snow fall	0.18
Coniferous woods from mean temperature 10°C to snow fall	0.14
Forest exfoliating in dry season	0.24
Forest exfoliating in rainy season	0.18

several figures of albedo according to season (Table 3.4). The difference in the albedo under wet and dry conditions has been studied in Thetford over a pine forest. Under low radiation conditions the difference was less than 1% except at large solar angles (Stewart, 1971). A similarly small difference has been established by Grulois (1968) over a deciduous broadleaved forest (Table 3.4).

Interception by forest biomass

Interception rate is essentially a function of height development of the total forest canopy, but other factors must be taken into account. If a complete, closed forest canopy is developed, the growth phase is the most important factor as the presence of leaves creates interception surfaces about 8 to 10 times the total surface of branches and trunks (Table 3.5). If the turbidity of the atmosphere (overcast days or clear sky) has an influence on the albedo coefficient, interception also will be affected (Table 3.4). The difference in interception between the foliated and leafless period is less than one would imagine, because the albedo of branches and trunks with mosses and lichens is about 0.31 at the Virelles IBP Site. This value must be considered together with the transparent properties of the foliage (Table 3.5). This has been confirmed by Roussel (1968) who observed about the same result over two years in an oak–beech–hornbeam forest in the east of France. He also found for light radiation, a transmission to soil of 0.03 in the foliated period, and the coefficient of foliage interception was 0.27 (entire foliated period).

In the beech forest of the Solling IBP Site, Kiese (1972) found over one year an interception of 0.815 (together with soil albedo) for clear sky and 0.82 for overcast days during the foliated period; 0.58 for clear-sky days and 0.526 for overcast days in the leafless period. Interception of foliage was 0.301 for clear sky and 0.369 for overcast days. In closed coniferous forests, extinction is very often higher than in deciduous forests because of the non-deciduous character and the more uniformly colored foliage through the whole year. Gay & Knoerr (1975) found an interception of 0.85 to 0.90 for *Pinus taeda* and a soil absorption of 0.06.

Extinction process through the canopy

Monsi & Saeki (1953) were the first to apply the Lambert–Beer Law to the extinction of radiation by monolayer communities:

$$I = I_o \cdot e^{-bx}, \tag{3.45}$$

where b is the extinction coefficient and x is the distance through medium with constant turbidy or leaf area index. This law is strictly valid for

Table 3.5. *Ten-day, monthly, and yearly values of incident global radiation (G); coefficients of reflexion (a), interception (i) and transmission (t) for global radiation (G)*

	Deciduous oak forest (April 1966–June 1967) (Virelles–Blaimont, Belgium) (Galoux, 1973)					Evergreen holly–oak forest (June 1973–Sept 1974) (Supramonte di Orgosolo, Sardinia, Italy) (Susmel et al., 1976)			
Month	Incident (G) cal cm^{-2}	(a)	(t) 12.0 m	(t) 1.20 m	(i)	Incident (G) cal cm^{-2}	(a)	(t) 1.30 m	(i)
April	8399.6	0.105		0.061	0.765	12828	0.130	0.035	0.835
May	15062.4	0.162		0.063	0.767	16740	0.134	0.030	0.836
June	13849.7	0.174		0.057	0.786	17850	0.138	0.025	0.837
July	12043.0	0.170		0.068	0.768	19508	0.140	0.024	0.836
August	12348.5	0.157		0.131	0.692	16306	0.138	0.025	0.037
September	9682.6	0.164		0.248	0.590	12210	0.134	0.030	0.836
October	4586.9	0.177	0.269	0.342	0.512	8153	0.124	0.040	0.836
November	2359.3	0.162	0.449	0.319	0.515	4921	0.118	0.050	0.832
December	1461.2	0.146	0.442	0.320	0.564	3735	0.110	0.060	0.830
January	2003.2	0.166	0.486	0.391	0.494	3906	0.109	0.059	0.831
February	3961.3	0.116	0.589	0.399	0.489	5432	0.117	0.050	0.832
March	6779.5	0.115		0.165	0.669	9300	0.124	0.040	0.836
April	10641.4	0.112		0.082	0.731				
May	13508.9	0.166							
June	13597.4	0.187							
Foliated period (8 May to 13 Nov.)	67220.5	0.178							
Leafless period	29329.2	0.122							
Annual	96549.7	0.161				130888	0.131	0.033	0.836

monochromatic radiation with parallel rays. Grulois (1967) obtained for the oak forest of Virelles IBP Site, during overcast days, the constants:

$$Y = 12.45 + 170.3 \cdot e^{-0.7728x} \tag{3.46}$$

where Y is the percentage of global radiation at a certain height and x is the distance, in meters, from the upper canopy surfaces. Kiese (1972) in beech forest of Solling IBP Site found $Y = 9.5 + 198.5 \cdot e^{-0.35x}$. Miller (1967) and Grulois (1967) established good correlations between the extinction coefficient of radiation and leaf slope and cosine of angle of incidence of the sun's rays. Landsberg, Jarvis & Slater (1973) found a relation between the extinction coefficient (k) and the sun's elevation β in coniferous forests. Fig. 3.9 gives the extinction coefficient as a function of (β) and as a function of cosec β, where k is $-0.51 + 4.1$ (cosec $\beta - 1$) ($\beta > \beta'$, morning against slope) and k is $0.16 + 1.3$ (cosec $\beta - 1$). ($\beta < \beta' =$ afternoon to slope). This indicates an

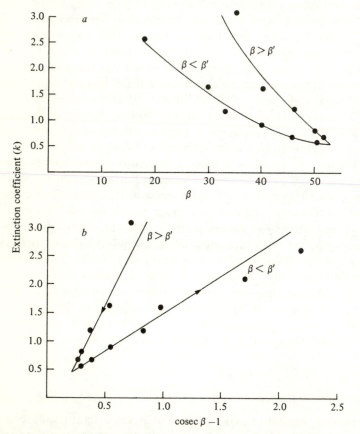

Fig. 3.9. *a*, the extinction coefficient (k) as a function of sun elevation (β); *b*, k as a function of cosec β. After Landsberg *et al.*, 1973.

interaction between the structure of the canopy, the slope of the ground and the incident direct beam solar radiation. The sensitivity of k to small changes in β in the morning rather than in the afternoon is due to the influence of the slight slope to the north-east on the penetration of radiation into the canopy, probably largely between the spire-shaped crowns. The results of Gay & Knoerr (1975) in *Pinus taeda* for spring and fall also suggest the influence of sun angle on interception processes through the canopy.

Transmission to the herbaceous layer or soil

Numerous measurements have been made which confirm the modifications of spectral composition of global radiation received at the soil surface (Sauberer & Hartel, 1959; Dirmhirn, 1964). Vezina (1961, 4, 5) used Bellani radiometers under *Pinus resinosa*. Sampling problems have appeared. During 104 snow days he found an average value of 0.08. The coefficient of transmission was smaller with clear sky (0.066) than for overcast days (0.102). In a closed deciduous forest at the Virelles IBP Site, Grulois & Schnock (1967) found through continuous measurements a relatively low variability from point to point (ranging from 0.30 to 0.33); the transmission coefficient was higher for overcast days than for clear sky. Integration in time considerably smooths the high variability of instantaneous values between points under the canopy. This has been confirmed by Gay, Knoerr & Braaten (1971) who found, in a pine plantation, a great variability of instantaneous values between plots but no significant differences of daily totals.

In the deciduous oak forest of Virelles, continuous measurements over one year (Grulois, 1968) gave an average value of 0.056 for clear-sky days and 0.082 for overcast days during foliated periods; 0.20 and 0.39 for the leafless period, respectively. The overall average values were 0.079 for the foliated period and 0.0358 for leafless period; these values are about the same as those obtained with circumglobal receiving surfaces of Bellani radiometers. Kiese (1972) confirms these general results in the beech forest of the Solling IBP Site: 0.467 for overcast days in the leafless period; in the foliated period 0.069 for clear-sky days and 0.095 for overcast days. For *Pinus taeda*, Gay & Knoerr (1975) found a transmission of global radiation to bare soil of 0.31 in the spring and 0.16 in the fall. For coniferous forests, Alexeyev (1963) and Mitscherlich, Künstle & Lang (1968) come to similar conclusions.

Flux of photosynthetically active radiation (PAR)

Knuchel (1914) was the first to make spectral analyses of light under forest canopies. But Alexeyev (1963) probably brought the largest contribution to the field of spectral studies in forest types in the vicinity of Leningrad. He

127

Table 3.6. *Ratio between light flux density and global radiation flux density. Monthly mean values:* $1 \, lux = k \cdot 1.10^{-5} \, cal \, cm^{-2} \, min^{-1} \, G$. *(Grulois & Vyncke, 1969)*

Month (1966)	k (Incident Radiation)	k' (Transmitted Radiation)	k'/k
June	1.865	12.635	6.775
July	1.859	12.486	6.716
August	1.869	11.934	6.385
September	1.889	12.055	6.382
Mean	1.870	12.276	6.565

conducted a considerable number of spectral analyses in numerous forest types and under various conditions. Various scientists made instantaneous measurements (Quantin, 1935; Roussel, 1953) of the energetic value of PAR within incident global radiation (G); many measurements at the level of vegetation made by ecologists give values around 0.5 (0.43–0.53). Such variations can be explained by the several conditions of instantaneous measurements (sun angle, air mass, clear or overcast sky, wave lengths considered). Long-term studies give values for temperate regions of a bit less than 0.5 (Wassink, 1968: 0.5; Yocum, Allen & Lemon, 1964: 0.47 for 0.4–0.7 μm). This value is less for clear sky than for overcast days, for small sun angle than for large sun angle.

Several measurements have been made comparing values of global radiation (G) and values given by photocells (lux). An energetic equivalence coefficient of lux $= k \cdot 10^{-5}$ cal cm^{-2} min^{-1}) may now be established. For clear sky and sun angles of 10° and 60°, Dogniaux (1964) gives a k for incoming global radiation, respectively, of 3.058 and 1.431. Brasseur & De Sloover (1973), found on 23 September that k was 1.5 at 0800 and 1.24 at 1300 h; Landsberg *et al.* (1973) give for the Feterosso Site (Sitka spruce in Scotland) the direct proportion PAR/G of 0.6 at 0730 and about 0.5 at 1200 h (average values for several days). At the Virelles IBP Site, for 32 days continuous measurements, the following values have been found: clear sky, 1.966 for incoming radiation above the canopy and 10.672 at the forest floor; overcast day, 1.789 above the canopy and 18.171 at the forest floor (Grulois & Vyncke, 1969). The average values of long-term continuous measurements for all days are given in Table 3.6. During the growth period, mean values of k are very homogeneous (1.870 for incoming radiation and 12.276 for transmitted radiation) (Table 3.6). The values of k for transmitted radiation give an idea of the spectral transformation of radiation through the canopy. There is 6.5 fold less light (PAR) for a broad-leaved forest in the transmitted than in the incoming radiation. Collected papers on light regimes, photosynthesis and productivity of the forest are to be found in Celniker (1967). Fig. 3.10 gives the spectral absorption, reflection and transmission of a leaf of *Populus deltoides*, and confirms the spectral properties of leaves of broadleaved tree species (Gates & Janke, 1965). Comparative measurements of light and global radiation under conifer stands are not numerous.

Instantaneous measurements of k have been made by Brasseur & De Sloover (1973) in the comparison of the selectivity of interception of a beech stand and a mature spruce forest. The values for transmitted radiation were 11.86 and 5.8, respectively, and 1.27 for an open area. The proportion of light in global radiation under a closed spruce stand is higher than that under a broadleaved tree species. Needles of *Picea abies* have a poorer permeability to light than the large beech leaves, and transmit relatively more light to the soil surface. Recent values for a Sitka spruce stand have been published by Jarvis (1976). On clear days within the canopy diffuse light is depleted in the near infrared relative to the composition of diffuse skylight above the canopy. The reflectivity is particularly large in the blue for *Picea pungens*; transmissivity in the visible spectrum varies from, 0.01 to 0.06 and

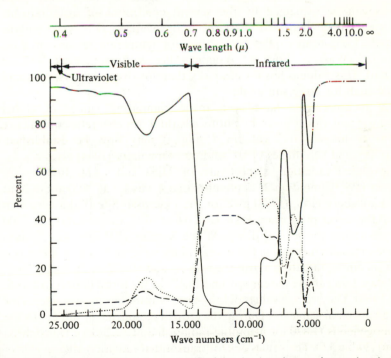

Fig. 3.10. Spectral absorption (———), reflectance (- - - -), and transmittance (·····) of *Populus deltoides* leaf as a function of the frequency of the radiation in wave numbers. A wave-length scale is given at the top. A wave number is the reciprocal of the wave length and is proportional to the frequency. After Gates & Janke, 1965.

129

from 0.15 to 0.50 in the near infrared. Consequently, absorptivity of needles is 0.97 in the visible and 0.58 in the near infrared.

Alexeyev (1970) established tables for PAR according to position in canopy gaps, sun elevation, and diameters of gaps. In the mediterranean evergreen oak forest of the Le Rouquet IBP Site, Methy (1974) found during overcast days that coefficients of transmission were 0.2 for near infrared, 0.15 for global radiation and 0.1 for PAR with clear sky; with sun flecks and shadow zones together, the coefficients of transmission are 0.27 for IR (0.7–3 μm), 0.23 for G and 0.18 for PAR. For clear sky, transmission of global radiation varies from 0.02 in shadows to 0.92 with sun flecks. But transmission of global radiation does not occur in two separate sets, one for shadow and one for the sun flecks. The distribution of the spectral bands is more integraded, because of the effect of penumbra. The maximum 1-cm long leaves of *Quercus ilex* at 10 m height have a diameter inferior to that of apparent diameter of sun (32 min). This has been thoroughly studied by Hutchison (1975) who worked on photographic assessment of forest structure (Fig. 3.11).

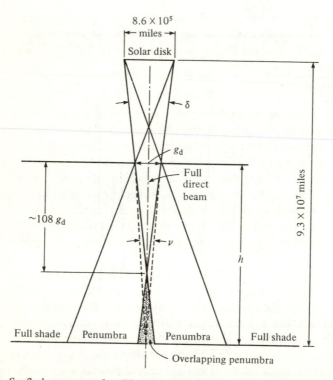

Fig. 3.11. Sunfleck geometry: δ, solid angle subtended by solar disk; ν, solid angle subtended by canopy gap; g_d, gap diameter (2h tan ν/2); h, height of gap above surface. After Hutchison, 1975.

Table 3.7. *Percentage contribution of each atmosphere layer containing 0.06 g cc^{-1} precipitable water to the total atmospheric radiation received at the ground. (Czepa & Reuter, 1950)*

Layer	Thickness (m)	Average temperature (°C)	Percentage contribution
1	87	9.2	72.0
2	89	9.0	6.4
3	93	8.6	4.0
6	108	7.4	1.2

Flux of long wave radiation

Downward atmospheric radiation

Atmospheric radiation (A) received just above the biospheric system depends essentially on water and carbon dioxide content and the temperature within the first 100 m of the low atmosphere. Brunt's formula is a tentative means to express these factors:

$$A = \sigma T^4 (a + b\sqrt{e}), \tag{3.47}$$

where T and e are in temperature and vapor pressure in air. Czepa & Reuter (1950) give relative contributions of several air layers to atmospheric downward radiation (Table 3.7). Atmospheric radiation flux (A) is nearly independent of vegetation, although high evaporative surfaces may increase the water content of air and have a small influence on atmospheric radiation flux density. Reflection of atmospheric radiation by vegetation surfaces is considered to be low value. Gates & Tantraporn (1952) have made some measurements of reflectivity of deciduous trees in the infrared to 25 μm and have found average values from 0.02 to 0.05, with up to 0.07 for poplars (Table 3.8).

Under the canopy, density of downward long-wave radiation often amounts to larger values during the day than those of atmospheric radiation (A) above the canopy. This is due to the temperature of foliage being higher than that of air, because of incoming short and long-wave fluxes. In fact, in any layer, the net long-wave radiation may be considered as the difference between the radiation emitted by the leaves and branches and the radiation received from leaves, branches and some atmospheric components outside that layer. This, for instance, appears in Gay & Knoerr's measurements on *Pinus taeda* (Fig. 3.12a), where the downward

131

Table 3.8. *Reflection coefficient of infrared wave radiation for deciduous trees. (Gates & Tantraporn, 1952)*

Angle of incidence 65°
(Wave length in μm)

	3.0	5.0	7.5	10.0	15.0	20.0	25.0
Acer saccharinum	0.03	0.04	0.045	0.055	0.05		
Populus alba					0.04	0.065	0.06
Populus deltoides	0.05	0.06	0.07	0.075	0.07		
Quercus robur	0.02	0.03	0.035	0.043	0.04		
Ulmus americana	0.022	0.024	0.024	0.024	0.025		

Angle of incidence 20°
(Wave length in μm)

	3.0	5.0	7.5	10.0	15.0	20.0	25.0
Acer saccharinum							
Populus alba	0.003	0.01	0.018	0.02	0.02	0.03	0.035
Populus deltoides	0.015	0.015	0.02	0.03	0.03	0.038	0.035
Quercus robur							
Ulmus americana	0.005	0.01	0.015	0.02	0.04	0.043	0.032

long-wave flux is 0.1 higher under the canopy than above the canopy. At the Virelles IBP Site, this measurement has been made for a number of days. Fig. 3.13 shows some results which indicate that during the day canopy downward radiation is higher than the atmospheric downward long-wave flux (amost 1.6 times in the afternoon). In July with short nights and clear sky this difference may remain positive for day and night. A warm canopy is a source of an important downward long-wave flux that contributes to all processes inside the forest and mainly to the soil heat regime (decomposition processes) (Fig. 3.13).

Upward radiation emitted by upper surfaces of vegetation

Terrestrial radiation. Radiation emitted upward by forest surfaces has been very rarely investigated. Gay & Knoerr (1975) were among the first to measure it over a *Pinus taeda* forest in 1965. During the course of the day, the flux varies according to canopy temperature, and is at its highest in the afternoon and at its lowest just before sunrise (Fig. 3.12*a,b*). Similar

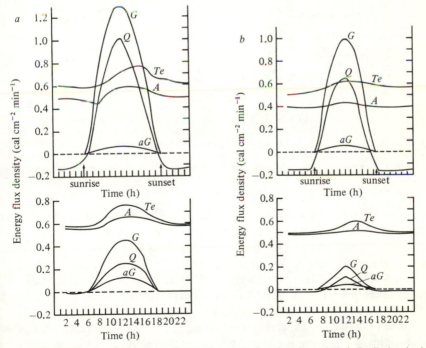

Fig. 3.12. *a*, Radiation budget components above the forest, spring, 1965; *b*, Radiation budget components above the forest, fall 1965. *G*, global radiation; *Q*, net radiation balance; *Te*, earth radiation; *A*, atmospheric radiation; *aG*, reflected global radiation. (Gay & Knoerr, 1975).

133

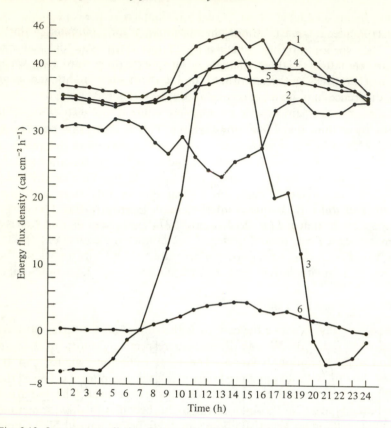

Fig. 3.13. Long-wave radiation fluxes and radiation balances above the canopy and at the forest floor (oak forest of Virelles IBP site). 1, $Te_{(19m)}$, upward long-wave radiation flux from canopy upper surfaces; 2, $A_{(19m)}$, atmospheric downward long-wave radiation to canopy upper surfaces; 3, $Q_{(19m)}$, radiation balance just above the canopy; 4, downward long-wave radiation flux from the surfaces of the canopy to the forest floor; 5, upward long-wave radiation flux from the soil surfaces; 6, $Q_{(1.2m)}$, radiation balance at the forest floor (1.2 m). (7.13.1967).

observations have been made at the Virelles IBP site (Fig. 3.14a). If we know the successive values of this flux, we may derive from the Stefan Law the surface temperatures. Research by Gay & Knoerr (1975), Galoux & Grulois (1968), Lorenz & Baumgartner (1970), Rauner & Ananjev (1971) are pertinent references. Gates & Papian (1971) devoted considerable effort to this problem and came to establish remarkable charts of leaf temperatures and an *Atlas of energy budgets of plant leaves*.

During the day with clear skies, leaf surfaces are warmer than the surrounding air. During the night it is generally the reverse. Gay & Knoerr (1975) found in 1965 over a *Pinus taeda* forest a maximum difference of +9.5°C at 1200 h in May. The largest difference found at the Virelles IBP

134

Site (oak forest) was 4.7°C, 1 May 1966, in the leafless period (Galoux & Grulois, 1968) and +6.1°C in July in the foliated period (Galoux, 1973). Above the spruce of the Ebersberger forest IBP Site, Lorenz & Baumgartner (1970) noticed differences of 1.3°C to 2.7°C. Rauner & Anajev (1971) give differences from +3 to 4°C. During the night before sunrise, differences are often negative and may be −1.5°C.

Flux density of the long-wave radiation emitted by soil surfaces is generally lower than that of canopy surfaces. This is illustrated by Gay & Knoerr (1975) for a *Pinus taeda* forest. They give a maximum difference of temperature (+2.7°C) between soil surface and air at 1.25 m. But after 1600 h, the difference is reversed (air warmer than soil). At the Virelles IBP Site, unpublished measurements made on a clear day in July show radiative temperature of soil surface always lower than air temperature at 2.00 m; so, we may conclude that a heat flow occurs from canopy to soil surface by diffusion.

Net radiation balance

Net radiation above the canopy

The radiation balance above the canopy is the sum of all the fluxes of short and long-wave radiation. We can distinguish, respectively, short-wave, long-wave and total radiation balances. The total radiation balance can be measured directly without any measurements of the components. This is the amount of radiation used by ecosystems in work or heat. The radiation balance remains negative throughout the night, becomes positive sometime after sunrise, and again negative about one hour before sunset. Night-time negative values represent net long-wave balance (Figs. 3.12a,b, 14a,b). The highest daily values measured are to be found in June in temperate regions (long photoperiod). Lützke (1966) found 308 cal cm^{-2} d^{-1} (over pine forest) and Rauner (1962) found 410 (broad-leaved forest in foliage near Moscow); Galoux (1973) found 352 (oak forest Virelles IBP Site), Gay & Knoerr (1975) found 540 (over *Pinus taeda*, North Carolina) and Strauss (1971a,b) found 492 (Ebersberg spruce forest).

The highest net radiation value is found at the lowest latitude. Tajchman (1967) proved that the vegetation cover, mainly through the coefficient of albedo, had a large influence on the values of net radiation balance. Three different vegetation covers situated in the same surroundings showed the following net radiation balance global radiation ratios: spruce forest, 0.699; potato field, 0.532; alfalfa, 0.514. The ratio has the following ranking for forests: spruce > beech > oak. Denmead (1969) found similar ratios for a *Pinus radiata* plantation and a wheat field. A good comparison of the net radiation balances over adjacent beech and spruce forests at the Solling IBP

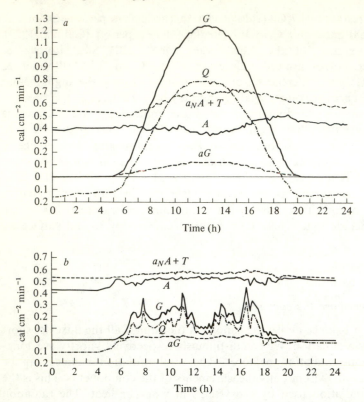

Fig. 3.14. Course of global radiation (*G*), atmospheric radiation (*A*), reflected atmospheric radiation ($a_N A$) and earth radiation (*T*), and net radiation balance (*Q*). Values and balances recorded each 10 min. *a*, Virelles–Blaimont forest at 19 m height, May 1966, sunny day. (Galoux & Grulois, 1968); *b*, Virelles–Blaimont forest at 19 m height, 27 April, 1966, overcast. (Galoux & Grulois, 1968).

Site has been made by Kayser & Kiese (1973). Fig. 3.15 shows that the spruce forest has a higher net radiation balance during the day and a lower one during the night.

Vertical profile of net radiation

Very few vertical profiles through forest stands have been measured. Baumgartner (1956, 7) was the first to achieve a radiation profile (Table 3.9) in a young spruce forest in Bavaria. Hager (1975) measured vertical net radiation profiles in the Ebersberger spruce forest. In the Solling beech forest (IBP site) at 26–8 m height with the first branches at 12 m, Kiese (1972) observed a larger balance value at a small distance below the highest living branches, than above the total canopy. This comes from an infrared

136

Table 3.9. *Profile of net radiation balance (Spruce forest, Bavaria, after Baumgartner 1956, 7)*

Height (m)		Day (cal cm^{-2})	(%)	Night (cal cm^{-2})	(%)
10.5	Above canopy	565	100	−22	100
5.0	Upper canopy level	555	98	−21	96
4.1	Sun crowns	223	39	−4	18
3.3	Shaded crowns	36	6	−2	9
0.2	Trunks	35	6	−1	5

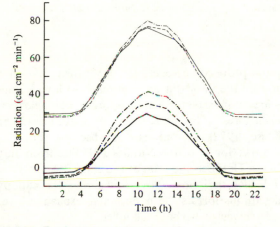

Fig. 3.15. Total global radiation from the upper hemisphere and net radiation in the Solling IBP Site. Germany July 1971. ——, grass; - - -, beech forest; −·−· spruce forest. Kayser & Kiese, 1973.

emission of these branches at high temperature in the middle of the day. Landsberg *et al.* (1973) give the course of the net radiation profile from 1300 to 2000 h in a Sitka spruce forest in Scotland. Fig. 3.16 expresses the evolution of the downward and upward fluxes. As sky radiation is higher with the approach of day, downward flux increases and upward flux decreases.

Net radiation at the forest floor

Values of net radiation above or under the herbaceous layer under the tree canopy depend strongly on the radiation exchange at the level of the

137

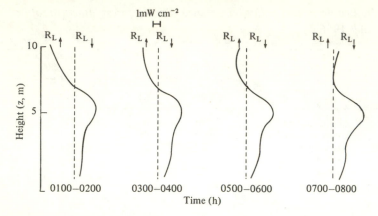

Fig. 3.16. Hourly average net radiation profiles in a spruce canopy from 0100 to 0800 h. Interrupted lines indicate zero radiation. R_L, net radiation. (Landsberg *et al.*, 1973).

canopy. The strong absorption property of vegetation cover for radiation is the reason that net radiation on the floor during daytime is so low. Values for clear sky, summer days are about 0.06 for a spruce forest (Baumgartner, 1956, 7); 0.31 for *Pinus taeda* forest (Gay & Knoerr, 1975) (Fig. 3.13*a,b*); 0.095 for beech forest (Kiese, 1972). Galoux (1973) observed that this proportion increases for the oak forest of the Virelles IBP Site from July onwards, with 0.105 to 0.263 in October and 0.68 at the end of October for yellowing foliage. In the fall, net radiation above and under the canopy approach each other. In deciduous forests the main active exchange layer during the winter is the soil surface or herbaceous layer.

Long term measurement of net radiation balances above and under forest canopies

Such measurements have been rare up to the present day. A relatively long period of measurement has been made at the IBP Site at Virelles above an oak forest. All the components are given in Fig. 3.17 which gives mean 10 − y values (1967). The general course of net radiation balance at the forest floor does not fluctuate much, a smooth light-decreasing curve being obtained from July to December. One can see the strong parallelism of the net radiation balance above the canopy with that of the global radiation. One can see also that the growth period is shifted in time in the course of the radiation flows. At the other IBP sites in Germany (Solling beech forest, Ebersberger spruce stand), long-term measurements have been recorded for net radiation balance. They generally indicate a higher net radiation balance

138

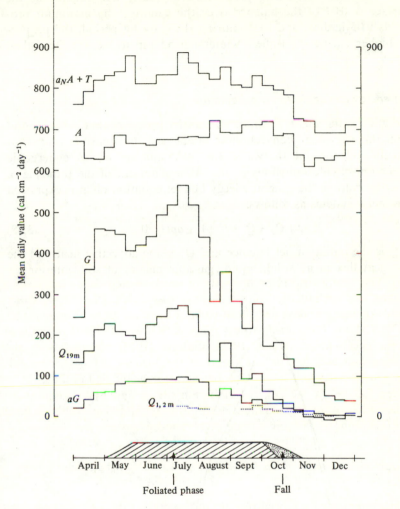

Fig. 3.17. Global radiation (G), reflected global radiation (aG), atmospheric radiation (A), reflected atmospheric radiation and terrestrial radiation ($a_N A + T$), net radiation balance above the canopy (Q_{19m}), net radiation balance on the forest floor ($Q_{1.2m}$). Mean values for 10-day periods (cal cm^{-2} day^{-1}) at Virelles–Blaimont, 1967 (Galoux, 1973).

during the vegetative period, in rather continental conditions compared to the more oceanic climate of the Virelles IBP Site (oak forest). Tajchmann (1972a) gives daily values of radiation in the canopy at the forest floor (Ebersberger forest site). In the IBP site of Bab (oak–hornbeam forest in Czechoslovakia), a complete yearly net radiation balance has been

139

reported (Smolen, 1975). The proportion of yearly net balance at the soil surface is 0.07 of the balance over the canopy. The minimum proportion is 0.05 in June and 0.065 during the growing period; the proportion is higher during the leafless winter period (up to 0.85 in December, Fig. 3.18).

Radiant energy transfer in forest ecosystems

Net radiation is the energy term for all transfer processes that organize and maintain the ecosystem: convection of latent heat and sensible heat; conduction through the soil; the water, the plant and air masses; endergonic photoreaction of chlorophyllous plants. As a statement of the principle of energy conservation, the general energy budget equation often is expressed for biospheric systems as follows:

$$Q + Q_L + Q_i + K + V + apG = 0, \qquad (3.48)$$

where Q is the radiation net balance and Q_L is the advective heat and the sensible heat flux in all solid, liquid, gaseous masses of soil, interior air

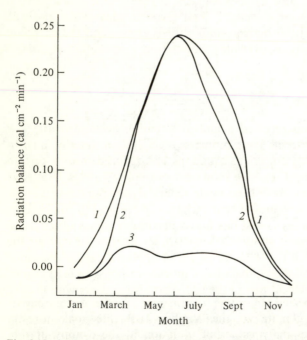

Fig. 3.18. Annual course of radiation balance at the oak–hornbeam forest, IBP site of Bab, Czechoslovakia (Smolen, 1975). *1*, above the stand; *2*, above an open area; *3*, in the forest stand.

masses, biomass; precipitation water (also with possible fusion or solidification latent heat). Q_i is expressed by:

$$Q_i = \left[\int_o^z \rho C_p (\delta T/\delta t)\delta z \right] \qquad (3.49)$$

where ρ is the density; c_p, heat capacity; T, temperature; z, height; V, $[Lk_v(\delta q/\delta z)]$, the latent heat of vaporization/condensation; q, specific humidity of air; L, latent heat of vaporization; k_V, coefficient of turbulent transfer for vapor; K, $[\rho c_p k_K(\delta\theta/\delta z)]$, the sensible heat in turbulent air; θ, potential temperature of air; k_K, coefficient of turbulent transfer for sensible heat; $a_P G$, $[\lambda k_C(\delta C/\delta z)]$, the net photosynthesis; λ, photochemical energy equivalent of carbon dioxide; k_C, coefficient of turbulent transfer for carbon dioxide; C, carbon dioxide concentration of the air.

Biomass surfaces are the most active interfaces for transfer of radiant energy. The rates of transfers depend essentially on the masses, the specific heat values, the conductibilities and surface-to-mass ratios. In a forest ecosystem, soil and sub-soil masses involved in transfers and exchanges can approach 300 g cm^{-2}, biomass up to 10 g cm^{-2} (often 2–4 g cm^{-2}), water flux through the system ranges from 50 to 300 g cm^{-2} year^{-1} (often 80–150 g cm^{-2} year^{-1}), air flow up to 5 tons cm^{-2} year^{-1}, and a chlorophyll mass of only 0.00024 g cm^{-2} year^{-1}. This gives an idea of the possible distribution of the radiation budget among these masses and the energetic input to ecosystem functions (warming, convection, evaporation, drainage, biological production).

Sensible heat in masses

In general, this term has received a limited attention because of the low values of these storage terms (soil, biomass, precipitation, interior air). In long-term studies, mostly empirical equations are deduced from a certain number of observations. Fig. 3.19 shows an example (Virelles IBP site), where hourly calculations have been made separately for biomass and soil. This figure shows the low values of heat exchange in biomass in comparison with values in soil. Grulois (1968) has found maximum values of 0.053 cal cm^{-2} min^{-1} for biomass at about 1200 h; 0.028 for interior air and for the soil. It often happens that the three fluxes are of opposite signs at the same time, because heat flux in the biomass has a quicker response to fluctuations of net radiation than the heat flux in the soil. This also appears with decade values (Fig. 3.19).

Total sensible heat flux in broadleaved forests may have maximum values of 30 to 34 cal cm^{-2} day^{-1} during a clear sky before complete foliation (April–May). Sensible heat flow in transpiration water, which is received as precipitation at low temperatures and evaporated at higher temperatures,

141

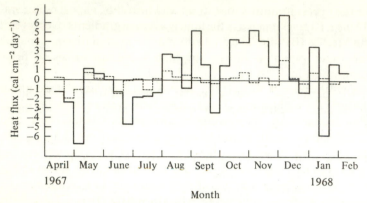

Fig. 3.19. Mean daily amounts for 10 day periods of heat flux in soil (Q_G, ----) and heat flux in biomass (Q_V, ---) at the Virelles oak forest (Galoux, 1973).

has been calculated for the Virelles IBP site to be about 120 cal cm^{-2} for the growing season.

Sensible heat and latent heat fluxes in turbulent air

Examples of the daily course of heat and vapor fluxes above forest ecosystems are now fairly numerous for spruce and pine forests for example, Baumgartner, 1956, spruce forest, Bavaria; Lützke, 1966 Scots pine forest near Berlin; Reifsnyder, 1967, *Pinus resinosa* forest USA; Tajchmann, 1967, Ebersberger spruce forest, Munich; Denmead, 1954 *Pinus radiata* forest; Black & McNaughton, 1971 Douglas fir forest (Vancouver, Canada); Storr *et al.*, 1970, spruce-fir forest, Alberta, Canada; Reimer & Desmaris, 1973, *Pinus banksiana* stand, Manitoba, Canada; Strauss, 1971, spruce forest Eberforst IBP Site, Germany; Tajchmann, 1972b, *Pinus sylvestris* forest, Freiburg, Germany; Stewart & Thom, 1973, *Pinus sylvestris* stand, Thetford, Great Britain; Gay, 1972, douglas fir forest, Cedar River, Washington; Gay 1973, *Pinus contorta* stand, Bend, Oregon. For broad-leaved deciduous forests there are data only for USSR oak forest (Rauner, 1958, 61); Grulois, 1968, and Galoux, 1973, oak forest Virelles IBP Site; Kiese, 1972, beech forest Solling IBP site.

Air temperature and vapor tension gradients above the canopy

As a result of radiation energy transfer in biomass, soil and air during the day, the temperature of the surrounding air becomes higher than air temperature a few meters above the forest, Vapor pressure in stomatal chambers and surrounding free air is higher than that of forced convection air above the stand. These

142

gradiens give the necessary forces to start air and heat flux, vaporization and vapor flux. Table 3.10 gives an idea of the hourly average gradients over several months at the Virelles IBP site. The gradients may be used in thermodynamic equations for water potential (notice the high values of forces to drive evapotranspiration) (Fig. 3.20).

Fig. 3.21 shows the energy budget components of a douglas fir stand under clear skies. During the night, while Q is negative, temperature of biomass and surrounding air decreases, the air mass descends and mixes with interior air and temperature inversion is established above the active surfaces. After sunset, turbulent heat and vapor fluxes start to develop rapidly. The maximum sensible heat flow occurs at 1000–1100 h for soil, at 1200 h for turbulent air and at 1500 h for evaporation. This last peak may find an explanation in the maximum water potential of air, which is to be found between 0100 h and 0400 h (see Table 3.10). These tendencies are observed in other similar studies. On overcast days the curves are very variable and represent low values. Similar courses of fluxes are found for the Ebersberger spruce forest (Tajchmann, 1967; Strauss, 1971), for a pine forest in the Rhine valley (Tajchmann, 1972b), and a pine forest near Berlin (Lützke, 1966). Kiese (1972) found an afternoon depression of V in the beech forest of the Solling IBP site (Fig. 3.22).

Five comparisons have been made by Gay & Holbo (1974) between the energy budgets of Scots pine and Douglas fir stands, although water availability at the five stands was not known. These data demonstrate a tendency for Douglas fir to transpire at a somewhat higher rate than Scots pine, perhaps because Douglas fir needles have lower stomatal resistance than the more xerophyllous Scots pine.

Masses, surfaces, gradients and exchanges in the forest profile

Profiles of mass exchanges and surfaces, temperature, radiation, vapor pressure, water potential of air, heat and vapor exchanges of the oak forest of the Virelles IBP site are given in Fig. 3.21 for 0100 h on a clear sky day (31 July 1967). While mass density increases in the direction of the soil, the surfaces are developed in two main layers, the tree and herbaceous layers. Global and radiation balance curves cross each other at a height of about 8 m, below foliage layers; downward and upward radiation fluxes also cross at about the same height. Sensible heat flow is proportional to mass density in the profile. The radiative surface temperature and air temperature curves cross at about 10 m (2/3 of total tree height). Vapor pressure in air increases from 7 m below tree foliage to the upper foliage surface at 16 m, and then decreases to 25 m. The course of water potential, the driving force for water cycling, shows a general increase from soil (-10 atmospheres) to air (-1300 atmospheres) at 7.16 and 25 m, respectively.

The vapor flux increases from soil to upper surfaces of foliage at 16 m.

Table 3.10. *Boundary layer above the forest canopy (between +16 and +25 m (°C/100 m) (Oak forest, Virelles, Galoux, 1973)*

Hour	2	4	6	8	10	12	14	16	18	20	22	24
a. Temperature gradient in air (1967) (°C/100 m)												
May	-3.6	-3.9	-3.7	0.4	2.4	3.1	3.3	3.6	0.8	-1.3	-3.7	-3.9
June	-3.3	-3.3	-1.5	2.3	4.5	4.6	4.1	3.6	1.7	-0.4	-3.7	-3.5
July	-5.6	-5.3	-4.7	1.7	5.8	7.3	6.3	5.0	2.0	-1.7	-4.8	-6.5
August	-4.5	-4.5	-3.0	0.3	2.6	4.1	3.4	1.6	1.5	-3.2	-4.7	-5.1
September	-3.2	-3.1	-2.8	-0.7	2.0	3.4	2.4	0.4	-1.0	-3.2	-3.4	-3.3
December	-0.3	-1.4	-1.4	-1.0	-1.4	-2.8	-1.4	-0.9	-0.7	-1.1	-1.5	-1.2
b. Vapor density gradient in air (1967) (g m^{-3}/100 m)												
May	-0.1	-0.1	0.2	1.1	0.3	0.0	0.6	-0.7	-0.4	0.1	0.6	0.4
June	0.7	0.6	0.9	2.3	2.8	3.8	4.0	3.9	3.8	2.2	0.9	1.1
July	-0.9	-0.9	0.2	2.8	3.2	3.7	3.6	4.3	2.0	0.8	0.0	0.1
August	-0.1	-0.5	-0.2	2.1	3.2	4.0	3.8	3.6	2.9	1.5	0.3	0.3
September	-0.8	-0.7	-0.3	1.1	2.1	3.0	3.9	3.4	2.2	0.8	0.1	0.4
December	-1.8	-2.0	-1.8	-1.5	-1.7	-1.0	-1.1	-0.9	-1.9	-1.8	-1.7	-2.0

*Oak forest Virelles-Blaimont. Research supported by the Fonds de la Recherche fondamentale collective, Belgium.

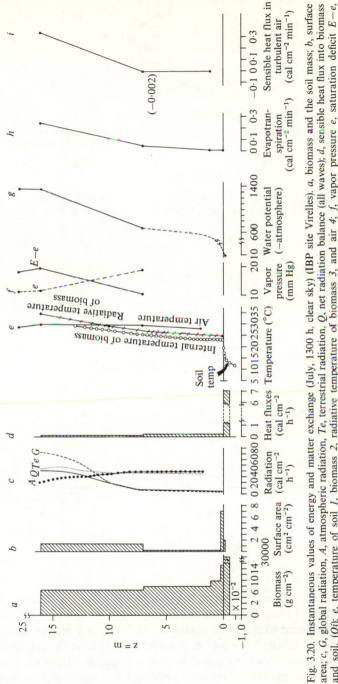

Fig. 3.20. Instantaneous values of energy and matter exchange (July, 1300 h, clear sky) (IBP site Virelles). *a*, biomass and the soil mass; *b*, surface area; *c*, *G*, global radiation, *A*, atmospheric radiation, *Te*, terrestrial radiation, *Q*, net radiation balance (all waves); *d*, sensible heat flux into biomass and soil. (*Qi*); *e*, temperature of soil *1*, biomass *2*, radiative temperature of biomass *3*, and air *4*; *f*, vapor pressure *e*, saturation deficit *E − e*, saturation vapor pressure; *g*, water potential, (*ψ*); *h*, evapotranspiration (*−V*); *i*, sensible heat flux in turbulent air (*−K*).

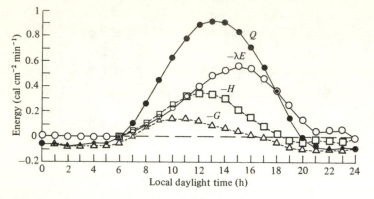

Fig. 3.21. Energy budget components under clear skies (29 July 1971): net radiation, Q; change in heat storage of soil and biomass, G; convection, H; latent energy, λE. (Gay, 1972).

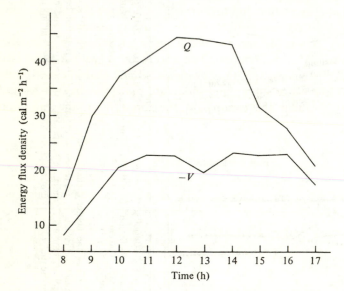

Fig. 3.22. Course of radiation balance (Q) and evapotranspiration (V) during 8 bright days in July for the beech forest of the IBP Solling site. (Kiese, 1972).

There is a thermodynamic active layer at some height in the tree foliage. From this layer to soil, a sensible heat flow comes from warm foliage and is oriented downward to soil. Above this active layer, at 7 m, the sensible heat flow is oriented upward. Rauner & Ananjev (1971) have demonstrated that the point of inversion of the sensible heat flux in air varies with structure according to species. This point is to be found at a lower relative height for

146

well-layered tree species like Scots pine than for highly concentrated foliage species as *Tilia* (Fig. 3.23).

Long-term measurements of sensible heat and latent heat fluxes

There are very few measurements on the entire growing season or annual cycle for these IBP sites: Ebersberger (spruce), Virelles (oak), Solling (beech), Triangle, USA (*Pinus taeda*). But simulation methods can supply such data. These are available, for other IBP sites, data which are not completely analyzed. The forest of Eberswald near Berlin also provides interesting long-term values. All these results are summarized in Table 3.11. Comparisons are difficult because of the variability of the radiant regimes over several years. Only two sets of measurements have been made in the same year. The most similar radiation regimes for May–September are Ebersberger forest (spruce) and Eberswald–Berlin (pine with beech). The monthly values of evaporation and turbulent heat flow are also comparable (V/Q being 0.66 and 0.65, respectively, for the growing season).

The oak forest of Virelles and the beech forest of Solling have similar radiation balances (May–September). The beech forest, with a good soil water supply and higher precipitation, evaporates more than the oak forest of Virelles, with low water supply and moderate precipitation (V, 0.84 and 0.65, respectively). Under the canopy, at the forest floor, the radiation

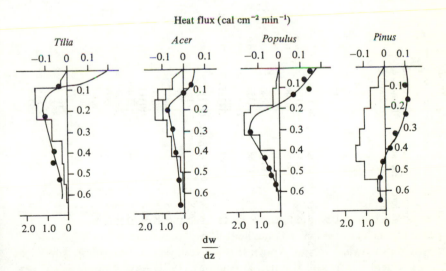

Fig. 3.23. Vertical course of heat flux (cal cm^{-2} min^{-1}) within the canopy for several tree-stands composed of *Tilia* sp., *Acer pseudoplatanus*, *Populus tremula*, and *Pinus sylvestris*) (mostly 1000 to 1400 h, ordinate; $z = 1 - \frac{z}{h}$ where h is stand height). (Rauner & Ananjev, 1971).

Table 3.11. *Radiation and heat energy transfers in four forest types (spruce, pine, beech, and oak)*

	Precipitation, P (mm)	Global radiation on a horizontal surface, G (cal cm^{-2})	Short and long-wave radiation "balance Q (cal cm^{-2})	Q/G	Sensible heat in soil and vegetation layer, Qi (cal cm^{-2})	Latent heat in evapotranspiration[a] (cal cm^{-2})	Sensible heat in turbulent air, K	V/Q	$-K/Q$	$-Qi/Q$
Spruce										
1 May – 30 September 1965 (Tachman, 1967)										
May	188	9033	6156	0.68	7	−4256	−1907	0.69	0.31	0.0
June	209	13029	9399	0.72	−142	−6675	−2582	0.71	0.27	0.02
July	155	11701	8737	0.75	−52	−5610	−3075	0.64	0.35	0.01
August	77	11432	8090	0.71	23	−4933	−3180	0.61	0.39	0.0
September	98	8768	5353	0.61	93	−3601	−1845	0.67	0.34	−0.01
Totals										
May–Sept.	724	53963	37735	0.70 (av)	−71	−25075	−12589	0.66 (av)	0.33 (av)	0.005 (av)
1 January – 31 December 1969 (Strauss, 1971)										
January	79.8		2154		37	−1975	−220	0.92	0.10	0.02
February	10.5		2226		48	−1988	−286	0.89	0.13	0.02
March	49.5		4049		34	−2440	−1640	0.60	0.41	0.01
April	57.2		8280		−144	−5517	−2619	0.67	0.32	−0.02
May	89.0		10314		−43	−8494	−1736	0.82	0.17	−0.00
June	153.4		8598		−69	−7314	−1215	0.85	0.14	−0.01
July	114.7		10636		−198	−8652	1795	0.81	0.17	−0.02
August	138.9		6203		−105	−5016	−1079	0.81	0.17	−0.02
September	14.3		6207		−42	−4668	−1503	0.75	0.24	−0.01
October	2.5		3633		96	−3698	−31	1.02	0.01	0.03
November	56.7		1035		132	−777	−390	0.75	0.38	0.13
December	78.1		1094		−31	−1032	−2	0.94	0.00	−0.03
Totals										
Jan–Dec	841.0	64429			−286	−51571	−12516	0.80 (av)	0.19 (av)	0.00 (av)
May–Sept	510.3	41958			−457	−34144	−7328	0.81 (av)	0.18 (av)	−0.06 (av)

Pine

1 April – 30 September 1967 (Lützke, 1969)

						(583.7 mm, 114% P)				
April	—	—	4612	—	−11	−2903	−1698	0.63	0.37	0.00
May	—	—	8350	—	−273	−4788	−3289	0.57	0.40	0.03
June	—	—	8988	—	−164	−6184	−2640	0.69	0.29	0.02
July	—	—	9020	—	−184	−5828	−3008	0.65	0.33	0.02
August	—	—	6939	—	−3	−4473	−2463	0.65	0.35	0.00
September	—	—	3953	—	70	−2962	−1061	0.75	0.27	−0.02
Totals										
April–Sept	—	62430	41862	0.67 (av)	−565	−27138 (460.7 mm)	−14159	0.65 (av)	0.34 (av)	0.01 (av)
May–Sept	—	—	37250	—	−554	−24235 (411.5 mm)	−12461	0.65 (av)	0.34 (av)	0.01 (av)

Beech

3 May – 20 October 1970 (Kiese, 1971)

May 3–31	—	10887.2	5457.8	0.50	−272.4	−3746.8	−1452.9	0.69	0.27	0.04
June	83.2	15624.0	8151.0	0.52	−180.0	−7008.0	−939.0	0.86	0.11	0.02
July	157.9	11563.0	6506.4	0.56	−96.1	−5682.3	−700.6	0.87	0.11	0.01
August	84.1	10332.3	5099.5	0.49	−77.5	−4188.1	−840.1	0.82	0.16	0.01
September	94.4	7041.0	2841.0	0.40	17.5	−2880.0	−156.0	1.01	0.05	0.00
October 1–20	—	3341.8	970.0	—	15.0	−1246.0	198.0	1.28	−0.20	0.00
Totals										
3 May – 20 Oct	609.7	58789.3	29025.7	0.49 (av)	−593.9	−24751.2 (421.3 mm)	−3890.6	0.84 (av)	0.14 (av)	0.02 (av)
3 May – Sept	515.3	55447.5	28055.7	0.51 (av)	−608.9	−23505.2 (401.8 mm, 78% P)	−4088.6	0.84 (av)	0.15 (av)	0.02 (av)

Table 3.11. *continued*

	Precipitation, P (mm)	Global radiation on a horizontal surface, G (cal cm^{-2})	Short and long-wave radiation[a] balance Q (cal cm^{-2})	Q/G	Sensible heat in soil and vegetation layer, Qi (cal cm^{-2})	Latent heat in evapotranspiration[a] (cal cm^{-2})	Sensible heat in turbulent air (cal cm^{-2})	-V/Q	-K/Q	-Qi/Q
Oak										
1 April – December 1967 (Galoux, 1973)										
April	57	10642	5128	0.48	-41	-5117	-1508	0.77	0.23	0.01
May	81	13509	6679	0.49	-54	-4433	-2084	0.67	0.32	0.01
June	16	13597	6592	0.48	-75	-4859	-3280	0.59	0.40	0.01
July	62	16801	8201	0.49	-62	-4234	-1197	0.79	0.22	-0.01
August	78	11797	5372	0.46	59	-2485	-661	0.79	0.21	-0.01
September	95	7836	3127	0.40	19	-1290	-203	0.94	0.15	-0.01
October	86	5032	1378	0.27	115	-563	362	5.80	3.73	-0.08
November	109	3163	97	0.03	104	11	117	-0.05	-0.57	-1.07
December	92	1429	-206	-0.14	78					-0.38
Totals										
April–May	676	83806	36368	0.32 (av)	143	-22970	-8454	1.28 (av)	0.59 (av)	-0.16 (av)
May–Sept	332	63540	29971 (47% G)	0.47 (av)	-113	-21128 (358 mm, 108% P)	-8730	0.70 (av)	0.29 (av)	0.00 (av)
January – December 1968 (Galoux, 19)										
January	97.2	2010.5	180.0	0.09	4.8	-277.4	96.8	1.50	-0.5	-0.03
February	72.6	3269.4	798.2	0.24	42.1					-0.05
March	69.2	6969.6	2888.1	0.41	-218.9					0.07
April	15.4	12929.0	5465.1	0.42	-71.4	-3914.4	-1479.3	0.72	0.27	0.013
May	63.7	11772.9	5850.7	0.49	-76.6	-4114.8	-1659.3	0.70	0.28	0.013
June	62.5	13148.1	6573.4	0.50	-85.4	-4108.4	-2379.6	0.62	0.36	0.013
July	156.1	13085.4	6977.8	0.53	0.5	-5520.2	-1458.1	0.79	0.21	0.00
August	66.4	9577.4	5308.9	0.55	-3.1	-4672.1	-633.7	0.88	0.12	0.00

Month									
September	136.9	7508.6	3603.5	0.48	83.9	−3425.1	−262.3	0.95	0.07	−0.03
October	42.6	4688.4	1707.1	0.36	−23.3					0.01
November	31.8	2466.5	22.9	0.01	179.8					−7.85
December	51.9	1652.0	141.2	0.08	132.4					−0.94
Totals										
Jan–Dec	866.3	89077.8	39516.9 (51% G)	0.44 (av)	−35.2	−21840.6 (376 mm, 77% P)	−6393.0	0.77 (av)	0.22 (av)	−0.73 (av)
May–Sept	485.7	55192.4	28314.3 (51% G)	0.51 (av)	−80.7					0.0 (av)

January – December 1969 (Galoux, unpublished)

Month									
January	64.3	1787.5	−197.4	−0.11	−47.6					
February	45.8	3468.5	846.1	0.24	88.7					
March	48.8	6292.6	2828.8	0.49	−144.4	−1829.7	−854.7	0.65	0.30	0.05
April	97.0	10965.4	5844.2	0.53	−204.1	−3881.6	−1758.5	0.66	0.30	0.04
May	90.9	13769.3	7552.3	0.55	−99.7	−5742.1	−1710.5	0.76	0.23	0.01
June	104.8	13841.8	7077.0	0.51	−170.7	−4869.7	−2036.6	0.69	0.29	0.02
July	54.7	13924.5	7477.2	0.54	−132.1	−5167.1	−2178.0	0.69	0.29	−0.02
August	113.9	11384.7	5683.1	0.50	75.0	−4115.7	−1642.4	0.72	0.29	−0.01
September	11.4	8705.7	3651.3	0.42	126.8	−2334.3	−1443.8	0.64	0.39	−0.03
October	4.4	6598.7	1736.8	0.26	80.4	−815.1	−1002.1	0.47	0.58	−0.05
November	129.5	2533.3	230.0	0.09	216.6					
December	58.8	1518.5	−122.0	0.08	275.5					
Totals										
Jan–Dec	824.3	94790.5	42607.4 (51% G)	0.45 (av)	64.4	−22228.9 (380.3 mm, 101% P)	−9011.3	0.71 (av)	0.277 (av)	0.006 (av)
May–Sept	375.7	61626.0	31440.9 (51% G)	0.51 (av)	−200.7					

[a] Values in parentheses represent percent of global radiation.

[b] Values in parentheses represent precipitation equivalent of latent heat of evapotranspiration and its percentage of precipitation.

balance is almost completely utilized by evaporation from the herb layer and soil surface. Some downward heat flow occurs during days with clear skies. At the Virelles IBP site, data for 72 days from July to September are available. The fraction of evaporation at the forest floor in the total evaporation above the forest varies from 12.2 to 22.6%, with 1.7% interception of precipitation on the herb layer. The latter percentage increases in the fall up to 30%. The fractions of total annual evaporation are 54.1% transpiration, 31.1% interception and 14.8% soil and herb layer evaporation. In the Ebersberger spruce forest, Tajchmann (1972b) found values of 62.0%, 33.2% and 4.7% for transpiration, interception and soil evaporation, respectively. The higher soil evaporation in the Virelles oak forest is probably due to the presence of an abundant herbaceous layer.

A detailed radiation and heat budget for the growing season is given in Fig. 3.24 for the oak forest at Virelles IBP Site. All the fluxes have been measured except some which are hypothetical. The direct sun radiation at the extra-atmospheric level has been estimated after the tables of Linacre (1969); the albedo has been taken from the latest estimates of Kondratyev (1969) (Möller for latitude 50° N); the fraction of terrestrial radiation reaching space was estimated from Kondratyev (1969); heat flow in precipitation has been estimated by calculation of daily amounts of precipitation and air temperature during and after showers; net ecosystem production (NEP) and heterotrophic respiration have been measured by growth analysis of the stand (Duvigneaud *et al.*, 1971) and continuous measurement of carbon dioxide soil flux by titration with potassium hydroxide (Froment *et al.*, 1971). The amount of energy for the Q_i and NEP terms in the general energy balance equation have been subtracted from the radiation balance before partitioning latent and turbulent heat fluxes.

Carbon dioxide balance studies

Schulze & Koch (1971), using carbon dioxide analysis in a transparent gas exchange chamber mounted in the tree crown, measured net photosynthesis in a long-term experiment in the beech forest of Solling. They distinguished sun and shade parts of the tree crown, because of their different photosynthesis regimes. In early summer (beginning of August), sun leaves produce one-third more photosynthate than shade leaves; net photosynthesis of sun leaves declines and by late September shade leaves produce twice as much as sun leaves (Fig. 3.25).

The annual balance of carbon dioxide gas exchange (mg CO_2 g^{-1} dry wt in leaves) is given in Table 3.12. The average net photosynthate produced by sun and shade leaves, from sprouting until leaf fall, is 9541.5 mg carbon dioxide g^{-1} dry wt, which is equal to 2.6 g carbon g^{-1} dry wt. Having a leaf mass of 3.3 tons dry wt ha^{-1}, assimilation is 8.6 tons carbon ha^{-1}. The

152

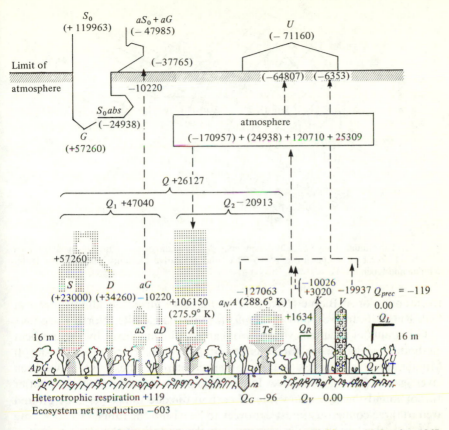

Fig. 3.24. Oak forest of Virelles–Blaimont. Energy balance from 25 May to 25 October 1967 (cal cm^{-2}). S_o, extra-atmospheric solar radiation on a horizontal surface (short waves); aS_o, extra-atmospheric solar radiation reflected by earth–atmosphere system; S_o abs, solar radiation absorbed by atmosphere; S, direct solar radiation on a horizontal surface; U, extra-atmospheric upward radiation (long waves); D, diffuse scattered radiation on a horizontal surface (short waves); G, global radiation on a horizontal surface $(S+D)$ (short waves); Te, terrestrial radiation (long waves); A, atmospheric radiation (long waves); aS, reflected solar radiation; aD, reflected diffuse radiation; aG, reflected global radiation; $a_N A$, reflected atmospheric radiation; apG, global radiation utilized in net photosynthesis; Q_1, short wave radiation balance $(G-aG)$; Q_2, long wave radiation balance $(A-Te)$; Q, short and long wave radiation balance $(G-aG+A-a_N A-Te)$; Q_G, sensible heat flux in soil; Q_V, sensible heat flux in vegetation; K, sensible heat turbulent flux; V, latent heat in evapotranspiration; Q_R, latent heat in water condensation; Q_b, advective sensible heat; Q_{prec}, sensible heat flux in precipitation water. Parameters of the stand (per ha): biomass, 156 ton; net primary production (ground), 14.6 ton. Exchange aerial surfaces (ha/ha): foliage (2 faces) of trees, 14; bark of trees, 2; herb layer, 2; litter, 1.5; total exchange surfaces (except litter 18 ha/ha). Figures in brackets are estimated values.

losses due to night respiration are 1.3 tons carbon ha^{-1}. With an average carbon content of 50% in beech wood, the increment of the stand is 2.1 tons carbon ha^{-1}. Recently Strain & Higginbotham (1976) have given an over-view of the results obtained in the two US IBP sites of the Eastern

153

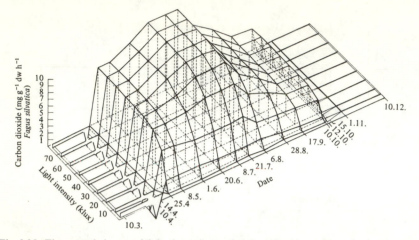

Fig. 3.25. The annual change of light dependence of carbon dioxide exchange in the sun crown of a beech forest at temperatures of 16–22°C and a relative humidity of 80–100%. (After Schulae and Koch, 1971).

Deciduous Forest Biome (*Liriodendron tulipifera* mixed deciduous forest at Oak Ridge, Tennessee, and the *Pinus taeda* Triangle Site, North Carolina).

The results of an intensive study of carbon dioxide fluxes by the gradient method in a coniferous forest is reported by Hager for the IBP site at Munich in this chapter. A few other coniferous sites also have been investigated (Jarvis, James & Landsberg, 1976; Jarvis, 1976). The latter author found the maximum rate of carbon dioxide influx per unit ground area to three coniferous forest canopies to lie within the range of 1 to 1.5 mg m^{-2} s^{-1}, the largest value occurring in the middle of the day. A rate of 3 mg m^{-2} s^{-1} was recorded for *Pinus radiata* in unstressed conditions. For Sitka spruce, the carbon dioxide flux amounts to an energy equivalent of 0.02 of the net radiation flux density, but can be as high as 0.15 in the early morning. The carbon dioxide flux from soil respiration ranges from 0.02 to 0.07 mg m^{-2} s^{-1}.

Annual course of ecosystem functions

The annual course of energy flow in a deciduous, broad-leaved oak forest ecosystem is given in Fig. 3.26 for three successive years. With the annual course of the sun, radiation is the energy source for all functions: these include heating of mass, the thermal balance of organisms that continuously lose heat through their interfaces with air, evaporation through membranes under the high values of water potential in the air, precipitation, drainage (cycling of water), warming of air and components on living surfaces through turbulent airflow, production and destruction of organic

Table 3.12. *The annual balance of carbon dioxide gas exchange (mg CO$_2$ g^{-1} dry wt) in beech leaves. (Schulze & Koch, 1971)*

	Sun leaves		Shade leaves		Percentage (sun leaves are 100%)
	CO$_2$ uptake	CO$_2$ release	CO$_2$ uptake	CO$_2$ release	
Bud and twig respiration (6 March – 9 April)	—	147.7	—	145.4	98.4
Bud and twig respiration (10 April – 19 April)	—	214.0	—	147.2	68.8
Net assimilation of the leaves (20 April – 21 October)	9855.1	—	9227.8	—	94.0
Respiration of leaves at night	—	1352.2	—	1237.0	91.5
Bud and twig respiration (22 October – 10 December)	—	75.5	—	158.6	210.0
Bud and twig respiration (11 December – 5 March)	—	25.0	—	25.0	—
Totals	9855.1	1814.4	9227.8	1713.2	93.4
Annual balance (mg CO$_2$/g dry wt)	8040.7		7514.6		
Net assimilation of leaves based on leaf surface area	4150.5		1729.3		42.4

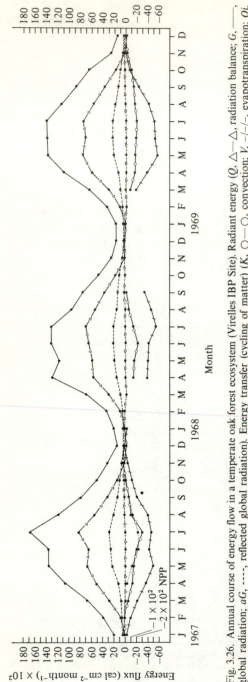
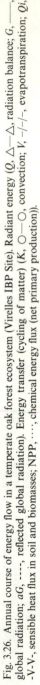

Fig. 3.26. Annual course of energy flow in a temperate oak forest ecosystem (Virelles IBP Site). Radiant energy (Q, $\triangle$—$\triangle$, radiation balance; G, ——, global radiation; aG, - - -, reflected global radiation). Energy transfer (cycling of matter) (K, $\circ$—$\circ$, convection; V, -/-/-, evapotranspiration; Qi, -V-V-, sensible heat flux in soil and biomasses; NPP, $\cdots$, chemical energy flux (net primary production)).

matter through photosynthesis and decomposition, which are all dependent on the sinusoidal daily and annual march of solar radiation input.

Sensible and latent heat in a spruce forest[*]

The Ebersberger forest is a seeded homogeneous spruce forest of about 400 ha of uniform age which was about 80 years old in 1972. The stand characteristics are: 800 stems/ha; 30 m average height of stand; zero point displacement parameter, 22.7 m (Jehle, 1968); roughness length, z_o, 3.1 m (Jehle, 1968).

Strauss (1971a) calculated the flux of sensible heat, L, and latent heat, V, according to the Sverdrup method for individual days and for the whole year (1969), using as reference level a plane in the average upper height of the stand. The data required to calculate the fluxes were measured by the sensors on a tower (Fig. 3.3) 114 times each hour, registered digitally, and reduced to hourly means. The fluxes directed towards the reference level were valued positively and the upward fluxes therefrom were valued negatively (Fig. 3.27). The fluxes for L and V are given in Fig. 3.28 and 3.29 and Table 3.13.

The total balance of energy fluxes for the year 1969, measured in kcal cm^{-2} yr^{-1}, as well as the percentage of the individual fluxes contained in net radiation Q, are shown in Table 3.13. Chemically fixed energy was not calculated. The monthly means of energy fluxes Q, L and V for the year 1969 are plotted in Fig. 3.28. B and P are not plotted; the value of P is negligibly

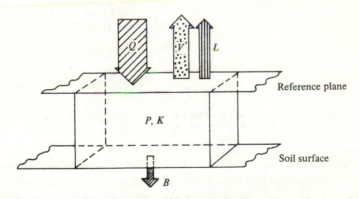

Fig. 3.27. Scheme of the energy exchange for the model volume of a stand. Q, net radiation; V, latent heat; L, sensible heat; B, heat flux into soil; P, into plants; K, chemical fixed energy.

[*]The values discussed in the following study were obtained under the guidance of Prof Dr A. Baumgartner in the Ebersberger forest IBP site near Munich latitude 48° 9′ N; elevation, 552 m).

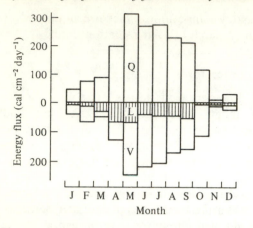

Fig. 3.28. Monthly means of the energy blance of a spruce forest in 1969 (Strauss, 1971*a*).

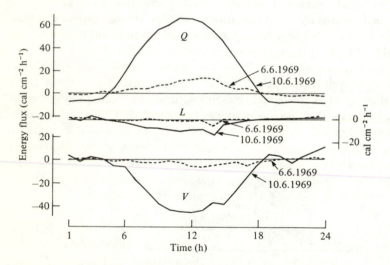

Fig. 3.29. Diurnal variation of the energy fluxes. *a*, net radiation, *Q*; *b*, sensible heat, *L*; *c*, latent heat, *V* on 6 June and 10 June, 1969. (Strauss, 1971*b*).

small, while the maximum value of *B* is 5% of net radiation. The diurnal variation of energy fluxes is depicted in Fig. 3.29. This figure shows fluxes *Q*, *L* and *V* on two days in June, with the sky being 10/10 cloudy for the whole day on 6 June 1969 and cloudless on 10 June 1969. In this conjunction it is interesting to note that at dawn net radiation becomes positive, while it becomes negative one hour before sunset. The reason is probably the long-wave emission of the forest which, because of greater surface temperature, is greater in the evening than in the morning. Moreover, at night, fluxes *V* and

158

Table 3.13. *Balance of energy fluxes for the entire year* (1969) *in the Ebersberger spruce forest*

	Q	V	L	B	P
kcal cm^{-2} yr^{-1}	54.6	-43.7	-10.3	-0.3	$+0.1$
%	100	80	19	1	

L can become positive, thus supplying energy to the forest (for example, by condensation).

Hager (1975) calculated the exchange coefficient for adiabatic stratification by applying both the Sverdrup method and the aerodynamic method, and calculated the fluxes of V and L from

$$F = Aa \cdot \text{grad } s, \qquad (3.50)$$

where F is the flux of any property in cm^{-2}.s^{-1}, Aa is the adiabatic exchange coefficient and grad s indicates the gradient of the property. Adiabatic exchange in the spruce forest was determined from

$$Aa_z = b \cdot u_z. \qquad (3.51)$$

In this equation b has the value 0.20345, and u_z is the wind speed at height z. Fig. 3.30 shows the regression of the turbulent exchange coefficient A across wind speed u.

Since it is difficult to measure very low wind speeds in the forest using cup anemometers, the measured wind speed u_z was converted into percentages of the wind speed of u_{40} (u_{40}, wind speed at a height of 40 m) (Table 3.14). The data in Table 3.14 allow derivation of the relation:

$$u_z = \frac{\% \text{ factor} \cdot u_{40}}{100}. \qquad (3.52)$$

With eqn 3.52 it is possible to calculate the adiabatic exchange coefficient, Aa, and at the same time the fluxes of L and V for very small wind speeds. Figs. 3.31 to 3.33 show the mean diurnal variations over a month for wind speed, air temperature and specific humidity in months. By means of eqn 3.52 and the flux eqn 3.50, the fluxes of L and V can be calculated for any height above the ground. If u_{40} is smaller than 150 cm s^{-1}, u_z in the forest must be calculated according to eqn 3.52. Assuming that the turbulent exchange A for momentum is also true for other properties of the air, the fluxes of other properties can be calculated in addition to L and V. Fig. 3.34 summarizes global radiation (sun and sky) above and within the Ebersberger spruce forest; Fig. 3.35 gives the distribution of mean net radiation in the forest.

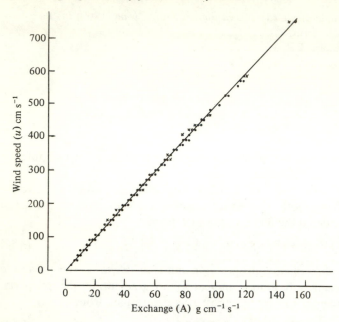

Fig. 3.30. Regression of the exchange coefficient, *A*, against wind speed, *u* (Hager, 1975). ●, observed & predicted; ○, predicted; ×, observed.

Table 3.14. *Correlation of wind speed,* u, *to height above the ground,* z, *in the Ebersberger spruce forest*

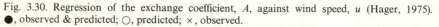

	height above the ground (m)							
z	0.5	2.5	5.0	7.5	10.0	15.0	20.0	40.0
u_z [in % of u_{40}]	1.9	4.0	5.8	4.9	3.7	1.6	0.8	100

Carbon dioxide balance in a spruce forest*

The Ebersberger spruce forest is an approximately 80-year-old stand of Norway spruce (*Picea abies* (L.) Karst.) within an even-aged and homogeneous 7000 ha tract of forest land. Vertical profiles of windspeed, wet and dry bulb temperatures, carbon dioxide concentration and short and long-wave radiation were measured continuously over long periods. (For a scheme of the instrumentation employed see Figs. 3.3 and 3.4). Results from

*The data and results presented here come from the IBP-woodlands site in the Ebersberger forest near Munich, and were gathered under the scientific direction of Prof. Dr A. Baumgartner. The project was funded by the Deutsche Forschungsge-meinschaft (DFG).

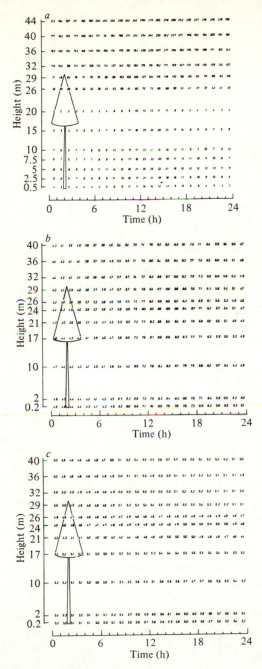

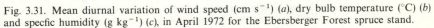

Fig. 3.31. Mean diurnal variation of wind speed (cm s^{-1}) (*a*), dry bulb temperature (°C) (*b*) and specfic humidity (g kg^{-1}) (*c*), in April 1972 for the Ebersberger Forest spruce stand.

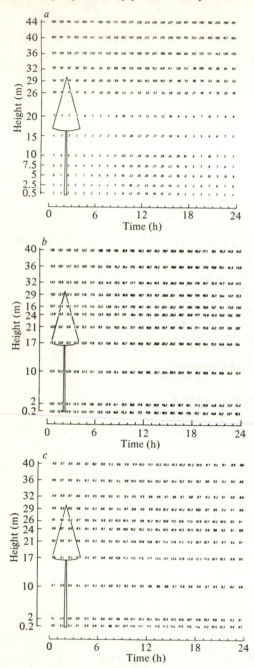

Fig. 3.32. Mean diurnal variation of wind speed (cm s^{-1}) (*a*), dry bulb temperature (°C) (*b*) and specific humidity (g kg^{-1}) (*c*) in August 1972 for the Ebersberger Forest spruce stand.

162

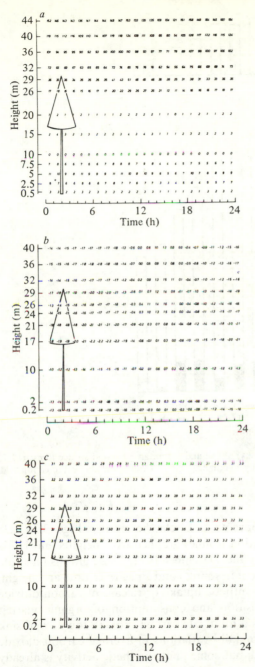

Fig. 3.33. Mean diurnal variation of wind speed (cm s^{-1}) (*a*), dry bulb temperature (°C) (*b*) and specific humidity (g kg^{-1}) (*c*) in December 1972 for the Ebersberger forest spruce stand.

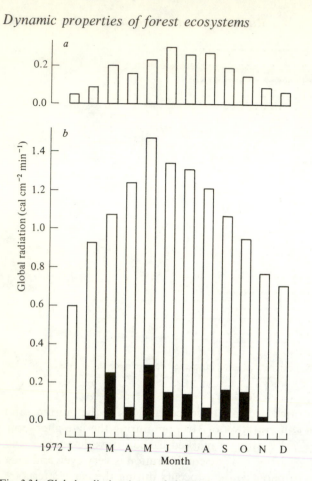

Fig. 3.34. Global radiation (sun and sky) above and within an 80-yr-old spruce forest (1972). *a*, monthly means; *b*, absolute monthly maxima. □, global radiation above the canopy; ■, global radiation 1.5 m above ground.

monitoring carbon dioxide concentrations can be seen in Figs. 3.36–3.38 which show average concentration isolines graphed over heights and hours for the months of April, August and December 1972. These graphs illustrate the effect of different carbon dioxide sources and sinks over time and height, which are due either to photosynthetic uptake or release of carbon dioxide by photorespiration, decomposition and consumption of organic matter. April 1972 (Fig. 3.36*a*) exhibits a moderate carbon dioxide source and sink activity and, therefore, no large daily variations in average carbon dioxide concentrations can be observed; but some photosynthetic activity is already indicated by the minimum concentrations around noon and during the afternoon. August 1972 (Fig. 3.26*b*) shows strong photosynthetic carbon

164

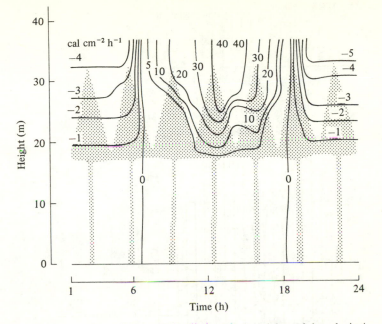

Fig. 3.35. Distribution of the mean net radiation of selected days of clear sky in August 1973 for the Ebersberger Forest, (Hager, 1975).

dioxide uptake, with a rapid decrease of concentrations in the morning and the absolute minimum concentration of 310 ppm during the late afternoon. December (Fig. 3.36c) lacks stronger marked carbon dioxide sources and sinks; the distribution of carbon dioxide concentration is very even and only a slight minimum due to photosynthesis occurs during the early afternoon hours.

After the evaluation of the turbulent exchange coefficients for different heights and hours by a modified aerodynamic method, carbon dioxide fluxes and their balances were calculated on an hourly basis. This yielded the carbon dioxide balance of the canopy stratum between 18.5 and 32.5 m height. The daily net carbon dioxide balance is graphed in Fig. 3.37 against daily net radiation and precipitation. Despite the strong variation of the daily balances, it is possible to observe nearly unchanged or only slightly decreasing monthly average carbon dioxide balances between April and June 1972. This is due to a change in primary productivity, resulting from the elongation and growth of the young spruce shoots during May and June. In July and August there occurs a strong rise in the carbon dioxide balances, with August 1972 being the most productive month, which is expressed in the carbon dioxide concentration isolines (see Fig. 3.36b). After August there is a strong drop in daily net carbon dioxide balances, with

165

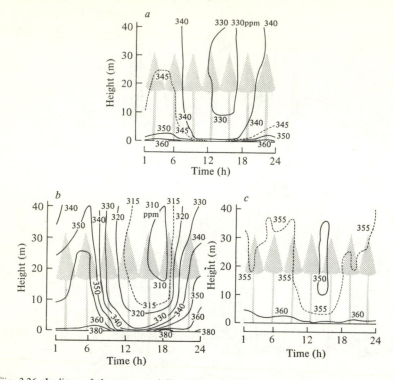

Fig. 3.36. Isolines of the average daily carbon dioxide concentration in their vertical distribution for *a*, April; *b*, August; *c*, December, 1972 for the Ebersberger spruce forest.

minimum values occuring during the month of December. The monthly net carbon dioxide balance remains positive in December, due to the ability of the spruce forest ecosystem to use favorable weather conditions for photosynthetic production even during the winter.

The average daily course of carbon dioxide balances of the spruce canopy can be seen in Fig. 3.38. An increase in carbon dioxide loss during the nighttime hours occurs from April until June, where a first maximum can be observed with a second maximum in September. Carbon dioxide gain by the canopy during daylight hours is moderate in April, shows a slight rise in May and a slight drop in June. After June a strong increase towards an August maximum takes place. Fall and winter months bring a rapid drop in daily carbon dioxide gains of the canopy. The numeric integration of Fig. 3.38 is shown in Table 3.15.

The sum of positive carbon dioxide balances represents an estimate of the daily gross assimilation rates, BAR, the sum of negative balances is an estimate of the respiration rate, RR, the difference of both represents the net assimilation rate, NAR, of carbon dioxide in the forest canopy. The

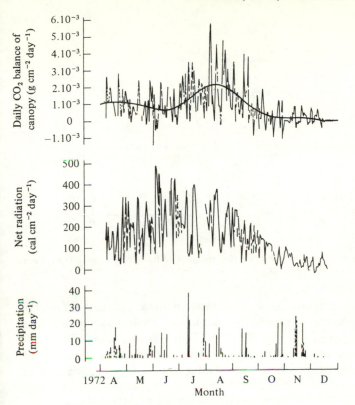

Fig. 3.37. *a*, daily net carbon dioxide balance of the canopy, ——— trend line of monthly mean; *b*, net radiation above the canopy; *c* daily precipitation during the year 1972 for the Ebersberger spruce forest.

absolute values of the monthly BAR show a steady increase from April until August, with the month of June as an exception. The observed dip in BAR in June is probably due to the growth of the young shoots in the spruce forest. After August a strong drop occurs from the month of September until November.

Respiration rates, RR increase to a first maximum in June (shoot growth) and reach a second maximum in August and September. Then the RR tapers off towards fall and winter. A similar behavior can be observed with NAR. For the whole year a net carbon dioxide consumption of the canopy stratum of 263.44 mg carbon dioxide cm^{-2} can be calculated, which is approximately equivalent to the production of 16 tons of organic dry matter ha^{-1} yr^{-1}.

This figure of the annual primary production of the forest canopy, which was obtained through the computation of turbulent carbon dioxide fluxes

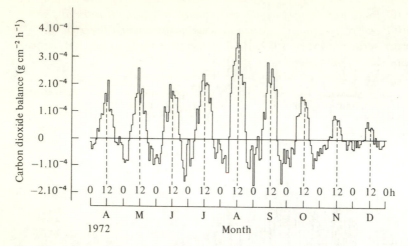

Fig. 3.38. Average daily course of carbon dioxide balances of the spruce canopy layer during the year 1972 for the Ebersberger forest.

and their balances, shows good agreement with a primary production evaluation by the harvesting method (von Droste, 1969), which gave a value of 15.5 tons of organic dry matter ha^{-1} yr^{-1} for a spruce stand in the same tract. This demonstrates that the computation of turbulent carbon dioxide fluxes and their balances can be a valuable tool for estimating organic matter production.

Radiant energy in the forest biomes

Budyko (1963, 74) was the first to describe the world-wide distribution of net radiation by maps that have been widely reproduced. We will try to refine the results of radiation exchange measurements at the level of the main forest biomes in the world, using a certain number of new measurements, mainly outside IBP sites, but also by interpreting data from IBP sites (Table 3.16).

It is well known that the largest values of net radiation are found over the oceans where cloud cover is frequent and where the losses to the cosmic space are low. This is of interest to us only in relation to coastal forests, which may have a higher net radiation than inland forests at the same latitude and altitude. The net radiation values on the continents scarcely amount to 100 kcal cm^{-2} yr^{-1} in equatorial rain forests. The minimum values for forest biomes may be about 15 kcal cm^{-2} yr^{-1} in the taiga north of the Arctic Circle. In such regions, even with continuous summer irradiation, positive net radiation throughout the year is small. The minimum winter value of net radiation shows a better correspondence with sun height

Table 3.15. *Daily and monthly sums and annual averages of gross and net assimilation and respiration of carbon dioxide in the canopy layer of the Ebersberger spruce forest in 1972. BAR, gross assimilation rate, RR, respiration rate; NAR, net assimilation rate*

		Average daily sums ($mg\ CO_2\ cm^{-2}\ d^{-1}$)	Monthly sums ($mg\ CO_2\ cm^{-2}\ month^{-1}$)	Percentages (%)
April	BAR	1.3056	39.17	100.00
	RR	0.1756	5.27	13.45
	NAR	1.1300	33.90	86.55
May	BAR	1.5380	47.69	100.00
	RR	0.6675	20.69	43.39
	NAR	0.8705	27.00	56.61
June	BAR	1.5089	45.27	100.00
	RR	0.8195	24.59	54.31
	NAR	0.6894	20.68	45.69
July	BAR	1.9550	60.62	100.00
	RR	0.4150	12.88	21.25
	NAR	1.5400	47.74	78.75
August	BAR	2.9784	92.33	100.00
	RR	0.8714	27.01	29.26
	NAR	2.1070	65.32	70.74
September	BAR	1.9740	59.21	100.00
	RR	0.9096	27.29	46.09
	NAR	1.0644	31.92	53.91
October	BAR	1.0650	33.00	100.00
	RR	0.7453	23.10	70.00
	NAR	0.3197	9.90	30.00
November	BAR	0.4817	14.45	100.00
	RR	0.2531	7.60	52.54
	NAR	0.2286	6.85	47.46
December	BAR	0.3153	9.77	100.00
	RR	0.2453	7.60	77.81
	NAR	0.0700	2.17	22.19
			Annual sum (weighted average of nine months; Oct–Dec given double weight) ($mg\ CO_2\ cm^{-2}\ yr^{-1}$)	(%)
Year	BAR	1.2487	457.02	100.00
	RR	0.5289	193.58	42.36
	NAR	0.7198	263.44	57.64

or short-wave incident radiation. In fact, the correlation between net annual radiation on continents and latitude all over the world is good and the regression line is given in Fig. 3.39.

Table 3.16. *Mean latitudinal distribution of annual net radiation totals over land, ocean, and the earth as a whole (kcal cm^{-2} yr^{-1}). After Budyko, 1974)*

Latitude (°N)	Net radiation			Latitude (°S)	Net radiation		
	Ocean	Land	Globe		Ocean	Land	Globe
70–60	23	20	21	0–10	115	72	105
60–50	29	30	30	10–20	113	73	104
50–40	51	45	48	20–30	101	70	94
40–30	83	60	73	30–40	82	62	80
30–20	113	69	96	40–50	57	41	56
20–10	119	71	106	50–60	28	31	28
10–0	115	72	105				
Earth as a whole	82	49	72				

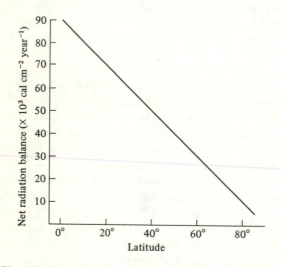

Fig. 3.39. Regression of annual net radiation balance in forest biomes on degree of latitude. $Q = -978.7 \cdot °Lat. + 89.999$; r, -0.962; R^2, 0.925; n, 17.

Equatorial forest biome

In tropical regions, the sun runs twice a year on the zenith and the course of solar and sky short-wave radiation is characterized by two minimum values (with small amplitude) every year. Total annual global radiation is about 150 kcal cm^{-2} and net (all wave) radiation about 80 kcal cm^{-2}. This value is high, because of low losses in long-wave radiation emitted by the earth-vegetation system, due to dense and regular cloud cover. For example, for

170

an identical solar and sky radiation, the value of net radiation may be higher over the equatorial forest than over the tropical dry forest with very high soil and vegetation temperature. The two maxima of net radiation occur in February–March–May and September–October–November, with monthly values mainly between 7000 and 8000 cal cm^{-2}. These maxima are not necessarily in correspondence with the two maxima of global radiation, because of variable incidence of cloud cover. Minimum monthly values may decrease to about 5000 cal cm^{-2}, sufficient for vegetative growth. By night, due to wetness and high evaporation rate, the temperature of surfaces is not particularly high and terrestrial radiation losses are strongly reduced (Fig. 3.40, Table 3.17).

Tropical forest biome

In the tropical forest biome, a course of net radiation similar to what has been described in equatorial forest biome occurs throughout the year, with two maximum and two minimum values. These maxima approach each other when latitude is higher. As the cloud cover decreases and the air is dry, very high values of global and net radiation may be found at first zenith position of the sun. These maximum values of net radiation may exceed 8000 cal cm^{-2} per month, which is a little more than that in the equatorial forest biome. The second maximum value of net radiation occurs later than the global radiation maximum. The hot surfaces of vegetation and soil cannot radiate enough through dry air to space to compensate for the high short-wave incoming energy, and net radiation values become higher than in the equatorial rain forest.

In the winter, monthly net radiation values may be three times lower than during the summer, because dry atmosphere radiation is low. Night-time net radiation may be strongly negative and frost may occur. Monthly minimum average values are about 3000 cal cm^{-2}. The highest night-time long-wave terrestrial radiations occur with the winter minimum of positive net radiation balance, and monthly values may be less than 3000 cal cm^{-2}. During summer-time, with cloudiness, night-time terrestrial radiation is not as large and radiation is lower than in winter-time. Yearly average short-wave incoming sum and sky radiation may reach 150–160 kcal cm^{-2} yr^{-1} as at the equator, with up to 90 kcal cm^{-2} during the growing season. The average annual value of net radiation is about 70–80 kcal cm^{-2} yr^{-1}.

Beyond the tropics, the one single maximum and one single minimum of global and net radiation are fixed by the maximum and minimum height position of the sun in the year. But in monsoon climates, two maximum values of global or net radiation are found during the year. This is the case for Chakia, India. The two periods with high night-time radiation losses are close to times of maximum values of global or net radiation (Figs. 3.40, 3.41, Table 3.17).

171

(a)

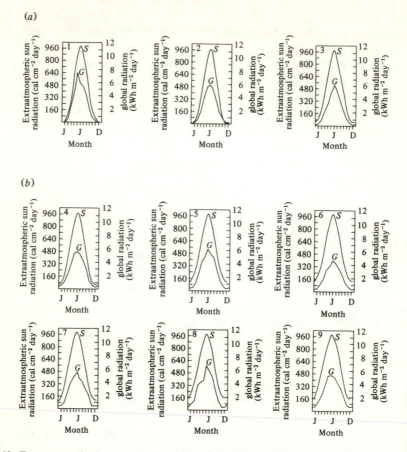

(b)

Fig. 3.40. Extra-atmospheric sun radiation and global radiation in the main forest biomes. *a.* Boreal. 1. Aklavik (69° N, 137° E (Collmann in Schulze, 1970)); 2. Onlu, Sodankyla (270 m; 66° 22′ N, 29° E (Havas, IBP 1973)); 3. Ultuna, 1963–72 (Hytteborn (1975)). *b.* Temperate. 4. Andersby, 1931–60 (30 m; 60° N, 17° E (Persson–Hytteborn, IBP 1973)); 5. Hestehaven, Odum 1966–71 (56° 18′ N, 10° 29′ E (Thamdrup–Petersen, IBP, 1973)); 6. Meathop, 1967–72 (45 m; 54° 12′ N, 2° 53′ E (Satchell, IBP, 1973)); 7. Hamburg (14 m (Collmann in Schulze, 1970)); 8. Bialowieza, 1968 (158 m; 52° 45′ N, 23° 52′ E (Falinsky, IBP, 1973)); 9. Virelles, 1964–9 (245 m;

Mediterranean forest biome

In the mediterranean forest biome, strong cloudiness occurs when the sun remains low on the horizon (wet winter); one single maximum value of net radiation is found at maximum sun height with cloud-free skies. This maximum is somewhat delayed from maximum zenith position of the sun. Monthly maximum values in summer-time of 9000 cal cm^{-2}, with dry air above, are probably among the highest measured in the world. At this

(c)

(d)

(e)

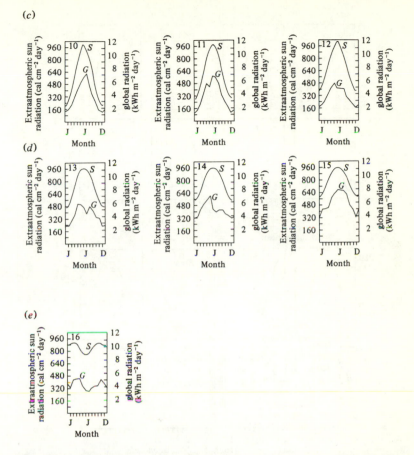

50° N (Galoux, 1973)); *c*. Mediterranean. 10. Rouquet (180 m; 49° N (Lossaint, 1973)); 11. Elba, Marciana–Marina, 1972–3 (43° N (Susmel & Sala (1975)); 12. Shigayama, 1968–72 (1790 m; 36° N (Kitazawa, 1973)); *d*. Tropical. 13. New Orleans (30° N (Collmann in Schulze, 1970)); 14. Chakia, 1971 (350 m; 25° N (Bandhu 1973)); 15. Honolulu (21° N (Collmann in Schulze, 1970)); *e*. Equatorial. 16. Kinshasha (4° N (Collmann in Schulze, 1970)). *S*, extra-atmospheric sun radiation. *G*, global radiation at the surface of the earth.

latitude, the monthly minimum value drops to 600 cal cm^{-2}; negative monthly values are not observed. The minimum night-time radiative value occurs at the same time as the maximum daily value – when surfaces are very hot and have very high radiation losses. Minimum night-time radiative losses occur during rainy and cloudy winters. Yearly global radiation is about 130–50 kcal cm^{-2} yr^{-1}, with 60–70 kcal cm^{-2} during the vegetative period and net radiation of about 60 kcal cm^{-2} (Figs. 3.40 and 3.41, Table 3.17).

Dynamic properties of forest ecosystems

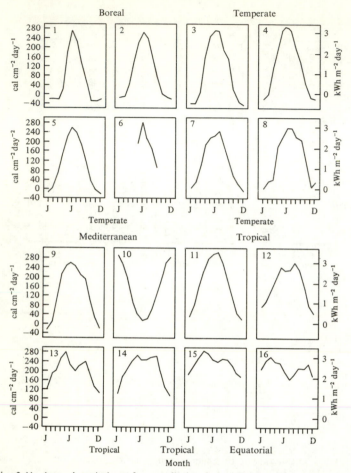

Fig. 3.41. Annual variation of net radiation balances (monthly mean values of daily sums, mostly from Kessler, 1973, but also from Strauss, 1971a, b, Kiese, 1972, Galoux, 1973) and, Baumgartner and Hager, 1973).
a. Boreal. 1. Verkoyansk 1964–70 (137 m, 67° 33′ N, 133° 23′ E); 2. Leningrad (60° N, 30° E).
b. Temperate. 3. Omsk 1964–70 (119 m, 54° 56′ N, 73° 24′ E); 4. Irkutsk 1964–70 (467 m, 52° 16′ N, 104° 21′ E); 5. Hamburg 1964–71 (14 m, 53° 38′ N, 10° 00′ E); 6. Solling 1970 (51° 15′ N); 7. Virelles 1967–9 (245 m, 50° 04′ N, 04° 22′ E); 8. Munich 1969 (45° 15′ N).
c. Mediterranean, 9. Vladivostok 1963–70 (80 m, 43° 07′ N, 131° 54′ E); 10. Aspendale 1964–70 (6 m, 38° 02′ S, 145° 06′ E); 11. Grendi 1964–8 (135 m, 35° 50′ N, 14° 26′ E).
d. Tropical. 12. New Delhi 1965–70 (212 m, 28° 35′ N, 77° 12′ E); 13. Poona 1964–6 (555 m, 18 32′ N, 73 51′ E); 14. Dakar 1965–70 (20 m 16° 44′ N, 17° 28′ W); 15. Khormaksar 1964–7 (3 m, 12° 49′ N, 45° 02′ E).
e. Equatorial. 16. Yangambi 1957–9 (485 m, 0° 49′ N, 24° 29′ E).

Temperate broad leaved and coniferous biomes

In temperate regions, the highest net radiation values occur with high short-wave radiation values during the summer-time. Maximum values often are

174

Table 3.17. *Global radiation and net radiation balance values at various sites in the main forest biomes. Descriptions of IBP Sites given in Chapter 11*

Site	Data from:	Altitude (m)	Country	Latitude	Longitude	Global radiation (yr)	Radiation balance (yr)	Global radiation (growing season)	Radiation balance (growing season)
						cal cm^{-2} yr^{-1}			
IBP sites									
Pasoh	J. A. Bullock	100	Malaysia	2° 59' N	102° 18' E	130140		130140	
Lubumbashi	F. Malaisse	1208	Zaïre	11° 29' S	27° 36' E	168265	93509		
Chakia	D. Bandhu	350	India	25° 20' N	83° E	155535			
Le Rouquet	Lossaint, Rapp	180	France	48° 56' N	1° 61' E	124282			
Watershed 1 Coweeta	W. T. Swank	706–988	N. Carolina, USA	35° 04' N	83° 26" W	149830		67412	
Watershed 18 Coweeta	C. D. Monk, F. P. Day	726–993	N. Carolina, USA	35° 03' N	83° 26' W	111129		63697	
Tigrovaya Balka	Molotovsky		USSR	37° 20' N	68° 30' E	109900			
Virelles	P. Duvigneaud, A. Galoux	245	Belgium	50° 04' N	4° 22' E	93131	41150	57198	27875
Shigayama	Y. Kitazawa	1790	Japan	36° 40' N	138° 30' E	117377		69984	
Thompson Res. Site. Seattle	S. P. Gessel, P. Sollins	210	Washington, USA	47° 23' N	121° 57' W		99280	70557	
Andersby	H. Hytteborn	30	Sweden	60° N	17° E	81000			
Linnebjer	F. Andersson	60	Sweden	55° 44' N	13° 18' E	90480		76300	
Kongolound	B. Nihlgard	120	Sweden	55° 59' N	13° 10' E	90000		70000	
Bialowieza	J. B. Falinski	158	Poland	52° 45' N	23° 52' E	97144		79107	
Koiwai	T. Satoo	360	Japan	39° 45' N	141° E	120085			
Meathop Wood	Satchell	45	United Kingdom	54° 12' N	2° 53' E	72018		62287	
Sinaia 1, 2	Popescu, Zeletin, Bindiu, Mocanu	950	Roumania	45° 23' N	23° 15' E	117000		83500	
Hestehaven	H. M. Thamdrup	11–28	Denmark	56° 18' N	10° 29' E	88362		65520	

(Table 3.17. *continued*)

Site	Data from:	Altitude (m)	Country	Latitude	Longitude	Global radiation (yr)	Radiation balance (yr)	Global radiation (growing season)	Radiation balance (growing season)
						cal cm^{-2} yr^{-1}			
Noe Woods	C. L. Loucks, G. J. Lawson	274	Wisconsin, USA	43° 02' N	89° 24' E		40000		
Walker Branch, Oak Ridge	Hutchison, Harris, Edwards, Reichle	265–360	Tennessee, USA	35° 58' N	84° 17' W	116000	74500		
Andrews, Exp. For.	C. C. Grier	430–670	Oregon, USA	44° 15' N	122° 20' W	124100	76500		78089
Solling	H. Ellenberg	430	West Germany	51° 45' N	9° 34' 0	60000	30000		
Munich	A. Baumgartner	552	West Germany	48° 15' N	11° 13' 0		54600		
Koinas Arkangelsk region	Rudneva, Tonkonogov, Dorochov		USSR	64° 40' N	47° 30' E		27000		
Central Forest Reserve	V. G. Karpov	200	USSR	56° 30' N	32° 40' E		35000		
Site 1 and 17	N. I. Kazimirov, R. M. Morozova	200	USSR	62° N	34° E	72100	31500	51400	
Tigrovaya Balka	Y. I. Molotovsky	200	USSR	37° 20' N	68° 30' E	109900			
Oulu	P. Havas	270	Finland	66° 22' N	29° E	75139	15630	46800	
Other sites									
Manaus		45	Brazil	3° 08' S	60° 01' W		73550		
Yangambi		487	Zäire	0° 49' N	24° 29' E		78089		
Belize		5	Br. Honduras	17° 32' N	88° 10' W		80950		
Dakar		20	Senegal	14° 44' N	17° 28' W		73710		
Khormaksar		3	Yemen	12° 49' N	45° 02' E		85510		
Poona		555	India	18° 32' N	73° 51' E		71743		
New Delhi		212	India	28° 35' N	77° 12' E		59929		
Shangai			China	31° 14' N	121° 27' E		49550		
Saigon		8	Vietnam	10° 47' N	106° 42' E		77300		
Aspendale		6	Australia	38° 02' S	145° 06' E		54106		
Qrendi		135	Malta	35° 50' N	14° 26' E		61057		
Roma		51	Italy	41° 54' N	12° 30' E		51100		
Lisbon		100	Portugal	38° 42' N	9° 08' W	120007	56150		
Supramonte di Orgosolo		1000	Italy (Sardinia)	40° 05' N	9° 08' E	130888			
Hamburg		14	Germany	53° 38' N	10° 00' E		37274		
Vladivostok		80	USSR	43° 07' N	131° 54' E		46636		
Irkutsk		467	USSR	52° 6' N	104° 21' E		43423		

delayed from maximum short-wave incident radiation. Maximum monthly values hardly exceed 8000 cal cm^{-2}, except in temperate continental regions. Opposite to the mediterranean regime, minimum night-time values of net radiation occur in the winter-time, mostly in February in the northern hemisphere with clear sky (polar air); monthly negative values occur frequently and may reach -500 cal cm^{-2}. Radiative losses are reduced to a minimum in the vicinity of oceans, where the atmosphere has a high vapor content. Total annual values of global radiation are about 85–110 kcal cm^{-2} (80% during the growing season); for net radiation, values are about 40–70 kcal cm^{-2} or half the value of the equatorial forest which has twice as long a growing season as the temperate deciduous forest. In monsoon climates, two maximum values of radiation are found, because of rainy and cloudy periods in coincidence with high sun position on the horizon. This is the case of Shigayama (Japan) with the high rainfall in June that depresses the radiation value and causes two maximum radiation values in May and July (Figs. 3.40 and 3.41, Table 3.17).

Boreal coniferous forest biome

In the boreal forest biome, the highest net radiation value occurs with maximum values of short-wave radiation similar to temperate regions. Although sun and sky radiation are not high, maximum monthly values may exceed these of the temperate deciduous biome, because of the long day length in high latitudes. So we may have almost 9000 cal cm^{-2} in June. But the length of the period with positive values is smaller; consequently, the annual course of the net radiation curve has a sharper form. Snow cover during winter-time, with a high albedo coefficient for short-wave radiation, tends to decrease radiation to a certain extent. This fact, associated with high radiation losses through dry air of polar anticyclone weather, is responsible for the low monthly net radiation values in winter that approach -800 cal cm^{-2}. Annual global radiation may vary from 80 to 70 kcal cm^{-2} yr^{-1} (40–50 kcal in the growing season), and annual net radiation values from 30 to 15 kcal cm^{-2} yr^{-1}, with high values in north Siberia (43 kcal cm^{-2} yr^{-1}) (Figs. 3.40 and 3.41, Table 3.17).

Water balances in the main forest biomes

Even though the study of water balances of forest ecosystems has been the subject of research for many years, there are not many complete balances for forests during the vegetative period. Most of the older data have come from the Soviet Union (Table 3.17); new advances have been made during this last decade, but on the whole, the available data are not numerous and are often incomplete. Interpretation and integration of these data, with a view to better comparative analyses of water balances for large forest biomes, remain very preliminary.

Very few data are available for the tropical and equatorial forest; the best data are for the boreal coniferous forest and, on a smaller scale, the temperate deciduous forests and the mediterranean oak forest. Therefore this synthesis is directed more towards the annual course of the water balance within the large forest biomes than towards the comparison of the absolute values obtained in each of these large vegetation formations. The results obtained during the IBP are presented in Table 3.19–3.24; they are complemented with information from the literature.

Water balance at the surface of continents

Table 3.18 (Budyko, 1974) gives a summary of the water balance at the surface of continents. In this table, rainfall in the Antarctic is not taken into account, except for the whole of the continents. The relationship between rainfall and evaporation is very different from one continent to another. Australia, and to a lesser extent Africa, is characterized by high evaporation rates (respectively, 89.4 and 62.3% of precipitation). This is because the equatorial forest biomes with a moderate evapotranspiration rate represent only 4% and 13% of the surface of these continents, respectively. On the other hand, in the same southern hemisphere South America, with 53% of its surface covered by the equatorial forest biome, has an evapotranspiration of 42.9% of precipitation. In Australia and Africa, that part of the net radiation balance used in evaporation is lowest (31.4 and 38.2%, respectively). Therefore, Australia and Africa are characterized, on the whole, by more arid climatic conditions than South America. Budyko's radiative index of aridity calculated for the different continents is also presented in Table 3.18.

In such a large territory as a continent, the vegetation cover is diverse. The limits of these large geobotanical zones correspond, according to Budyko (1974), to isolines of the radiative index of aridity. The most humid conditions correspond to the lowest values of this index, and the deserts and semi-deserts correspond to the highest values. Thus, isoline 0.3 defines the limit of tundra and forests, isoline 1.0 forest and grassland, and isoline 2.0 grassland, deserts and semi-deserts. Within the forested regions of the earth, the values of the net radiative balance determine the geographical distribution of the five large forest biomes.

Equatorial forest

In the equatorial forest, precipitation is very great (1500 to 2000 mm yr^{-1}) and more or less well distributed during the course of the year. The soil is

Table 3.18. *Water balance on the continents after Budyko, 1974*

Continent	Precipitation (mm yr^{-1})	Evaporation (mm yr^{-1})	Evaporation (%)	Runoff (mm yr^{-1})	Runoff (%)	Radiative index of aridity $\left(\dfrac{\text{net radiative balance}}{\text{latent ht of vaporization} \times \text{precipitation}}\right)$
Europe	640	390	60.9	250	39.1	1.036
Asia	600	310	51.7	290	48.3	1.211
North America	660	320	48.5	340	51.5	1.030
South America	1,630	700	42.9	930	57.1	0.730
Africa	690	430	62.3	260	37.7	1.675
Australia	470	420	89.4	50	10.6	2.532
Total land surface	730	420	57.5	310	42.5	

179

constantly wet. As water supply is not a limiting factor, evapotranspiration will depend mainly on the quantity of available energy. An important part of the net radiation balance is that used in evaporation (87.3% in Manaos for the Amazonian Forest, and 73.1% in Saigon for the equatorial Monsoon forest in South Vietnam).

In the evergreen rain forests, the annual course of evaporation practically parallels the net radiative balance, and shows two maxima at the equinoxes (Fig. 3.42*a*). In the Monsoon forests (Fig. 3.42*b*), characterized by a period of drought at the beginning of spring, evaporation shows an important decrease that reaches its minimum at the end of the drought period when the water supply in the soil is very low. Evaporation increases again at the beginning of the monsoon, during which its level becomes steady at about 80 mm per month. In the El Verde forest, on Puerto Rico Island, rainfall is clearly higher than the average for an equatorial area (Fig. 3.42*c*). Evaporation slightly exceeds the runoff (*V*, 55.4%), and its annual course follows closely that of rainfall, except in August where minimum evaporation coincides with maximum rainfall.

It is well established that in equatorial forests, *sensu stricto*, the runoff function is continuous and regularly feeds the uninterrupted flow of streams during the course of the year. Table 3.19 shows the annual values of the water balance for a few equatorial stations; IBP sites are marked with an asterisk.

Tropical forest

In tropical areas, the components of the water balance can vary widely from one region to another, and they restrict the extension and the nature of various types of vegetation. On the average, between 1000 and 1500 mm rainfall occurs annually, but the distribution is very irregular; indeed, during one or two drought periods a marked water deficit will occur. At the beginning of the main drought period, the well-saturated soil from preceeding months of rainfall dries up very quickly because the saturation deficit or water potential is relatively high. The amplitude of the water deficit increases with the length of the drought period. Runoff is restricted to rainy periods when the soil water supply is restored, and the flow of streams is very irregular. For most streams, discharge is practically zero in the drought period. Table 3.20 summarizes the data on water balance in various natural and artificial stands in tropical forest regions.

Mediterranean vegetation

In the biome of the evergreen sclerophyllous forest of the mediterranean region, the average annual water supply shows an important geographical variation of rainfall (400–1300 mm) and also sharp fluctuations from year to

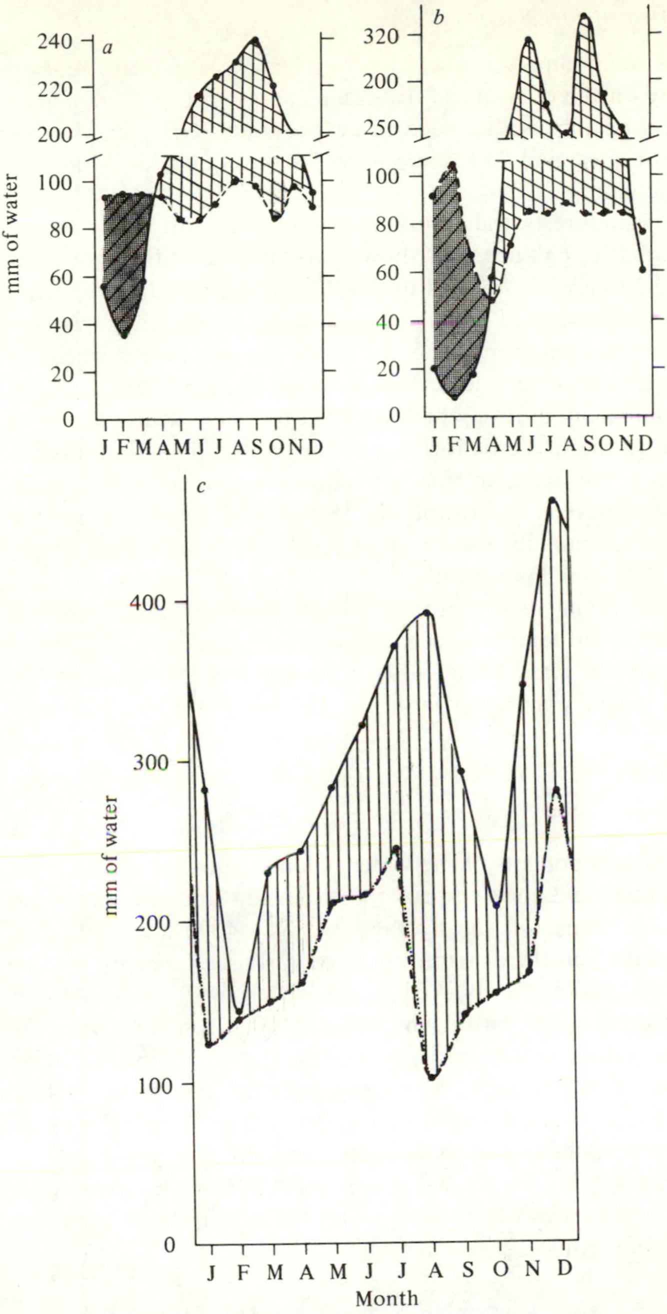

Fig. 3.42. Annual course of the water balance for three stations located in the evergreen rain forest zone: *a*, Manaos (Brazil), from Budyko (1958); *b*, Saigon (South Vietnam) from Budyko (1958); and *c*, El Verde (Puerto Rico) from Odum *et al.* (1970). (——), rainfall; (----), actual evapotranspiration; (····), potential evapotranspiration.
▩, water deficit; ▥, drainage; ▧, soil water recharge; ▨, soil moisture depletion.

Table 3.19. *Water balance of various equatorial rain forest types*

Forest type	Site (*denotes IBP site)	Year	Precipitation (P, mm)	Actual evapotranspiration				Runoff			Through-fall (% P)	Stem flow (% P)
				Interception (% P)	Transpiration (% P)	Soil evaporation (% P)	Sum (% P)	Drainage (% P)	Surface runoff (% P)	Sum (% P)		
Sub-equatorial forest (Huttel, 1975)	Yapo (*) (Côte d'Ivoire)	1969–71	1950.0	22.0	—	—	73.0	—	<1.0	27.0	—	→78.0←
	Banco (*) (Côte d'Ivoire)	—	—	—	—	—	—	—	—	—	—	—
Tropical rain forest (Odum, Moore & Burns 1972)	Plateau	1969–71	1800.0	10–12	—	—	64.0	—	<1.0	36.0	—	→88–90←
	Talweg	1969–71	1800.0	10–12	—	—	66.0	—	<1.0	34.0	—	→88–90←
	El Verde (Puerto-Rico)	—	3759	26.4	—	—	55.4	—		44.6	55.6	18.0
Alstonia scholaris plantation (Banerjee, 1973)	Arabari Range (Bengal)	10/1/70–9/30/71	1622.8	21.3	45.4	35.0	100.0	0	0	0	—	—
Alstonia scholaris plantation (Ray, 1970)	Arabari Range (Bengal)	—	—	21.6 to 36.3	—	—	—	—	—	—	48.7–59.1	13.4–22.9
Shorea robusta plantation (Ray, 1970)	Arabari Range (Bengal)	—	—	16.5 to 35.4	—	—	—	—	—	—	59.3–74.2	5.3–10.1
Shorea curtisii forest (Low, 1972)	Sungai Lui Catch (W. Malaysia)	9 months	—	36.0	—	—	—	—	—	—	—	—
Tropical moist forest (McGinnis et al., 1969)	(Panama)	—	—	17.0	—	—	—	—	—	—	—	—
Tropical wet forest (McColl, 1970)	(Costa Rica)	—	—	—	—	—	—	—	—	—	94.5	—

→values← are sums of colums 12 and 13.

Table 3.20. *Water balance of various tropical forest types*

Forest type	Site	Year	Precipitation (P, mm)	Interception (% P)	Actual evapotranspiration			Runoff			Through-fall (% P)	Stem flow (% P)
					Transpiration (% P)	Soil evaporation (% P)	Sum (% P)	Drainage (% P)	Surface runoff (% P)	Sum (% P)		
Pine Forest (Smith, 1974)	Lat. 33° 24′ S Long. 150° E (Australia)	10/17/68–4/28/71	861.6	18.7	—	—	—	—	—	—	78.3	<3
Eucalypt Forest Smith, 1974	Lat. 33° 24′ S Long. 150° E (Australia)		861.6	10.6	—	—	—	—	—	—	86.4	<3
Pinus longifolia plantation (Dabral et al. 1968)	Dehra Dun (India)	Summer monsoon, 1960	—	22.1	—	—	—	—	—	—	74.3	3.6
Tectona grandis plantation (Dabral & Rao, 1968)	Dehra Dun (India)	Summer monsoon, 1960	—	27.0	—	—	—	—	—	—	69.7	3.3
Shorea robusta plantation (Dabral & Rao, 1969)	Dehra Dun (India)	Wet period, 1961	2159	38.2	—	—	—	—	—	—	54.6	7.2
Acacia catechu plantation (Dabral et al. 1969)	Dehra Dun (India)	Wet period, 1961	2159	28.5	—	—	—	—	—	—	67.3	4.2
Eucalyptus globulus plantations Thomas, Chanara Sekhar & Haldorai, 1972	Nilgiris water sh. (India)	—	1300	—	26.7	—	—	—	—	—	—	—
Tropical forest (Hopkins, 1960)	Mpanga Res. For. (Uganda)	Two wet periods	1130	33.6	—	—	—	—	—	—	66.4	—

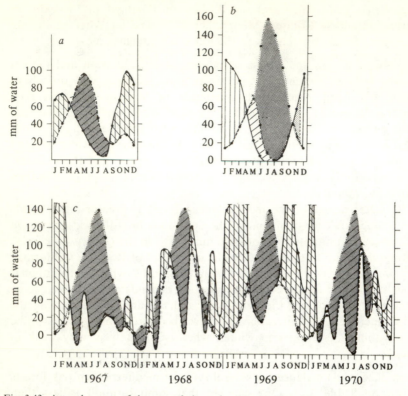

Fig. 3.43. Annual course of the water balance for three stations located in the mediterranean vegetation biome: *a*, Lisbon, Portugal, from Budyko (1958); *b*, Rocklin, California, from Major (1967); *c*, Le Rouquet, France, from Ettehad *et al.* (1973). The symbols used are as for Fig. 3.42.

year (1628 mm in 1969 and 719 mm in 1970 in the Rouquet Station near Montpellier, Ettehad, Lossaint & Rapp, 1973). What makes this biome different from the temperate deciduous forest is the seasonal distribution of precipitation. Most of the rainfall occurs in autumn and spring, whereas the summer months are characterized by a more or less marked and long drought. Showers are more periodic, the intensity of rainfall is less, and there are fewer rainy days. Every year the mediterranean forest goes through a period of summer drought, varying between 3 and 7 months (April to October). The soil-water supply is rapidly exhausted in spring (April and May), and the wilting point is reached every year for a relatively long period. Water availability is the main limiting factor of the intensity and phenology of vegetation.

The course of annual, seasonal, and monthly fluctuations of the main components of the water balance is shown in Figs. 3.43*a,b* and *c*. These data were gathered from three stations: Lisbon in Portugal (*a*), Rocklin in California (*b*) and le Rouquet-Montpellier in France (*c*). After examining these graphs closely, one comes to the following conclusions:

Following closely the course of incident global radiation and the net radiative balance, potential evapotranspiration increases very quickly in March and April, reaching its maximum in June or July and then regularly decreasing until the end of the year. Where the water potential of the air reaches its maximum in months with little rainfall, the water deficit shows up as soon as April or May. At this time the water demand is no longer satisfied by the atmospheric water supply, and the available soil-water supply is quickly exhausted. Actual evapotranspiration is then limited by rainfall.

The water deficit of the mediterranean forest is greater and extends over a longer period than that of the deciduous broadleaved forest. On the average, it amounts to about 430 mm (46%) in the *Quercus ilex* forest in Rouquet and 530 mm (62%) in the chapparal of Rocklin, California.

Runoff is practically restricted to winter months, when water input is rather strong and actual evapotranspiration depends on available energy at the site. Considering the periodicity and the amount of rainfall, the runoff function is most irregular. The river bed, broad and stony, can be filled within a few hours in winter, whereas the flow is very weak in summer.

A complete water balance for several types of mediterranean oak forests is shown on Table 3.21. One can see that annual runoff is less than evapotranspiration. On a percentage use basis of water supply, the ever-green mediterranean forest is not very different from that of a temperate deciduous forest. This can be explained by the fact that the runoff factor in the mediterranean biome is restricted to a short winter period with high water and weak energy availability.

Temperate forest

The deciduous forest receives, on an average and according to the location, between 500 and 1500 mm rainfall a year. A slight minimum occurs in the summer in areas under oceanic influence and in the winter in areas with a more continental character. Therefore, the supply of soil water through precipitation depends on the frequency of cyclonic rains. Precipitation is nearly always sufficient in the Appalachian Forest (eastern US) and in the temperate forests of western Europe.

In the areas where continental influence is stronger, rainfall can not recharge the soil water supply adequately to satisfy the demand; there is then a water deficit that tends to increase and expand in years with little

Table 3.21. *Water balance of various mediterranean forest types*

Forest type	Site (*denotes IBP site)	Year	Precip- itation (P, mm)	Intercep- tion (% P)	Actual evapotranspiration			Runoff			Through- fall (% P)	Stem flow (% P)
					Transpi- ration (% P)	Soil eva- poration (% P)	Sum (% P)	Drainage runoff (% P)	Surface runoff (% P)	Sum (% P)		
Quercus ilex forest (Ettehad et al., 1973)	Le Rouquet (*) France	1967–70	987.7	30.7		→20.3←	57.0	41.4	0	41.4	65.1	4.2
	La Madeleine (*) (France)	1966–70	636.6	33.7		→30.9←	64.6	26.6	0	26.6	61.2	5.0
Foothill woodland (Major, 1967)	Rocklin (California)	—	569.0	—		—	56.9	—	—	43.1	—	—
Chaparral (Rowe, Storey & Hamilton 1951)	North Fork (California)	1934–8	1149.6	5.0		→37.2←	42.2	57.8	Trace	57.8	80.0	15.0
Chaparral (Rowe, Storey & Hamilton 1951)	Tanbark Flat (California)	1942–5	528.6	11.1		—	—	—	—	—	80.5	8.4

→values← are sums of columns 6 and 7.

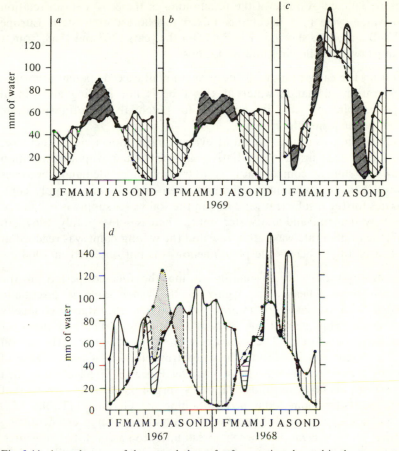

Fig. 3.44. Annual course of the water balance for four stations located in the temperate forest biome: *a*, Paris, France, from Budyko (1958); *b*, Kongalund (Sweden) Norrvidinge and Sodervidings 1969–71 for precipitation, Svalov 1931–60 for actual evapotranspiration according to Aslyng (1965) courtesy of Nihlgard (IBP, 1973); *c*, Munich, Germany, from Strauss (1971); and *d*, Virelles–Blaimont, Belgium, from Schnock (1970). Symbols as in Fig. 3.42.

summer rainfall. The annual course of the water balance in temperate deciduous forests is shown in Fig. 3.44*a,b,c* and *d*. The selected stations include Paris (France), Kongalund (Sweden), Munich (Germany) and Virelles (Belgium) and show a water shortage that is more or less typical of location and season (generally between May and September). During this growing period, evapotranspiration depends directly on the available water supply, whatever the available quantity of energy. In winter dormancy, on the contrary, rainfall is excessive and energy limits evapotranspiration; water surplus is removed through drainage towards the water table and by surface flow into streams.

187

A more detailed example of the functioning of the ecosystem in relation to the water cycle in a deciduous mixed oak forest of western Europe (Virelles–Blaimont) is shown in Fig. 3.44*d* for the years 1967 and 1968; from it one can draw the following interpretations:

> During the early months of the year, rainfall exceeds potential evapo-transpiration (demand); water at this time is not limiting and actual evapotranspiration can be equated to potential evapotranspiration. Excess water forms the runoff.

> With spring warming, potential evapotranspiration quickly increases with the increase in the net radiative balance. Actual evapotranspiration follows a similar course until evapotranspiration (demand) becomes greater than precipitation water input; then the soil water supply compensates for the insufficient rainfall and the soil water supply is exhausted to satisfy demand and the water deficit increases (May–July 1967). In 1967, the rainfall rate was rather low and the wilting point was reached at the end of July, whereas the phenomenon was not observed in 1968.

Evapotranspiration occurs mainly during the foliated period in the deciduous forest, whereas recharging of the water table is essentially associated with a season of minimal radiant energy. Man has locally substituted artificial plantations of conifers (mainly spruce and pine) for natural hardwood forests. The consequences of this change reverberate on the water balance by increasing the evapotranspiration at the expense of recharging ground water. Benecke & Van Der Ploeg (1975) have established that, for 1969, actual evapotranspiration of spruce plantations exceeds that of beech forests by about 100 mm (IBP Solling site, Germany) (Tables 3.22, 3.23). The coniferous stand intercepts the rainfall to a large extent during the entire year, whereas the beech forest intercepts very little in winter. Transpiration by vegetation and drainage to the water table are less in the spruce than in the beech forest (Tables 3.22, 3.23). Values for water balances in deciduous temperate forests and in artificial conifer stands in temperate deciduous forest areas are relatively numerous. Tables 3.22 and 3.23 do not cover all the published results. Additional information may be obtained from Rutter (1968), Mitscherlich (1971) and Molchanov (1960, 1971).

Boreal coniferous forest

The boreal coniferous forest (taiga) is well adjusted to a very long and very cold winter, and to a soil that remains frozen nearly the entire year. Precipitation, a part of which is snow, can vary on an annual basis between 400 and 700 mm. Maximum precipitation occurs during summer months. The taiga has a relatively short vegetative period. The soil water is only available from spring onwards (snow melt) when the soil is well saturated

Table 3.22. *Water balance of various temperate deciduous forest types*

Forest type	Site (*denotes IBP site)	Year	Precipitation (P, mm)	Actual evapotranspiration				Runoff			Through-fall (% P)	Stem flow (% P)
				Interception (% P)	Transpiration (% P)	Soil evaporation (% P)	Sum (% P)	Drainage (% P)	Surface runoff (% P)	Sum (% P)		
Mixed oak forest (Schnock, 1970)	Virelles (*) (Belgium)	1964-8	965.9	17.0		35.5	52.5	47.0	0	47.0	76.8	6.7
Oak-hornbeam forest (G. Schnock & A. Galoux (unpublished))	Ferage (*) (Belgium)	1964-8	898.0	30.7	—	—	—	—	—	—	65.6	3.7
Oak-hornbeam forest (G. Schnock, R. Dalebroux & A. Galoux (unpublished))	Virelles (*) (Belgium)	1964-8	994.0	16.6	—	—	—	—	—	—	80.3	3.1
Oak-hazel forest (G. Schnock & A. Galoux (unpublished))	Ferage (*) (Belgium)	1964-8	898.0	28.3	—	—	—	—	—	—	68.5	3.2
Oak-hornbeam forest (Intribus, 1975)	Bab (*) (Czechoslovakia)	11/71-10/72	771.9	17.8	—	—	—	—	—	—	78.6	3.6

189

(Table 3.22 continued)

Forest type	Site (*denotes IBP site)	Year	Precipitation (P, mm)	Actual evapotranspiration				Runoff			Through-fall (% P)	Stem flow (% P)
				Interception (% P)	Transpiration (% P)	Soil evaporation (% P)	Sum (% P)	Drainage (% P)	Surface runoff (% P)	Sum (% P)		
Oak-hornbeam forest (Ukrecky, Smolik & Lanar, 1974)	Brno (*) (Czechoslovakia)	4/71–10/71	38.5	24.5	—	—	—	—	—	—	75.5	—
Mixed deciduous forest (White & Carlisle, 1968)	Meathop (*) (United Kingdom)	12/66–11/67	1554.4	14.2	—	—	—	—	—	—	77.3	8.5
Oak–ash–lime forest (Molchanov, 1971)	Tellermanovsky (*) (USSR)	1952–61	513.0	14.4	48.5	21.5	84.4	5.2	10.4	15.6	—	—
Oak (Q. daschorochensis forest (Cepel, 1967)	Near Istamboul (Turkey)	5 yr	1032.7	20.0	—	—	—	—	—	—	69.1	10.9
Oak-pine forest (Bodeux, 1954)	Campine (Belgium)	1950	849.6	26.8	—	—	—	—	—	—	65.5	7.7
Old birch coppice (Noirfalise, 1959)	Ottignies (Belgium)	1945–6 (one year)	876.9	29.8	—	—	—	—	—	—	63.9	6.3

Beech forest (Benecke & van der Ploeg, 1975)	Solling (Germany)	1/1/69 10/31/69	896.3	11.8	35.6	47.4	55.7	0	55.7	88.2
Beech forest (Lemee, 1974)	Fontainebleau* (France)	8/19/70– 8/16/73	554.0	—	—	—	—	—	—	74.4	—
Beech–hornbeam forest (Aussenac, 1970)	Nancy (France)	5/66–6/67	724.2	17.0	—	—	—	—	—	76.0	7.0
Beech forest (Eidmann, 1959)	(Germany)	1952–7	1216.0	7.6	—	—	—	—	—	75.9	16.5
Beech (F. orientalis) forest (Cepel, 1967)	Near Istamboul (Turkey)	5 yr	1032.7	17.4	—	—	—	—	—	67.1	15.5
Beech forest (Delfs, 1967)	Sauerland	4 yr	1150.0	7.6	—	—	—	—	—	75.8	16.6
Oak forest on plateau (Galoux, 1963)	Virelles* (Belgium)	1961	921.9	—	—	—	—	—	—	84.2	—
Thermophilous oak forest on south slope (Galoux 1963)	Virelles* (Belgium)	1961	921.9	—	—	—	—	—	—	74.6	—
Maple–ash–elm forest on north slope (Galoux, 1963)	Virelles* (Belgium)	1961	921.9	—	—	—	—	—	—	86.8	—

Table 3.23. *Water balance of coniferous artificial stands in temperate deciduous forest areas*

Forest type	Site (* denotes IBP site)	Year	Precipitation (P, mm)	Interception (% P)	Actual evapotranspiration				Runoff			Through-fall (% P)	Stem flow (% P)
					Transpiration (% P)	Soil evaporation (% P)	Sum (% P)	Drainage runoff (% P)	Surface runoff (% P)	Sum (% P)			
Spruce forest (Benecke et al., 1975)	Solling* (Germany)	1/1/69–10/31/69	896.3	26.7	→29.1←		55.8	46.0	0	46.0	—	73.3
Spruce forest (Strauss, 1971a,b)	München* (Germany)	1969	841.0	—	—	—	87.6	—	—	—	—	—
Spruce forest (Aussenac, 1970)	Nancy (France)	5/65–6/67	750.5	34.4	—	—	—	—	—	—	63.8	1.8
Spruce forest (Eidmann, 1959)	Germany	1953–8	1216.0	25.9	—	—	—	—	—	—	73.3	0.7
Spruce forest Mature spruce forest (Delfs 1967)	Sauerland Hartz	4 yr 4 yr	1150.0 1356.0	25.9 36.0	— —	— —	— —	— —	— —	— —	73.4 63.2	0.7 0.8
Pine forest (Aussenac, 1970)	Nancy (France)	5/65–6/67	750.5	30.9	—	—	—	—	—	—	65.7	1.6
Pinus sylvestris plantation (Rutter, 1967)	— (England)	4/58–3/59	710.0	33.8	46.5	5.6	85.9	—	—	14.1	52.1	14.1
Pinus sylvestris plantation pool stage High Wood (Lützke & Simon, 1975)	Berlin (DDR)	10/1/67–9/30/71	626.2 627.0	33.1 28.5	49.6 39.6	15.8 19.0	98.5 87.0	— —	— —	3.8 13.2	64.2 71.0	2.7 0.5
Grand Fir plantation (Aussenac, 1970)	Nancy (France)	5/65–6/67	750.5	42.3	—	—	—	—	—	—	56.7	1.0

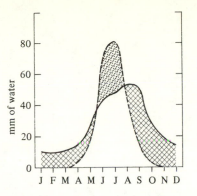

Fig. 3.45. Annual course of the water balance at the station of Turukhansk located in the boreal coniferous forest biome (from Budyke, 1958). Symbols as in Fig. 3.42.

and evapotranspiration increases regularly until July and then decreases rapidly (Fig. 3.45). This decrease sharpens at the end of the summer in areas where the soil-water supply has dried up. For the Turukhansk station (Siberia), evapotranspiration consumes about three quarters of the net radiation balance (calculated with Budyko's 1974 graph).

Even if rainfall exceeds evapotranspiration during summer months, runoff is non-existent for this period. Runoff begins with the melting of snow (March, April) and reaches its maximum in April, when melt water cannot penetrate a partially thawed soil and runs over the deeper frozen layers. In the summer, when rainfall is at its highest, runoff is restricted because of increased evaporation. Water balances of boreal forests have been made mainly for the Russian area of the euro-siberian region. Table 3.24 gives a few values from the work of Molchanov (1971).

Among the five large forest biomes that have been studied, the wet equatorial forest has the most favorable water balance. Large and regular rainfall provides a continuous and sufficient supply of soil water and the water deficit is practically zero or very weak. With a favorable radiation balance as well, the growth conditions for plants are ideal throughout the year and the yearly average productivity can be as high as 20 tons ha^{-1} yr^{-1} (Lieth, 1973). In equatorial monsoon forests, a slight water deficit can take place before the monsoon, due to reduction of the soil water supply.

In dry tropical forests, the radiative balance is always very high, but the water balance periodically shows a deficit. The importance of this deficit increases according to the frequency and the duration of dry seasons. As water is a limiting factor during those periods, vegetation develops forms that are better adapted to the drought and shows a halt in growth. These different adaptations have an influence on the average yearly productivity of approximately 6 tons ha^{-1} yr^{-1}.

193

Table 3.24. *Water balance of various boreal coniferous forest types*

Forest type	Site (* denotes IBP site)	Year	Precipitation (P, mm)	Actual evapotranspiration				Runoff			Through-fall (% P)	Stem flow (% P)
				Interception (% P)	Transpiration (% P)	Soil evaporation (% P)	Sum (% P)	Drainage (% P)	Surface runoff (% P)	Sum (% P)		
Northern taiga (Molchanov, 1971) Pine forest and mixed pine–spruce forest of various ages	(*) Lat. 63° 30' N Long. 40° 7' E	1948-9	525	6.3–24.4	20.6–33.5	10.5–17.0	43.6–70.5	1.7–47.6	4.8–27.8	29.5–56.4	—	—
Middle taiga (Molchanov, 1971) Spruce forests	(*) Lat. 60° 30' N Long. 34° 35' E	—	600	26.3–27.7	16.8–17.2	10.8	53.8–57.7	42.1–44.0	2.2	44.3–46.2	—	—
Mixed birch–spruce forests	Lat. 60° 30' N Long. 34° 35' E	—	600	29.7–30.5	33.5–40.3	10.0–11.7	74.8–80.8	17.0–23.0	2.2	19.0–25.2	—	—
Birch forest	Lat. 60° 30' N Long. 34° 35' E	—	600	5.5–17.7	14.7–24.2	12.7–14.8	35.0–54.5	40.8–54.5	4.7–10.5	45.5–65.0	—	—
Southern taiga (Molchanov, 1971) Spruce forests	(*) Lat. 59° N Long. 39° E	—	730	17.5–20.3	35.1–48.5	9.6–11.0	63.6–78.4	1.1–12.4	20.5–24.0	21.6–36.4	—	—
Birch forests	Lat. 59° N Long. 39° E	—	730	10.8–11.0	32.2–34.1	20.0	63.0–65.1	10.9–13.0	24.0	34.9–37.0	—	—

In a similar way to the dry tropical forest, mediterranean vegetation has a favorable radiation balance and a water supply that is periodically in deficit. During warm and dry summers, the soil-water supply is quickly exhausted and the water deficit can last from three of seven months. The wilting point is reached every year. Potential and actual evapotranspiration for *Quercus ilex* oak forests can amount to 430 mm yr^{-1} and up to 530 mm yr^{-1} for chaparral in California. The average yearly productivity of these oak forests is approximately 8 tons ha^{-1} yr^{-1} (Lieth, 1973).

Except during dry years, the temperate deciduous forest is regularly watered during the vegetative period. Rainfall recharges the soil at short intervals of a few days and considerably reduces the water deficit. For Virelles–Blaimont oak forest it does not exceed 60 mm yr^{-1}. In such forests, either water or energy, in turn, are the limiting factors during the growing season. The wilting point is not reached every year, as does occur in a mediterranean climate. The growth conditions are favorable, however, and make possible average annual productivities of 10 tons ha yr^{-1}.

Covered with snow in winter and overflowed in the spring, the soil of boreal coniferous forests still has a sufficient water supply during a large part of the vegetative period. Only at the end of the vegetative period does the water deficit occur. In these forests, the available energy limits the metabolic processes. Under such harsh growing conditions, the average annual productivity reaches only 5 tons $ha^{-1}yr^{-1}$.

When we examine the data of Tables 3.18 to 3.24, we notice that a variability exists for the main components of the water balance between the different forest types. Nature, age, composition and structure of the observed stands, rainfall distribution, exposure, instrumentation used and duration of the measurement period are the main factors contributing to these discrepancies. While it is difficult to establish general tendencies, at least the following points can be made:

In nearly all the studied stands, actual evapotranspiration consumes more than 50% of rainfall (41–100%). But no significant differences appear between the large forest biomes. Although only one result has been available up to now (Benecke, 1975), it seems that the substitution of conifers with high interception rates for temperate deciduous broadleaved forests contributes to reduction of underground water reserves (drainage is 55.7% for the beech forest and 46.0% for the spruce stand at the Solling).

In spite of the great variability of the interception of rainfall by the canopy (5–42%), the evergreen conifer stands (mainly spruce) intercept more than deciduous forests.

In most cases, stemflow does not exceed 10% of rainfall. Stemflow is generally weak for conifer species but can, however, reach 15–20% for the

195

temperate deciduous beech forests and even exceed this for certain tropical species. The nature of the bark, the location of the tree and the exposure of the site are the main factors responsible.

If runoff is less than half the rainfall, it occurs at different periods in the different biomes: permanent flow in equatorial forests; restricted to the rainy summer period in tropical forests; essentially a winter phenomenon in mediterranean and temperate deciduous forests (although it can also appear during the vegetative period after rainy periods or heavy summer showers); particularly during snow melting in the spring in the taiga.

References

Alexeyev, V. A. (1963). Quelques problèmes de propriétés optiques de la forêt. Recueil. Problèmes d'écologie et de physiologie des plantes forestières. *Akad. For. Kirov* (*Leningrad*), pp. 47–79. (In Russian; translated into French by M. Nitsch, Phytotron, Gif-sur-Yvette, France).

Alexeyev, V. A. (1970): The additional photosynthetically active radiation received by the vegetation of the lower storys through gaps in the canopy. *Fitocenologija i biogeocenologija temno hvojnoj tajgi*. Nauka Publishing House, Leningrad. pp. 19–31.

Andrews, R. E. & Neuman, E. I. (1969). Resistance to water flow in soil and plant. III. Evidence from experiments with wheat. *New Phytol.* **68**, 1051–8.

Aussenac, G. (1970). Action du couvert sur la distribution au sol des précipitations. *Ann. Sci. For.* **27**, 383–99.

Bandhu, D. (1974). International Woodlands Workshop 1BP-PT 1973. Göttinger Bodenkunliche Berichte 30.

Banerjee, A. K. (1973). Computing transpiration and soil evaporation from periodic soil moisture measurements and other physical data. *Indian For.* **99**, 82–91.

Baumgartner, A. (1956). Untersuchungen über den Wärme und Wasserhaushalt eines jungen Waldes. *Ber. deuts. Wetterdienstes*, **5**, 4–53.

Baumgartner, A. (1957). Beobachtungswerte und weitere Studien zum Warmehaushalt eines jungen Waldes. *Wissenchaftliche Mitteilungen Meteorologischen Institut, Universität München*, **1**, No. 4.

Baumgartner, A. (1965). The heat, water and carbon dioxide budget of plant cover: methods and measurements. In *Proceedings Montpellier Symposium, UNESCO, Paris*, pp. 495–512.

Baumgartner, A. (1969). Meteorological approach to the exchange of CO_2 between the atmosphere and vegetation, particularly forest stands. *Photosynthetica*, 3, 127–149.

Benecke, P. (1974). Energy and matter exchange processes – Water. Report from Working Group on Energy and Water Balance. *Göttinger Bodenkundliche Berichte* 30. pp. 18–46.

Benecke, P., & Van Der Ploeg, R. R. (1975). Nachhaltige Beeinflussung des Landschaftswasserhaushaltes durch die Baumartenwahl. *Forstarchiv*, **46**, 97–102.

Berz, G. (1969). Untersuchungen zum Wärmehaushalt der Erdoberfläche und zum bodennahen atmosphärischen Transport. Meteorologisches Institut, Universität München, 16.

Birbeback, R. & Birbebak, R. (1964). Solar radiation characteristics of tree leaves. *Ecology*, **45**, 646–9.

Black, T. A., Gardner, W. R. & Thurtell, G. W. (1969). The prediction of evaporation, drainage, and soil water storage for a bare soil. *Proc. Soil Sci. Amer.* **33**, 655–60.

Black, T. A. & McNaughton, K. G. (1971). Average Bowen-ratio methods of calculating evapotranspiration applied to a Douglas-fir forest. *Boundary-Layer Meteorol.* **2**, 466–75.

Bodeux, A. (1954). Recherches écologiques sur le bilan de l'eau sous la forêt et la lande de Haute Campine. *Agricultura*, II(2):(1), 1–80.

Brasseur, F. & De Sloover, J. R. (1973). L'extinction de l'irradiance et de l'éclairement dans deux peuplements forestiers de Haute Ardenne. *Bull. Soc. Roy Bot. Belg.* **106**, 219–36.

Briggs, G. E. (1967). *Movement of Water in Plants*. F. A. Davis Company, Philadelphia, Pennsylvania.

Budyko, M. I. (1958). *The heat balance of the earth's surface* (translated by Nina A. Stepanova). Office of Technical Services, US Department of Commerce, Washington, D.C.

Budyko, M. I. (1963). *Atlas Teplovogo Balansa Vara*. Moscow.

Budyko, M. I. (1974). *Climate and Life*. Academic Press, International Geophysics Series No. 18.

Celniker, J. L. (1967). The light regime, photosynthesis and productivity of the forest. *Svetovoji rezim, fotosintezi, produktivnost' lesa*. Nauka Publishing House, Moscow.

Cepel, V. N. (1967). Interzeption (= Niederschlagsverdunstung im Kronenraum) in einem Buchen-, einem Eichen- und einem Kiefernbestand des Belgrader Waldes bei Istanbul. *Forstwiss. Centralbl (Hamburg)*, **5**, 301–14.

Cowan, I. R. & Milthorpe, F. L. (1965). Plant factors influencing the water status of plant tissues. pp. 137–89. In *Water Deficits and Plant Growth*, ed. T. Kowlowki.

Czepa, O. & Reuter, H. (1950). Beitrag der effektiven Ausstrahlung in Bodennähe bei klarem Himmel. *Archiv für Meterologie, Geophysics und Bioklimatologie*, B. **2**, 250–8.

Dabral, B. G. & Rao, B. K. S. (1968). Interception studies in chir and teak plantations, New Forest. *Indian For.* **94**, 541–51.

Dabral, B. G. & Rao, B. K. S. (1969). Interception studies in Sal (*Shorea robusta*) and Khair (*Acacia catechu*) plantations. *Indian For.* **95**, 314–23.

Deirmerdjian, D. & Sekera, Z. (1954). Global radiation resulting from multiple scattering in a Rayleigh atmosphere. *Tellus*, **6**; 382–98.

Delfs, J. (1965). Interception and stem flow in stands of Norway spruce and beech in West Germany. pp. 179–185. In *International symposium on forest hydrology*. Pergamon Press, Oxford.

Denmead, O. T. (1969). Comparative micrometerology of a wheat field and a forest of Pinus radiata. *Agric. Meteorol.* **6**, 357–71.

Dirmhirn, I. (1964). *Das Strahlungsfeld im Lebensraum*. Akademische Verlagsgesellschaft, Frankfurt am Main.

Dogniaux, R. (1964). *L'éclairage naturel et ses applications*. Editions SIC, Bruxelles.

Duvigneaud, P., Denaeyer-De Smet, S., Ambroes, P. & Trimperman, J. (1971). Recherches sur l'écosysteme forêt. Biomasse, productivité et cycle des polyéléments dans l'ecosystème 'chênaie caducifoliée'. Essai de phytogéochimie forestière. *Institut Roy. Sci. naturelles de Belgique. Mémoire No. 164.*

Dyer, A. I & Maher, F. I. (1965). The 'EVAPOTRON'. An instrument for the measurement of eddy fluxes in the lower atmosphere. *SCIRO, Technical paper No. 15.* Melbourne, Australia.

Edwards, N. T. (1974). A moving chamber design for measuring soil respiration. *Oikos*, **25**, 97–101.

Eidmann, F. E. (1959). Die Interception in Buchen-und Fichtenbeständen. Ergebnis mehrjähriger Untersuchungen im Rothaargebirge (Sauerland). *Publication de l'Association Internationale d'Hydrologie Scientifique, Gentbrugge.*

Ettehad, R., Lossaint, P. & Rapp, M. (1973). Recherches sur la dynamique et le bilan de l'eau des sols de deux écosystèmes méditerranéens à chêne vert. *Editions du Centre National de Recherches Scientifiques, (Paris)*, **40**, 199–289.

Falinsky, J. B. (1974). International Woodlands Workshop IBP-PT 1973. Göttinger Godenkundliche, Berichte **30**.

Froment, A. *et al.* (1971). La chênaie mélangée calcicole de Virelles-Blaimont en Haute Belgique, pp. 635–66. In *Productivité des ecosystèmes forestiers, Actes Colloque Bruxelles, 1969.* Ecologie et conservation, **4**, UNESCO, Paris.

Galoux, A. (1963). Budgets et bilans dans l'écosystème forêt. *Lejeunia, revue de Botanique*, **21**.

Galoux, A. (1973). La chênaie mélangée calcicole de Virelles-Blaimont. Flux d'énergie radiante, conversions et transferts dans l'écosystème (1964–7). *Travaux Station de Recherches des Eaux et Forêts, Groenendaal-Hoeilaart*, **A 14.**

Galoux, A. (1974). Ecosystem, open thermodynamic system. The Oak Forest of Virelles-Blaimont. *Göttinger Bodenkundliche Berichte*, **30**, 131–49.

Galoux, A. & Grulois, J. (1968). La chênaie de Virelles-Blaimont. Echanges radiatifs et convectifs en phase vernale. *Travaux Station de Recherches des Eaux et Forêts, Groenendaal-Hoeilaart*, **A 13.**

Gates, D. M. (1965). Energy, plants and ecology. *Ecology*, **46**, 1–13.

Gates, D. M. & Janke, R. (1965). The energy environment of the alpine tundra. *Oecol. Plant.* **1**, 39–62.

Gates, D. M. & Tantraporn, W. (1952). The reflectivity of deciduous trees and herbaceous plants in the infrared to 25 microns. *Science*, Washington **115**, 613–61.

Gates, D. M. & Papian, LaVerne, (1971). *Atlas of energy budgets of plant leaves.* Academic Press, New York.

Gay, L. W. (1972). Energy flux studies in a coniferous forest ecosystem, pp. 243–53. In *Proceedings of research on coniferous forest ecosystems.* Bellingham, Washington.

Gay, L. W. (1973). Energy exchange studies at the earth's surface. I. Energy budgets of desert, meadow, forest and march sites. *Department of Atmospheric Sciences, Technical report 73–1.* Oregon State University, Corvallis.

Gay, L. W. & Holbo, H. R. (1974). *Studies of the forest energy budget.* WRRI–24. Water Resources Research Institute, Oregon State University, Corvallis.

Gay, L. W. & Knoerr, K. R. (1975). *The forest radiation budget.* Bulletin 19. Duke University School of Forestry and Environmental Studies, Durham, North Carolina.

Gay, L. W. Knoerr, K. R. & Braaten, M. O. (1971). Solar radiation variability on the floor of a pine plantation.. *Agric. Meteorol.* **8**, 39–50.

Gietl, G. (1974). Energy and matter exhange processes-heat and vapor. Report from working group on energy and water balance. Göttinger Bodenkundliche Berichte **30**, 18–46.

Grulois, J. (1967). La chênaie de Virelles-Blaimont. Extinction du rayonnement global, tropismes et paramètres foliaires. *Bull. Soc. Roy bot. Belg.* **100**, 139–51.

Grulois, J. (1968). La chênaie de Virelles-Blaimont. Réflexion, interception et transmission du rayonnement de courtes longueurs d'onde: variations au cours d'une année. *Bull. Soc. Roy. Bot Belg.* **102**, 13–25.

Grulois, J. & Schnock, G. (1967). La chênaie de Virelles-Blaimont. Rayonnement global sous le couvert en période défeuillée. *Bull. Inst. R. Sci. Nat. Belg.* **43**.

Grulois, J. & Vyncke, G. (1969). Relation entre les éclairements lumineux et énergétiques incidents et transmis sous forêt en phénophase feuillée. *Oecol. Plant.* **43**, 27–46.

Hager, H. (1974). Energy and matter exchange processes. Carbon dioxide. Report from Working Group on Energy and Water Balance. *Göttinger Bodenkundliche Berichte* **30**, 18–46. Göttingen.

Hager, H. (1975). *Kohlendioxydconzentrationen-Flüsse und-Bilanzen in einem Fichten-hochwald.* Wissenschaftliche Mitteilung **26**, Meteorologisches Institut, Universität Munchen.

Harlow, W. M. (1931). The identification of pines of the United States, native and introduced, by needle structure. *Bull. NY State College of Forestry*, Tech. Pub. 32.

Havas, P. (1974). International Woodlands Workshop IBP-PT 1973. Göttinger Bondenkundliche Berichte **30**.

Higginbotham, K. O. (1974). The influence of canopy position and age of leaf tissues on the growth and photosynthesis of loblolly pine. Ph.D. Dissertation. Duke University, Durham, North Carolina.

Hopkins, B. (1965). Vegetation of the *Olokemeji forest reserve*, Nigeria. The micro-climates with special reference to their seasonal changes. *J. Ecol.* **53**, 125–38.

Hopkins, B. (1960). Rainfall interception by a tropical forest in Uganda. *East Afr. Agric. J.* **25**, 255–8.

Hutchison, B. A. (1975). Photographic assessment of deciduous forest radiation regimes. US Department of Commerce, National Oceanic and Atmospheric Administration. ATDL Contribution File No. 75/3. 164 pp.

Huttel, C. (1975). Recherches sur l'Ecosystème de la forêt subéquatoriale de basse Côte d'Ivoire. La Terre et la Vie. *Rev. Ecol. Appl.* **29**, 192–202.

Hytteborn, H. (1975). Deciduous woodland at Andersby, Eastern Sweden. Above-ground tree and shrub production. *Acta Phytogeogr. Suec.* **61**, 96.

Inoue, E. (1965). On the CO_2-concentration profiles within crop canopies. *J. Agric. Meteorol. Tokyo*, **20**, 137–40.

Intribus, I. (1975). Water balance factors in the ecosystem of an oak–hornbeam stand at the object in Bab. *Research Project IBP Progress Report*, **2**, 337–51. Slovak Academy of Sciences, Bratislava.

Jarvis, P. G. (1976). Exchange properties of coniferous forest canopies, pp. 90–98. In *XVI Congress of International Union of Forest Research Organizations, Norway, Division II.*

Jarvis, P. G., James, G. B. & Landsberg, J. J. (1976). Vegetation and the atmosphere. **2**, 171–240. In *Coniferous Forest*, ed. Monteith J. L. Academic Press, London.

Jehle, A. (1968). Die Parameter des vertikalen logarithmischen Windprofils über verschiedenen Pflanzenbeständen. Diplomarbeit F. Meteorologie, Universität München.

Kayser, C. & Kiese, O. (1973). Energiefluss and -umsatz in ausgewählten Ökosystemen des Sollings, pp. 484–91. *Deutscher Geographentag Kassel, Tagungsbericht und wissenschaftliche Abhandlungen.*

Kayser, C. & Kiese, O. (1974). Energy and matter exchange processes – Radiation. Report from Working Group on Energy and Water Balance. *Göttinger Bodenkundliche Berichte*, **30**, 18–46. Göttingen.

Kessler, A. (1973). Zur Klimatologie der Strahlungsbilanz an der Erdoberfläche.

Tages-und Jahresgänge in den Klimaten der Erde. *Erkunde, Arch. Wissen. Geogra.* **27**, 1–10.

Kiese, O. (1972). Bestandsmeteorologische Untersuchungen zur Bestimmung des Wärmehaushalts eines Buchenwaldes. *Berichte des Instituts für Meteorologie und Klimatologie der Technischen Universität Hannover.*

Kinerson, R. S. (1975). Plant surface area-respiration relationships of a forest community.

Kitazawa, Y. (1974). International Woodlands Workshop IBP-PT 1973. Göttinger Bodenkundliche Berichte, **30**.

Knuchel, H. (1914). Spektrophotometrische Untersuchungen im Walde. *Mitt. Schweiz. Anst. Forstl. Versuchswes*, **11**, 1–94.

Koch, P. (1972). Utilization of Southern Pines, **I**. *USDA Forest Service, Agric. Handbook No. 420.*

Kondratyev, K. Ya. (1969). *Radiation in the Atmosphere.* International Geophysics Series, Vol. 12. Academic Press, New York.

Kramer, P. J. (1969). *Plant and water relations: a modern synthesis.* McGraw-Hill Book Company, New York.

Kramer, P. J. & Kozlowski, T. T. (1960). *Physiology of trees.* McGraw-Hill Book Company, New York.

Krinov, E. L. (1953). Spectral reflectance of natural formations. Laboratoria Acromethodov Akad. Nauk URSS. *National Council of Canada, Comm. T. 439.*

Landsberg, J. J., Jarvis, P. G. & Slater, M. B. (1973). The radiation regime of a spruce forest, pp. 411–8. In *Plant response to climatic factors, proceedings of the Uppsala symposium.* UNESCO, Paris.

Lemee, G. (1974). Recherches sur les écosystèmes des réserves biologiques de la forêt de Fontainebleau. IV. Entrée d'éléments minéraux par les précipitations et transfert au sol par le pluviolessivage. *Oecol. Plant.* **9**, 187–200.

Leonard, R. E. (1968). Albedo of a red pine plantation. Diss. Abstr. 28 B (10) (3951). Ann Arbor, Michigan.

Lettau, H. (1939). Atmosphärische Turbulenz. Akademische verlagsgesellschaft m.b.h., Leipzig.

Lieth, N. (1973). Primary production: Terrestrial ecosystems. *Hum. Ecol.* **1**, 303–32.

Linacre, E. T. (1969). Net radiation to various surfaces. *J. Ecol.* **6**, 61–73.

Lorenz, D. & Baumgartner, A. (1970). Oberflächentemperatur und Transmission infraroter Strahlung in einem Fichtenwald. *Arch. Meteorol. Geophys. Bioklimatol.* **B 18**, 305–24.

Lossaint, P. (1974). International Woodlands Workshop IBP-PT 1973. Göttinger Bondenkundliche Berichte, 30.

Low, K. S. (1972). Interception loss in the humid forested areas (with special reference to Sungai Lui catchment, West Malaysia). *Malay. Nat. J.* **25**, 104–11.

Lützke, R. (1966). Vergleichende Energieumsatzmessungen im Walde und auf einer Wiese. *Arch. Forstwes.* **15**, 995–1015.

Lützke, R. & Simon, K. H. (1975). Zur Bilanzierung des Wasserhaushalts von Waldeständen auf Sandstandorten der Deutschen Demokratischen Republik, pp. 806–7. In *Beiträge für die Forstwirtschaft*, **1**, Akademie-Verlag, Berlin in Algemeine Forstschrift, 37, Munchen.

Major, J. (1967). Potential evapotranspiration and plant distribution in western states with emphasis on California, pp. 93–126. In *Ground level climatology.* American Association for Advance of Sciences, Washington.

McColl, J. G. (1970). Properties of some natural waters in a tropical wet forest of Costa Rica. *Bioscience*, **20**, 1096–100.

McGinnis, J. T., Golley, F. B., Clements, R. G., Child, G. I. & Duever, M. J. (1969). Elemental and hydrologic budgets of the Panamian tropical moist forest. *Bioscience*, **19**, 697–700.

Methy, M. (1974). Interception du rayonnement solaire par différents types de végétation dans la zone méditerranéenne. Thèse de l'Université des Sciences et Techniques du Languedoc.

Miller, P. C. (1967). Leaf temperatures, leaf orientation and energy exchange in Quaking aspen (*Populus tremuloides*) and Gambell's oak (*Quercus gambelii*) in Central Colorado. *Oecol. Plant.* **2**, 241–70.

Mitscherlich, G. (1971). Wald Wachstum und Umwelt, Bd 2, Waldklima und Wasserhaushalt. J. D. Sauerländers Verlag. Frankfurt am Main.

Mitscherlich, G., Moll, W., Künstle, E. & Maurer, P. (1966). Ertragskundlichokölogische Untersuchungen im Rein-und Mischbestand. IV: Niederschlag, Stammablauf und Bodenfeuchtigkeit, *Allg. Forst-Jagdzg*, **137**, 1–13.

Mitscherlich, G., Künstle, E. & Lang, W. (1967). Ein Beitrag zur Frage der Beleuchtungsstärke im Bestande. *Allg. Forst. Jagdz.* **138**, 213–23.

Molchanov, A. A. (1960). The Hydrological Role of Forests (Translated by A. Gourevitch, 1963), Isräel Program Scientific Translations, Jerusalem.

Molchanov, A. A. (1971). Cycles of atmospheric precipitation in different types of forests of natural zones of the USSR. Ecology Conservation **4**, pp. 49–69. In *Productivity of forest ecosystems*. Proceedings of the Brussels Symposium, UNESCO, Paris.

Monsi, M. & Saeki, T. (1953). Uber den Lichtfaktor in den Pflanzengesellschaften und seine Bedeutung fur die Stoffproducktion. *Japanese J. Bot.* **14**, 22–52.

Monteith, J. L. (1965). Evaporation and environment, pp. 205–34. In *Symposium of Soc. Exp. Biology*, **19**. Cambridge University Press, Cambridge.

Monteith, J. L. (1972). *Survey of instruments for micrometeorology*. IBP Handbook 22. Blackwell, Oxford.

Murphy, C. E. (1974). Energy and matter exchange processes – Modeling of the physical environment. Report from Working Group on Energy and Water Balance. *Göttinger Bodenkundliche Berichte* **30**, Göttingen.

Murphy, C. E. & Knoerr, K. R. (1972). *Modeling the energy balance processes of natural ecosystems*. EDFB/IBP–72/10. Oak Ridge National Laboratory, Oak Ridge, Tennessee.

Murphy, C. E. & Knoerr, K. R. (1970). A general model for energy exchange and microclimate of plant communities, pp. 786–97. In *Proceedings 1970 Summer Computer Simulation Conference*. Simulation Councils Inc., LaJolla, California.

Noirfalise, A. (1959). Sur l'interception de la pluie par le couvert dans quelques forëts belges. *Bull. Soc. R. For Belg. (Bruxelles)*, **10**, 433–9.

Odum, T. H., Moore, A. M. & Burns, L. A. (1970). Hydrogen budget and compartments in the rain forest, pp. 105–22. In *A tropical rain forest*, **3**. US Atomic Energy Commission, Washington, DC.

Penman, H. L. (1948). Natural evaporation from open water, bare soil and grass. *Proc. R. Soc.* **A193**, 120–45.

Penndorf, R. (1956). Luminous reflectance visuel albedo of natural objects. *Bull. Am. Meteorol. Soc.* **37**, 142–4.

Person, H. & Hytteborn, H. 1974. International Woodlands Workshop IBP-PT 1973. Göttinger Bodenkundliche Berichte, 30.

Prandt, L. (1959). Führer durch die Strömungslehre Fr. Vieweg, Braunschweig.

Quantin, A. (1935). L'évolution du tapis végétal à l'étage de la chênaie dans le Jura méridional. *Commun. Stat. Int. Géobot. Méditerr. Alp.* **37**, 1–382.

Rauner, Y. L. (1958). Some results of heat budget measurements in a deciduous forest. *Izvestisa Akademii Nauk SSSR, Seriia Geografceskaja*, **5**, 79–86.

Rauner, Y. L. (1961). On the heat budget of a deciduous forest in winter. Izvestisa Akademii Nauk SSSR, Serija Geografceskaja 4, 83–90. Translation by A. Nurklik, 1964. Canadian Department of Transport, Meteorological Branch. (Meteorological translation **11**, 60–77).

Rauner, Y. L. & Ananjev, I. P. (1971). Merkmale der atmosphärischen Turbulenz in Waldbeständen. (Charakteristiki atmosfernoj turbulentnosti Vustorijack lesa, Izvestija Akademii Nauk SSSR, Serija Geografceskaja, 2, Moskva 1971). Ubersetsung Eidgenossische Anstalt fur das forstliche Versuchswesen, Birmensdorf, ZH. 23 pp.

Ray, M. P. (1970). Preliminary observations on stem flow, etc., in *Alstonia scholaris* and *Shorea robusta* plantations at Arabari, West Bengale. *Indian For.* **96**, 482–93.

Reifsynder, W. E. (1967). Forest meterology. The forest energy balance. *Int. Rev. For. Res.* **2**, 127–75.

Reimer, A. & Desmaris, R. (1973). Micrometeorology energy budget methods and apparent diffusivities for boreal forest and grass sites at Pinawa, Manitoba, Canada. *Agric. Meteorol.* **11**, 419–36.

Roussel, L. (1953). Recherches théoriques et pratiques sur la répartition en quantité et en qualité de la lumière dans les milieux forestiers, influence sur la végétation. *Ann. Ec. Eaux For. (Nancy)*, **13**, 295–400.

Rowe, P. B., Storey, H. C. & Hamilton, E. L. (1951). Some results of hydrologic research. US Department of Agriculture Forest Service, Californian Forest and Range Experiment Station, Miscellaneous publications, **1**(5).

Rutter, A. J. (1967). Evaporation dans les forêts. *Endeavour*, **26**, **97**, 39–43.

Rutter, A. J. (1968). Water consumption by forests, pp. 23–84. In *Water deficits and plant growth*, **2**. Academic Press, New York.

Satchell, J. E. (1974). International Woodlands Workshop IBP-PT 1973. Göttinger Bodenkundliche Berichte, **30**.

Sauberer, F., Hartel, O. (1959). *Pflanze und Strahlung*. Akademische Verlagsgesellschaft, Geest und Portig, Leipzig.

Schnock, G. (1970). Le bilan de l'eau et ses principales composantes dans une chênaie mélangée calcicole de Haute-Belgique (Bois de Virelles-Blaimont). Thèse Université Libre de Bruxelles, Bruxelles.

Schulze, R. (1970). Strahlenklima der Erde. Dr. Dietrich Steinkopff Verlag Darmstadt. 217 pp.

Schulze, E. D. & Koch, W. (1971). Measurement of primary productivity with cuvettes, pp. 141–56. In *Productivité des écosystèmes forestiers*. UNESCO, Paris.

Shachori, A. Y., Stanhill, G. & Michaeli, A. (1965). The application of integrated research approach to the study of effects of different cover types on rainfall disposition in Carmel mountains in Israel, pp. 479–87. In *Methodology of plant eco-physiology: proceedings of the Montpellier symposium*. UNESCO, Paris.

Sinclair, T. R. (1972). A leaf photosynthesis submodel for use in general growth models. Eastern Deciduous Forest Biome Memo Report 72–14. Oak Ridge National Laboratory, Oak Ridge, Tennessee.

Sinclair, T. R., Allen, L. H., Jr. & Stewart, D. W. (1971). A simulation model for crop–environmental interactions and its use in improving crop productivity, pp. 784–94. In *Proceedings 1971 Summer Computer Simulation Conference*. Simulation Councils Inc., LaJolla, California.

Sinclair, T. R., Murphy, C. E., Jr. & Knoerr, K. R. (1975). Development of less-complex models for simulating vegetative photosynthesis and transpiration based on a soil-plant-atmosphere model.

Slatyer, R. O. (1967). Plant–water relationships. Academic Press, London.
Smith, M. K. (1974). Throughfall, stemflow and interception in pine and eucalyptus forest. *Aust. For.* **36**, 190–7.
Smolen, F. (1975). The radiation balance at the research site of Bab. *Research Project Bab, IBP Progress report II*, pp. 369–81.
Stewart, D. W. & Lemon, E. R. (1969). A simulation of net photosynthesis of field corn. Microclimate Investigations Interim Report. 69–3, USDA, ARS, Cornell University, Ithaca, New York.
Stewart, J. B. (1971). The albedo of a pine forest. *Quart. J. Roy. Meteorol. Soc.* **97**, 561–4.
Stewart, J. B. & Thom, A. S. (1973). Energy budgets in pine forests. *Quart. J. roy. meteorol. Soc.* **99**, 154–70.
Storr, D., Tomlain, J., Cork, H. F. & Munn, R. E. (1970). An energy budget study above the forest canopy of Marmot Creek. *Water Resour. Res.* **6**, 705–16.
Strain, B. R. & Higginbotham, K. O. (1976). A summary of gas exchange studies on trees undertaken in the US IBP Eastern Deciduous Forest Biome, pp. 91–102. In *XVI Congress of international union of forest research organization, Norway.*
Strauss, R. (1971*a*). Energiebilanz and Verdunstung eines Fichtenwaldes im Jahre 1969. Universität München, Meteorologisches Institut, München.
Strauss, R. (1971*b*). Energiehaushalt eines Fichtenwaldes, 1969. Meteorologisches Institut der Universität München. *Wiss. Mitt.* **21**, 17–19.
Susmel, L., Capelli, M., Viola, F. & Bassato, G. (1975). Autoecologia del pino radiato al Grighini (Sardegna centro-occidentale). *Estratto dagli annali del centro di economia montana delle Venezie.* **9**, 1968–9.
Susmel, L., Viola, F. & Bassato, G. (1976). Ecologia della Lecceta del Supramonte di Orgosolo (Sardegna Centro-orientale). *Annali del Centro di Economia montana delle Venezia, Padova,* **10**.
Sverdrup, H. (1936). The eddy conductivity of the air over a smooth snow field. *Geofysika* Publ. **11**, 7.
Swinbank, W. C. (1951). The measurement of vertical transfer of heat and water vapor by eddies in the lower atmosphere. *J. Meteorol.* **8**, 135–45.
Tajchmann, S. (1967). Energie-und Wasserhaushalt verschiedener Pflanzenbestände bei München. *Wissenschaftliche Mitteilungen Meteorologisches Institut Universität München* **12**.
Tajchmann, S. (1972*a*). The radiation and energy balances of coniferous and deciduous forests. *J. Appl. Ecol.* **9**, 359–77.
Tajchmann, S. J. (1972*b*). Messungen zum Wärmehaushalt über einer Kiefernschonung im Trockengebiet des Oberrheines. *Allg. Forst Jagdzg.* **143**, 35–8.
Tanner, C. B. (1964). Basic instrumentation and measurements for plant environment and micrometeorology. Soils Department Bulletin 6, College of Agriculture, University of Wisconsin.
Thamdrup, H. M. & Peterson. (1974). International Woodlands Workshop IBP-PT 1973. Göttinger Bodenkundliche Berichte, **30**.
Thomas, P. K., Chandrasekhar, K. & Haldorai, B. (1972). An estimate of transpiration by *Eucalptus globulus* from Nilgiris Watersheds. *Indian For.* **98**, 168–72.
Tombesi, L. (1971). Su alcuni aspetti fondamentali della fertilita, bilanci energetici, idrologici e nutritivi delle culture, pp. 205–48. In *Annali dell'Istituto sperimentale per la nutrizione delle piante, Roma.* Vol. II.
Uchijima, Z. (1970). Carbon dioxide environment and flux within a corn crop canopy, pp. 179–96. In *Proceedings of the IBP–PP technical meeting, Trebon.* September 14–21, 1969.

Ukrecky, I., Smolik, Z. & Lanar, M. (1975). Preliminary evaluation of the precipitation balance in the floodplain forest near Lednice in Moravia, Ecosystem study on foodplain forest in South Moravia. PT–PP–IBP, Institut of the IBP, Brono, Czechoslovakia, Report No. 4, 287–96.

Von Droste, B. (1969). Struktur und Biomasse eines Fichtenbestandes auf Grund einer Dimensionsanalyse an oberirdischen Baumorganen. Dissertation Univ. München.

Von Paller, H. (1972). Kohlendioxydverteilung, -Ströme und – Bilanzen in einem Fichtenwald. Diplom. Arbeit, München.

Van den Honert, T. H. (1948). Water transport in plants as a cantenary process. *Discuss. Faraday Soc.* **3**, 146–53.

Van Miegroet, M. (1965). Die Lichttransgression und die Lichtreflexion bei Blattern einiger Laubbaumarten. *Schweiz. Z. Forstwirtsch.* **7**, 556–89.

Vezina, P. E. (1961). Variations in total solar radiation in three Norway spruce plantations. *For. Sci.* **7**, 257–64.

Vezina, P. E. (1964). Solar radiation available below and over a snow pack in a dense pine forest. *Agric. Meteorol.* **1**, 54–65.

Vezina, P. E. (1965). Solar radiation available below thinned and unthinned balsam fir canopies. *Canadian Department Forestry, Forest Research Branch*, No. 63, Q, 8. 12 pp. Multigraphié.

Vezina, P. E., & Boulter, D. W. K. (1966). The spectral composition of near ultraviolet and visible radiation beneath forest canopies. *Can. J. Bot.* **44**, 1267–84. 84.

Waggoner, P. E. & Reifsnyder, W. E. (1968). Simulation of temperature, humidity and evaporation in a leaf canopy. *J. Appl. Meteorol.* **7**, 400–7.

Wassink, E. C. (1968). Light energy conversion in photosynthesis and growth of plants. Functioning of terrestrial ecosystems at the primary level, pp. 53–64. In *Proceedings of the Copenhagen Symposium.* UNESCO, Paris.

White, E. J. & Carlisle, A. (1968). The interception of rainfall by mixed deciduous woodland. *Q. J. For.* 310–20.

Wittaker, R. H. & Woodwell, G. M. (1971). Measurement of net primary production of forests, pp. 159–77. In *Productivité des Ecosystèmes Forestiers.* Actes du Colloque Bruxelles, 1969. UNESCO, Paris.

Yocum, C. S., Allen, L. H. & Lemon, E. R. (1964). Photosynthesis under field conditions. VI: Solar radiation balance and photosynthetic efficiency. *Agron. J.* **56**, 249–53.

Zuffa, J. (1975). Global radiation under the conditions of the oak–hornbeam stand in course of the vegetation period, pp. 391–404. In *Research Project Bab*, IBP Progress Report II.

4. Water relations and hydrologic cycles

R. H. WARING, JAMES J. ROGERS, & W. T. SWANK

Contents

How do ecosystems control the flow and storage of water, and how does water control the functioning of ecosystems? As a result of the International Biological Programme, we are in a better position to answer these questions than ever before. A systematic attempt to quantify water budgets for a variety of vegetations and climates has resulted in a data base for testing both old and new hypotheses. The ecosystem objective of linking the hydrologic system to carbon and mineral cycling has forced a revaluation of the processes controlling water movement. This perspective extends to understanding how water flows from one terrestrial unit to another and into aquatic systems.

This chapter has two major objectives. The first is to describe the processes or groups of processes affecting water movement and storage, e.g. the details of structure and function. The second objective is to demonstrate the general application of the hydrologic processes by assembling them into a detailed computer simulation model and applying this model to three extremely different kinds of forested watersheds where streamflow data were available.

Throughout this presentation we have eliminated all formal mathematics. We rely heavily on figures to summarize mathematical relations graphically, and on some box and arrow diagrams to illustrate relations between various processes. Pertinent literature is cited for those who want the mathematical details. The second objective requires the use of the master accountant, the computer, to keep track of water within the ecosystem. Computer simulations facilitate interpretation of behavior but also subsume many details.

Fig. 4.1 provides an overview of the structure of a watershed hydrologic system. It summarizes the main components and processes controlling

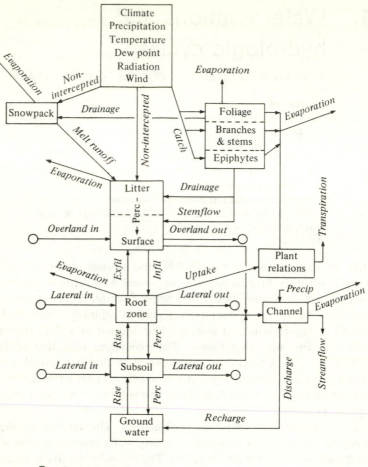

Fig. 4.1. The structure of a watershed hydrologic system showing the main components and processes controlling water movement.

water movement. These processes have a major effect on the structure and functioning of forest and aquatic ecosystems. Each component or process is the subject of individual sections.

The climatic variables indicated in Fig. 4.1 drive the entire system. Of particular importance are humidity and radiation, variables too often ignored but essential to estimate adequately transpiration, evaporation, and the energy content of snow, litter, and soil.

The hydrologist may see refinements in Fig. 4.1 unnecessary for a simple hydrologic budget. For example, the biological system is subdivided into several compartments above and below ground. For ecosystem purposes an essential distinction is made between the green canopy, important in fixing carbohydrates and the intercepting non-foliar surfaces. The process of evaporation operates from both. The canopy, however, derives water also from the root zone and stem wood of the tree; the amount supplied affects the actual rate of transpiration.

Some compartments, such as litter, hold very little water. Nevertheless, the moisture content of the litter is important because it affects decomposition through its control on the activity of micro-organisms. The snow-pack is hydrologically important in many areas yet completely absent in others. When present it moderates the temperature of the soil, in turn affecting water, nutrient uptake, plant growth, and canopy development.

The below-ground portion of the hydrologic system involves at least two subdivisions of the soil: (a) the root zone which can be subdivided depending upon rooting depth of different plants; and (b) the subsoil below the rooting zone. Processes controlling overland flow and lateral movement through the soil will be discussed in detail.

It should be noted that the size of a compartment which affects the water stored is often in part defined by biological criteria, such as rooting depth, leaf area, or sapwood volume. For example, water stored in the sapwood compartment plays a part in determining stomatal control and growth as will be treated in some detail in the section on plant water relations. These are critical distinctions for ecosystem analyses.

The diagram represents only a single terrestrial ecosystem with simple linkages shown to streams, lakes, and to the ground water. These linkages, characterized by properties of the vegetation and soil, control the amount of water and its routing through the system. Routing forms the basis for examining the interrelation of one terrestrial unit to another.

Finally, to evaluate whether the essential properties of the hydrologic system have been described, we will contrast a variety of watersheds with differing climates, soils and vegetation using simulation models developed independently by the US Deciduous and Coniferous Forest Biomes.

Structure and function of a watershed hydrologic system

The structure we develop is somewhat different from the conventional way of depicting the hydrologic system. The eventual objective of linking to erosion and carbon-mineral cycling has caused us to view the hydrologic system from an ecosystem perspective. Thus, we emphasize the importance of interactions between biological components and the hydrologic system, and spatial variation of these processes within a watershed.

Dynamic properties of forest ecosystems

Stratification into terrestrial and aquatic units

Spatial variation of hydrologic processes within a watershed may be caused by differences in soils, vegetation, geology, physiography and climate. Differences may exist in both terrestrial and aquatic portions of the watershed.

Where significant spatial variation does exist we can stratify a watershed into a set of homogeneous terrestrial units and a set of homogeneous aquatic (or channel) units. The following considerations are important and should be considered in addition to those commonly used in stratification of watersheds such as soil type, slope, and aspect.

Environmental stimuli of soil moisture, soil and air temperature, light, soil fertility, and mechanical stress have been shown to be significant factors in the distribution and growth of forest flora and vegetation (Cleary & Waring, 1969; Waring, 1969; Waring & Youngberg, 1972; Emmingham & Waring, 1973). These stimuli produce quantifiable plant responses including phenology, carbon dioxide exchange, plant moisture stress, stomatal resistance, and foliar nutrition (Waring, Reed & Emmingham, 1972; Reed & Waring, 1974). Stimuli and responses then form ecological indices which locate ecosystems in an environmental grid and permit prediction of such characteristics as productivity and species composition (Waring *et al.* 1972). An example of this grid is Fig. 4.2, which shows the distribution of selected conifers in the Siskiyou Mountains in southern Oregon in relation to a temperature–growth index and plant moisture stress (Waring *et al.* 1972). Such an ecological assessment of environment served to stratify experimental watersheds in the Coniferous Forest Biome.

Given stratification of a watershed into terrestrial and aquatic units, on an ecological basis, we are in a position to study the behavior of the watershed from an ecosystem viewpoint. That is, we will be able to interpret watershed behavior from the behavior of the smaller interacting units. We can in turn interpret the behavior of each unit in terms of the vegetation, physical characteristics and inputs to that unit.

Climate variation over a watershed

Before discussing the hydrologic processes on each unit we should first discuss the climatic environment and how this varies over a watershed. We are generally interested in: (1) amount and kind of precipitation, (2) air temperature, (3) vapor pressure, (4) short and long wave radiation, and (5) wind movement. We will limit our discussion to some methods of estimating important variables, and problems associated with extrapolating point measurements to other areas. We will not discuss the microclimate within a stand, and refer the reader to Geiger (1965) and to Chapter 3 of this volume for information.

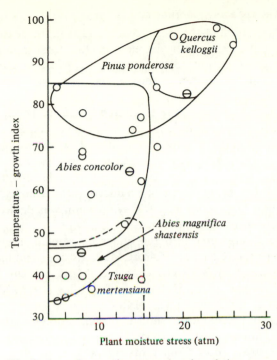

Fig. 4.2. Distribution of natural regeneration in relation to gradients of moisture and temperature defined by plant response indices (Waring *et al.*, 1972). Indices relate to response by 1–2 m tall Douglas fir. The temperature index sums the fractions of potential growth for each day during the growing season, assuming maximum potential with 25°C air and 20°C soil temperature. The moisture stress index represents predawn values measured in September at the peak of the summer drought.

Amount and kind of precipitation

The processes controlling precipitation are understood and the reader is referred to Eagleson (1970) for a careful review which includes cloud physics. It is important to have an appreciation of the phenomena and to be aware of the spatial and temporal variability of precipitation which may occur over a watershed.

The occurrence, total duration and total amount of storm precipitation seems to be largely stochastic and treatable by probabilistic methods (Eagleson, 1970). Stochastic models for point precipitation from summer convective storms have been developed by Grace & Eagleson (1966), Sariahmed & Kisiel (1968), Duckstein, Fogel & Kisiel (1972), Duckstein, Fogel & Thames (1973), and Smith (1974). It may be possible to treat synoptic scale storms in a similar manner.

The internal structure of a given storm may be largely deterministic but is dependent on the type of storm (Fogel & Duckstein, 1969; Eagleson, 1970;

209

Smith, 1974). Eagleson (1970) states that in general the difference between mean rainfall over an area about the storm center and what is measured at the storm center: (1) increases with decrease in the total rainfall, (2) decreases with increasing duration, (3) is greater for convective and orographic precipitation than for cyclonic, and (4) increases with increasing storm area.

Topography can have significant effects on precipitation. Precipitation usually increases with elevation because of orographic cooling and a precipitation–elevation relation is frequently needed when using point rainfall data. Eagleson (1970) indicates this effect may only extend to about 1500 above the general terrain, but local topography can have significant influence (Burns, 1953; US Army Corps of Engineers, 1956; Linsley, 1958). A World Meteorological Organization (1972) symposium was dedicated to discussion of precipitation in mountain areas.

The above considerations enter into the design of rainfall monitoring networks (Rodriguez-Iturbe & Mejia, 1974*a,b*). Methods of extending point data to areas are discussed by Rodriguez-Iturbe & Mejia (1974*b*), Rodda (1970), Mandeville & Rodda (1970), Hutchinson & Walley (1972), Shaw & Lynn (1972), and Wei & McGuinness (1973).

Separation into type of precipitation, snow and rain, is done on the basis of observation or by temperature. Eagleson (1970) suggested using screen air temperature to separate snow from rain with a dividing line of 1.1–1.7°C. Anderson (1968) suggested using wet bulb temperature, which can be estimated from air and dew point temperature, with a dividing line of 0.5°C.

Temperature and vapor pressure

Ambient temperature typically decreases with increasing elevation at a rate between 5 and 8°C km^{-1}. This gradient is referred to as the ambient lapse rate of temperature (Eagleson, 1970), and can be safely used only if the mountain localities being compared have approximately similar environments and differences between them in character and profile of the surface are small. Local topography and climatic conditions may influence temperatures within a watershed. For example, the presence of night-time cold air drainage or formation of inversion layers may result in a positive ambient lapse rate near the ground. This could cause warmer temperatures at higher elevations than at lower elevations and the formation of thermal belts. Large bodies of water may also influence local temperatures. An example of how the lapse rate can vary is shown in Fig. 4.3. The differences shown are due to transition from plains to the foothills and to influences of glaciers in the upper part of the valley. Variations within the day are due to different rates of cooling of various parts of the valley. The open areas of the plains and foothills cool more rapidly than the higher elevations (from Kuzmin, 1961).

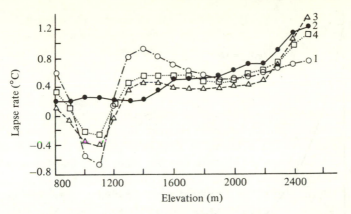

Fig. 4.3. Changes in mean annual lapse rate at (1) 7 a.m., (2) 1 p.m., (3) 9 p.m., and in (4) the mean daily air temperatures with elevation along the Zeraushan River Valley, USSR, for 1933. From Kuzmin, 1961.

The water holding capacity of the air is generally dependent on the temperature and pressure. We can expect vapor pressure to vary in ways similar to temperature. Since this is strongly influenced by elevation, we can expect vapor pressure to decrease with altitude. Kuzmin (1961) reports several relations of which one is shown in Fig. 4.4.

Radiation

The total radiation incident on any surface is the sum of (1) direct short wave radiation from the sun, (2) diffuse short wave radiation from the sky, (3) reflected short wave from nearby surfaces, (4) long wave radiation from atmospheric emission, and (5) long wave emitted from nearby surfaces.

For the northern hemisphere, Houghton (1954) reports mean values for direct short wave of 0.12 ly min^{-1}, diffuse short wave of 0.11 ly min^{-1} and atmospheric long wave of 0.52 ly min^{-1}. For ecological purposes the short wave component is generally the most important. The long wave component becomes important in some water balance studies such as snow melt.

The short wave radiation on a plane surface at any instant is dependent on the slope, aspect, and latitude of the surface, the declination of the sun, time of day, and transmissivity of the atmosphere. Given this information, the direct short wave radiation on any slope may be estimated for any time interval using theoretical methods (Garnier & Ohmura, 1968, 70; Buffo, Fritschen & Murphy, 1972); Short wave radiation can vary significantly with topography (Fig. 4.5).

A problem arises in determining the atmospheric transmissivity. Garnier & Ohmura (1970) reported a method for estimating atmospheric transmis-

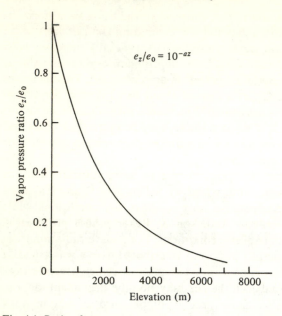

Fig. 4.4. Ratio of vapor pressure, e_z, at elevation, z, to that at sea level, e_o, for $a = 0.2$.

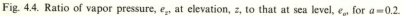

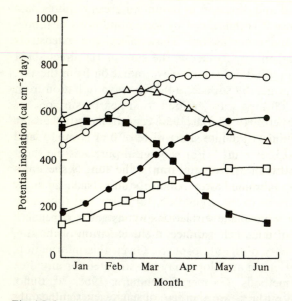

Fig. 4.5. Isograms of daily values of direct solar radiation at latitude 40°N and slopes of 60°E (●—●), 90°E (□—□), 30°S (○—○), 60°S (△—△), and 90°S (■)—■) for atmospheric transmissivity of 0.90. From Buffo *et al.*, 1972.

sivity from observed global and diffuse short wave radiation on a horizontal surface. They also report a method for estimating diffuse short wave radiation on slopes. On clear days, transmissivity generally increases with altitude because the optical path length through the atmosphere is reduced while the moisture content and dust pollution are decreased. This increases clear sky short wave radiation at a rate of about 0.012 ly/min per 100 m (Kuzmin, 1961). Since cloudiness usually increases with altitude in mountain areas, the gain under clear sky conditions may be more than offset and total short wave radiation on horizontal surfaces may remain constant or decrease with altitude (Kuzmin, 1961).

Evaluation of long wave radiation income is more difficult. Various methods proposed have been summarized by Sellers (1965) and Eagleson (1970). Some methods are applicable only to clear sky conditions while others consider effects of clouds. Anderson & Baker (1967) proposed a method for use under all atmospheric conditions. Long wave radiation from atmospheric emission is generally uniformly distributed over a watershed. It is not significantly affected by topography except in extremely steep terrain, when back radiation from adjacent slopes may become important.

Wind

Wind speed increases with altitude only in the free atmosphere. The determination of wind speed and direction in mountain areas requires data from direct observation since it is influenced by local conditions of topography, exposure, heating and other factors. These effects are extensively discussed by Geiger (1965).

Hydrology of terrestrial strata

Here we will discuss the hydrology of individual terrestrial units from an ecosystem viewpoint. In the next sections, we will link these units to each other and to aquatic units.

In words, the water content of a terrestrial unit is the sum of the water stored (1) on the foliage, branches and stems, (2) in the snowpack, (3) in the litter, (4) on the soil surface, (5) in the vegetation, (6) in the soil root zone, and (7) in the subsoil. The structure and relations were shown in Fig. 4.1. We will discuss each part of the water balance in the following sections.

Interception by foliage, branches and stems

In vegetated regions, plant surfaces are the first obstacles encountered by precipitation. As a storm begins, rain strikes the foliage, branches, and stems or falls directly through canopy openings to the forest floor. The latter route is minor in fully closed forest stands. During the initial stages of a storm, much of the precipitation is stored on the canopy or upon the stems. As a storm continues and these surfaces reach their capacities, excess water drains to the forest floor. Evaporation of intercepted precipitation may take place throughout a storm but is of primary importance after precipitation ceases. Thus, interception is the combination of processes determining water storage on tree foliage, branches, and stems. The change in water storage on foliage, branches and stems is the sum of (1) intercepted precipitation, minus (2) drainage from canopy and stem drip, (3) stem flow to the ground, and (4) evaporation from vegetation surfaces. Intercepted precipitation may be rain, snow, rime or dew.

Interception has been studied for nearly a century (Hoppe, 1896) and has been shown to account for losses ranging from 10 to 35% of annual precipitation (Kittredge, 1948; Zinke, 1967). The principle process causing interception loss is evaporation from water stored on plant surfaces. Leonard (1967) discussed and depicted the theoretical relationships between interception loss processes during a storm (Fig. 4.6), and pointed out the need for improvement in measurement techniques for separating processes. Other descriptions of the rainfall interception process have been provided by Horton (1919), Grah & Wilson (1944), Rowe & Hendrix (1951), and Rutter, Kershaw, Robins & Morton (1971). The process of snow interception by trees is described by Hoover & Leaf (1967) and Miller (1967). We do not consider the interception processes for other forms of water deposited on plant surfaces although they may be of local importance. For example, rime has been recognized as a contribution to the hydrological balance of conifers (Berndt & Fowler, 1969) and hardwoods (Gary, 1972); the importance of dew in the water balance of conifers has also been demonstrated for some localities (Kittredge, 1948; Fritschen & Doraiswamy, 1973).

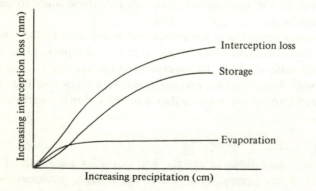

Fig. 4.6. Theoretical relationship between interception loss and precipitation by components during the initial stages of a storm. After Leonard, 1967.

In the following sections, we will address the problems of understanding and estimating interception components.

Storage and evaporation processes. The surface water storage capacity of a forest is related to the surface area represented by foliage, branches, and stems. The configuration, orientation, and texture of the surfaces also influence storage capacities. Since the surface water holding characteristics of each tree component vary, it is conceptually desirable to treat each component separately.

The storage capacity of forest vegetation has been estimated by extrapolating regressions of interception loss on precipitation back to the intercept on the *y*-axis and also by experimental techniques (Leyton, Reynolds & Thompson, 1967). For a plantation of Scots pine (*Pinus sylvestris* L.), Rutter (1963) found that leafy shoots retained an amount of water approximately equal to the dry weight of foliage. Storage values of foliage, branches, and stems were estimated to be 0.8, 0.3, and 0.25 mm, respectively, Voigt & Zwolinski (1964) estimated that storage capacities for red pine (*Pinus resinosa* Ait.) and white pine (*Pinus strobus* L.) were 0.8 and 0.5 mm, respectively, with about 40% of the water retained on the stems. Swank (1972) estimated 1.8 mm storage capacity for a Douglas fir (*Pseudotsuga menziesii* (Mirb.) Franco) stand with about 36% retained by the stem. In a predictive model of rainfall interception for Douglas fir, Rutter, Morton & Robins (1975) used 2.1 mm storage capacity for the stand with 43% contributed by the stems. Canopy storage for Scots pine, Norway spruce (*Picea abies* (L.) Karst.), grand fir (*Abies grandis* (Dougl.) Lindl.) and a deciduous stand of *Fagus* and *Carpinus* was estimated as 3.0, 3.1, 3.8, and 1.9 mm, respectively (Aussenac, 1968) and a value of 1.5 mm was estimated for Norway spruce (Leyton *et al.* 1967). Zinke (1967) summarized interception storage values for conifers and hardwoods; when taken collectively with other values in the literature, average rainfall storage capacities for conifers are about 1.9 mm and for hardwoods about 1 mm. Snow storage for conifers averages about 3.8 mm (Zinke, 1967).

Large deviations from these average values occur by season and stage of stand development. The distribution of storage by tree components also changes with stand development. Surface area of branches and stems continues to increase with stand age and comprises a substantial proportion of the total intercepting surface in mature stands (Fig. 4.7). Values of foliar surface (both sides) per unit of ground area for closed deciduous forests of the eastern United States range between 6 and 12, branch area 1.2–2.2, and stem area 0.3–0.6 (Whittaker & Woodwell, 1967). Values reported for a young eastern white pine plantation were 17.8, 2.3, and 0.4 for foliage (all surfaces), branch, and stems areas, respectively (Swank & Schreuder, 1973).

Evaporation is a process operating from all surfaces which hold water and are exposed to the atmosphere. Evaporation from the litter and soil

215

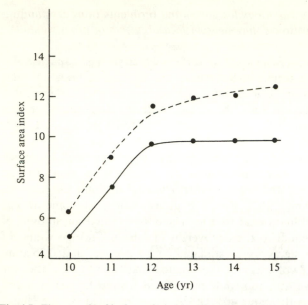

Fig. 4.7. Time trends of leaf area index (——) and total above ground surface area index (----) for a plantation of eastern white pine (*Pinus strobus*) from age 10 to 15 yr. From Swank & Schreuder, 1973.

surfaces are discussed in a separate section. Evaporation of intercepted water is an energy-dependent process and is predicted from measurements of net radiation, temperature, humidity, and wind speed (Penman, 1956; Monteith, 1965).

Currently, interception models are based on a combination of physical processes and empirical relationships. A model of the evaporation of intercepted rainfall (Rutter *et al.*, 1971) was modified and applied to both deciduous broad-leaved and evergreen coniferous stands (Rutter *et al.*, 1975). Good agreement was found when interception loss values calculated by model simulation were compared with observed values. Evaporation of water retained on tree stems was estimated to be 20–30% of the total loss from trees in leaf, and 30–40% for leafless trees (Rutter *et al.*, 1975). A simulation model of evaporation of intercepted rainfall has also been used to show that precipitation intercepted by vegetation evaporates at a greater rate than transpiration from the same type of vegetation in the same environment (Murphy & Knoerr, 1975).

Empirical relationships for drainage and stemflow. Summaries of the results from many traditional interception studies were made by Helvey & Patric (1965) for mature hardwood forests of the eastern United States and by

216

Table 4.1. *Summary of equations for computing throughfall and stemflow for coniferous[a] and hardwood[b] forests from measurements of gross rainfall*

Species	Average equations (gross rainfall in cm)		Interception loss (cm)
	Throughfall	Stemflow	
Red pine	0.87p− 0.04	0.02p	0.15
Loblolly pine	0.80p − 0.01	0.08p − 0.02	0.15
Shortleaf pine	0.88p − 0.05	0.03p	0.14
Ponderosa pine	0.89p − 0.05	0.04p − 0.01	0.13
Eastern white pine	0.85p − 0.04	0.06p − 0.01	0.14
Average (pines)	0.86p − 0.04	0.05p − 0.01	0.14
Spruce–Fir–Hemlock	0.77p − 0.05	0.02	0.26
Mature mixed hardwoods			
Growing season	0.90p − 0.03	0.041p − 0.005	0.10
Dormant season	0.91p − 0.015	0.062p − 0.005	0.05

[a] Conifer equations from Helvey (1971).
[b] Hardwood equations from Helvey & Patric (1965).

Helvey (1971) for conifers in North America. More than 50 different forest stands are included in these summaries and the stands represent closed conditions rather than thinned, pruned, or other manipulated conditions. In these standard interception studies, interception loss is derived from the difference in measured values of gross precipitation and throughfall plus stemflow. The general form of prediction for throughfall, stemflow, and interception loss for individual storms is a linear regression with gross precipitation as the independent variable. A composite of the summaries is given in Table 4.1 with estimated interception loss assuming 1 cm of precipitation. Annual or periodic throughfall and stemflow can be computed by solving the equations for annual (or periodic) gross rainfall and multiplying the constant term by the number of storms in which gross rainfall equals or exceeds the constant term. Total rainfall delivered in storms which are smaller than the constant terms in the equations, is added to the difference between precipitation − (throughfall + stemflow) to obtain interception loss. These equations represent average values from a variety of forest stands and a range of constants exist. Interception loss is greatest in the spruce–fir–hemlock type, intermediate in pine, and least in broad-leaved deciduous forests. Because surface area index also is greatest in the spruce–fir–hemlock type (Burger, 1925), intermediate in pine species (Swank & Schreuder, 1974), and least in deciduous forests (Whittaker & Woodwell,

1967), the surface area of forests provides an important denominator in modeling interception processes. Other extensive summaries of interception studies are provided by Kittredge (1948), Molchanov (1960), and Zinke (1967).

There have been few reports of the quantitative relationship between interception processes and specific vegetation characteristics. Attempts to relate interception loss quantitatively to stand basal area have been inconclusive (Rogerson, 1967; Kittredge, 1968), but some results suggest a relationship with canopy density (Rothacher, 1963; Clegg, 1963). Grah & Wilson (1944) showed a close correlation between interception loss and foliage weight for Monterey pine, and Swank (1972) described models for estimating interception components from live crown weight for Douglas fir. Several studies have demonstrated a strong relationship between stemflow and coniferous tree measurements (Wicht, 1941; Leonard, 1961; Rutter, 1963; Swank, 1972).

From this examination of the interception process, it is clear that periodic interception loss of rainfall can be regarded primarily as a function of periodic precipitation, evaporation during the storm, and the number of times storage capacity is filled within a specific period. Equally clear is the fact that a wide range of storage capacities must exist for forest communities since the quantity and proportion of evaporating surfaces are highly variable.

Snowpack

Snowpack dynamics are complex but fairly well understood. The major reference volume is the report of the Cooperative Snow Investigations Program by the US Army Corps of Engineers (1956). Eagleson (1970) summarized this and more recent work and Kuzmin (1961) summarized Russian work.

An understanding of snowpack dynamics requires familiarity with the physical nature of the snowpack as well as with the energy and water balance processes. We will briefly review each of these.

Snowpack structure. A new snowpack has low density because of snowflake structure. However, it undergoes metamorphosis with time resulting from (1) exchange of heat at snow surface by radiation convection and condensation and at ground surface by conduction, (2) compaction under its own weight, (3) rain or melt water percolation through the snowpack, (4) wind, and (5) variations within the snowpack of temperature and water vapor (US Army Corps of Engineers, 1956; Eagleson, 1970).

The density of new snow varies from about 0.06 to 0.34 g cm^{-3} (US Army Corps of Engineers, 1956; Eagleson, 1970) increasing with both wind and temperature. Variation of density with temperature is shown in Fig. 4.8.

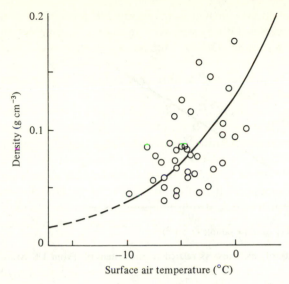

Fig. 4.8. The density of new fallen snow as related to air temperature. From US Army Corps of Engineers, 1956.

Average density of the snowpack generally increases with time, decreasing after each new snowfall. It generally increases with depth with local variations due to buried snow crusts, ice lenses and planes. Snow crystals gradually assume a granular form due to percolating water and convection of heat and water along temperature gradients in the pack.

Physical characteristics of the snowpack which are generally related to density include thermal conductivity, liquid water holding capacity, and thermal diffusivity. Relations for these are reported by the US Army Corps of Engineers (1956), Eagleson (1970), Anderson (1968), and Eggleston, Israelson & Riley (1971). The relation of liquid water holding capacity to density is shown in Fig. 4.9.

As metamorphosis proceeds, the snowpack tends to become homogeneous with respect to density, temperature, liquid water content and grain size. When runoff from the snowpack occurs, the greater part of the snow has been brought to the melting point and the latent heat of fusion has been added. When the pack is at the melting point it is called isothermal. The pack is said to be ripe, or ready to transmit and discharge any water which enters at the surface, when it is isothermal and contains all the water it can hold against gravity. Only the upper surface changes when this state is reached. These changes are day-time thaw which produces melt water and night-time freezing which forms a surface crust.

Another physical characteristic affecting snowpack energy balance and melt rate is the albedo or reflectivity of the snow surface. Albedo is generally

219

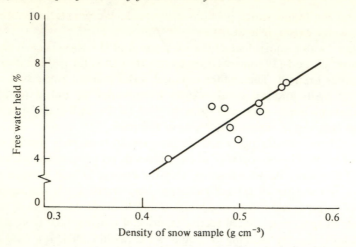

Fig. 4.9. Water holding capacity of ripe snow as related to snow density. From US Army Corps of Engineers, 1956.

related to the state of metamorphosis of the snow surface. This is reflected in Fig. 4.10 in which albedo is shown to decrease with accumulated temperature after snowfall.

Energy and water balance of the snowpack. In studying heat and water balance processes of snowpacks it is most useful to distinguish between the

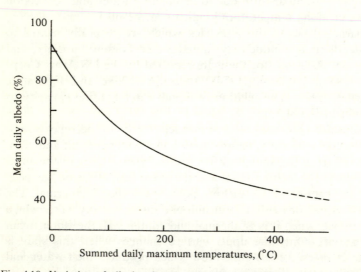

Fig. 4.10. Variation of albedo of snowpack with accumulated temperature index. From US Army Corps of Engineers, 1956.

ice portion and the liquid water portion of the pack. We generally assume that if liquid water exists, it is at 0°C.

The change in water equivalent of the ice portion of the snowpack is the sum of (1) snowfall, and (2) condensation from vapor to solid (sublimation, if negative) minus (3) melt. The change in liquid water content is the sum of (1) rainfall, (2) melt, and (3) condensation from vapor to liquid (evaporation, if negative), minus (4) runoff from the pack. If the melt is negative it represents the freezing of liquid water within the pack.

A number of heat-transfer processes are involved in snowpack dynamics and melting of snow. The importance of each depends on geography, season and climate. The energy balance assumes that the change in heat storage of the snowpack is the sum of (1) net radiation heat transfer, (2) release of latent heat of vaporization by condensation or, if negative, removal by sublimation or evaporation, (3) convective transfer of sensible heat from air, (4) gain of latent heat by freezing of liquid water within pack (or melt if negative), (5) conduction of heat from underlying ground, (6) advection of heat by a gain (precipitation) or loss (runoff) of water (Anderson, 1968). Fig. 4.11 summarizes the energy and water balances of a snowpack. We will briefly discuss each component of the energy balance.

With the exceptions of freezing of liquid water within the pack, and conduction of heat from the ground all components of the energy balance can be assumed to take place in a thin surface layer of snow (Anderson, 1968). This is important in determining the surface temperature of the snowpack since the net radiation and transfer of sensible and latent heat are dependent on surface temperature.

The transfer of latent heat to the surface is a turbulent exchange process. The direction of transfer is controlled by the vapor pressure gradient. If the air vapor pressure is greater than the surface vapor pressure, moisture is transferred to the surface. This results in condensation and release of latent heat to the surface. If the gradient is reversed, the direction of moisture and heat transfer is reversed. If the snow surface is at 0°C it is generally assumed the phase change is from vapor to liquid or vice versa and uses the latent heat of vaporization (597.3 cal g^{-1}). If it is below 0°C the phase change is from vapor to solid and uses the latent heat of sublimation (677 cal g^{-1}). This process can be important in areas where warm, moist, turbulent air occurs over snowpacks, since 1 cm of condensate can produce about 7.5 cm of melt, or 8.5 cm of runoff (Anderson, 1968; Eagleson, 1970).

The transfer of sensible heat to the surface is also a turbulent exchange process. The direction of transfer is controlled by the temperature gradient. If the air temperature is warmer than the snow surface there is a direct heat transfer from the air to the snow and vice versa. The quantities of melt by convection from warm turbulent air can be similar to those from condensation melt.

221

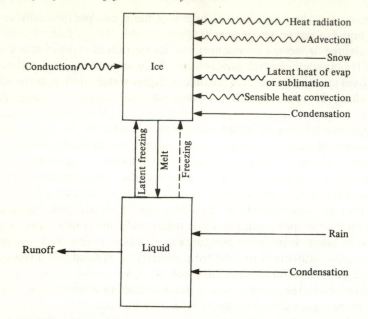

Fig. 4.11. The energy and water balance components of snow showing major processes affecting them. Solid lines denote water transfer and wavy lines energy transfer.

Rain is warmer than the snowpack and hence there is a transfer, by advection, of heat from rain to the snowpack. Since the specific heat of water is 1 cal g^{-1}, the amount of heat added is easily calculated from the amount (cm) and temperature (°C) of the rain. The temperature of rain can be assumed to be air temperature. Since the latent heat of fusion is 80 cal g^{-1} one can see that the melt-producing capacity of rain is small. For example, 4 cm of rain at 20°C must be added to a ripe snowpack to produce 1 cm of melt. However, when the snowpack is below freezing the rain would add its small heat content and also give up the latent heat of fusion (80 cal g^{-1}) when it froze within the pack. This is a quite significant source of heat and will quickly bring a cold snowpack to an isothermal or ripe condition.

Net radiation heat transfer includes both short wave and long wave radiation balances. In forested areas this is complicated by the presence of forest cover. The solar radiation incident above the canopy is attenuated by the foliage. This must be considered in determining what is incident at the snow surface (*see* p. 231). The albedo of the snow surface then determines the portion reflected, and the balance is absorbed.

Part of the long wave radiation from the atmosphere, discussed in an earlier section, is absorbed by the forest cover. However, the forest cover also emits long wave radiation, and is usually assumed to behave as a black body radiating at air temperature. Hence, the portion of the snowpack covered by forest is assumed to receive radiation from the canopy, while that in the open receives it from the atmosphere. The snowpack is usually assumed to absorb all incident long wave radiation. It in turn emits long wave radiation as a black body at the temperature of the snow surface.

Conduction of heat from the ground generally is the result of release of heat stored in the soil during the summer months. This may produce melt from the bottom of the snowpack. The process is controlled by the temperature gradient from the soil–snow interface. The amounts are generally small and usually neglected.

Melt which occurs in the surface of a cold snowpack percolates down where it refreezes and releases the latent heat of fusion. This warms the pack until it eventually becomes isothermal. Additional melt then is held as liquid water until the water holding capacity of the pack is reached and runoff from the bottom of the pack occurs.

Simulation models of the snowmelt process have been reported by Anderson (1968), Eggleston *et al.* (1971), and Leaf & Brink (1973). The model by Leaf & Brink (1973) is generally restricted to sub-alpine conditions. The latter two have broader applicability, include more processes and require more climatic data than Leaf & Brink's model.

Litter and soil surface

Water balance processes in the litter layer are important in litter decomposition and nutrient movement. Hence it is necessary that this be dealt with in hydrologic models of an ecosystem.

The change in water content of the litter and soil surface is the sum of (1) snowpack runoff, (2) drip from the foliage, stems and branches, (3) direct precipitation throughfall, (4) lateral flow into the layer, and (5) exfiltration from the soil up into the litter; minus (6) evaporation, (7) lateral flow out of the layer, and (8) percolation from the bottom of the litter layer into the soil as infiltration. For water balance purposes we assume that any stemflow moves directly into the soil. Any roots in the litter are included in the root zone.

Infiltration is the process of entry into the soil of water made available at its surface (Philip, 1969). Percolation is the process by which water moves downward under gravitational forces through the litter, soil, or other porous media. Lateral flow is the movement of water under gravitational forces generally parallel to the slope. Although infiltration is primarily a soil related phenomenon, we discuss it here because of its importance in determining what happens in the litter layer.

223

There have been several studies of litter water balance. Many of these are reviewed by Helvey & Patric (1965), Zinke (1967), and Helvey (1971). A detailed study is reported by Plamondon, Black & Goodell (1972), but such detailed studies are rare literature.

During a storm, water falling on the litter will gradually increase the litter moisture level to field capacity. Additional water will then percolate through to the bottom of the litter and infiltrate into the soil unless the capacity of the soil to absorb water is exceeded. Then additional rainfall could raise the litter moisture to saturation. Lateral flow can occur through the litter under the force of gravity after field capacity is reached. This may appear as overland flow in areas where litter is sparse.

The forest floor may be divided into several layers, based on stages of decomposition. There are several methods of nomenclature for this. For purposes of discussion we will divide the forest floor into two main layers: (1) an upper horizon in which the origin of most material is identifiable to the naked eye, and (2) a lower horizon in which the origin of material is not recognizable to the naked eye. The hydrology of the two layers is different.

The upper layer is generally composed of leaves, branches, and logs. These have the capacity to hold water on their surface and absorb a certain amount of water. The lower layer on the other hand behaves more like the soil.

Since the moisture content of fine fuels (< 8 cm diameter) in the upper layer is significant in determining flammability, these have been studied extensively in developing fire danger ratings (Schroeder & Buck, 1970; Deeming *et al.*, 1974). In general, litter tends to approach a moisture content in equilibrium with its surroundings. The rate at which it comes to equilibrium depends on temperature, humidity, structure of the material, surface weathering, density and so on (Deeming *et al.*, 1974).

The water-holding capacity of the surface horizon essentially depends on the surface area of the material, analogous to storage on foliage, stems and branches. Data reported by Clary & Ffolliott (1969) indicate that upper horizons of Ponderosa pine litter can hold about 175% of dry weight at field capacity. In the lower layers water is held by capillary force and the capacities increase. Clary & Ffolliott (1969) report moisture contents of 210% for the lower layer of Ponderosa pine litter at field capacity.

For the entire litter layer Helvey (1971) reports an average capacity for conifer litter of 215% at field capacity and about 330% at saturation. For hardwood litter Helvey & Patric (1965) report about 160% at field capacity and 220% at saturation. The minimum moisture content for both hardwood and conifer litter was about 30% (Schroeder & Buck, 1970).

Evaporation from litter storage is another area in which literature is limited. Most work in the area has been done in development of the fire danger rating technology (Deeming *et al.*, 1974).

224

Moore & Swank (1975) developed a process model for litter water content and evaporation. They used data reported by Helvey (1964) to test the model and results are shown in Fig. 4.12.

Plamondon *et al.* (1972) found that excess water percolates very rapidly through litter. However, movement into litter and percolation may be restricted if the litter is very dry initially and hydrophobic conditions exist (Meeuwig, 1971; Debano & Rice, 1973) or if matting occurs (Pierce, 1967). We will neglect hydrophobic problems here. Hence the rate at which percolating water can infiltrate into the soil determines whether an excess is available for lateral flow through the litter (Pierce, 1967).

In examining infiltration under natural conditions it is helpful to divide soil porosity into two parts – a channel system and a capillary system (R. M. Dixon, 1971; Dixon & Peterson, 1971). A channel system exists in both the surface and subsurface and includes a network of large soil pores which drain and fill by gravity. The channel system also includes micro-topographical characteristics of the soil surface and pores created by clay shrinkage, cracking, roots, earthworms and other soil organisms. The

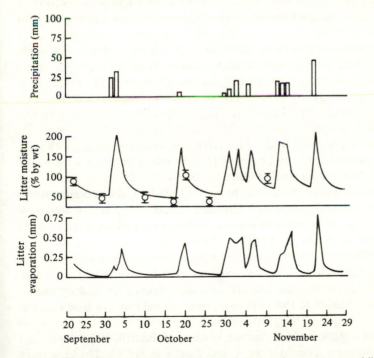

Fig. 4.12. Measured precipitation and simulated litter moisture content and litter evaporation for an 80-day period in 1961. Experimental estimates of water content (from Helvey, 1964) with 67% confidence intervals shown. From Moore & Swank, 1975.

225

capillary system includes textural and structural pores within the soil mass which drain and fill mainly as a result of capillary forces. The channel system is embedded in the capillary system and shares a common boundary at the soil surface and along the walls of the large pores. R. M. Dixon (1971) reports that a good channel system can increase intake rates by an order of magnitude or more over that observed on similar soils lacking such. This is substantiated by other studies where root development and vegetative cover combined to produce high infiltration rates (Rauzi, Fly & Dyksterhuis, 1968; Verma & Toogood, 1969; Meeuwig, 1970; Rauzi & Smith, 1973). Vegetation removal or soil compaction leads to significantly reduced infiltration capacities (Dyrness, 1972).

There are two general approaches to the treatment of infiltration. The first and most common is to neglect the channel system and consider soil as an isotropic, homogeneous porous medium. The second attempts to consider the channel system. The first approach led to the development of a mathematical theory for soil water movement and infiltration through homogeneous soils or soils of two or more homogeneous, horizontal layers (Bear, Zavlavsky & Irmay, 1968; Philip, 1969; Childs, 1969). The infiltration equations developed from this theory can be used to predict infiltration rates in laboratory tests where the channel system has been destroyed, but are rarely accurate under natural field conditions. The second approach is, at present, beyond deterministic mathematical analysis. Hence one must rely on empirically or physically based conceptual models for estimating infiltration, such as those developed by Green & Ampt (1911), Gardner & Widstoe (1921), Kostiakov (1932), Horton (1933), Philip (1954, 69), and Holtan (1961, 1969, 1971). These have been reviewed by Childs (1969), and Whisler & Bouwer (1970).

The infiltration process can also be affected by water temperature (Klock, 1972), frozen soils (Haupt, 1967; Sartz, 1973; Dingman, 1975), and shrinkage cracks (Blake, Schlichting & Zimmerman, 1973).

When the infiltration capacity of the soil is exceeded, the excess water in the litter may begin to move laterally downslope under the influence of gravity. Conditions which may cause this to happen in forests are reviewed by Pierce (1967).

Some water may be held in small depressions as detention or static storage. The excess water moves overland as sheet flow or in small rivulets. This eventually converges into the lowest order channels formed by micro-topography provided that the flow does not infiltrate before reaching the channel.

The physical nature and mechanics of the overland flow processes are described elsewhere by Chow (1964), and Eagleson (1970). The equations describing the process are also discussed by Wooding (1965), Grace & Eagleson (1966), Woolhiser, Hansen & Kuhlman (1970), and Overton (1971).

226

The application of the equations to natural watersheds is complicated by the presence of the litter layer and other low ground cover such as rocks, and low vegetation. However, Simons, Li & Stevens (1975) have recently developed a method for modeling lateral flow through or over litter which takes into account the resistance to flow from the litter and other low cover.

Plant water relations

Plants exert a unique influence upon the hydrologic cycle by extracting water from the soil below the zone affected by evaporation. This in turn influences the amount remaining for seepage and streamflow. Water moves through plants as a liquid in the conducting system and by diffusion across cell membranes. From the mesophyll tissue within the leaf, water is transformed into the vapor phase passing out through the stomata or the leaf cuticle to the atmosphere. The driving force for the movement of water through plants is a fall in water potential initiated in the leaves as a result of the loss of water by evaporation.

Specifically, this section describes (1) how roots extract water from the soil, (2) how the internal water balance of vegetation changes diurnally and seasonally and (3) how plant water relations interact with the environment to control transpiration. From this discussion, important structural and physiological characteristics of the vegetation are identified that aid in assessing the water balance of vegetation and its influence upon the hydrologic cycle.

Root zone water. In mature forests, plant roots usually have reached their maximum depth. Disturbing the vegetation results in untapped reservoirs of water until roots reoccupy the area. In assessing water uptake by vegetation it is thus necessary to have some knowledge of root distribution.

The distribution of tree roots is affected by the depth of soil and by the soil water regime. In areas where precipitation is adequate, trees often develop rather shallow root systems. In such ecosystems a brief period without precipitation may stress the vegetation more than in areas where drought is more common and roots occupy a greater volume of soil.

The pattern of water withdrawal from the root zone is similar for many plants. If the soil is fully charged, plants first remove water from the upper layers until it becomes difficult to extract (Woods, 1965). Thereafter lower layers are sequentially drawn upon. The pattern is similar for agricultural crops (Hsiao *et al.*, 1976), conifers and hardwoods (Krygier, 1971), and desert shrubs (Caldwell & Fernandez, 1975). For forest plants, the total volume of water available in the rooting zone is approximately that held between 0.1 and 15 bars soil water potential.

Measurements of plant water potential made before dawn usually represent an equilibrium between the water available in the soil and the internal balance of the plant. This assumes that the stem is not frozen and that transpiration is not occurring. Under such conditions, curves similar to that presented in Fig. 4.13 result as soil water is withdrawn (Sucoff, 1972; Hinckley & Ritchie, 1973). Typically, a large fraction of the available water can be removed without causing a change in plant water potential. Assuming the rooting depth to represent the zone occupied by at least 90% of the root biomass, Running, Waring & Rydell (1975), defined the inflection point for an abrupt change in plant water potential to correspond with a depletion of 80% of the available water. Nocturnal uptake was restricted exponentially as the remaining water was extracted until at about −30 bars essentially no uptake was possible. Accordingly, only a fraction of the total soil profile need be recharged to permit plant water potentials to recover to values of −3 to −5 bars at night.

If plants vary greatly in their rooting depth they should be grouped into appropriate classes to better estimate root water extraction and its consequences upon the community water budget.

Internal storage of water. There is often a lag between transpirational losses from the foliage and uptake by the roots as indicated by sap flow at night

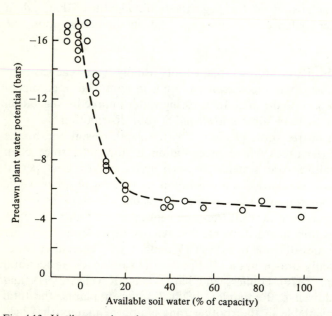

Fig. 4.13. Until more than three-quarters of the available water is depleted, red pine (*Pinus resinosa*) trees recover to the same predawn water potential. Thereafter, an abrupt decrease begins until all water is depleted. After Sucoff, 1972.

when there is no transpiration (Lassoie, 1973). A lag indicates water is not merely passing through a rigid conducting system but is being withdrawn from somewhere within the tree. Two kinds of tissue are involved: one shows volumetric change and is made up of extensible living cells; the other is predominantly non-living sapwood which shows no dimensional fluctuation as its water content changes.

The extensible tissue reserve, represented by foliage, phloem, cambium and new xylem cells has significance diurnally because it can provide supplemental water to maintain normal transpiration for one to several hours (Jarvis, 1975). Sapwood, on the other hand, contains a much larger volume of extractable water which is less readily available. In conifers where the sapwood exhibits up to a 50% change in volume of water, a mature forest may store 6 cm per hectare (Waring & Running, 1976). Different species of trees have varying proportions of sapwood to leaf area (A. F. G. Dixon, 1971; Grier & Waring, 1974) and therefore different amounts of sapwood. Although the amount of water withdrawn from the sapwood on a given day may be relatively small, under sustained demand the reserve can be emptied in a few weeks (Fig. 4.14) and recharge may be even more rapid (Fries, 1943). The sapwood of conifers can withstand repeated breakage of water capillaries because the vascular elements have valves (border pits) which seal off one cell from another when steep water potential gradients

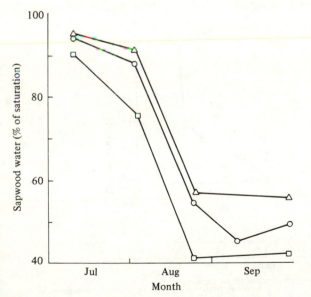

Fig. 4.14. Seasonal course of water extraction from different sapwood zones of Sitka spruce (*Picea sitchensis*) grown in Scotland. △—△, outer 2 cm; ○—○, mid 2 cm; □—□, inner 2 cm. After Chalk & Bigg, 1956.

occur (Gregory & Petty, 1973). This also permits gas under high pressure to redissolve and the water column to be re-established following disruption by freezing or desiccation (Hammel, 1967). In general a forest of mature trees has greater internal reserves to draw upon than a younger forest with equal leaf area. This results in the older forest being less susceptible to climatic extremes affecting the water budget.

Transpiration. The process of transpiration refers to water loss through the stomata or cuticle of the leaves. Transpiration is directly proportional to the leaf area and the water vapor gradient between the air and leaf surfaces. It is inversely proportional to the canopy resistance, encompassing stomatal, cuticular resistance, and a boundary layer resistance. Usually the biological portion of these resistances are pooled as leaf resistance or its reciprocal, conductance.

The water flux from a canopy can be estimated with knowledge of the surface resistance using the Penman–Monteith equation (Monteith, 1965). If the surface area of the foliage is multiplied by the surface resistance of the canopy, an estimate of the average foliage resistance is obtained, subtracting any boundary layer resistance (Stewart & Thom, 1973). Fig. 4.15 illustrates the correspondence between these two expressions.

The Penman–Monteith equation, although complex, does not require specific knowledge of foliage temperatures because in the process of combining both mass and conductive transfer of heat and water vapor the

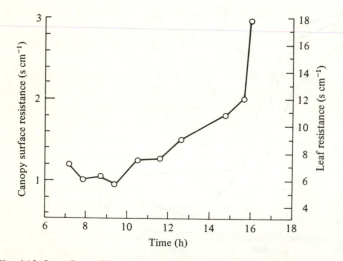

Fig. 4.15. In a Scots Pine (*Pinus sylvestris*) plantation canopy surface resistance increased throughout the afternoon corresponding to a stomata resistance change from 6 s cm^{-1} to more than 18 s cm^{-1}. Values calculated for projected surface, not total leaf area. LAI = 6. After Stewart & Thom, 1973.

temperature terms cancel. This makes the equation particularly useful in forests with broad leaf vegetation, for such leaves exhibit large boundary layer resistances resulting in leaf temperature well above ambient. The needle-shaped leaves of conifers, in contrast, have a small boundary layer resistance (usually less than 0.5 cm s^{-1}) and are thus within a few degrees of ambient temperature (Gates, 1968).

It is possible for broad leaf plants to exhibit some stomatal control without a reduction in transpiration. In fact, if the leaf temperature is elevated sufficiently, transpiration may even be increased in spite of partial stomatal closure (Gates, 1968).

Stomatal control by light, temperature, and humidity. Stomata respond to critical levels of light, temperature, and humidity, as well as to internal water deficits. The first three factors are mainly external in that they operate upon the guard cells of the stomata and need not affect the internal water balance of the leaf or other tissue.

Most plants have a minimum level of light which is necessary to trigger opening of the stomata. This is often equivalent to the carbon dioxide compensation point. Plants differ in this respect and also in the time it takes to respond to changes in light conditions. In general, fast growing pioneer species usually require greater amounts of light to open than slow growing advanced successionary species. The former are also less responsive to sudden changes in light than the latter (Woods & Turner, 1971).

Light attenuation is difficult to estimate under a variable canopy but tends to follow Beer–Lambert's law in relation to increasing foliage density (Fig. 4.16). Lower foliage in the canopy can be expected to be more sensitive and to require somewhat less light for stomata to open.

Increasing the temperature may result in further opening of stomata, at least in some species (Schulze *et al.*, 1973). On the other hand, temperature causing freezing of the guard cells or any part of the conducting system will result in stomatal closure (Drew, Drew & Fritts, 1972). Indirectly, low root temperatures may affect stomata by inhibiting water uptake, resulting in a water deficit within the leaves (Bababola, Boersma & Youngberg, 1968; Havranek, 1972).

Low humidity, or more precisely a high evaporative demand can affect guard cells of the stomata and cause partial closure (Lange *et al.*, 1971; Watts, Neilson & Jarvis, 1976).

These external environmental controls may help in explaining differences in water use by forests. Climates characterized by high evaporative demand may have forests that exhibit stomatal control even with adequate soil water. This means that a higher proportion of the solar energy is dissipated as sensible or conductive heat loss than by way of evaporation. Similarly,

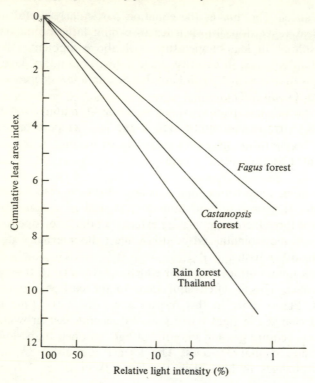

Fig. 4.16. Light attenuation in relation to leaf area follows a negative exponential curve as predicted by Beer-Lambert's Law. Kira, Shinozaki & Hozumi, 1969.

light limitations due to cloudy weather or a shaded position may result in less transpiration than expected by an analysis of evaporative demand alone. For modeling purposes, threshold and limits can be established to identify conditions that may be expected to alter the canopy resistance abruptly (Waring & Running, 1976).

How plant water potential and internal water deficits affect stomata. The evaporative demand, soil water supply, and resistance to water flow through the conducting system interact to affect the water balance of leaves. If insufficient water is supplied to the leaves to meet the evaporative demand, the immediate result will be an increase in the tension on the water column throughout the conducting tissue. This is indicated by an increase in the water potential gradient. Additional conducting tissue may be brought into use with the net result that water will flow faster and that uptake will balance demand. In such a situation there will be little volumetric change in extensible tissue, stomata will remain open, and the water potential may, at

232

least in some deciduous species, fall linearly with increasing transpiration (Landsberg *et al.*, 1975) as illustrated in Fig. 4.17.

If, however, the internal resistance becomes too great to supply water at the necessary rate, the deficit will be met first by withdrawal of water from the extensible tissue near the evaporative sites. As water is withdrawn it becomes more difficult to obtain additional water and untapped reserves in tissue further from the evaporative sites must be utilized (Gibbs, 1958; Doley, 1967; Schnock, 1972; Jarvis, 1975). The water potential gradients necessary to extract water from different tissue vary but curves such as those presented in Fig. 4.18 are typical of many forests (Hellkvist, Richards & Jarvis, 1974). Often a diurnal change of 10% (Clausen & Kozlowski, 1965; Gary, 1971) in the relative water content is observed in the foliage of trees. Under extreme conditions more than a 40% change in volume or relative water content occurs in some plants (Huck, Klepper & Taylor, 1970). As water becomes more difficult to obtain, the guard cells of the stomata are themselves depleted and this causes at least partial stomatal closure. The point where water becomes limiting can often be related to a particular water potential gradient. On forest samplings of *Pseudotsuga*

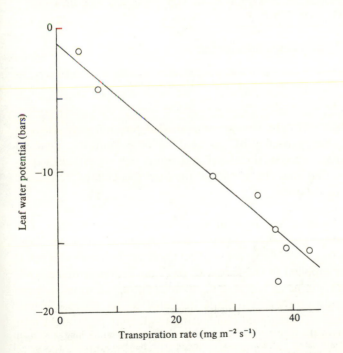

Fig. 4.17. In apple trees, increased transpiration may be sustained through an increased water potential gradient. After Landsberg *et al.*, 1975.

233

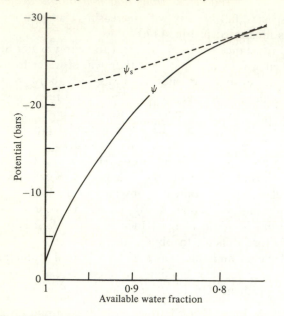
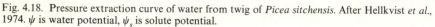

Fig. 4.18. Pressure extraction curve of water from twig of *Picea sitchensis*. After Hellkvist *et al.*, 1974. ψ is water potential, ψ_s is solute potential.

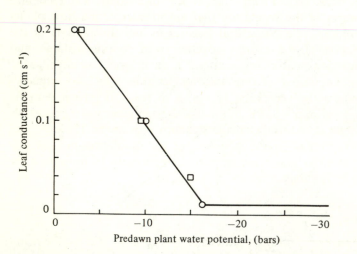

Fig. 4.19. Predawn plant water potential appears to control the maximum opening of stomata as illustrated by the relationship with leaf conductance of both hardwoods (Dinger, personal communication) and conifers (Running, 1976). Conductance, $1/r$, is calculated for all leaf surfaces.

234

menziesii and *Pinus ponderosa* the value is between −18 and −20 bars (Running, 1976; Lopushinsky, 1969). There are, of course, internal adjustments by the plants with height and season as a result of increased osmotic concentrations in the leaves (Richter, Halbwachs & Holzner, 1972).

A useful relationship has been established between predawn water potential and the maximum leaf conductance for both hardwoods and conifers (Fig. 4.19), (Running, 1976; Dinger, personal communication). This suggests that the water uptake by roots at night fully rehydrates the extensible tissue and that the relative water content of foliage controls the initial leaf conductance. If the evaporative demand during the day remains low, leaf conductance remains essentially constant (Running, 1976). On the other hand, a high evaporative demand may result in internal water deficits and lead to at least partial stomatal closure. A family of curves can be drawn or the average leaf conductance estimated for different kinds of days and initial conditions (Running *et al.*, 1975).

If root uptake become inadequate to balance losses through transpiration, more and more water is withdrawn from the sapwood. Most of the water is initially withdrawn from the interior zone with the outer 0.5 cm remaining saturated (Chalk & Bigg, 1956). Extraction may continue until only about 40% of the void spaces are filled with water. The water potential gradient may continue to increase to −50 or −60 bars which corresponds with lethal limits for conifer seedlings.

Key plant characteristics. From the above discussion we recognize certain key features of the vegetation that aid in interpreting how plants affect the hydrologic cycle. Structural features include an analysis of the vertical distribution of the canopy together with rooting volume and sapwood storage. Seasonal changes in leaf area, if large, should also be estimated. Some knowledge of threshold temperature, light and humidity effects are helpful if the vegetation is shown to react to typical climatic patterns. Monitoring of pre-dawn and mid-afternoon plant water potentials is desirable in order to indirectly calculate canopy resistances. These general features are incorporated in the structure of the simulation models discussed in later sections and were employed to estimate transpiration by individual trees by Running *et al.*, 1975).

Soil water balance

The physical processes controlling water movement in the soil are the result of strong surface and capillary forces with which the porous media hold water against the force of gravity. The soil water potential is a measure of the energy status of water in soil. This varies primarily with soil water content but is also influenced by temperature and chemical composition of the

235

water. The hydraulic conductivity is a measure of the ability of a soil to conduct water in response to gradients in soil water potential. It also varies strongly with soil water content. Both soil water potential and hydraulic conductivity are strongly influenced by physical characteristics of the soil such as texture, structure and pore size distribution. An example of soil water potential and hydraulic conductivity relation with water content is shown in Fig. 4.20 and 4.21.

Water moves from areas of high potential to areas of low potential at a rate dependent on the potential gradient and the conductivity. The direction of movement is controlled by the gradient. The conductivity is highest when the soil is saturated and can decrease by many orders of magnitude as the soil moisture decreases to field capacity (0.1 bars tension). This is because of the increased tortuosity of the path of water movement through the soil, as water must flow along the thin films on the soil particles.

Here we again need to recognize the possible existence of a channel system within the capillary system, as discussed previously. Evaporation and root uptake processes occur in the capillary system. Infiltration and percolation may occur in either the capillary or channel system. Storage may occur in either system but storage in the channel system would generally be very temporary. Lateral flow is a large scale form of redistribution. However, redistribution generally occurs within the capillary system or from the channel system to the capillary under the influence of potential gradients.

Forest soils are rarely isotropic and homogeneous, but are generally layered. Movement of moisture in layered soils is complicated by large changes in conductivities between strata. The boundaries between layers

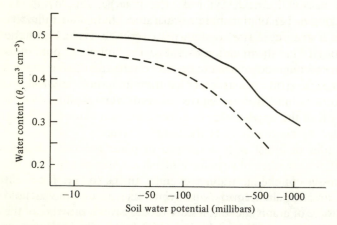

Fig. 4.20. Volumetric water content as a function of soil water potential for two surface horizons of a Honeywood silt loam. From Elrick, 1968.

236

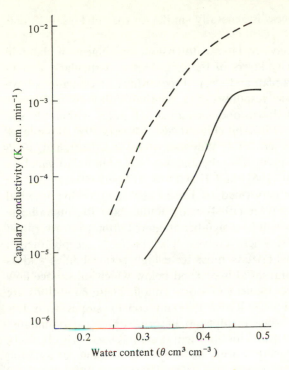

Fig. 4.21. Capillary conductivity as a function of water content for two surface horizons of Honeywood silt loam. From Elrick, 1968.

may be sharp or gradual. Infiltration into layered soils has been studied by Hanks & Bowers (1962), Miller & Gardner (1962), Miller (1969), Whisler, Watson & Perrens (1972), and Aylor & Parlange (1973). We find a significant difference in behavior between sharp and gradual stratification, particularly when a stratum of low conductivity overlies a coarse stratum of high conductivity. If the stratification is sharp the upper stratum must nearly saturate before significant water can move into the coarse stratum. In any case, the coarse stratum may never saturate in a very deep soil. For the reverse situation, infiltration to depths proceeds very rapidly with the eventual saturation of the upper layer. In either case, percolation to deeper layers is controlled by conductivity of the limiting layer.

The permeabilities of forest soils developed in place generally decrease with depth. Furthermore, they always have higher combined conductivities in a direction parallel to the slope than at right angles to the slope. This difference with direction is termed anisotropy. Its causes are discussed by Bear *et al.* 1968, and include the effects of biological activity as well as the formation of litter and humus layers. This almost assures that over-land flow is a rare occurrence on forest watersheds (Freeze, 1972*b*)

Hence the infiltration process is generally similar to that of Figs. 4.22 and 4.23.

This situation is conducive to lateral downslope movement of water in the upper, high conductivity layers of the soil. This is particularly true in forested areas with high rainfall and steep slopes. Many investigations have found that the hydrologic response of a forested hillslope to rain is dominated by the lateral downslope movement of water within the soil mineral layers (Whipkey, 1967a,b; Hewlett & Hibbert, 1967; Dunne & Black, 1970; Nutter & Hewlett, 1971; Weyman, 1970, 3; & Megahan, 1972; Hornbeck, 1973; Hewlett, 1974). The phenomenon has been investigated on artificial slopes by Hewlett (1961) and Hewlett & Hibbert (1963).

Simulation studies were performed by Freeze (1972a,b) for hypothetical slopes and Stephenson & Freeze (1974) using actual data. Results indicate that subsurface lateral flow will be a significant contribution to storm runoff only under the conditions where convex hillslope feeds steeply incised channels. The surface soil horizons must have high permeabilities and as permeabilities decrease a threshold is reached below which subsurface flow cannot be an important component of storm runoff. These conditions are illustrated in Fig. 4.24. Freeze found that on convex slopes with low permeabilities and on all concave slopes, the storm runoff was dominated by direct runoff of precipitation on transient wetlands near the channels. These wetlands expand during storms as surface saturation occurs from

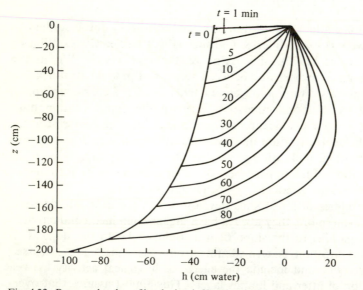

Fig. 4.22. Pressure head profiles during infiltration into a heterogeneous porous medium with saturated conductivity of 0.5 cm min^{-1} at the top uniformly decreasing to 0.05 cm min^{-1} at the bottom (-200 cm). From Whisler *et al.* 1972.

238

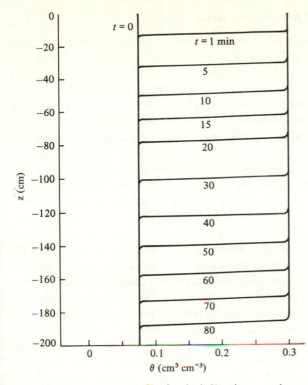

Fig. 4.23. Water content profiles for the infiltration case shown in Fig. 4.22. From Whisler *et al*. 1972.

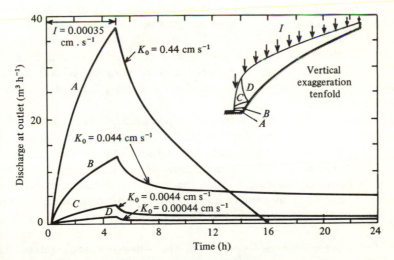

Fig. 4.24. Simulated stream hydrographs of runoff generated by subsurface stormflow at outlet of upstream source area. Source area has convex, 7.5% hillslope with maximum soil thickness of 100 cm. Results for four saturated conductivities are shown for a rainfall event of five hours duration and intensity (I) of 0.00035 cm/s. The inset show the cross-section of hillside and position of water table for each case at end of five hours, From Freeze, 1972*b*.

below due to infiltration toward very shallow water tables, and not from lateral subsurface flow. The behavior is compatible with the variable source area concept (Hewlett & Hibbert, 1967).

Freeze also concludes that a watershed just permeable enough to sustain subsurface flow to a channel, without formation of wetlands, will exhibit the minimum storm runoff response. In some cases this permeability may be so low that subsurface lateral flow just satisfies the evaporation demands from the channel.

For water balance purposes, it is useful to divide the soil into at least two layers: a root zone and a subsoil. Biological processes influence water movement in the root zone but purely physical processes are acting in the subsoil.

Root zone. The change in water content of the root zone is the sum of (1) infiltration, (2) rise of water from the subsoil, (3) subsurface lateral flow in, minus (4) evaporation, (5) uptake by roots, (6) percolation into subsoil, (7) subsurface lateral flow out, and (8) exfiltration. We have previously discussed infiltration, subsurface lateral flow, root uptake, and percolation, and will briefly discuss evaporation, rise, and exfiltration.

The drying of soil by evaporation has been studied by Lemon (1956), Philip (1956), Gardner (1959), and Kimball & Jackson (1971). There are three recognizable stages in the drying of soil. During the first stage, there is a rapid loss of moisture while capillary flow to the surface is adequate to meet evaporation demands. During the second stage, the surface dries and there is a rapid drop in the rate of loss of moisture because the rate of movement to the surface is controlled by intrinsic hydraulic characteristics of the soil. During the third stage the rate of loss is very low and is controlled by adsorptive forces at the liquid-solid interfaces. A litter layer can significantly reduce soil temperature and evaporation of soil water. Tomanek (1969) summarizes much of the work on effects of litter on evaporation from soil.

Water can move upward in response to potential gradients within the soil. This process is termed capillary rise. It most often occurs in response to water being either removed by evaporation and uptake processes, or added by the upward movement of a saturated zone due to a net lateral flow in. The process of water movement from the soil to the surface is termed exfiltration. It can be significant in maintaining a wet evaporating surface when a saturated zone is present close to the surface.

Subsoil. The change in water content of the subsoil is the sum of (1) percolation from the root zone, (2) subsurface lateral flow in, (3) rise from regional groundwater, minus, (4) rise to the root zone, (5) subsurface lateral flow out, and (6) percolation (or seepage) to the regional groundwater. We previously discussed these processes with the exception of rise from and seepage to regional groundwater.

The interaction between the subsoil on a small forested watershed and the regional groundwater is dependent to a large extent on the geology of the area, the permeability of the bedrock beneath the subsoil, and the characteristics of the regional groundwater aquifers. Since we are restricting ourselves to small watersheds within a larger basin, we will not deal with capillary rise to the subsoil from regional groundwater, since in mountainous areas this probably is not significant. However, percolation can occur through the bedrock at a rate limited by the permeability of the bedrock. Freeze & Witherspoon (1966, 7, 8) and Freeze (1969*a,b*) provide comprehensive discussion and analysis of regional groundwater.

If the bedrock has low permeability, a saturated zone can form in the subsoil. This is termed perched groundwater because it exists or is 'perched' above the regional groundwater. It is often the source of small springs or seeps which may dry up in extended periods of drought.

Summary

This completes our discussion of the hydrology of an individual terrestrial unit. In summary, precipitation entering a unit passes first through a canopy consisting of foliage, branches, and stems. Some is held and evaporated and the balance directly or indirectly reaches the ground surface. It may be stored in a snowpack which eventually melts and delivers water to the surface. At the surface it may be held in the litter and evaporate or percolate through the litter to the surface where it may infiltrate. Excess water on the surface may runoff through the litter as overland flow, but this is rare in forested watersheds. Once in the soil, water can percolate downward and move laterally in response to potential gradients. Percolation can deliver water to regional groundwater. Within the root zone, water in excess of the wilting point is available for uptake by plants. This water may not immediately leave the system by transpiration but can be stored in the sapwood and extensible tissue reserves of the vegetation. Transpiration withdraws water from these reserves, reducing the conducting capabilities of the vascular system, which in turn creates a water deficit in the foliage and reduces transpiration by affecting stomata.

Coupling of terrestrial and aquatic units to describe a watershed system

Overland and subsurface lateral flow provides the basis for coupling terrestrial units to other terrestrial units as well as to aquatic units. Surface topography determines the path of overland flow and is an indication of the probable paths of subsurface flow. The actual paths of subsurface flow may be modified by the local geology. Flow paths through the aquatic system are more easily determined.

Knowledge of the flow paths can be used to route both overland and subsurface flow from upper units to lower units and into the stream or aquatic units. This has been done in simulation studies by Simons *et al.* (1975), US Forest Service (1972) and Rogers (1973).

Comparison of total watershed behavior

We have described the water balance processes of individual terrestrial units. We are able to describe the path of movement of excess water between individual units and into the aquatic system. By putting this all together mathematically, in the form of a general hydrologic model, we will be able to reproduce and better explain a number of phenomena which we have observed in practice. This would contribute substantially to our theoretical understanding of the hydrology and water relations of watershed ecosystems. We will discuss two efforts in this direction – PROSPER and ECOWAT.

Comparison of the PROSPER and ECOWAT hydrology models

By comparing PROSPER and ECOWAT, we hope to illustrate the structure and capabilities of the models, and also clarify the major difference in order to place each in perspective.

PROSPER

PROSPER was developed within the Eastern Deciduous Forest Biome and is described by Goldstein, Mankin & Luxmoore (1974). The objective in development of the model was to simulate water relations in the soil–plant–atmosphere system emphasizing the evapotranspiration processes. Initial work was done in a deciduous forest watershed. Hence, PROSPER models only a portion of the hydrologic system depicted in Fig. 4.1; specifically the foliage, root zone and plant water relations components for a simple hydrologic unit. The current implementation does not consider internal storage in the sapwood and is generally for the deciduous forest area where it was developed. However this could be changed.

Swift *et al.* (1975) modified PROSPER slightly and used it to simulate evapotranspiration from both mature and clearcut, deciduous forest watersheds, and from a young pine plantation. We will review the results (p. 248),

242

ECOWAT

ECOWAT is a result of extensive refinements and revision of a model originally developed by Rogers (1973). It incorporates component models for plant water relations developed within the Coniferous Forest Biome. ECOWAT models all of the terrestrial water balance processes depicted in Fig. 4.1 and described on p. 241 with the exception of groundwater. It can model many units simultaneously, and routes excess water between units to the aquatic system. (It does not yet route flows through the aquatic system.) It can model units with either coniferous or deciduous vegetative cover. The current version of ECOWAT is documented by Rogers (1975) and we briefly describe the component models used in the following sections.

Canopy water balance component. The model which we use is similar to that developed by Rutter *et al.* (1975). We assume that the water holding capacity of the canopy is a function of the leaf, branch and stem area, as well as vegetation type (conifer or deciduous). We also assume that the capacity is the same for rain and snow, under the assumption that while more snow may be caught, the excess will eventually fall off before significant loss occurs. The model keeps track of amounts of rain and snow caught by the canopy, passing directly through without striking the canopy, and excess dripping from the canopy as well as evaporative losses from the canopy storage. We assume evaporation proceeds at the potential rate until storage is exhausted as determined by either Penman–Monteith equation, or Fick's Law (method is user selected).

Snowpack component. A good snowpack model requires consideration of both the water and energy balances of the snowpack. The model we use is based primarily on the work of Anderson (1968), with some modifications to allow for pack settlement, density changes, albedo changes, and water holding capacities.

Briefly, we consider the energy balance of the thin surface layer of the snowpack. Heat gained in this layer produces melt which, together with any rain, percolates into the pack. If the pack has a heat deficit, the water will freeze and release heat which warms the pack, eventually bringing it to an isothermal condition. When the pack is isothermal additional water will fill the water holding capacity and the excess will runoff through the bottom of the pack. A heat loss from the surface layer will cool the pack, and excess water will freeze to satisfy any heat deficit until all excess water is frozen and the pack develops a heat deficit.

Litter water balance. The litter water balance operates very similarly to the canopy water balance. It recognizes the several drying stages of litter. The first is where moisture evaporates at the potential rate, the second where the rate of evaporation is limited by an increasing surface resistance, and the third where the hygroscopic properties are important and moisture transfer may be in either direction. The model gives amounts evaporated from storage, passing through the litter, and passing through openings in the litter without being caught by the litter.

Surface water balance component. This component deals with the processes of overland flow, detention storage and infiltration. It is closely coupled with the soil water balance component. We currently assume that anything in excess of the detention storage capacity flows out of the unit. A provision for real time routing will be incorporated later. Holtan's equation for infiltration as used in the USDAHL–70 model (Holtan & Lopez, 1971) is the basis for estimating infiltration. Routing of overland flow between units and to the stream is also handled by this component.

Soil water balance component. This component incorporates the processes of percolation, lateral flow, and non-capillary flow. Uptake is incorporated in the plant water relations component. We subdivide the soil horizon into up to four layers, depending on the rooting depth and soil properties. We assume that all units have the same number of layers, but not necessarily of the same thickness.

We assume that the soil water conductivity–soil moisture relationship is a log–log relationship. This could be replaced, but we use this relation because data are rarely available under most conditions. We assume that all water in excess of the capillary water holding capacity (which may be the saturated water holding capacity) flows laterally down to the next unit. Water left then moves at a rate determined by the conductivity. It is divided into percolation down to the next layer, and lateral flow out to the next unit, on the bases of slope and conductivities of both layers. The method for doing this is derived from work of Zavlavsky & Rogowski (1969). We iterate to get the average conductivity for each time step. Lateral flow between units and to the stream is also handled in this component.

Plant water relations. This component incorporates the processes of uptake, transpiration, internal storage, moisture stress, leaf resistance and surface resistance. It is applicable to deciduous cover and coniferous cover with the primary difference being that we assume deciduous cover has no effective internal storage of water. In other words deciduous plants essentially depend on uptake to meet transpiration demand, while coniferous plants have the capability of satisfying demand from internal storage, primarily in the sapwood.

The method used is essentially an adaptation of an individual tree model developed by Running *et al.* (1975). We first estimate plant moisture stress as a function of temperature and readily available moisture. We then estimate leaf resistance and surface resistance. For coniferous species we then take up water from the soil to satisfy the internal storage, or sapwood water deficit. Uptake proceeds first with readily available water from the top layers downward. The water holding capacity of the sapwood is a function of the leaf area and tree height. For deciduous species we determine transpiration demand and then uptake to meet the demand if possible, and the actual transpiration is then equal to uptake. For coniferous species we meet transpiration demand from sapwood storage if possible. Transpiration demand is determined on the basis of the Penman–Monteith equation or a modification of Fick's Law, both of which include surface resistance terms.

Discussion

It should be obvious that ECOWAT is directed toward a different type of application from PROSPER. PROSPER could be a component of ECOWAT, replacing the existing water relations routine. PROSPER can also be used to study water relations on a small site where other components not included in the model, such as snow, are not important. ECOWAT can be directed toward applications on both larger watersheds where all components and heterogeneity are important; and on small sites, where snow and other components are important.

Applications of the models

In order to evaluate the models we made initial applications of ECOWAT in three areas and PROSPER in one area. Both were applied to several vegetation cover conditions in each area, in order to evaluate the utility of the models for studying and predicting total watershed behavior. In general, the cover conditions studied in each area included (1) the original undisturbed vegetation, (2) a reduction in original vegetation density as a result of thinning or clearing, and (3) a change in type of vegetation cover.

Description of the watersheds

The areas in which the models were applied were (1) the Coweeta Hydrologic Laboratory in North Carolina, (2) the H J Andrews Experimental Forest in Oregon, and (3) the Beaver Creek Experimental Watershed in Arizona. These represent three sites with significant differences in climate, vegetation, soils, and geology.

245

The Coweeta Hydrologic Laboratory. The Coweeta Hydrologic Laboratory is located in western North Carolina. It was established in 1933 and contains a network of 31 gaged watersheds on which a series of vegetation alteration experiments have been made. These are described by Hewlett & Hibbert (1967), Swank & Miner (1968), and Swank & Helvey (1970).

In general, the area is located in the high precipitation (2% snow) region of western North Carolina. It is characterized by deep, permeable soils underlain by granite bedrock, steep slopes, and an original cover type of oak–hickory.

Our tests were made on Watershed 18 using climatic and streamflow data for the period May 1971, through April 1973. Watershed 18 is a 12.5-ha control watershed with mature oak–hickory forest which has not changed in over 40 yr. The growing season leaf area index is 5 (one sided) and dormant season is 0.5. It has a north facing slope with elevation ranging from 721 to 977 m. Precipitation for 1971–2 water year (May–April) totaled 199 cm and 1972–3 was 234 cm. Complete climatic data are available.

Soils data were obtained from Luxmoore (1972). Albedo values were taken from Swift (1972), and interception storage data for canopy and litter from Helvey (1967). The watershed is very homogeneous and was represented by one response unit with no stratification.

The H J Andrews Experimental Forest. The H J Andrews Experimental Forest is a 6000–ha watershed in the western Cascades of Oregon. The drainage is characterized by steep topography, with about one-fifth of the study area consisting of more gentle slopes or benches. Elevation varies from 457 m to more than 1523 m. Mean forest air temperature varies from 2°C in January to 18°C during the summer months. Annual precipitation ranges from 225 cm at lower elevations to 350 cm at the highest ridges. Highest elevations are characterized by extensive snowpack during the winter, while rain predominates at lower elevations.

Watershed 10 is 10.2 ha and is located on the edge of the H J Andrews Experimental Forest. Elevations on the watershed range from 430 m at the stream gaging station to about 670 m at the highest point. Slopes on the watershed average about 45% but frequently exceed 100%.

The study site is located in an area underlain by volcanic tuff and breccia. Soils of the watershed are derived from these materials and range from gravelly, silty clay loam to very gravelly clay loam. The <2–mm fraction of these soils ranges from 20% to 50% clay and contains gravel amounting to 30–50% of the soil volume. The forest floor ranges from 3 to 5 cm thick and is classified as a duff–mull.

The present overstory vegetation is dominated by a 60- to 80-m-tall, 450-year-old stand of Douglas fir (*Pseudotsuga menziesii*) containing small islands of younger age classes. Distribution of understory vegetation reflects topography and slope-aspect on this watershed. Dry ridge-tops and south-

246

facing slopes have an understory composed primarily of chinkapin (*Castanopsis chrysophylla*), Pacific rhododendron (*Rhododendron macrophyllum*), and salal (*Gaultheria shallon*). More mesic parts of the watershed support an understory of vine maple (*Acer circinatum*), rhododendron, and Oregon grape (*Berberis nervosa*, with a well-developed intermediate canopy of *Tsuga heterophylla*. Subordinate vegetation of the moist areas along the stream and on north-facing slopes is primarily vine maple and sword fern (*Polystichum munitum*). The average leaf area index is 25 to 30 (two sided) for all sites. Tree basal area averaged 76 m² ha.

The climate of Watershed 10 is typical for the western Oregon Cascades. Average annual precipitation is 230 cm per year with over 75% of the precipitation falling as rain between October and March. Snow accumulations on the watershed are not uncommon, but seldom last more than two weeks. Based on two years' data, the average daytime temperature for July is 21°C and for January is 0°C. Observed extremes have ranged from a high of 41°C in August to a low of − 20°C in December. Complete climatic data are available.

Stream discharge varies from around 0.23 litre sec in the summer to about 140 litre sec during winter freshets. The uppermost forks are intermittent during the summer months. Mean width of the stream channel ranges from 0.25 m in the upper reaches to 0.75–1.0 m at the base of the watershed.

The Beaver Creek watershed. The 110 000-ha Beaver Creek watershed is located in Central Arizona. It was established in 1957 and now contains a network of about 44 watersheds ranging in size from about 10 ha to about 6400 ha. About one-third of the watershed is ponderosa pine and the balance Utah and Alligator juniper.

In the pine type, elevations vary from about 2073 m to 2439 m. Underlying bedrock consists of igneous rock of volcanic origin. Soils are developed on basalts and cinders, and are mostly silty clays and silty clay loams less than 0.75 m deep. January temperatures average − 3°C and July temperatures average 19°C. Precipitation averages 63.5 cm annually with 64% falling during winter, October–April. Winter snow regimes are quite variable and range from a continuous snowpack in heavy snowfall years, to an intermittent pack in light snowfall years. Average annual streamflow has varied from 0.5 to 41.6 cm with an overall average of 13.5 ± 1.3 cm with 93% coming during winter (October–April).

The forest is composed of a mixture of ponderosa pine (*Pinus ponderosa* Laws.), Gambel oak (*Quercus gambelli* Nutt.) and alligator juniper (*Juniperus deppeana.* Steud). Of a combined basal area of 26 m² ha⁻¹ pine occupies 21 m², oak 4 m², and juniper 1 m². A more detailed description is available in Brown *et al.* (1974).

Leaf area data were lacking on Beaver Creek, as were solar radiation,

247

wind and good dew point data. We estimated solar radiation, used a constant wind speed and estimated dew point from available humidity data of unknown quality. Good data should be available for later years. Our leaf area estimate of 4.0 (all sides) is very tentative and we hope to improve this using relationships of foliage mass to sapwood area developed by Grier & Waring (1974).

Comparison of reduction in vegetation density

A reduction in cover density may take two forms (1) uniform reduction by thinning the vegetation in a uniform way, and (2) non-uniform reduction by removing clumps, strips or patches of the vegetation so as to create openings in the original cover. The two forms may be combined. The significant difference is that a light uniform thinning may not create openings in which the soil is unoccupied by roots, while non-uniform thinning does if the openings are large enough. This can result in significant differences in transpiration. A very heavy thinning will create openings and of course both can be carried to the extreme of clearing of all vegetation.

Other effects include (1) reduction in surface area of foliage, branches and stems, (2) increased penetration of solar radiation through the canopy to the snow or soil surface, (3) possible effects on distribution of snowpack, and (4) possible disturbance or exposure of the soil surface which could lead to reduced infiltration capacity (Goodell, 1967).

One-year-old clearcut at Coweeta. Both PROSPER and ECOWAT were used to simulate a one-year-old regrowing clearcut on Coweeta Watershed 18. This clearcut was represented as a stand of sprout and seedling regrowth with growing season leaf area index of 0.75 (one sided) and dormant season of 0.5. Simulations for the original hardwood forest cover provide a base for comparison. Results are given in Table 4.2 with ECOWAT runs being *e–f* and *i–j* for original hardwoods and clearcut, respectively; and PROSPER runs being *g–h* and *k–l*, respectively. Not all evaporation and transpiration results were available for the PROSPER runs. The differences in total precipitation should be noted when comparing ECOWAT and PROSPER runs.

For the original oak–hickory conditions the maximum error in simulated streamflow was 8.1% (11 cm) with all others within 2%.

Three clearcut treatments on north- or east-facing Coweeta watersheds produced first year streamflow increases ranging from 36.0 to 41.3 cm. The increases determined by simulation for water years 1972 and 1973 were 36.0 and 38.0 cm, respectively, for ECOWAT; and 36.5 and 42.2 cm, respectively, for PROSPER. Differences in climatic conditions and the shape, aspect and elevation of the treated watersheds prevent direct comparison of the

Table 4.2. *Summary of the results of simulation runs under different conditions (all values in cm)*

Watershed		No. and condition	Water year	Precipitation		Evaporation				Infil-tration	Transpiration					Total evapotrans-piration	Streamflow	
				Rain	Snow	Canopy	Litter	Snow	Total		A	B	C	D	Total		Predicted	Actual
Beaver Creek	*a.*	Original	1965	48	50	10.6	1.2	2.5	14.3	69.7	33.7	8.8	8.3	6.4	57.2	71.5	26.0	33.1
	b.	Thinned	1973	45	74	13.7	1.5	3.9	19.1	70.3	28.6	7.8	8.3	4.9	49.5	68.6	56.3	55.4[a]
	c.		1973	45	74	10.6	1.6	3.0	15.2	67.8	23.3	5.8	6.4	1.4	36.9	52.1	67.6	64.3
	d.	Shrubs	1973	45	74	9.7	1.6	3.6	14.9	67.5	18.8	5.8	6.0	0.4	31.0	45.9	71.7	70.1[a]
Coweeta	*e.*	Hardwood	1972	192	3	24.4	1.4	0.0	25.8	168.0	53.7	4.4	4.3	—	62.4	88.2	103.0	104.3
	f.		1973	235	5	24.2	1.3	0.0	25.3	205.0	51.0	6.6	0.4	—	58.0	83.3	147.0	136.0
	[b]*g.*	Hardwood	1972	199	0	c	c	c	24.5	c	c	c	c	—	66.6	91.1	103.6	104.3
	[b]*h.*		1973	234	0	c	c	c	22.8	c	c	c	c	—	66.0	88.8	138.0	136.0
	i.	Clearcut	1972	192	3	8.6	2.4	0.0	11.0	182.0	41.9	0.0	0.0	—	41.9	52.9	139.0	142.5[d]
	j.		1973	235	5	8.5	2.2	0.0	10.7	218.0	34.3	0.0	0.0	—	34.3	45.0	185.0	e
	[b]*k.*	Clearcut	1972	199	0	c	c	c	c	c	c	c	c	—	c	c	140.1	142.5[d]
	[b]*l.*		1973	234	0												180.2	e
	m.	Pine	1972	192	3	30.9	1.4	0.0	32.3	163.0	73.4	4.2	0.0	—	77.6	109.9	84.4	84.1[d]
	n.		1973	235	5	30.4	1.3	0.0	31.7	200.0	65.3	6.3	1.8	—	74.2	105.9	122.0	117.7[d]
	[b]*o.*	Pine	1972	199	0	c	c	c	34.2	c	c	c	c	—	82.2	116.4	83.4	84.1[d]
	[b]*p.*		1973	234	0	c	c	c	31.9	c	c	c	c	—	76.5	108.4	121.1	117.7[d]
H J Andrews	*q.*	Original	1973	119	48	32.9	0.6	-1.5	32.0	134.0	30.0	7.5	5.0	—	42.5	74.5	82.7	80.1
	r.	Clearcut	1974	206	98	22.3	0.7	-2.2	20.8	269.0	19.0	7.5	7.1	—	33.6	54.4	267.0	258.5
	s.		1973	119	48	2.3	3.3	-2.4	3.2	163.0	18.8	0.1	—	—	18.9	22.1	130.0	
	t.	Upper 37% cut	1973	119	48	21.0	1.7	-1.7	21.0	145.0	25.9	5.0	2.7	—	33.5	54.5	98.0	e
	u.	Lower 10% cut	1973	119	48	29.9	0.8	-1.7	29.0	137.0	28.9	6.6	4.5	—	40.1	69.1	87.9	e

[a] Estimates based on results from other watersheds at Beaver Creek (Brown *et al.* 1974).
[b] Results of simulations with PROSPER as given in Swift *et al.* (1975).
c Not available.
[d] Estimates based on results from other watersheds at Coweeta as presented in Swift, Swank *et al.* (1975).
e No estimate available.

249

simulation results with measured treatment effects. Water year 1973 was a record year for precipitation at Coweeta.

The ECOWAT simulations showed reduction in evaporation of 58% for both years and in transpiration of 33 to 41%. The reduction in evaporation is due to reduced storage capacity of foliage, branches and stems. The reduction in transpiration is proportionally less than that of evaporation because increased exposure of the remaining foliage results in much higher rates of transpiration per unit of leaf area (Landsberg *et al.*, 1975).

Heavy thinning at Beaver Creek. We used ECOWAT to simulate measured treatment effects on the 120 ha Watershed 17 which is typical of the pine watersheds on Beaver Creek. This watershed was thinned by group selection in 1969. 75% of the initial 27.5 m^2 basal area was removed by thinning leaving even-aged groups with average basal area of 7 m^2 ha for all species. All Gambel oaks over 38 cm dbh except den trees were removed, leaving 1 m^2 ha. All junipers were removed. Slash was windrowed. We used a leaf area index of 4.0 (all sides) for 1965 and of 2.0 for 1973. Pre- and post-treatment conditions were simulated for the water year October 1972 to September 1973, and pre-treatment conditions for water year 1965. For simulation purposes the watershed was stratified into three response units. The results are shown as runs *a–c* in Table 4.2. The model underestimated flow by 21% for water year 1965 and overestimated flow by 5% for water year 1973.

The estimated increase streamflow due to thinning for water year 1973 is 8.9 cm. The simulation showed an increase of 11.3 cm or 20%. This was due to a 20% reduction in evaporation and a 25% reduction in transpiration. These reductions are attributable to reduced storage capacity of foliage, branches and stems, and increased surface resistance to transpiration.

Almost all transpiration at H J Andrews and Coweeta is taken from the top soil layer. The pine stand at Beaver Creek, however, is dependent on lower layers to meet transpiration demand as shown in Fig. 4.25. Uptake proceeds first from the top 20 cm which is soon exhausted and withdrawal from lower layers begins. Rain from summer storms recharges the top layer in July and August and this then provides part of the demand.

Both the Coweeta and H J Andrews site rarely have snowpacks which persist. However, Beaver Creek can have a persistent pack and Fig. 4.25 represents a record year for both precipitation and snow. The peak in infiltration in May represents the melting of the snow pack.

Partial cutting at H J Andrews. We simulated effects of clearing different parts of a watershed on the H J Andrews Watershed 10. Original conditions were simulated for water years 1973 and 1974 and compared to measured

250

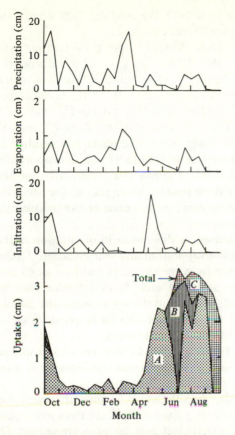

Fig. 4.25. Actual precipitation, and simulated evaporation from foliage, infiltration and root uptake from *A* 0 to 20 cm, *B* 20 to 40 cm, and *C* 40 to 100 cm at 14-day intervals for Beaver Creek Watershed 17, October 1972 through September 1973. The simulation represents a thinned stand of Ponderosa pine with total leaf area of 2 and average height of 32 m.

results. We then simulated effects of (1) clearcutting the entire watershed, (2) clearcutting the upper slopes, and (3) clearcutting the lower slopes.

The watershed was stratified into six units for purposes of simulation. Essentially the north and south slopes were divided into upper (dry ridgetops), middle (mesic) and lower (moist) zones. All flow from the upper units passes through the lower units before reaching the stream.

Clearcutting a unit was represented by a deciduous stand with leaf area of 0.75 in summer and 0.5 in winter. This was the same as in the Coweeta clearcut. Clearcutting of the upper units affected 37% of the total watershed area while clearcutting the lower units affected 10% of the area.

The results for the original conditions are given as runs *q* and *r* in Table 4.2. The predicted streamflow was high by 1.6 cm (2.0%) for water year 1973

251

and high by 8.5 cm (3.3%) for water year 1974. We used the 1973 simulation (run *q*) as the base for comparing treatment effects.

The results of simulations of complete clearcut, upper slope clearcut, and lower slope clearcut are given in Table 4.2 as runs *s*, *t* and *u*, respectively. The total clearcut resulted in a 47.3 cm (57.2%) increased streamflow with a reduction in evaporation of 90% and in transpiration of 55.5%. Clearcutting the upper 37% of the watershed resulted in a 15.3 cm (18.5%) increased streamflow with an overall reduction in evaporation of 34.4% and in transpiration of 21.2%. Clearing the lower 10% of the watershed resulted in 5.2 cm (6.3%) increased streamflow with an overall reduction in evaporation of 9.4% and in transpiration of 5.6%. Cutting the lower 10% of the watershed resulted in a larger relative response with respect to the area cut, than cutting the top 37%. Hence, streamflow response is not necessarily directly proportional to the area cut.

The difference in response was caused by differences in orientation of the areas cut which caused differences in evaporation and transpiration. Under the original conditions, transpiration from the two upper units was 43 and 41 cm; and from the two lower units, 49 and 40 cm. Under the cut condition, transpiration was 20 and 18 cm from the upper units and 20 and 19 cm from the lower units. Hence the lower units had a larger decrease in transpiration (and in evaporation) than the upper units.

This emphasizes the need to stratify watersheds into response units in areas where heterogeneity of soils and topography has significant effects on the vegetation and the hydrologic response of those units.

It should be noted that in order to simulate a change in vegetation density, it is only necessary to change the values of no more than five inputs to ECOWAT for each unit. These are (1) leaf area for growing season, (2) leaf area for dormant season, (3) stem area, (4) average tree height, and (5) litter capacity. Of these the leaf areas have by far the greatest effect and demonstrate the need for obtaining good estimates of this variable. Changes in stem area and litter capacity have some effect but are minor compared to changes in leaf area.

Comparison of changes in vegetation type

Changes in cover type can have many forms and be the result of natural (successional), catastrophic (fire), timber harvest, or other management activities such as revegetation, seeding, planting, or brush control activities. In any case, they can have more significant effects on water balance processes than simple reduction in leaf area. Effects can be due to changes in (1) amount and seasonal distribution of leaf area, (2) density and depth of rooting, (3) season and duration of active growth, (4) albedo and aerodynamic characteristics of the vegetation. These and others are discussed by

Goodell (1967), Swank & Miner (1968), and Swank & Douglas (1974).

Conversion to conifers at Coweeta. We simulated conversion of an oak–hickory forest to white pine at Coweeta using both ECOWAT and PROSPER. The conversion was represented by a stand of 16-year-old white pine with growing season leaf area index of 12 (surface with stomata) and dormant season of 6. The simulations for the original forest provide a base for comparison. Results are given in Table 4.2 with ECOWAT runs being *e–f* and *m–n* for original hardwoods and conifer, respectively; and PROSPER runs being *g–h* and *o–p*, respectively. Not all evaporation and transpiration results were available for PROSPER runs, and total precipitation was different for the ECOWAT and PROSPER runs.

The simulation results are compared to measured results from a watershed with a white-pine plantation on a north-facing slope adjacent to the simulated Watershed 18. Simulation showed decreases in streamflow for water years 1972 and 1973 of 18.6 and 25.0 cm, respectively, for ECOWAT; and 20.2 and 16.9 cm, respectively, for PROSPER. The measured decreases were 20.2 and 18.3 cm, respectively.

The ECOWAT simulations showed increases in evaporation of 25% for both years and in transpiration of 24 to 28%. PROSPER showed evaporation increases of 40% and transpiration increases of 16 to 23%. The increases in total evapotranspiration were 25 to 27% for ECOWAT and 22 and 28% for PROSPER.

The increase in evaporation is due to increased foliar storage capacity during the dormant season when hardwoods are normally bare. Summer transpiration by pine and oak–hickory was similar except that pine transpiration rates increased about one month earlier than hardwoods. Simulation indicates more than half of the increase in total evapotranspiration from pine was due to dormant season transpiration.

Conversion to deciduous shrubs and grass at Beaver Creek. We also simulated a conversion from mature ponderosa pine to grasses, forbs, shrubs and oak sprouts on Beaver Creek Watershed 7 using ECOWAT. This represented the current condition on Beaver Creek Watershed 12 which was cleared in 1966. The conversion was represented as a deciduous stand with growing season leaf area of 2 (one sided) and dormant season of 0.5. The results are shown in Table 4.2 with the original condition as run *b* and the conversion as run *d*.

The simulation shows an increase in streamflow of 15.4 cm (27%) compared with an estimated increase based on measured results of 14.7 cm (27%). Evaporation decreased by 22% (4.2 cm) and transpiration decreased by 37% (18.5 cm). This was due to reduction in foliar storage capacity, increased surface resistance, and reduced leaf area in winter.

253

Conclusions

We believe that models which represent the processes that control water movement and storage, in a way consistent with the structure and function of terrestrial ecosystems, simulate the hydrologic regimes of these ecosystems under a variety of conditions. The last objective provided convincing evidence to support our belief. The processes described and incorporated into the ECOWAT and PROSPER models gave very encouraging results. Fig. 4.26 compares predicted versus observed annual streamflow for sites on which ECOWAT was tested. These predictions were made without benefit of prior calibration of the model. With both models the annual totals of simulated streamflow agreed closely with measured streamflow for the three areas under different vegetation and climatic conditions.

Streamflow from a watershed is influenced by both events in past months and current climatic and vegetative conditions. However, vegetative conditions are also dependent on events in past months as well as the current hydrologic and climatic conditions. The ability of the models to estimate day-to-day changes in the hydrologic regime can be improved by in-

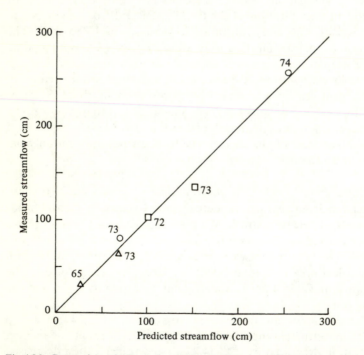

Fig. 4.26. Comparison of measured annual streamflow versus streamflow predicted by ECOWAT. The line represents perfect prediction. The r^2 is 0.995. □, Coweeta, Watershed 18; △, Beaver Creek, Watershed 17; ○, HJ Andrews, Watershed 10.

corporating models for the short and long term response of the vegetation on the watershed to both the treatment and to the current hydrologic and climatic conditions.

Both simulations revealed research needs. The models are sensitive to the leaf area and particularly to leaf conductance, surface resistance and moisture stress. These types of data are generally lacking for vegetation on forest watersheds.

Better data, both environmental and structural, should improve both resolution of predictions and general understanding. We encourage further integrative watershed ecosystem studies. These provide a complete set of data which we use both to improve our understanding of structure and function, and completely test and evaluate models which synthesize our knowledge and understanding.

References

Anderson, E. A. (1968). Development and testing of snowpack energy balance equations. *Water Resources Res.* **4**, 19–37.

Anderson, E. A. & Baker, D. R. (1967). Estimating incident terrestrial radiation under all atmospheric conditions. *Water Resources Res.* **3**, 975–88.

Aussenac, G. (1968). Interception des precipitations par le convert forestier. *Ann Sci. Forestières*, **25**, 135–56.

Bababola, O., Boersma, L. & Youngberg, C. T. (1968). Photosynthesis and transpiration of Monterey pine seedlings as a function of soil water suction and soil temperature. *Plant Physiol.* **43**, 515–21.

Bear, J., Zavslavsky, D. & Irmay, S. (1968). *Physical Principles of Water Percolation and Seepage.* UNESCO, Paris.

Berndt, H. W. & Fowler, W. B. (1969). Rime and hoarfrost in upperslope forests of eastern Washington. *J. Forest.* **67**, 92–5.

Blake, G., Schlichting, E. & Zimmerman, U. (1973). Water recharge in a soil with shrinkage cracks. *Proc. Soil Sci. Soc. America*, **37**, 669–73.

Brown, H. E., Baker, M. B. Jr., Rogers, J. J., Clary, W. P., Kovner, J. L., Larson, F. R., Avery, C. C. & Campbell, R. E. (1974). *Opportunities for increasing water yields and other multiple use values on ponderosa pine forest lands.* US Department of Agriculture Forest Service Research Paper RM–129. Rocky Mountain Forest and Range Experiment Station, Fort Collins, Colorado.

Buffo, J., Fritschen, L. J. & Murphy, J. L. (1972). *Direct solar radiation on various slopes from 0 to 60 degrees north latitude.* US Department of Agriculture Forest Service, Pacific Northwest Forest and Range Experiment Station Research Paper PNW–142.

Burger, H. (1925). Holz-Laub-und Nadeluntersuchungen. *Schweiz. Z. Forstwesen,* 266–310.

Burns, J. I. (1953). Small scale topographic effects on precipitation distribution in San Dimas Experimental Forest. *Trans Amer. geophys. Union*, **34**, 761–8.

Caldwell, M. M. & Fernandez, O. A. (1975). Dynamics of Great Basinshrub root systems. In *Physiological adaptations of desert organisms*, ed. N. F. Hadley. Dowden, Hutchinson & Ross, Stroudsburg, Pennsylvania.

Chalk, L. & Bigg, J. M. (1956). The distribution of moisture in the living stem in Sitka spruce and Douglas-fir. *Forestry*, **29**, 5–21.

Childs, E. C. (1969). *An Introduction to the physical basis of soil water phenomenon.* John Wiley & Sons, New York.

Chow, V. T. (1964). *Handbook of applied hydrology.* McGraw-Hill Book Company, New York.

Clary, W. P. & Ffolliott, P. F. (1969). Water holding capacity of ponderosa pine forest floor layers. *J. Soil Water Conserv.* **24**, 22–3.

Clausen, J. J. & Kozlowski, T. T. (1965). Use of the relative turgidity technique for measurement of water stresses in gymnosperm leaves. *Canad. J. Bot.* **43**, 305–16.

Cleary, B. D. & Waring, R. H. (1969). Temperature collection of data and its analysis for the interpretation of plant growth and distribution. *Canad J. Bot.* **47**, 167–73.

Clegg, A. G. (1963). Rainfall interception in a tropical forest. *Caribbean Forester*, **24**, 75–9.

Debano, L. F. & Rice, R. M. (1973). Water repellent soils: their implications in forestry. *J. Forest.* **71**, 220–3.

Deeming, J. E., Lancaster, J. W., Fosberg, M. A., Furman, R. W. & Schroeder, M. J. (1974). *National fire-danger rating system.* US Department of Agriculture Forest Service, Rocky Mountain Forest and Range Experiment Station Research Paper RM–84.

Dingman, S. L. (1975). *Hydrologic effects of frozen ground: literature review and synthesis.* US Army Corps of Engineers: Cold Regions Research and Engineering Laboratory, Species Report 218.

Dixon, A. F. G. (1971). The role of aphids in wood formation. *J. appl. Ecol.* **8**, 165–79.

Dixon, R. M. (1971). Role of large soil pores in infiltration and interflow. In *Proceedings of the Third International Seminar for Hydrology Professors.* Urbana, Illinois.

Dixon, R. M. & Peterson, A. E. (1971). Water infiltration control: A channel system concept. *Proc. Soil Sci. Soc. America*, **35**, 968–73.

Doley, D. (1967). Water relations of *Eucalyptus marginata* Sm. under natural conditions. *J. Ecol.* **55**, 597–614.

Drew, A. P., Drew, L. G. & Fritts, H. C. (1972). Environmental control of stomatal activity in mature semiarid site ponderosa pine. *Arizona Acad. Sci.* **7**, 85–93.

Duckstein, L., Fogel, M. M. & Kisiel, C. C. (1972). A stochastic model of runoff producing rainfall for summer type storms. *Water Resources Res.* **8**, 410–21.

Duckstein, L., Fogel, M. M. & Thames, J. L. (1973). Elevation effects on rainfall: a stochastic model. *J. Hydrol.* **18**, 21–35.

Dunne, T. & Black, R. D. (1970). Partial area contributions to storm runoff in a small New England watershed. *Water Resources Res.* **6**, 1296–311.

Dyrness, C. T. (1972). *Soil surface conditions following balloon logging.* US Department of Agriculture Forest Service, Pacific Northwest Forest and Range Experiment Station Research Note PNW–182.

Eagleson, P. E. (1970). *Dynamic hydrology.* McGraw-Hill Book Company, New York.

Eggleston, K. O., Israelson, E. K. & Riley, J. P. (1971). *Hybrid computer simulation of the accumulation and melt processes in a snowpack.* Utah Water Research Laboratory, College of Engineering, Utah State University, Logan.

Elrick, D. E. (1968). The microhydrological characterization of soils. In *Water in the unsaturated zone*, ed. P. E. Rijtema & H. E. Wassink, pp. 311–7. International Association of Scientific Hydrology Publication No. 83.

Emmingham, W. H. & Waring, R. H. (1973). Conifer growth under different light environments in the Siskiyou Mountains of southwestern Oregon. *Northwest Sci* **47**, 88–99.

Fogel, J. M. & Duckstein, L. (1969). Point rainfall frequencies in convective storms. *Water Resources Res.* **5**, 1229–37.

Freeze, R. A. (1969*a*). *Regional groundwater flow: Old Wives Lake Drainage Basin, Saskatchewan.* Inland Waters Branch, Canadian Department of Energy, Mines and Resources, Science Series No. 5.

Freeze, R. A. (1969*b*). The mechanism of natural groundwater recharge and discharge. 1. One-dimensional, vertical, unsteady, unsaturated flow above a recharging and discharging groundwater flow system. *Water Resources Res.* **5**, 153–71.

Freeze, R. A. (1972*a*). Role of subsurface flow in generating surface runoff. 1. Base flow contributions to channel flow. *Water Resources Res.* **8**, 609–23.

Freeze, R. A. (1972*b*). Role of subsurface flow in generating surface runoff. 2. Upstream source areas. *Water Resources Res.* **8**, 1272–83.

Freeze, R. A. & Witherspoon, P. A. (1966). Theoretical analysis of regional groundwater flow. 1. Analytical and numerical solutions to the mathematical model. *Water Resources Res.* **2**, 641–56.

Freeze, R. A. & Witherspoon, P. A. (1967). Theoretical analysis of regional groundwater flow. 2. Effect of water-table configuration and sub-surface permeability variation. *Water Resources Res.* **3**, 623–34.

Freeze, R. A. & Witherspoon, P. A. (1968) Theoretical analysis of regional groundwater flow. 3. Quantitative interpretations. *Water Resources Res.* **4**, 581–90.

Fries, N. (1943). Zur Kenntnis des Winterlichen Wasserhaushalts der Laubbaume. *Svensk Botanisk Tidskrift,* **37**, 241–65.

Fritschen, L. J. & Doraiswamy, P. (1973). Dew: an addition to the hydrologic balance of Douglas-fir. *Water Resources Res.* **94**, 891–4.

Gardner, W. R. (1959). Solutions of the flow equation for the drying of soils and other porous media. *Proc. Soil Sci. Soc. America,* **23**, 183–7.

Gardner, W. & Widstoe, J. A. (1921). Movement of soil moisture. *Soil Sci.* **11**, 215–232.

Garnier, B. J. & Ohmura, A. (1968). A method of calculating the direct shortwave radiation income of slopes. *J. appl. Meteorol.* **11**, 796–800.

Garnier, B. J. & Ohmura, A. (1970). The evaluation of surface variations in shortwave radiation income. *Solar Energy,* **13**, 21–34.

Gary, H. L. (1971). Seasonal and diurnal changes in moisture contents and water deficits of Engelmann spruce needles. *Bot. Gazette,* **132**, 327–32.

Gary, H. L. (1972). Rime contributes to water balance in high-elevation aspen forests. *J. Forest.* **70**, 93–7.

Gates, D. M. (1968). Transpiration and leaf temperature. *Annu. Rev. Plant Physiol.* **19**, 211–38.

Geiger, R. (1965). *The climate near the ground.* Harvard University Press, Cambridge, Mass.

Gibbs, R. D. (1958). Patterns in the seasonal water content of trees. In *The physiology of forest trees,* ed. K. V. Thimann, pp. 43–69. Ronald Press, New York.

Goldstein, R. A., Mankin, J. B. & Luxmoore, R. J. (1974). *Documentation of PROSPER. A model of atmosphere-soil-plant water flow.* Eastern Deciduous Forest Biome Report EDFB/1BP–73/9, International Biological Program, Oak Ridge National Laboratory, Oak Ridge, Tennessee.

Goodell, B. C. (1967). Watershed treatment effects on evapotranspiration. In *Forest* hydrology, ed. W. E. Sopper & H. W. Lull, pp. 477–82, Pergamon Press, New York.

Grace, R. A. & Eagleson, P. S. (1966). *The synthesis of short-time-increment rainfall sequences.* MIT Department of Civil Engineering, Hydrodynamics Laboratory Report 91, Cambridge, Mass.

Grah, R. F. & Wilson, C. C. (1944). Some components of rainfall interception. *J. Forest.* 42, 890–8.

Green, W. H. & Ampt, G. A. (1911). Studies on soil physics: I. Flow of air and water through soils. *J. agric. Sci.* 4, 1–24.

Green, R. E. & Corey, I. C. (1971). Calculation of hydraulic conductivity: A further evaluation of some predictive methods. *Proc. Soil Sci. Soc. America*, 35, 3–8.

Gregory, S. C. & Petty, J. A. (1973). Value actions of bordered pits in conifers. *J. Exp. Bot.* 24, 763–7.

Grier, C. C. & Waring, R. H. (1974). Conifer foliage mass related to sapwood area. *Forest Sci.* 20, 205–6.

Hammel, H. T. (1967). Freezing xylem sap without cavitation. *Plant Physiol.* 42, 55–66.

Hanks, R. J. & Bowers, S. A. (1962). Numerical solution of the moisture flow equation for infiltration into layered soils. *Proc. Soil Sci. Soc. America*, 26, 530–4.

Haupt, H. F. (1967). Infiltration, overland flow and soil movement on frozen and snow-covered plots. *Water Resources res.* 3, 145–61.

Havranek, W. (1972). Uber die Bedeutung der Bodentemperatur fur die Photosyntheses und Transpiration junger Forstpflanzen und die Stoffproduktion an der Waldgrenze. *Angewandte Botanik*, 46, 101–16.

Hellkvist, J., Richards, G. P. & Jarvis, P. G. (1974). Vertical gradients of water potential and tissue water relations in Sitka spruce trees measured with the pressure chamber. *J. Ecol.* 11, 637–68.

Helvey, J. D. (1964). *Rainfall interception by hardwood forest litter in the Southern Appalachians.* US Department of Agriculture Forest Service, Southeastern Forest Experimental Station Research Paper SE–8.

Helvey, J. D. (1967). Interception by eastern whitepine. *Water Resources Res.* 3, 723–9.

Helvey, J. D. (1971). A Summary of rainfall interception by certain conifers of North America. In *Proceedings of the International Symposium for Hydrology Professors. Biological effects in the hydrological cycle*, pp. 103–13. Purdue University, Lafayette, Indiana.

Helvey, J. D. & Patric, J. H. (1965). Canopy and litter interception by hardwoods of Eastern United States. *Water Resources Res.* 1, 193–206.

Hewlett, J. D. (1961). Soil moisture as a source of base flow from steep mountain watersheds. *US Department of Agriculture Forest Service, Southeastern Forest Experiment Station Paper* 132.

Hewlett, J. D. (1974). Comments on letters relating to 'Role of subsurface flow in generating surface runoff. 2. Upstream source areas' by R. A. Freeze. *Water Resources Res.* 10, 605–6.

Hewlett, J. D. & Hibbert, A. R. (1963). Moisture and energy conditions within a sloping soil mass during drainage. *J. geophys. Res.* 68, 1080–7.

Hewlett, J. D. & Hibbert, A. R. (1967). Factors affecting the response of small watersheds to precipitation in humid areas. In *Forest hydrology*, ed. W. E. Sopper & H. W. Lull, pp. 275–90. Pergamon Press, New York.

Hinckley, T. M. & Ritchie, G. A. (1973). A theoretical model for calculation of xylem sap pressure from climatological data. *American Midland Naturalist*, 90, 56–69.

Holtan, H. N. (1961). *A concept for infiltration estimates in watershed engineering.* US Department of Agriculture Agricultural Research Service 41–51.

Holtan, H. N. (1969). Hydrologic research for watershed engineering. *J. Hydrol.* **8**, 207–16.

Holtan, H. N. (1971). A formulation for quantifying the influence of soil porosity and vegetation on infiltration. In *Proceedings of the International Seminar for Hydrology Professors. Biological effects in the hydrological cycle.* Urbana, Illinois.

Holtan, H. N. & Lopez, N. C. (1971). USDAHL–70 model of watershed hydrology. *US Department of Agriculture Technical Bulletin* 1435.

Hoover, M. D. & Leaf, C. F. (1967). Process and significance of interception in Colorado Sub-alpine forest. In *International Symposium on Forest Hydrology,* pp. 213–24. Pergamon Press, New York.

Hoppe, E. (1896). Regenmessung unter Baumkronen. *Mitteilungen aus der Forstlichen Versuchswesen Osterreichs* **21**, 1–75. (Precipitation measurements under tree crowns. US Forest Service Translations 291 by A. H. Krappe).

Hornbeck, J. W. (1973). Stormflow from hardwood-forested and cleared watersheds in New Hampshire. *Water Resources Res.* **9**, 346–54.

Horton, R. E. (1919). Rainfall interception. *Monthly Weather Review,* **47**, 603–23.

Horton, R. E. (1924). Determining the mean precipitation on a drainage basin. *New England Water Works Association Journal,* **38**, 1–43.

Horton, R. E. (1933). The role of infiltration in the hydrologic cycle. *Trans. Amer. geophys Union,* **14**, 446–60.

Houghton, H. G. (1954). On the annual heat balance of the Northern Hemisphere. *J. Meteorol.* **11**, 1–9.

Hsiao, T. C., Fereres, E., Avecedo, E. & Henderson, D. W. (1976). Water stress and dynamics of growth and yield of crop plants. Chapter 17. In *Water and Plant Life–Problems and Modern Approaches,* ed. O. L. Lange, L. Kappen, & E. D. Schulze. Springer Verlag, Berlin.

Huck, M. G., Klepper, B. & Taylor, H. M. (1970). Diurnal variations in root diameter. *Plant Physiol.* **45**, 529–30.

Hutchinson, P. & Walley, W. J. (1972). Calculation of a real rainfall using finite clement techniques with altitudinal corrections. *Bull. int. Ass. sci. Hydrol.* **17**, 259–72.

Jarvis, P. G. (1975). Water transfer in plants. In *Heat and Mass Transfer in the Environment of Vegetation. 1974 Seminar of the International Centre for Heat and Mass Transfer, Dubrovnik.* Scripta Book Company, Washington, D.C.

Kimball, B. A. & Jackson, R. D. (1971). Seasonal effects on soil drying after irrigation. In *Hydrology and Water Resources in Arizona and the Southwest, Proceedings of the 1971 meetings of Arizona Section AWRA and Hydrology Section,* pp. 85–94. Arizona Academy of Science.

Kira, T., Shinozaki, K. & Hozumi, K. (1969). Structure of forest canopies as related to their primary productivity. *Plant and Cell Physiol.* **10**, 129–42.

Kittredge, J. (1948). *Forest Influences* (first ed.), pp. 99–114. McGraw-Hill Book Company, New York.

Klock, G. O. (1972). Snowmelt temperature influence on infiltration and soil water retention. *J. Soil Water Conserv.* **27**, 12–14.

Kostiakov, A. N. (1932). On the dynamics of the coefficient of water percolation in soils and on the necessity of studying it from a dynamic point of view for purposes of amelioration. *Translations of the 6th Conference of the International Society Soil Science* Russian Part A, pp. 17–21.

259

Krygier, J. T. (1971). Comparative water loss of Douglas-fir and Oregon white oak. Ph.D. Thesis, Colorado State University.

Kuzmin, P. P. (1961). *Melting of snow cover*. Israel Program for Scientific Translations, 1972. 290 pp.

Landsberg, J. J., Beadle, C. L., Biscoe, P. V., Butler, D. R., Davidson, B., Incoll, L. D., James, G. B., Jarvis, P. G., Martin, P. J., Neilson, R. E., Powell, D. B. B., Slack, E. M., Thorpe, M. R., Turner, N. C., Warrit, B. & Watts, W. R. (1975). Diurnal energy, water and CO_2 exchanges in an apple (*Malus pumila*) orchard. *J. appl. Ecol.* **12**, 659–84.

Lange, O. L., Losch, R., Schulze, E. D. & Kappen, L. (1971). Responses of stomata to changes in humidity. *Planta*, **100**, 76–86.

Lassoie, J. P. (1973). Diurnal dimensional fluctuations in a Douglas-fir stem in response to tree water status. *Forest Sci.* **19**, 251–5.

Leaf, C. F. & Brink, G. E. (1973). Computer simulation of snowmelt within a Colorado subalpine watershed. *US Department of Agriculture Forest Service, Rocky Mountain Forest and Range Experiment Station Research Paper RM–99*.

Lemon, E. R. (1956). The potentialities for decreasing soil moisture evaporation loss. *Proc. Soil sci Soc America*, **20**, 120–5.

Leonard, R. E. (1961). Interception of precipitation by northern hardwoods. *US Forest Service, Northeastern Forest Experiment Station Paper 159*.

Leonard, R. E. (1967). Mathematical theory of interception. In *International Symposium on Forest Hydrology*, pp. 131–6. Pergamon Press, New York.

Leven, A. A., Meurisse, R. T., Carleton, J. O. & Williams, J. A. (1974). Land response units–an aid to forest land management. *Proc. Soil Sci. Soc. America*, **38**, 140–4.

Leyton, L., Reynolds, E. R. C. & Thompson, F. B. (1967). Rainfall interception in forest and moorland. In *International Symposium on Forest Hydrology*, pp. 163–78. Pergamon Press, New York.

Linsley, R. K., Jr. (1958). Correlation of rainfall intensity and topography in northern California. *Trans. Amer. geophys. Union*, **39**, 15–18.

Lopushinsky, W. (1969). Stomatal closure in conifer seedlings in response to leaf moisture stress. *Bot. Gazette*, **130**, 258–63.

Luxmoore, R. J. (1972). Some Coweeta soil water characteristics. *Eastern Deciduous Forest Biome Memo Report 72–129, Oak Ridge National Laboratory, Oak Ridge, Tennessee*.

Luxmoore, R. J. (1973). Application of the Grier and Corey method for computing hydraulic conductivity in hydrologic modeling. *Eastern Deciduous Forest Biome Report EDFB/IBP–73/4, Oak Ridge National Laboratory, Oak Ridge, Tennessee*.

Mandeville, A. N. & Rodda, J. C. (1970). A contribution to the objective assessment of a real rainfall amounts. *J. Hydrol.* (NZ) **9**, 281–91.

Meeuwig, R. O. (1970). Infiltration and soil erosion as influenced by vegetation and soil erosion in northern Utah. *J. Range Management*, **23**, 185–8.

Meeuwig, R. O. (1971). Infiltration and water repellency in granitic soils. *US Department of Agriculture Forest Service. Intermountain Forest and Range Experiment Station Research Paper INT–111*.

Megahan, W. F. (1972). Subsurface flow interception by a logging road in mountains of central Idaho. In *Proceedings of the National Symposium on Watersheds in Transition*, pp. 350–6. American Water Resources Association & Colorado State University, Fort Collins.

Miller, D. E. (1969). Flow and retention of water in layered soils. *US Department of*

Agriculture Agricultural Research Service Conservation Research Report 13. 28 pp.

Miller, D. E. & Gardner, W. H. (1962). Water infiltration into stratified soil. *Proc. Soil Sci. Soc. American*, **26**, 115–9.

Miller, D. H. (1967). Sources of energy for thermodynamically-caused transport of intercepted snow from forest crowns. In *International symposium on Forest Hydrology*, pp. 201–11. Pergamon Press, New York.

Molchanov, A. A. (1960). *The hydrological role of forests*. Academy of Science USSR Institute of Forestry. (Translation from Russian by Israel Program for Scientific Translations, 1963).

Monteith, J. L. (1965). Evaporation and environment. In *The state and movement of water in living organisms*, ed. G. E. Fogg, pp. 205–36. XIX Symposium of the Society of Experimental Biology 1964. Cambridge University Press.

Moore, A. & Swank, W. T. (1975). A model of water content and evaporation for hardwood leaf litter. Paper presented at Symposium on Mineral Cycling in Southeastern Ecosystems, Augusta, Georgia, 1–3 May 1974.

Murphy, C. E., Jr. & Knoerr, K. R. (1975). The evaporation of intercepted rainfall from a forest stand: an analysis by simulation. *Water Resources Res.*, **11**, 273–80.

Nutter, W. L. & Hewlett, J. D. (1971). Streamflow production from permeable upland basins. In *Proceedings Third International Seminar for Hydrology Professors*. Urbana, Illinois.

Overton, D. E. (1971). Modeling watershed surface water flow systems. In *Proceedings Third International Seminar for Hydrology Professors*, Urbana, Illinois.

Penman, H. L. (1956). Evaporation–an introductory survey. *Netherlands J. agric. Sci.* **4**, 9–29.

Philip, J. R. (1954). An infiltration equation with physical significance. *Soil Sci.* **77**, 153–7.

Philip, J. R. (1956). Evaporation and moisture and heat fields in the soil. *J. Meteorol.* **14**, 354–66.

Philip, J. R. (1969). Theory of infiltration. In *Advances in hydroscience*, ed. V. T. Chow, Vol. 5, pp. 215–305. Academic Press, New York, 305 pp.

Pierce, R. S. (1967). Evidence of overland flow on forest watershed. In *Forest hydrology*, ed. W. E. Sopper & H. W. Lull, pp. 247–52. Pergamon Press, New York.

Plamondon, P. A., Black, T. A. & Goodell, B. C. (1972). The role of hydrologic properties of the forest floor in watershed hydrology. In *National Symposium on Watersheds in Transitions*, pp. 341–8. American Water Resources Association and Colorado State University, Fort Collins, Colorado.

Rauzi, F., Fly, C. L. & Dyksterhuis, E. J. (1968). Water intake on mid-continental rangelands as influenced by soil and plant cover. *US Department of Agriculture Agricultural Research Service Technical Bulletin* 1390.

Rauzi, F. & Smith, F. M. (1973). Infiltration rates: three soil with three grazing levels in northeastern Colorado. *J. Range Management*, **26**, 126–9.

Reed, K. L. & Waring, R. H. (1974). Coupling of environment to plant response: a simulation model of transpiration. *Ecology*, **55**, 62–72.

Richter, H., Halbwachs, G. & Holzner, W. (1972). Saugspannungsmessungen in der Krone eines Mammutbaumes (*Sequoiadendron giganteum*). *Flora*, **161**, 40–420.

Rodda, J. C. (1970). On the questions of rainfall measurement and representativeness. In *Proceedings of the World Water Balance Symposium, Reading UK Publication No. 92. A.I.H.S.*

Rodriguez-Iturbe, I. Mejia, J. M. (1974a). The design of rainfall networks in time and space. *Water Resources Res.* **10**, 713–27.

Rodriquez-Iturbe, I., & Mejia, J. M. (1974b). On the transformation of point rainfall to areal rainfall. *Water Resources Res.* **10**, 729–35.

Rogers, J. J. (1973). Design of a system for predicting effects of vegetation manipulation on water yield in the Salt-Verde Basin. Ph.D. Thesis dissertation. University of Arizona, Tucson.

Rogers, J. J. (1975). Ecosystem analysis of forested watersheds: Documentation of a general water balance model. Coniferous Forest Biome, Corvallis, Oregon. In the press.

Rogerson, T. L. (1967). Throughfall in pole-sized loblolly pine as affected by stand density. In *International Symposium on Forest Hydrology*, pp. 187–90. Pergamon Press, New York.

Rothacher, J. (1963). Net precipitation under Douglas-fir forest. *Forest Sci.* **9**, 423–9.

Rowe, P. B. (1941). Some factors of the hydrology of the Sierra Nevada foothills. *Trans. Amer. geophys. Union*, **22**, 90–100.

Rowe, P. B. & Hendrix, T. M. (1951). Interception of rain and snow by second-growth ponderosa pine. *Trans. Amer. geophys. Union*, **32**, 903–8.

Running, S. W. (1976). Environmental control of leaf water conductance in conifers. *Canad. J. for. Res.* **6**, 104–12.

Running, S. W., Waring, R. H. Rydell, R. A. (1975). Physiological control of water flux in conifers: a computer simulation model. *Oecologia*, **18**, 1–16.

Rutter, A. J. (1963). Studies in the water relations of *Pinus sylvestris* in plantation conditions. *J. Ecol.* **51**, 191–203.

Rutter, A. J., Morton, A. J. & Robins, P. C. (1975). A predictive model of rainfall interception in forests. II. Generalization of the model and comparison with observations in some coniferous and hardwood stands. *J. Appl. Ecol.* **12**, 367–80.

Rutter, A. J., Kershaw, K. A. Robins, P. C. Morton, A. J. (1971). A predictive model of rainfall interception in forests. 1. Deviation of the model from observations in a plantation of Corsican Pine. *Agricult. Meteorol.* **9**, 367–83.

Sariahmed, A. & Kisiel, C. C. (1968). Synthesis of sequences of summer thunderstorm volumes for the Atterbury Watershed in the Tuscon area. In *Proceedings of the International Association of Scientific Hydrology Symposium, Use of Analo Digital Computers in Hydrology*, Vol. 2, pp. 439–47, Tuscon.

Sartz, R. S. (1973). Effect of forest cover removal on depth of soil freezing and overland flow. *Soil Sci. Soc. Amer. Proceedings*, **37**, 774–7.

Schnock, G. (1972). Contenu en eau d'une phytocenose et bilan hydrique de l'ecosysteme: Chenaie de virelles. *Oecology Plantarum*, **78**, 205–26.

Schroeder, M. J. & Buck, C. C. (1970). Fireweather, *US Department of Agriculture Forest Service Agricultural Handbook* 360.

Schulze, E. D., Lange, O. L., Kappen, L., Buschbom, U. & Evenari, M. (1973). Stomatal responses to changes in temperature at increasing water stress. *Planta*, **110**, 29–42.

Sellers, W. D. (1965). *Physical climatology*. University of Chicago Press, Chicago.

Shaw, E. M. & Lynn, P. P. (1972). Areal rainfall evaluation using two surface fitting techniques. *Bull. int. Ass. sci. Hydrol.* **27**, 419–33.

Simons, D. B., Li, R. M. & Stevens, M. A. (1975). Development of models for predicting water and sediment routing and yield from storms on small watersheds. *Report prepared for US Department of Agriculture Forest Service, Rocky Mountain Forest and Range Experiment Station, Civil Engineering Department, Engineering Research Center, Colorado State University, Fort Collins.*

Smith, R. E. (1974). Point processes of seasonal thunderstorm rainfall. 3. Relation of point rainfall to storm areal properties. *Water Resources Res.* **10**, 424–6.

Stephenson, G. R. & Freeze, R. A. (1974). Mathematical simulation of subsurface flow contributions to snowmelt runoff, Reynolds Creek Watershed, Idaho. *Water Resources Res.* **10**, 284–94.

Stewart, C. M., Tham, S. H. & Rolfe, D. L. (1973). Diurnal variations of water in developing secondary stem tissue of Eucalypt trees. *Nature, Lond.* **242**, 479–80.

Stewart, J. B. & Thom, A. S. (1973). Energy budgets in pine forest. *Quart. J. roy. met. Soc.* **99**, 154–70.

Sucoff, E. (1972). Water potential in red pine: Soil moisture, evapotranspiration crown position. *Ecology,* **53**, 681–6.

Swank, W. T. (1972). Water balance, interception, and transpiration studies on a watershed in the Puget Lowland region of Western Washington. Ph.D. Disseration, University of Washington, Seattle.

Swank, W. T. & Douglas, J. E. (1974). Streamflow greatly reduced by converting deciduous hardwood stands to pine. *Science, Wash.* **185**, 857–9.

Swank, W. T. & Helvey, J. D. (1970). Reduction of streamflow increases following regrowth of clearcut hardwood forests. In, *Proceedings IASH–UNESCO–Symposium on the results of research on representative and experimental basins,* pp. 346–61, Wellington, New Zealand.

Swank, W. T. & Miner, N. H. (1968). Conversion of hardwood-covered watersheds to white pine reduces water yield. *Water Resources Res.* **4**, 947–54.

Swank. W. T. & Schreuder, H. T. (1973). Temporal changes in biomass, surface area and net production for a *Pinus strobus* L. forest. In *IUFRO Biomass Studies. Working Party Mensuration Forest Biomass,* pp. 171–82. College of Life Sciences and Agriculture, University of Maine at Orono.

Swank, W. T. & Schreuder, H. T. (1974). Comparison of three methods of estimating surface area and biomass for a forest of young eastern white pine. *Forest Sci.* **20**, 91–100.

Swift, L. W. (1972). Effect of forest cover and mountain phycography on the radiant energy balance. D.F. Thesis, Duke University, Durham, North Carolina.

Swift, L. W., Jr., Swank, W. T., Mankin, J. B., Luxmoore, R. J. & Goldstein, R. A. (1975). Simulation of evapotranspiration from mature and clearcut deciduous forest and young pine plantation. Paper presented at the Symposium on Evaporation and Transpiration from Nature Terrain, 1974 (*Eastern Deciduous Forest Biome Memo Report* 73–79).

Tomanek, G. W. (1969). Dynamics of mulch layer in grassland ecosystems. In *The Grassland Ecosystem: A Preliminary Synthesis,* ed. R. L. Dix & R. G. Beidleman, pp. 225–40. Range Science Department Science Series No. 2, Colorado State University, Fort Collins.

US Army Corps of Engineers. (1956). *Snow Hydrology.* North Pacific Division, Portland, Oregon.

US Forest Service. (1972). *The resources capability system. A user's guide.* Watershed System Development Unit, Berkeley, California.

Verma, T. R. & Toogood, J. A. (1969). Infiltration rates in soils of the Edmonton area and rainfall intensities. *Canad. J. Sci.* **49**, 103–9.

Voigt, G. K. & Zwolinski, M. J. (1964). Absorption of stemflow by bark of young red and white pines. *Forest Sci.* **10**, 277–82.

263

Waring, R. H. (1969). Forest plants of the eastern Siskiyous: Their environmental and vegetational distribution. *Northwest Sci.* **43**, 1–17.

Waring R. H., Reed, K. L. & Emmingham, W. H. (1972). An environmental grid for classifying coniferous forest ecosystems. In *Research on Coniferous Forest Ecosystems*, ed. J. F. Franklin, L. J. Dempster & R. H. Waring, pp. 79–92. US Department of Agriculture Forest Service, Porland, Oregon.

Waring, R. H. & Running, S. W. (1976). Water uptake, storage, and transpiration by conifers: a physiological model. In *Water and Plant Life-Problems and Modern Approaches*, ed. O. L. Lange, L. Kappen & E. D. Schulze, Springer-Verlag, Berlin.

Waring, R. H. & Youngberg, C. T. (1972). Evaluating forest sites for potential growth response of trees to fertilizer. *Northwest Sci.* **46**, 67–75.

Watts, W. R., Neilson, R. E. & Jarvis, T. G. (1976). Photosynthesis and stomatal conductance in a Sitka Spruce Canopy. *J. appl. Ecol.* **13**, 623–31.

Whipkey, R. Z. (1967a). Theory and mechanics of subsurface stormflow. In *Forest Hydrology*, ed. W. E. Sopper & H. W. Lull, pp. 255–9, Pergamon Press, New York.

Whipkey, R. Z. (1967b). Storm runoff from forested catchments by subsurface routes. *Proceedings of the Leningrad Symposium on Floods and Their Computation*, Vol. II, pp. 773–9.

Whisler, F. D. & Bouwer, H. (1970). Comparison of methods for calculating vertical drainage and infiltration for soils. *J. Hydrol.* **10**, 1–19.

Whisler, F. D., Watson, K. K. & Perrens, S. J. (1972). The numerical analysis of infiltration into heterogeneous porous media. *Soil sci. Soc. Amer., Proceedings*, **36**, 868–74.

Wei, T. C. & McGuinness, J. L. (1973). Reciprocal distance squared method, a computer technique for estimating areal precipitation. US Department of Agriculture *Agricultural Research Service Publication*, ARS–NC–8.

Weyman, D. R. (1970). Throughfall on hillslopes and its relation to the stream hydrograph. *Bull. int. Ass. sci. Hydrol.* **15**, 23–25.

Weyman, D. R. (1973). Measurements of the downslope flow of water in a soil. *J. Hydrol.* **20**, 267–88.

Whittaker, R. H. & Woodwell, G. M. (1967). Surface area relations of woody plants and forest communities. *Amer. J. Bot.* **54**, 931–9.

Wicht, C. L. (1941). An approach to the study of rainfall interception by forest canopies. *J. South African for. Ass.* **6**, 54–70.

Wooding, R. A. (1965). A hydraulic model for the catchment stream problem. *J. Hydrol.* **3**, 254–82.

Woods, D. B. & Turner, N. C. (1971). Stomatal response to changing light by four tree species of varying shade tolerance. *New Phytol.* **70**, 77–84.

Woods, F. W. (1965). Tritiated water as a tool for ecological field studies. *Science, Wash.* **147**, 148–9.

Woolhiser, D. A., Hansen, C. L. & Kuhlman, A. R. (1970). Overland flow on rangeland watersheds. *J. Hydrol. (N.Z.)* **9**, 336–56.

World Meteorological Organization. (1972). *Distribution of Precipitation in Mountainous Areas*. WMO/OMM 326, Geneva, Switzerland.

Zavslavsky, D. & Rogowski, A. S. (1969). Hydrologic and morphologic implications of anisotropy and infiltration in soil profile development. *Soil Sci. Soc. Amer., Proceedings*, **33**, 594–9.

Zinke, P. J. (1967). Forest interception studies in the United States. In *Forest Hydrology*, ed. W. E. Sopper & H. W. Lull, pp. 137–60, Pergamon Press, New York.

5. Soil processes

B. ULRICH, P. BENECKE, W. F. HARRIS, P. K. KHANNA &
R. MAYER

Contents

Dynamics of water

Dynamics of water in this context means considering a section of the water cycle (Fig. 5.1), namely the fate of precipitation water that happens to fall on some kind of an ecosystem. This section deals with the processes the water is subjected to during its passage from the canopy to the bottom of the soil profile from where it may seep deeper to some aquifer thus recharging the ground water. As such it may be pumped by water plants or re-emerge in wells and contribute to rivers and streams.

The scope of this section is represented in Fig. 5.2, which shows a model of a forest ecosystem. This model, furthermore, is helpful in explaining and visualizing the processes that are essential for the water turnover in ecosystems. The dashed line represents interfaces between an ecosystem and neighboring systems: the atmosphere, the underground and possibly adjacent ecosystems. Arrows crossing this line are termed as input or output, arrows between the subdivisions (or compartments) of the ecosystem are interactions, normally fluxes of water, matter or energy, within the ecosystem.

Input and output variables act as driving forces because they trigger flow processes by changing the hydraulic potential at the boundary. Evaporation lowers the potential, thus establishing a hydraulic gradient that causes the

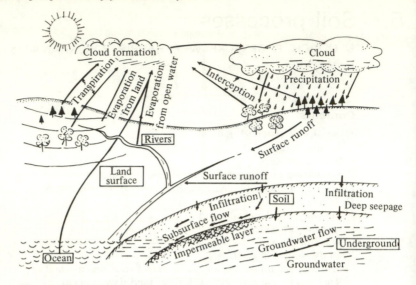

Fig. 5.1. The water cycle. Benecke, 1976.

soil water to move towards the evaporating surface. Inversely, rain arriving at the soil surface enhances the potential hereby giving rise to water movement into the soil. Both processes start from the boundary and propagate subsequently into deeper layers of the soil profile. Water movement in the soil in general is associated with changes in the soil water content, although there are exceptions. The velocity of water movement in the narrow soil pores is rather slow. It decreases with increasing depletion of the soil water. Even under conditions without substantial reduction of its availability to plants the bulk movement of soil water may be negligibly small. The capability of the soil to adjust the flow velocity of the soil water corresponding to the water saturation is comparable to the function of a hydraulic switch that acts in favor of the oxygen and water supply for the vegetation. This background makes it worthwhile to take a closer look at the dynamics of water in ecosystems.

Driving forces

Input variables (precipitation, throughfall, stem flow)

In ground water-free and level areas, precipitation (rain, snow, hail, fog) is the only input variable by which an ecosystem gains water. Lateral flow of ground water, surface and subsurface flow of water – including the un-

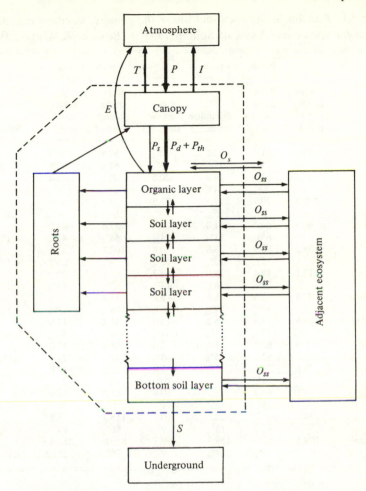

Fig. 5.2. Compartment model of a soil-plant ecosystem. *P*, precipitation, *E*, evaporation, *T*, transpiration; *I*, interception; P_s, stemflow; P_d, canopy drip; P_{th}, throughfall; O_s, surface runoff; O_{ss}, subsurface flow; *S*, deep seepage. The arrows indicate possible directions of the water fluxes. Benecke, 1976.

saturated form – are other possible sources. Restricting interest to precipitation, it is important to notice that rain is fractioned during the passage through the canopy (Fig. 5.2):

$$P = I + P_{th} + P_d + P_s,\tag{5.1}$$

where *P* is precipitation, *I* is interception, P_{th} is throughfall, P_d is canopy drip, and P_s is stemflow.

Table 5.1. *Rainfall in the open and under the canopy, stemflow and inter-ception for spruce and beech at Solling 1968–9 (Benecke & Mayer, 1970)*

		Beech, 120 years			Spruce, 95 years		
	Rain in the open (mm)	Rain under canopy (mm; %)	Stemflow (mm; %)	Inter-ception (mm; %)	Rain under canopy (mm; %)	Inter-ception (mm; %)	Stemflow
May	96.2	71.9	12.1	12.2	59.5	36.7	
		75%	12%	13%	62%	38%	<1%
June	125.5	88.0	16.9	20.6	80.1	45.4	
		70%	14%	16%	64%	36%	<1%
July	109.7	70.9	17.3	21.5	59.7	50.0	
		65%	15%	20%	54%	46%	<1%
August	98.6	64.2	11.0	23.4	57.6	41.0	
		65%	11%	24%	58%	42%	<1%
September	138.3	98.9	18.8	20.6	102.7	35.6	
		72%	13%	15%	74%	26%	<1%
October	112.1	79.4	18.0	14.7	92.1	20.0	
		71%	16%	13%	82%	18%	<1%
November	32.8	32.6	8.6	−2.6	35.4	−2.6	
		99%	27%	−26%	108%	−8%	<1%
December	29.6	39.4	5.2	−15.0	41.2	−11.6	
		133%	18%	−51%	139%	−39%	<1%
January	75.6	62.2	11.4	2.0	58.0	17.6	
		82%	15%	3%	77%	23%	<1%
February	79.8	70.0	2.0	7.8	63.8	16.0	
		88%	2%	10%	80%	20%	<1%
March	76.1	63.2	12.5	0.4	61.2	14.9	
		83%	16%	1%	80%	20%	<1%
April	134.1	109.7	24.5	−0.1	115.3	18.8	
		82%	18%	0%	86%	14%	<1%
Total	1108.4	850	158.3	99.7	826.6	281.8	
		76%	14%	9%	75%	25%	<1%

Part of the rain is immediately changed into an output variable called interception (*see* Chapter 4). The rain, less the amount of interception, reaches the soil surface in three ways: throughfall, by canopy drip or by stemflow. Throughfall means that portion of the rain that falls through the openings of the canopy and remains unaffected by the above-ground vegetation. Depending on the closure of the canopy, the rain-exposed surface area and its properties such as roughness and wettability, another portion of the rain will not immediately find its way to the soil surface, but will first wet the above-ground vegetation and only then either drip from leaves, needles or twigs or run down the stem. Accordingly, rain has changed significantly after passing through the above-ground vegetation compartments. It is diminished by the amount of interception and reaches the soil surface in a more or less irregular spatial distribution. An example is given in Table 5.1, which shows results from readings from 30 rain gauges under beech and another 30 under spruce. The variability is considerable and is in most cases higher under spruce. The spatial variability under beech is further strongly increased by stemflow, which is negligibly small under spruce. Beech takes an outstanding position in this respect, as can be seen from Fig. 5.3 (after Kittredge, 1948). Data collected from the same beech stand as in Table 5.1 show an even higher portion of the rain running down the stem (Fig. 5.4), though only the summer months (June–September) are included. About 12% of the rain above the canopy runs down the stem

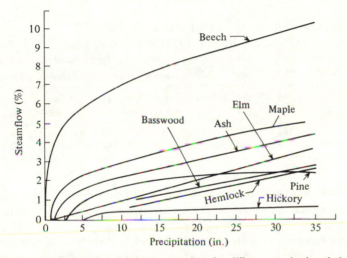

Fig. 5.3. Percentage of precipitation stemflow for different species in relation to precipitation per storm. After Kittredge, 1948.

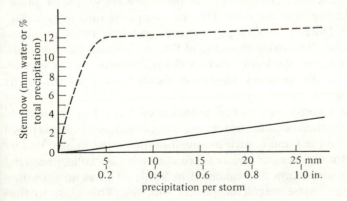

Fig. 5.4. Stemflow in mm (full line) and in percent (interrupted line) of the precipitation per storm for a 120-yr beech stand for June–September, 1969–72. (Solling, W. Germany).

269

Table 5.2. *Soil-water–tension response to the distance from the stem; (data are averaged over time (June to October 1968), depths and plots, in cm water column (Benecke & Mayer, 1971))*

	Part of the area of crown projection near to the stem (cm)	Canopy drip (remaining part of crown projection (cm)	Outside crown projection (cm)
Beech	−42.3	−58.8	−43.6
Spruce	−94.2	−64.8	−47.6

Table 5.3. *Rainfall (mm) in the open (R_o), and in the forest (R_f), Odaigahara, Japan, 1922. (After Hirata (1929) in Penman, 1963)*

		April	May	June	July	August	September	October	Total
No fog	R_o	56	27	307	26	167	49	100	732
	R_f	42	21	209	19	118	36	68	512
Fog	R_o	265	182	56	469	100	230	606	1909
	R_f	293	167	55	517	81	216	450	1780

and enters the soil, generally in a narrow region at one side of the base of the stem. Here less than 1% of the total area receives more than 10% of the rain.

Many authors have dealt with the spatial rainfall distribution beneath the canopy, particularly in order to find an adequate measuring system (Delfs *et al.*, 1958; Weihe, 1968; Ziemer, 1968). From soil moisture measurements (Table 5.2; Eschner, 1967) it appears that there are gradients along radial lines originating from the stem. They have opposite directions under beech and spruce. The reason obviously lies predominantly in the arrangement of the branches. The rising branches of the beeches lead the water on their smooth surfaces to the stem, whereas the falling branches of spruce lead the rainwater to the periphery, where it drops off, leaving practically no water for stemflow.

Rainfall has been considered as one precipitation form. Fog is another form, though of limited significance in the water budget. Fog may be accumulated on the trees and contribute significantly to the budget under two conditions. Border areas of forests exposed to fog may collect enough fog to produce extra stemflow and canopy drip. Table 5.3 gives an example. The other condition is freezing temperatures with fog. This leads to the formation of rime on the trees that later may melt and, thus, produce canopy drip and stemflow, characteristically during bright weather.

Far more important is the contribution of snow to the water budget of ecosystems. The behavior of snow is quite different from that of rain. Far larger amounts may be intercepted during the snowfall as would be the case with rain. But the 'intercepted' snow evaporates not at all or only to a small degree. Most of it in general is transferred to the soil surface in various ways. Either it subsequently melts or it slips in whole portions to the ground. Thus snow interception is not a genuine interception with subsequent evaporation but a temporary storage which sooner or later is converted with its major portion to an 'input variable' to the soil and only with a minor portion to an 'output variable' in the sense of interception.

Another important difference as compared with rain is the accumulation of snow above the soil surface prior to entering the soil. By considering the snow layer as a temporary storage compartment, it could be treated in the same way as a soil compartment (Fig. 5.1.). But unfortunately the principles of water movement and water release are not the same. This makes it extremely difficult to balance the input, changes of the amount of snow and output either by release at the bottom or by evaporation from the top. (The interested reader can find plentiful literature on this subject. Examples are Garstka (1964), de Quervain & Gand (1967), Mitscherlich (1971), Brechtel & Zahorka (1971), Brechtel (1970, 1971) and Heiseke (1974).

Output variables (transpiration, evaporation, run-off, subsurface flow, seepage)

There are numerous output variables which make it desirable to have a scheme or an equation in order to avoid confusion. Fig. 5.2 is such a scheme that shows by means of the arrows crossing the dashed line in outside direction whence the output variables originate and whither they go. Another common way to put them into a meaningful order is given by the water-balance equation:

$$\underset{\text{Input}}{P} = \underset{\text{Output}}{I + E + T + O_s O_{ss} + S +} \underset{\text{Storage and depletion}}{\Delta R}, \quad (5.2)$$

where P is precipitation, I is interception, E is evaporation, T is transpiration, O_s is surface run-off, O_{ss} is subsurface flow, S is deep seepage, and ΔR is change in storage.

It must be emphasized that all of the components in eqn 5.2 are time-dependent variables, i.e. they possess a 'dynamic' character. Thus, the equation makes sense only if each of its components represents the cumulative (or integrated) value over a certain time interval. In addition, the terms O_{ss} and ΔR need to be integrated over soil depth. Except for the interception all the remaining output terms of the water-balance equation depend on the soil, thus emphasizing the focal role that the soil plays in water dynamics and exchange.

271

Transpiration – sometimes labelled productive evaporation because of its close ties to production of biomass – is a major output variable depending mainly on the vapor pressure gradient and the energy available at the transpiring surfaces and the soil water availability. Table 5.4 shows results of the evapotranspiration of a beech and a spruce stand (both timber size) under the moderate, cool-humid conditions of the mountainous central part of West Germany (Solling). Interception may be of the same order of magnitude as the evapotranspiration. Both together are the total evaporation. The evaporation from the soil surface – here combined with transpiration – plays a minor role.

In discussing transpiration in a more general way a basically simple but conceptually important view is to consider the system soil–plant–atmosphere as a hydraulic continuum. This unified system has been called 'SPAC' by J. R. Philip (in Hillel, 1971). Within this system the water transfer obeys the same law which is principally expressed as Darcy's law (eqn 5.6) and can be stated in more general terms by saying that the rate of water movement is proportional to a gradient as driving force by a transfer-coefficient. The gradient represents ultimately an energy gradient which in this system is realized as a pressure gradient. The initial point is in the soil in the form of the soil-water suction (negative pressure) and the terminal point in the atmosphere as vapor pressure. The overall pressure difference is in general very large in day-time (hundred or even several hundreds of bars) and small during the night.

Most of this enormous pressure difference is used up between the walls of the substomatal cavities and the external air where the water is converted from the liquid to the vapor state (Fig. 5.5).

It is remarkable that the role of the plant in the water exchange is rather passive as long as the evaporative demand of the atmosphere can be met (Penman, 1963; Hillel, 1971). As long as there is sufficient water supply in the root zone, plants transpire at a rate that is dominated by the energy available at the transpiring surfaces rather than by plant physiological requirements. Only from the moment the soil-water availability to plants becomes limited do plants begin to regulate the transpiration flux through their stomata. Thus, soil-water availability becomes a crucial question in the water exchange of ecosystems. From experience it is known that, once the soil water tension in the root zone exceeds 15 bars, most plants will wilt irreversibly (Slatyer, 1967). But evidently it is not the binding force that limits the water availability, but the greatly restricted mobility and accessibility of the soil water. It would take <1 cal to release 1 cm^3 of water held with 15 bars as compared to 580 cal necessary for its evaporation. With water supply becoming limited, the water conductivity drops rapidly. Steep hydraulic gradients around the roots may nevertheless allow for adequate water flow to the roots and the growing roots themselves continue to invade

Table 5.4. *Water balance terms of a beech stand (B, 120 years) and an adjacent spruce stand (F, 90 years) in the mountainous central part of West Germany (Solling). After Benecke & van der Ploeg (1976)*

Year		Rain in the open (mm)	Rain beneath canopy (mm)	Evapo-transpiration (mm)	Deep seepage (mm)	Net change of water storage in the soil (0–180 cm) (mm)	Net change of water stored in snow caves (mm)	Interception (mm)	Interception (%)	Total evaporation (mm)	Total evaporation (%)
1968	B	746.1	639.1	254.4	382.5	+16.8	+19	107.0	14.3	361.4	48.4
(May–December)	F		529.9	362.8	145.8	+2.3	+19	216.2	30.0	579	77.6
1969	B	1064.0	912.0	307.0	582.9	−9.8	+32.0	152.0	14.3	459.0	43.1
	F		743.7	383.9	368.9	−10.1	+1.0	320.0	30.1	704.2	66.2
1970	B	1479.1	1206.3	261.0	972.6	+23.7	−51	272.8	18.4	533.8	36.1
	F		1152.6	260.6	890.5	+21.6	−20	326.5	22.1	586.9	39.7
1971	B	809.7	622.7	311.0	303.9	+7.8	0.0	187.0	23.1	498.0	61.5
	F		555.3	310.5	232.2	+12.6	0.0	254.4	31.4	564.9	69.8
1972	B	910.4	716.0	245.0	343.0	+128.2	0.0	194.4	21.3	439.4	48.3
	F		605.0	307.2	308.4	−10.5	0.0	305.5	33.5	612.7	67.3
Mean	B	1065.8	864.3	281.0	550.6			201.6	18.9	482.6	45.3
(without 1968)	F		764.2	315.6	450.0			301.7	28.3	617.2	57.9

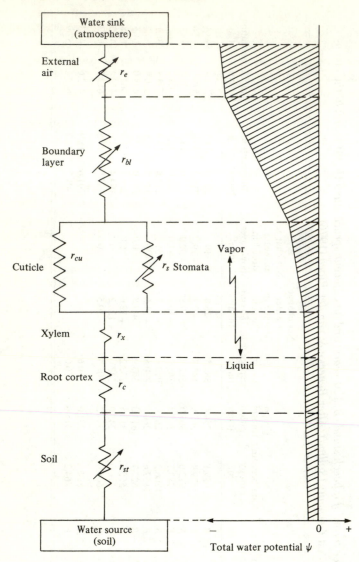

Fig. 5.5. Left: illustrating the various soil, plant and atmosphere resistances to the transpiration stream. While no resistance is fixed, those shown with an arrow are particularly variable. Right: showing the decreasing total potential of water as it moves from soil through a plant to the atmosphere (not to scale). After Rose, 1966.

new regions with less exhausted water supply. But even this combined effect becomes eventually insufficient. This point is marked by a tension of about 15 bars in the bulk soil, which means that the tension of the soil water may be much greater in the immediate neighborhood of the roots.

274

The word evaporation is often used in a somewhat ambiguous way. In a physical sense it means, of course, the transition of liquid water into vapor, regardless of whether plants are involved or not. Since there are three forms of evaporation in ecosystems it is good practice to distinguish between transpiration, evaporation from the soil surface and interception. Transpiration is associated with plant physiological processes. Interception is a somewhat inaccurate notation for an evaporative output. It simply means the amount of intercepted rainfall wetting the standing crop and subsequent evaporation after cessation of the storm. The third form is often simply called evaporation, too, thus giving rise to some ambiguity. More precise would be evaporation from the soil surface.

Interception is the only output variable independent of the soil. It depends on stand properties like age, surface roughness, wettability, leaf area index, closure of the canopy, thinning, and storm characteristics like yield, duration and distribution. Furthermore, it changes with season, i.e. with energy supply. Figs. 5.6 and 5.7 show the effect of interception in relation to yield per storm, based on data for the summer months of four years. The multiple dependency is reflected in a considerable scattering of the data points. Fig. 5.8 is derived from Figs. 5.6 and 5.7. It shows that

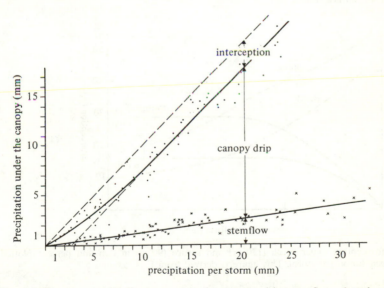

Fig. 5.6. The upper solid line represents the canopy drip+stemflow plotted against the precipitation per storm in a beech stand for June–September 1969–72. The vertical distance to the upper interrupted line corresponds to the amount of interception. The lower solid line shows the stemflow as in Fig. 5.4. The lower interrupted line intersects the abscissa at the point equal to the total interception capacity, whereas the point where the lower interrupted line joins the upper solid line gives the value of minimum precipitation per storm to fill the total interception capacity.

275

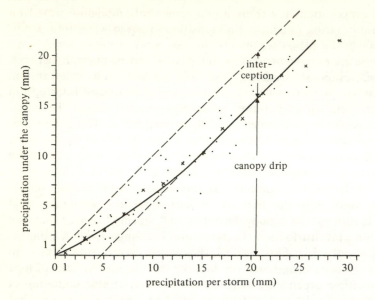

Fig. 5.7. Same as Fig. 5.6 for a spruce stand June–September 1969–72. Stemflow hardly occurred in the spruce stand.

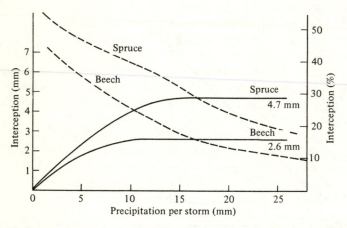

Fig. 5.8. Interception under beech (120 yr) and spruce (90 yr) in mm and percent plotted against precipitation per storm. This figure is derived from Figs. 5.6 and 5.7.

interception gradually approaches a maximum value with increasing yield per storm. The maximum value – 2.6 mm for beech and 4.7 for spruce – may be called interception capacity. It requires about 10 mm (beech) and 12 mm (spruce) of continuous rain to reach the maximum. Expressed in terms of percentage, a decrease is observed with increasing yield per storm,

276

starting out with more than 50% for spruce. For the example demonstrated here there is an average interception of 19% (beech) and 28% (spruce) based on the annual precipitation results (Table 5.4).

It is evident from the literature that the effect of interception differs over an extremely wide range, i.e. the reported data just serve as an example and are valid only for the associated site properties. One final remark deals with the ecological effect of interception. Sometimes it is labeled 'unproductive' evaporation suggesting its uselessness for ecosystems. This may not be entirely correct, since for obvious reasons it can be expected to reduce transpiration (Rijtema, 1965). First and predominantly, energy is used up and second the vapor pressure gradient is likely to be reduced. Van Bavel (in Rijtema, 1965) reports that the total evaporation from wet and dry Sudan grass was observed to be the same, indicating that the transpiration can be reduced by the amount of interception.

Evaporation from the soil surface occurs in two principal ways, either as 'litter-interception' with water supply by canopy drip or by preceding capillary rise of soil water to the surface and evaporation from there. Little is known about the quantity of litter-interception. Taking into account that hydraulic gradients exist, which tend to facilitate infiltration into deeper layers, that most of the radiation energy is shielded off by the canopy, that air movement is more or less absent and that smaller vapor pressure gradients exist as compared to canopy conditions, one would expect the litter-interception to be much smaller than canopy interception. Naturally, the interception capacity of the soil surface, meaning the capacity of keeping rainfall water at or near the surface and thus preventing it from infiltration will be higher under forests with an organic layer on top of the soil than in the case of a mineral soil surface. If this would indicate a higher surface interception in the presence of a litter layer, an inverse relation would hold for the evaporation with preceding capillary rise.

Lacking capillary pathways, the capillary rise is highly effectively interrupted in the organic top layer as can be concluded from countless observations. A mineral soil surface on the other hand may allow large quantities of ascending soil water to evaporate at the surface, provided the capillary rise keeps pace. Since the soil-water conductivity drops sharply with growing water deficit, the capillary rise, too, will drop, unless a lasting water source, like ground water, prevents the build-up of a deficit. Fig. 5.9 gives an example showing the rates of capillary rise to the surface as a function of the depth of a ground-water table (Giesel, Renger & Strebel, 1971). A ground-water level at 40 cm below surface allows the evaporation of 10 mm/day, whereas only 1 mm/day moves to the surface if the ground water is 130 cm in depth.

So far we have dealt with evaporative output variables; some non-evaporative ones remain – surface and subsurface run-off and deep seepage.

277

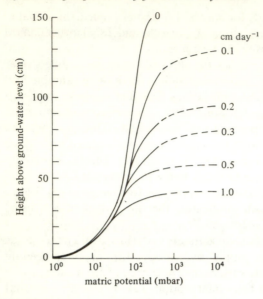

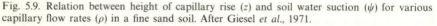

Fig. 5.9. Relation between height of capillary rise (z) and soil water suction (ψ) for various capillary flow rates (ρ) in a fine sand soil. After Giesel *et al.*, 1971.

All the precipitation water not contributed to interception reaches and infiltrates the soil as long as the input rate does not exceed its 'infiltrability' (Hillel, 1971). Once this happens, the water begins to accumulate on the soil surface and may give rise to surface run-off and possibly erosion and flooding in the valleys, provided the rainstorm yields enough water and the area is sloped. Surface run-off is relatively unlikely under forest cover, mainly because of the organic top layer that itself provides a high storage capacity for water and furthermore shields the underlying mineral soil. Without this protective cover the mineral soil may be puddled by the direct impact of the rain drops which, in turn, may sharply decrease the infiltrability. Preventing or at least greatly reducing the danger of surface run-off is one of the important protective functions of forests.

As long as the infiltrability of the soil exceeds the input rate, all of the rain water enters the soil. Because all or almost all of the pore space is of capillary dimensions, the capillary forces from now on play a dominating role. They counteract the gravitational forces, tending to keep the water in place. But this is, in general, achieved only partly, because the capillary forces themselves usually exhibit differences at different points of the soil, thus giving rise to the establishment of gradients. As is explained in subsequent sections, the sum of the capillary forces and the gravitational forces at a particular point form the hydraulic potential, and it is the gradient of the hydraulic potential that causes the water to move.

278

According to Darcy's law, the water flow in the soil is in the direction of decreasing hydraulic potentials. Its rate depends further on the soil-water conductivity or permeability. To make things more complicated it must be added that the conductivity is not constant, but varies with the water saturation of the soil, becoming lower with decreasing water content.

These brief remarks explain when subsurface flow will occur. It requires sloped areas, soils with decreasing conductivity in deeper layers and a high water saturation. In level soils the infiltrating rain increases the moisture content in a horizontal layer evenly, thus allowing the build-up of hydraulic gradients only in a vertical direction. But in sloped soils the gravitational potential drops in the same layer due to its inclination. This causes the soil water to move laterally within the layer. The movement may take any direction between almost vertical and parallel to the surface. Subsurface flow may become the only nonevaporative output variable in sloped areas with shallow soils over impermeable bedrock or clay. Such an area is particularly suited for catchment investigations, since the total run-off can be measured in weirs at the exit of the valley. Fig. 5.10 is an example that shows the order of the seasonal changes.

In a physical sense deep seepage is to be distinguished from subsurface flow only by its vertical direction. Its ecological as well as its economic significance lies in the fact that this water moves more and more out of the reach of the roots thus being lost to the plants but possibly gained for the

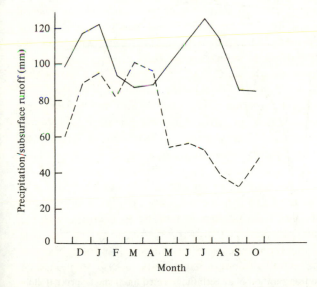

Fig. 5.10. Precipitation (full line) and subsurface run-off (interrupted line) on a monthly basis, averaged over 15 yr (1951–65). Catchment: 80 ha, with spruce. After Friedrich, Liebscher, Rudolph & Wagenhoff, 1968.

supply of water plants by recharging ground-water aquifers. It should be pointed out that deep seepage has the least priority of all output variables, this is to say, all the others have to be satisfied first, before deep seepage takes place. This accounts for the generally large seasonal amplitude, of which Fig. 5.11 gives an example. Under humid conditions the seasonal changes are superimposed by short range amplitudes caused by heavy or continued rainfall. Since deep seepage and subsurface flow are closely connected with storage and depletion and with internal processes, the discussion is kept short here but is continued in subsequent sections.

Storage and depletion, spatial variation in soil

Besides the driving forces at the boundary of an ecosystem that have been considered so far there are internal forces that cause the water to move into regions of water depletion by the output variables. These forces are mainly of capillary nature. Almost all of the soil pores are of capillary dimension ranging from $<0.2\ \mu$ to $>50\ \mu$ in diameter. Fig. 5.12 shows the diameter distribution and the corresponding capillary forces (in suction units) of three soils with typical textures. It should be noted by the way that the total

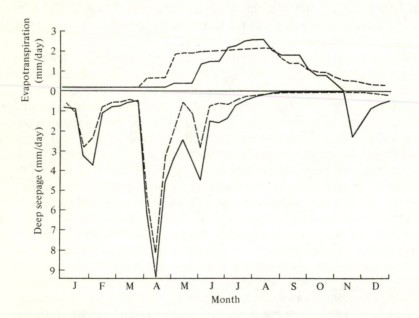

Fig. 5.11. Deep seepage and transpiration of a beech (interrupted lines) and a spruce (full lines) stand (120 and 90 yr, respectively) in 1969. The total rainfall was 1064 mm, the canopy drip under beech (including stemflow) 912 mm and under spruce 744 mm, the seepage under beech 583 mm and under spruce 369 mm.

280

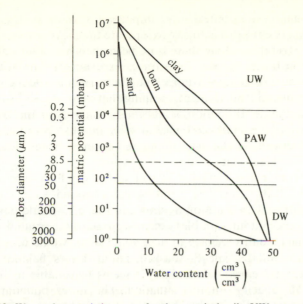

Fig. 5.12. Water characteristic curves for three typical soils. UW, unavailable water or 'dead water'; PAW, plant-available water held by the soil against gravity with its lower limit dependent on the depth of a zero-tension-plane; DW, water that drains off within two or three days, free drainage downwards provided.

average pore space of a soil is slightly less than 50% by volume. There are big differences. Very compact soils have less than 30%, loose soils up to 70% and peat soils even about 90% and more pore space.

The capillary nature of the pore space is most fundamental for the hydraulic function of soils. As long as air and water are both present in the soil, menisci will be present, too, and according to capillary laws (eqn 5.5b), the associated forces are inversely proportional to the curvature of the menisci which, in turn, is closely tied to the pore diameter (Fig. 5.12). The phenomenon of capillarity is supplemented by adsorption of water molecules by the internal surface of the soil, i.e. the walls of the pores. Adsorption forces are much stronger than capillary forces and the water held in that way is practically immobile (and not available to plants).

These briefly outlined principles may help to understand the concept of 'field capacity', roughly meaning that portion of the pore volume in which water is held against the action of gravity. Field capacity is further subdivided (Fig. 5.12) into plant-available-water (PAW) and 'dead water' (UW), the latter being stored in pores with diameter $<0.2\ \mu$ corresponding to a suction of >15 bars. Once a soil dries out beyond this mark in the root zone, most of our crops will irreversibly wilt. This has lead to the notation of permanent wilting point.

281

More difficult, and in this context more important, is to define the lower boundary of 'field capacity', this is to say to separate the moving water from the resting water. Strictly speaking there is no permanently resting water in a soil as can, for instance, be seen from tracer experiments with tritiated water. These experiments (Blume, Munnich & Zimmermann, 1966) confirm the concept of downward displacement, meaning that all of the soil water is displaced downward under the action of infiltrating water. Nonetheless, it is meaningful in terms of the hydroecological functioning of soils to make the above distinction between 'field capacity' and 'self-draining water'. Fig. 5.13 may help to explain these relationships.

Two forces determine the behavior of soil water, capillary forces and gravitational forces. If one considers a soil having a ground-water table at some depth, the potential distribution as shown in Fig. 5.13 would develop, provided all input and output variables except seepage to the ground water are zero. After some time this seepage to the ground water would cease, and a static equilibrium will become established, i.e. all forces balance each other and no water movement takes place. It seems reasonable to base a definition for 'field capacity' on this situation. The lower boundary (in mbar) would be equal to the height above the water table (in cm) of the point under consideration. Keeping this point fixed in the soil and allowing the water table to sink would mean, according to this definition, that the field capacity decreases, thus becoming dependent on the depth of a ground-water table. This approach can be extended to ground-water-free soils, because the rate of downward movement decreases strongly with increasing suction. Once this rate drops below ~0.1 mm/day with further decreasing tendency, it becomes a negligible quantity. Averaging roughly, a ground-water depth of about 3 m corresponds to this condition. Hence, a ground-water-free soil has about the same field capacity as a soil with a 3 m-deep ground-water table. This holds for soils with free drainage. Layers

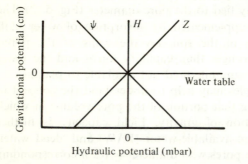

Fig. 5.13. Soil water suction (matric potential), ψ; gravitational potential, Z; hydraulic potential, H ($\psi + Z$) in a soil containing a free water table under the condition of static equilibrium.

with low conductivity have a effect similar to higher ground-water tables, this effect being greater, the shallower the layers are located and the higher their total resistance is to water movement.

It is for the outlined reasons that the field capacity in Fig. 5.12 has two boundaries at the lower end marking the interval within which according to the individual conditions the lower boundary of the field capacity is located.

Internal processes

Water potentials

Water held in capillary systems like the soil is subjected to at least two forces; gravitational and capillary forces. Forces stemming from pressure or osmotic differences may occur under corresponding conditions.

The impact of these forces on an infinitesimally small volume element of water held at a certain position in a porous system is reflected in its energy status, and the potential is nothing else than the amount of internal energy relative to a reference level. Thus, the definition of a soil-water potential requires first the definition of an energy reference level. As such, a pool of pure water under a pressure of 1 bar and at a specified elevation is commonly used. Water molecules being in the surface layer of such a pool are assigned zero potential. This implies that the elevation of a horizontal plane where the gravitational potential is zero can be chosen arbitrarily. Generally, this is done in connection with setting the origin of a coordinate system, relative to which the location of any point in the flow region can be defined.

Two simple experiments may illustrate the effect of a potential difference. Firstly, a vertical column of dry soil is allowed to touch with its bottom the 'zero-pool'. Water would immediately start to move into the column against gravity. Secondly, the same experiment is performed but with an initially completely water saturated soil column. This time the column would partially drain. Both experiments eventually would turn out to have the same result by showing a potential distribution as displayed in Fig. 5.13 (hysteresis neglected). Since water is always flowing in the direction of lower potentials – most obviously demonstrated by any river or creek following decreasing gravitational potentials by flowing downhill, the capillary system of dry soil evidently possesses a lower effective potential than the pool. Fig. 5.12 explains that this is a consequence of the steeply increasing binding forces when soil water is withdrawn. In other words, the capillary potential can be expressed as a function of the water content, becoming the more negative the more water is withdrawn.

In contrast to the conditions in an open river bed where only gravitational forces act, the effective potential in a capillary system is made up of

283

two potentials. It is the sum of the gravitational and the capillary potential which is frequently called the hydraulic potential (Fig. 5.13). Since in the first experiment with an initially dry soil column the high negative capillary potential far outweighs the small positive increment of the gravitational potential, the hydraulic potential drops strongly with height and the water keeps moving upward until there is no further hydraulic gradient, i.e. difference in hydraulic potential with height. This is the condition of Fig. 5.13. Inversely, in a completely water saturated soil column the capillary potential – except for very special conditions that do not matter here – is zero throughout the column. So initially only the gravitational potential distribution determines the direction and magnitude of the hydraulic potential gradients, as in an open river bed. Water starts moving downward. This means withdrawal of soil water that in turn is bringing capillary forces into action, which increasingly counteract the gravitational forces until an equilibrium of forces is established, again as in Fig. 5.13.

Since forces and potentials have both been used their relationship is shown in eqn 5.3.

$$-\frac{\partial P}{\partial s} = F_s \tag{5.3}$$

The force, F_s, is the first derivative of the potential P with respect to s, i.e., the potential gradient is a force vector. Soil water potentials are discussed in detail in almost any soil physics textbook, as for example in Rose (1966) or Hillel (1971).

Fundamental equations and principles, soil water characteristic, hydraulic conductivity, (Darcy's law, equation of continuity, flow equations)

Only a few basic relationships need to be kept in mind for a fundamental understanding of the hydrological processes that have been discussed so far, as well as for the mathematical simulation of the water exchange of ecosystems, which is the subject of the next section.

The law of capillarity states that water rises in a cylindrical capillary tube until an equilibrium is established between the surface forces at the liquid:solid:air contact and the weight of the risen water column in the tube. Hence

$$2r\pi\sigma \cos \theta = r^2\pi\zeta hg, \text{ and} \tag{5.4a}$$

$$h = 2\sigma \cos \theta / r\zeta g, \tag{5.4b}$$

where h is the length of the water column, r the capillary radius, σ the surface tension (73 dyn/cm), θ the wetting angle (generally taken as zero), ζ the water density and g the gravitational acceleration. Since the water 'hangs' in the capillary, the negative pressure or tension directly beneath the meniscus is equal to the upward directed surface forces. Hence, if ψ is the tension per unit square area

$$\psi\, r^2\, \pi = 2r\pi\sigma \text{ for } \theta = 0, \text{ and} \qquad (5.5a)$$

$$\psi = 2\sigma/r \; (\text{dyne/cm}^2). \qquad (5.5b)$$

Frequently the tension ψ is not expressed in the pressure unit dyne/cm^2 but in cm water column. Substituting eqn 5.5a into eqn 5.4a shows that this is in accordance with the underlying equilibrium of forces. Eqn 5.5b established the inverse proportionality between the tension of the soil water and the pore diameter. As has already been pointed out the force necessary to withdraw a unit volume of water from the soil depends on the size of the pores that hold the water. This relationship is used to determine experimentally the soil water characteristic as shown in Fig. 5.12.

Since soil pores are not cylindrical tubes, the term 'diameter' indicates an idealization. The fact that real soil pores are of rather irregular shape is reflected in a generally considerable hysteresis of the soil water character-istic curve $\psi = f(\theta)$ ($\psi =$ capillary potential, $\theta =$ water content by volume). To determine this function, an initially water saturated, undisturbed soil core was used, from which the water was gradually withdrawn (desorption curve). One could equally well use an initially dry soil core and allow it to adsorb water gradually. But a different curve in general results. Giesel *et al.* (1973) give an example (Fig. 5.14). The adsorption curve always exhibits a

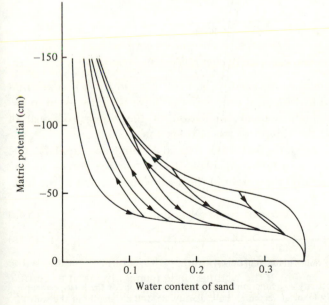

Fig. 5.14. Relations between the matric potential ψ and the water content of a sand (in Giesel *et al.*, (1973) with data from Vachaud & Thony (1971)).

lower water content at the same capillary potential or suction. Infinitesimally many 'scanning' curves may connect the two boundary' curves to each other depending on the wetting and drying history of a soil. This phenomenon of hysteresis is discussed for instance by Vachaud & Thony (1971), Poulovassilis (1973) and Beese & van der Ploeg (1976).

Closely connected with the water saturation of a soil, besides the capillary potential, is the conductivity for water. The forces trying to keep the water in place naturally resist its movement. Furthermore, with decreasing water content the pathways for the water to flow are narrowed. A steep drop of conductivity is the consequence as shown for example in Fig. 5.15.

Attempts to express the conductivity analytically as a function of the capillary potential (which stays for the diameter of the pores by the law of capillarity (eqn 5.5b) have more or less failed so far. Relations that have been discussed in the literature (Rijtema, (1965), Nielsen, Biggar & Erh (1973)) generally involve some constants that must be determined experi-

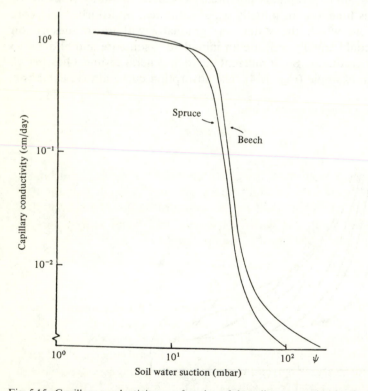

Fig. 5.15. Capillary conductivity as a function of the soil water suction for the least permeable horizons (very stony sandy loam of very high bulk density, 85–140 cm depth) of the soils of Figs. 5.17 and 5.18. Benecke, 1976.

mentally and have only limited suction ranges of validity. This, too, may be attributed to the fact that the porous system of a certain soil is rather specific and irregular and hardly apt for shaping a geometrical model to deal with it theoretically. The conductivity for soil water comes into one of the most fundamental laws in soil water dynamics:

$$q = -k(\theta)\frac{dH}{ds}\left(\frac{cm^3}{cm^2t}\right), \qquad (5.6)$$

which is known as Darcy's law (q is the amount of water flowing per unit time perpendicular to a square unit area in the direction of the decreasing hydraulic potential H, s is an arbitrary space coordinate, $k(\theta)$ is the conductivity coefficient which can be expressed as a function of the water content θ or as a function of the capillary potential ψ, since ψ, too, is a function of θ).

Darcy's law may be specified for horizontal and vertical water movement. Observe that:

$$H = \psi + Z \qquad (5.7)$$

where Z is the gravitational potential, and

$$q = -k(\theta)\frac{d\psi}{dx} \qquad (5.8)$$

for horizontal water movement, and

$$q = -k(\theta)\left(\frac{d\psi}{dz} + 1\right), \qquad (5.9)$$

for vertical water movement (x is a horizontal and z a vertical coordinate).

Darcy's law describes a constant flux q of water through the soil. The more general and frequent case in the unsaturated zone, however, is a variable flux that combines water movement and water storage or depletion. For this the equation of continuity applies:

$$\frac{dq}{dx} = -\frac{d\theta}{dt}, \qquad (5.10)$$

(q as defined in Darcy's law and θ is the volumetric water content). It says that the change of flux along some distance x is equal to the change of water content during that time interval, and can be best understood if one considers the flux through a (small) unit volume of soil during a unit time interval. A full derivation is given for instance in Kirkham & Powers (1972). This flux may not be in the x-direction but in any direction. In this case the three space components of the flux are considered. Substituting q by the corresponding right-hand expression of Darcy's law gives the general soil

Dynamic properties of forest ecosystems

water flow equation for three-dimensional water movement:

$$\frac{\partial}{\mathrm{d}x}\left(k_x\frac{\partial\psi}{x}\right)+\frac{\partial}{\partial y}\left(k_y\frac{\partial\psi}{\partial y}\right)+\frac{\partial}{\partial z}\left(k_z\frac{\partial\psi}{\partial z}\right)+\frac{\partial k_z}{\partial z}=\frac{\partial\theta}{\partial t}. \tag{5.11}$$

If the flux is vertical, the first two terms on the left-hand side disappear:

$$\frac{\partial}{\partial z}\left(k_z\frac{\partial\psi}{\partial z}\right)+\frac{\partial k_z}{\partial z}=\frac{\partial\theta}{\partial t}. \tag{5.12}$$

In this form the flow equation is frequently used, since a lot of field investigations are carried out on level sites. The simulation of the water exchange of an ecosystem, described in the next section, is also based on this equation.

Simulation of evapotranspiration and percolation

Without entering a general discussion on modeling and simulation, this section aims at showing how the principles and laws that have been considered so far can be used to build a model. Nonetheless a quotation from de Wit & Arnold (1976) may serve as a general definition. 'A system is part of reality that contains interrelated elements, a model is a simplified representation of a system, and simulation may be defined as the art of building mathematical models...'.

The basic idea in the present consideration is to model and simulate what is going on with the water in the soil and to solve for evapotranspiration and percolation. Eqn 5.12 is the central part of the simulation, but will be changed to the finite-difference form for the following reasons: (1) numerical methods allow a higher degree of physical accuracy of the model with respect to the represented ecosystem and the boundary conditions and (2) explaining the building of eqn 5.12 with finite-differences makes the problem highly transparent. Another important reason in favor of numerical methods are the present-day computer facilities, that allow a relatively effortless handling of the computations.

If in Fig. 5.13 rain were allowed to enter the soil, this would initiate two processes that can be visualized best by going back to the compartment model (Fig. 5.2). The rain is considered a flux through the surface of the top compartment. It enters the compartment and hence increases the water content and consequently the hydraulic potential (as can be concluded from Fig. 5.12). This in turn establishes a hydraulic potential gradient between the first and the next deeper compartment which initiates a water flux from the top compartment to the next deeper one. Consequently the water content increases in the second compartment and water starts to flow from the second into the third compartment. Next time the fourth compartment is included and so on. Eventually, considering a certain time interval

288

$t_i - t_{i-1} = \Delta t_i$, each compartment receives an input flux through its upper boundary plane and releases an output flux through its lower boundary plane. If the incoming flux is greater than the outgoing flux, water will be stored in that compartment. Conversely, if the outgoing flux exceeds the incoming flux, depletion of the originally stored water occurs. Eventually it may happen that both fluxes are equal and then no change of the water content of that compartment takes place though a (in this case constant) flux is passing through the compartment (steady-state-flux). At the end of the time interval Δt_i some or all of the compartments have a different water content as compared to time $t = t_{i-1}$. This new water content distribution sets the initial conditions for the next time interval Δt_{i+1} at the end of which there again will be another water content distribution. So time step after time-step is governed by the outlined rules.

Boundary conditions that must be known are the input function (time rate of water input into the soil) and the drainage conditions at the bottom of the soil profile. Furthermore the soil water characteristic (water content: water tension relationship) and the conductivity function (water content: water conductivity relationship) are to be known for each of the soil layers separately. To establish eqn 5.12 in the finite difference form three consecutive compartments of arbitrary location may be considered (Fig. 5.16). Taking the geometric centers, P, Q and R as representative for the corresponding compartments, and applying Darcy's law gives approximately:

$$q_{P,Q} = -k(\psi_u) \frac{H_Q - H_P}{z_Q - z_R}, \tag{5.13a}$$

for the flow through the upper plane of the middle compartment and

$$q_{Q,R} = -k(\overline{\psi}_e) \frac{H_R - H_Q}{z_R - z_Q}, \tag{5.13b}$$

for the simultaneous flow through the lower plane of the middle compartment, where q is the flow of soil water per unit time and unit square area in vertical direction ($cm^3 . cm^{-2} . day^{-1}$), k the conductivity (cm/day^{-1}), ψ_u is ($\psi_P + \psi_Q$)/2, the average capillary potential taken to be valid at the upper plane (cm), ψ_e is ($\psi_Q + \psi_R$)/2, the average capillary potential taken to be valid at the lower plane (cm), H is $\psi + Z$, hydraulic, capillary and gravitational potential respectively (cm), z the vertical coordinates (cm), and P, Q, R are points in space representing the geometrical centers of the compartments.

The difference of the flows between the upper and the lower boundary or, better, the change of flow along the vertical distance of the middle compartment must according to the law of conservation of matter be equal to the change of water content within the same compartment. This already has

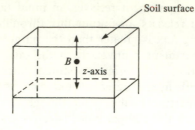

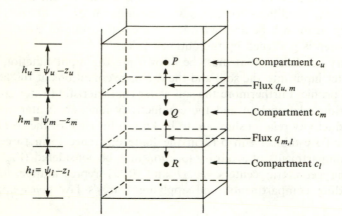

Fig. 5.16. Compartment model of a soil profile with three representative compartments. van der Ploeg, 1974.

been stated by eqn 5.10, the law of continuity. In difference form

$$-\frac{\Delta\theta}{\Delta t} = \frac{q_{P,R} - q_{Q,R}}{z_u - z_e} = \frac{q_u - q_e}{z_u - z_e} = \frac{\Delta q}{\Delta z}, \tag{5.14}$$

where θ is the volumetric water content, and u and e refer to the upper and lower boundary planes of the middle compartment respectively. Combining eqns 5.14 and 5.13a and 5.13b gives

$$-\frac{\Delta\theta}{\Delta t} = \frac{-k(\overline{\psi}_u)(H_Q - H_P) + k(\overline{\psi}_e)(H_R - H_Q)}{(\Delta z)^2}. \tag{5.15}$$

$$(z_Q - z_P = z_R - z_Q = z_u - z_e = \Delta z)$$

Eqn 5.15 corresponds to eqn 5.12 and is the working equation in simulating the water dynamics in level soils with only vertical water movement. Only the general steps have been outlined here. A more detailed discussion can be found in the papers of van der Ploeg (1974) and van der Ploeg & Benecke (1974a, b), where the special conditions at the boundaries of the flow region, that cannot be discussed here, are dealt with. Fig. 5.17 shows the results of a simulation period of about a month. Evapotranspiration is neglected in the case of beech, because of the season. The response of the soil-water tension to the water input (rainfall beneath the canopy – including stemflow) is damping out and delayed with increasing depth. Also the dynamics of redistribution are well reflected in the figure. The deep seepage – or deep percolation – is shown at the bottom of the figure. For comparison the same simulation was carried out for a spruce stand at 400 m distance from the beech stand. Fig. 5.18 shows the results. This time evapotranspiration had to be included. An iterative procedure was used by first assuming daily evapotranspiration rates that were built into the model by adding 'sink' terms in different soil depths according to the root distribution (see the subsequent discussion). The daily evapotranspiration rates were corrected, until the simulated and the measured tension values agreed satisfactorily. The 'sink' term just mentioned is added to eqn 5.12 to give

$$\frac{\partial}{\partial z}\left(k_z\frac{\partial H}{\partial z}\right) + A(z, t) = \frac{\partial \theta}{\partial t}, \tag{5.16}$$

which is equation 2 in Nimah & Hanks, 1973*a*.

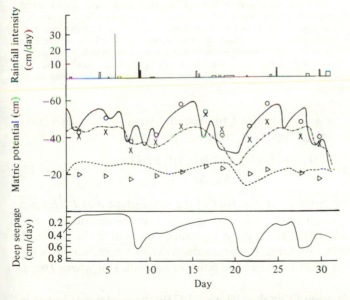

Fig. 5.17. Simulation of the soil water tension at different depths (———, 15 cm, –·–·–45 cm, ----, 135 cm) as compared to measured values ○, at 15 cm; ×, at 45 cm; △, at 135 cm (mean values of about 40 tensiometers per depth) for a beech stand in May 1969. Rainfall intensity and distribution beneath the canopy (including stemflow) are shown at the top of the figure. The total rainfall amounts to 12.27 cm in May, 1969. Deep seepage as determined by the model is depicted at the bottom of the figure. It totals 9.27 cm. Evaporation is neglected. After van der Ploeg & Benecke, 1974*a*, deep seepage added.

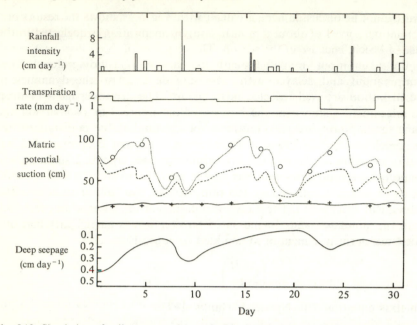

Fig. 5.18. Simulation of soil water tension as in Fig. 5.17 in May 1969, for a spruce stand. Under spruce poor agreement at 15 cm depth between simulated and measured values was observed when evaporation was neglected (- - - - -). But with the shown transpiration rate, agreement was good at 15 cm (....,). Observations at 15 cm depth, ○; at 175 cm, +. Here the total rainfall is 10.47 cm (due to higher interception as compared to the beech stand in Fig. 5.15), the total evaporation is 6.25 cm and the total deep seepage 5.41 cm. Both stands are comparable with respect to soil and meteorological conditions. After van der Ploeg *et al.*, (1975) deep seepage added.

In Fig. 5.18 $A(z, t)$ was determined by trial and error. Nimah & Hanks attempt a more analytical approach, defining $A(z, t)$

$$A(z, t) = \frac{[H_{root} + (RRES.z) - H(z, t)].RDF(z).k_z}{\Delta x.\Delta z}. \tag{5.17}$$

H_{root} is the hydraulic potential in the root at the soil surface and *RRES* is a root resistance term equal to $1 + Rc$. Rc is a flow coefficient in the root system, assumed to be 0.05, $H(z, t)$ is the hydraulic potential in the bulk soil. *RDF* is a root distribution term, k_z the conductivity in z-direction, x is the distance between the plant root and the point in the bulk soil where $H(z, t)$ is measured.

In a further publication (Nimah & Hanks, 1973*b*) the model was tested in the field to predict water content profiles, evapotranspiration, water flow from or to the water table, root extraction and root-water potential at the surface under transient conditions. The test was carried out with alfalfa. Reasonable agreement was found, though the authors point out that agreement was poor immediately after irrigation and best 48 h later. Other authors who worked on root models are, for example, Gardner, 1960 and Feddes & Rijtema, 1972.

Literature on modeling and simulating ecosystems is growing fast, thus showing the increasing interest in this method to gain deeper, analytical insight into the functioning of ecosystems. Attitudes towards this technique are mixed. But neither scepticism nor limitless enthusiasm are dominating. Many results show that, despite the simplifications that necessarily have to be made, specific aspects or parts of ecosystems can be represented by dynamic models. Often the validity of such models is limited to restricted ranges of specified parameters. But despite some reservations and in view of the enormous possibilities and, last but not least, the lack of an equivalent alternative, modeling and dynamic simulation of ecosystems are likely to become a prominent tool in ecology. This certainly is particularly true for investigating the dynamics of water.

Dynamics of chemical elements

The dynamics of chemical elements are understood as element fluxes within the soil, including the exchange with its environment (soil input and soil output), their spatial distribution and time behavior, and the processes governing them. In discussing any soil flux or process, it is essential to define the scale with respect to space (microscopic, ecosystem, landscape, global) and to time (day, month, year, or any other time unit). We deal with the soil as part of the forest ecosystem and with the long-range transport of elements through it. Eventually short-range transport on a microscopic scale has to be considered to explain the internal processes.

The long-range transport of most elements does not take place as single atoms or simple compounds but rather as constituents of a material flux. This leads to the distinction between the transport medium (e.g. water, litter) and the element transported with that medium. The dynamics of an element is very dependent on the dynamics of the transport medium, while the flux of the transport medium is, in most cases, very little affected by the element itself. This fact has important implications on the experimental approach for the assessment of element cycling; i.e. the element flux may be determined by measuring the flux of the transport medium and the element concentration in the medium.

Pathways of chemical input

This section covers different pathways of input to the soil usually found in a forest ecosystem. It will be shown that the estimation of element cycling requires a complete inventory of the input fluxes to the ecosystem.

293

Dynamic properties of forest ecosystems

Input with precipitation

Methods. The element input to the soil coupled with different forms of precipitation (rain, snow, dew, dust etc.) is measured close to the soil surface below the forest canopy. The methods used to collect precipitation are, in principle, the same as those employed in precipitation chemistry in non-forested areas: collectors of different forms and sizes that are continuously open may be used to sample bulk precipitation which includes wetfall (or wet deposition) and dry fallout (or dry deposition). Wetfall may be obtained separately from dry fallout by use of collectors that remain open only during rainfall.

Under the forest canopy it is difficult to avoid contamination of the precipitation (throughfall) sample by all kinds of litter. Therefore, most investigators try to exclude all dry particulate matter from the precipitation samples by covering the collectors with a screen and filtering the sample before analyzing it. All dry particulate matter is included as a part of the litter. In many forests (especially those with broad-leaved trees) consider-able amounts of rain input reach the forest floor along the stems. In these cases stemflow is collected separately by troughs or spiral gutters along the trunks. The technique of sampling precipitation inside the forest has to take into account the heterogeneity of that flux, in its amount as well as in its element content, under the canopy. Usually a great number of collectors is required to obtain representative mean amounts and concentrations.

Components of precipitation input to soil

The difficulty in understanding the dynamics of the element input to the soil by precipitation arises from the fact that this element flux is the sum of several single fluxes. The main contributors, each with its own dynamics, are:

1, the incident precipitation, or wetfall, to the forest canopy;
2, the dry fallout, or dry deposition, of particulate matter to the for-est canopy;
3, impaction of small particles, mainly aerosols, by the vegetation, includ-ing dissolution of gaseous substances like SO_2, NH_3 or nitrous oxides in the intercepted water film covering plant surfaces, subsequently washed down by rain. This is sometimes called 'filtering effect'; and
4, leaching of metabolites and salts from the vegetation.

Input by impaction is not always distinguished from dry deposition; yet impaction on wet leaves may well occur during rainfall.

There are a great number of publications dealing with the natural and industrial sources of the elements present in the atmosphere (*see* Junge, 1963; Singer, 1970; Munn & Bolin, 1971). Research has been concentrated

294

on the chemistry of the main industrial pollutants in the atmosphere, like sulfur (*see*, Eriksson, 1963; Urone & Schroeder, 1969. Prince & Ross, 1972) and nitrogen (*see* Eriksson, 1952; Jost, 1969).

Air chemistry defines as aerosol solid or liquid materials dispersed in air with a particle size smaller than 20 μm regardless of their origin and chemical composition. Particles with a radius larger than 20μm have a very short life time in the atmosphere owing to fast sedimentation and are, therefore, of importance for areas only very close to the emitters of this type of particles. The aerosols are further classified as Aitken particles (≤ 0.1 μm radius), large particles (0.1–1.0 μm radius), and giant particles (≥ 1.0 μm radius).

Aerosols have their origin in maritime sources like sea spray (main constituents are chloride, sodium, potassium and magnesium ions), in mineral dust (carbonates, silicates), chemical reaction products from gases and water vapor in the atmosphere, smoke from industry and from natural fires (sulphur dioxide, sulphate and nitrate ions), and flowering plants producing pollen.

The existence of a filtering effect of forests on aerosols was demonstrated in a great number of investigations, where aerosol concentration inside and outside a forest stand was measured by the use of air chambers or collector foils (*see* Neuberger, Hosler & Kocmond, 1967). However, these techniques are suitable only for an estimation of the filtering efficiency of different types of vegetation, but do not furnish the absolute amount of element desposition by impaction. The values from collectors of different design are not comparable (Stafford & Ettinger, 1971).

The increase in the element load of precipitation passing through the forest canopy is caused not only by impaction, but is also due to leaching. Since both fluxes cannot be measured independently, only their sum can be calculated as the difference between elements in precipitation below canopy (throughfall and stemflow) and above canopy (wetfall and dryfall). Therefore, many investigators measuring the element fluxes with precipitation in forests did not distinguish the two sources of elements, even though they found evidence that impaction and leaching may contribute to varying extents to the concentration of the different elements (Madgwick & Ovington, 1959; Denaeyer-de Smet, 1962, 9; Miller, 1963; Grunert, 1964; Carlisle, Brown & White, 1966; Rapp, 1971; Bernhard-Reversat, 1975).

There have been attempts to develop methods for estimation of impaction. Theoretical considerations by Chamberlain (1960) refer to grasslike surfaces as sinks for atmospheric substances. The input to forests by impaction has been studied by White & Turner (1970). In a mixed deciduous forest in Britain (Meathop Wood) they found the input with dry fallout including impaction exceeding the input with incident precipitation in the case of sodium, potassium and magnesium (Table 5.5). Extremely

295

Table 5.5. *Annual atmospheric input of elements in a mixed deciduous forest in Meathop Wood, Britain. After White & Turner* (1970)

| | Annual atmospheric input (kg ha^{-1} yr^{-1}) | | | | |
	Sodium	Potassium	Calcium	Magnesium	Phosphorus
By incident precipitation (wet deposition)	50.5	4.0	8.8	6.8	0.34
By interception airborne particles (dry deposition including impaction)	125.2	6.3	4.2	16.2	0.12

high values for sodium and magnesium were probably due to high aerosol concentrations from sea spray present in this region. Nihlgard (1970) tried to reach the same goal by the use of plastic nets as aerosol impactors. He found the contribution of aerosols to the element content in precipitation beneath the canopy increasing in the order

$$N < Mn, \ K < P, \ S < Ca, \ Cl < Mg, \ Na.$$

Based on measurements of precipitation input to a beech forest over several years, Mayer & Ulrich (1974) derived values for the filtering of the forest canopy from seasonal and annual element balances. It was then possible to calculate the total atmospheric input as the sum of bulk incident precipitation plus impaction (filtering). Table 5.6 gives the mean annual fluxes for a central European beech forest (Ulrich *et al.*, 1976).

In contrast to the input fluxes treated so far, the contribution of leaching to the element input to soil does not originate from sources outside the forest ecosystem, but is part of the internal turnover of the vegetation by which elements taken up from the soil may return to it (Stenlid, 1958; Tukey, 1970). Total amounts of elements have not been measured but have been calculated from the element balance as the difference between precipitation input to soil and total atmospheric input (bulk precipitation plus impaction). Table 5.6 gives mean annual values for the leaching in a beech forest. The figures show that within this ecosystem leaching makes an important contribution to the precipitation input to soil only in the case of potassium and manganese.

Accurate estimates of the impaction and leaching in evergreen forests are not yet available, because of the lack of distinct periods of inactive physiological stages during the seasons when internal turnover (with possible leaching) of metabolites is largely reduced. As a result it should be pointed out that in mineral cycling studies the precipitation input to soil as well as to the forest canopy should possibly be measured at the test site.

Table 5.6. *Annual atmospheric input and leaching of elements from the canopy in a German beech forest (Solling). Values are averaged over the period May 1969 to April 1972. After Ulrich et al. (1976)*

	Annual value (kg ha^{-1} . yr^{-1}) (in brackets: error of the mean)										
	Sodium	Potassium	Calcium	Magnesium	Iron	Manganese	Aluminium	Chlorine	Sulfur	Phosphorus	Nitrogen
Bulk incident precipitation	7.2 (0.3)	3.7 (0.2)	14.4 (1.0)	2.4 (0.1)	0.9 (0.05)	0.2 (0.3)	1.1 (0.7)	16.0 (0.6)	24.1 (1.0)	0.8 (0.1)	22.6 (1.0)
Impaction (filtering)	5.5 (1.1)	13.0 (2.6)	11.4 (2.2)	1.4 (0.3)	0.3 (0.1)	1.7 (0.4)	0.5 (0.2)	13.4 (2.0)	22.2 (3.3)	0.0	1.2 (0.4)
Total atmospheric input (Bulk incident precipitation + impaction)	12.7 (1.1)	16.7 (2.6)	25.8 (2.4)	3.8 (0.3)	1.2 (0.1)	1.9 (0.5)	1.6 (0.7)	29.4 (2.1)	46.3 (3.5)	0.8 (0.1)	23.8 (1.1)
Precipitation input to soil	12.9 (0.4)	24.3 (0.9)	28.3 (0.9)	4.0 (0.3)	1.1 (0.4)	3.0 (0.1)	1.6 (0.1)	30.8 (1.1)	47.6 (1.2)	0.6 (0.1)	23.8 (0.6)
Leaching of canopy (Precipitation input to soil– total atmospheric input)	0.2 (1.2)	7.6 (2.8)	2.5 (2.6)	0.2 (0.4)	–0.1 (0.4)	1.1 (0.5)	0.0 (0.7)	1.4 (2.4)	1.3 (3.7)	–0.2 (0.2)	0.0 (1.2)

297

Where assessment or accurate estimations of dry fallout and impaction are not possible, approximate assumptions should be used. In no case input by impaction should be neglected without checking its relative importance; otherwise the element budget for the forest ecosystem may be erroneous.

Dynamics

The dynamics of individual fluxes contributing to the element load in forest precipitation has been studied intensively by meteorologists, ecologists and plant physiologists. Leaching of elements from vegetation is determined by factors inside the forest ecosystem and by environmental factors. Investigations have led to the hypothesis that leaching of cations is primarily passive, involving ion exchange from the cell walls and cuticular membrane surfaces and diffusion into the leaching solution following a concentration gradient (Mecklenburg, Tukey & Morgan, 1966; Yamada, Rasmussen, Bucovac & Witter, 1966). Protons as exchange counterparts may be provided by the precipitation (carbonic acid from atmospheric carbon dioxide, sulfuric acid from sulfur dioxide and trioxide). The exchanged cations then form carbonates or sulfates which are removed with the precipitation or are precipitated upon the leaf surfaces.

The hypothesis of passive leaching could explain why the amount of elements leached is fairly independent of the volume of leaching solution: a long-lasting rainfall of low intensity is more effective in leaching than a short high intensity rainfall (Witter & Teubner, 1959; Tukey, 1970). It explains, furthermore, why changing temperature, radiation and energy conditions – major factors influencing physiological activity – had little effect upon the leaching of cations (Mecklenburg *et al.*, 1966).

Leaching is more intensive at the beginning of a precipitation period (Stenlid, 1958) and from mature leaves than from young leaves (Tukey, 1970). In deciduous forests leaching of potassium was found to increase before leaffall (Denaeyer-de Smet, 1962). Intensive leaching was observed when trees were tagged with radioactive cesium (Witherspoon, 1964; Witkamp & Frank, 1964; Waller & Olson, 1967); leaching of radioactive calcium was less intensive (Thomas, 1969).

There is some evidence that elements from the atmospheric input may be adsorbed or taken up by vegetation (Witter & Bukovac, 1969). These elements may be precipitated on leaf surfaces, bark, epiphytes and mosses, or taken up by leaves and translocated into the plant, thus being protected from precipitation washout. In the first case, the elements reach the soil with litterfall. The magnitude of adsorption of elements by vegetation is probably small for most elements compared with the amounts carried with precipitation in forests and is, thus, negligible. It may, however, play an important role in the case of nutrient elements like nitrogen and phosphorus (Ovington, 1962) and for many heavy metals like lead, mercury, zinc and cadmium (Lagerwerff, 1972; Ernst, 1972; Little, 1973; Ruhling & Tyler, 1973; Schlesinger, Reiners & Knopman, 1974; Holl & Hampp, 1975). For these elements the concentration in the precipitation under the canopy is lower than in incident precipitation during the vegetation period (nitrogen and phosphorus; *see* Mayer & Ulrich, 1974; Khanna, 1975), or during all seasons (heavy metals).

The element contribution to soil input by incident precipitation, dry fallout and impaction is mainly controlled by factors present outside the forest ecosystem. Amount of precipitation and its distribution in space and time depend upon the regional meteorological conditions. Processes leading to deposition of atmospheric substances are the formation of clouds and precipitation (rainout) including chemical reactions, rain scavenging (washout) and sedimentation. Sedimentation can be predicted for particles of known radius (Junge, 1963). But for particles smaller than about 2 μm radius residence time becomes very long and fall velocity very small so that particles of this size are more likely to be washed out or removed from the atmosphere by impaction. These two latter processes are much more complicated because chemical processes and flow dynamics of the transport media, precipitation and air, are involved.

In a number of investigations a general correlation between element concentration in the incident precipitation and the amount and type of precipitation has been attempted. The average concentration shows a decrease with increasing amount of precipitation (Madgwick & Ovington, 1959; Junge, 1963). This relationship may readily be explained by two facts.

1. Evaporation of the rain droplets during rainfall increases the element concentration in the droplet.
2. The higher the liquid water content of a cloud (larger rainfalls) the lower is the element concentration (dilution effect) attained by rainout processes. At the same time, evaporation of the droplets during rainfall is reduced because of high humidities.

The concentration (c) against precipitation (h) plot from data reported by Junge (1963) for several elements can be approximated by a curve of the form:

$$c = a \cdot h^{-0.3}, \tag{5.18}$$

where a is a constant. The same relationship holds for the stemflow (Ulrich, 1969) and for the canopy drip (Mayer, 1972a). Seasonal variations in the element input to soil with precipitation are very distinct especially in deciduous forests and in all regions with climatic seasonality. The different elements often show a characteristic behavior (Madgwick & Ovington,

1959; Henderson *et al.*, 1971; Ulrich, Mayer, Khanna & Prenzel, 1973; Khanna, 1975).

Models for the dispersion and deposition of atmospheric substances with regional or continental validity have been developed (Granat, 1972, 4, 5; Rodhe, 1972; Bolin & Persson, 1975; Granat & Söderlund, 1975). Application of these models requires a knowledge of meteorological data such as the probability of occurrence of rain and rain duration and of the vertical transfer of substances at a given place. It is probable that few of these data are available for most regions, and therefore computation of a dispersion/deposition model will give only approximate results.

Gas adsorption by soil

This way in input may be of importance only for elements which are present in the gaseous phase in the atmosphere above the soil surface. These are primarily oxygen, nitrogen, and carbon (as carbon dioxide). Large amounts of carbon and oxygen are exchanged between atmosphere and soil due to photosynthesis and respiration of plants and microorganisms. The processes involved in this exchange are treated in this volume in the chapters on carbon dioxide balances, root dynamics, and decomposition (Chapters 3, 5 and 9). Other gas exchanges are less important, only adsorption of nitrogen may be of significance in some forest ecosystems. It is known that atmospheric nitrogen is absorbed and transformed by several soil microorganisms which may be free-living (e.g. *Azotobacter, Clostridium*, some blue-green algae (Jensen, 1965)) or associated with the roots of plants (symbiotic nitrogen fixation; Nutman, 1965). But only the bacteria in the root nodules of legumes (*Rhizobium*) seem to contribute significantly to the nitrogen input to the soil. Nitrogen gains of more than 200 kg/hectare/year have been reported for soils under legume crops (Allison, 1965). Several genera of non-leguminous shrubs and trees (e.g. *Alnus*) have root nodules with symbiotic microorganisms which may fix atmospheric nitrogen.

Within the scope of the IBP woodland sites the tropical forest ecosystems often contain leguminosae, but data on the gaseous fixation of nitrogen in these ecosystems have not yet been reported.

Nitrogen fixation due to the activity of non-symbiotic microorganisms probably plays only a minor role in the nitrogen cycle of most ecosystems (Campbell & Lees, 1967), and data are not reported for the IBP woodland sites. It should be pointed out that considerable methodical difficulties are encountered in all investigations dealing with that problem, for measuring techniques may interfere with the delicate equilibrium in the activity of microorganisms involved.

Gas adsorption by trees (Roberts, 1974) and by soils (Prather, Miyamoto & Bohn, 1973) may be considerable when atmospheric pollutants are

300

present in high concentrations, but these conditions are beyond the scope of the present synthesis.

Litterfall

An important element input to the forest floor is coupled with litterfall which includes leaves, fruits, flowers, bud scales, twigs and branches, grass and detritus from above-ground animals. This input is part of the internal turnover of the forest vegetation, by which elements taken up by the roots return to the soil. As mentioned in the previous chapters, this is not true for oxygen and carbon, which are adsorbed as gases from the atmosphere, and perhaps for small quantities of other elements, which are deposited from the atmosphere by wetfall, by dryfall or by impaction, and which were neither washed down from the canopy by precipitation nor translocated into the shoots. Only in areas with a high level of air pollution the latter element input may be of importance, and may perhaps exceed the internal turnover fraction in litter in the case of some heavy metals and trace elements.

The element flux coupled with litterfall is measured continuously by collecting the litter in traps or in funnel-type collectors with defined surface area. Methods are described and discussed by, e.g. Duvigneaud, Denaeyer-de Smet & Marbaise (1969) and Saito & Shidei (1972). Care has to be taken that no elements are leached (e.g. potassium) or accumulated (e.g. iron) from the litter by rain when deposited in the collector. Therefore, collectors have to be sampled in short intervals. The different components of litter may be determined by separating them from the bulk material collected. Fractions of larger size, with very heterogeneous distribution on the forest floor, are determined by collecting them from larger areas (several square meters), primarily cleaned from coarse litter, after a defined period of time. The element content in the different fractions is determined by chemical analysis. The element flux is obtained by multiplication of the element content (mass of element per unit mass of litter) and the amount of litter reaching the forest floor (mass per unit area and unit time).

The element input to soil with litter is dominated by the dynamics of the litterfall and by the element content of the different litter fractions. In deciduous forests, the flux becomes small during the winter months, but seasonal variations are also found in evergreen forests (Duvigneaud & Denaeyer-de Smet, 1970; Ando, 1970; Pavlov, 1972; for other citations see following section). In various forest ecosystems of the world the element content of freshly fallen litter shows seasonal characteristics for most elements. Maximum phosphorus and nitrogen contents were found in the beginning of the vegetation period, maximum potassium and magnesium contents during summer and fall, maximum calcium, silicon, iron and aluminium contents during fall and winter. These general conditions were

301

observed in New Zealand hard beech (*Nothofagus truncata*) forests by Miller (1963), in central European mixed beech (*Fagus sylvatica*) and Scots pine (*Pinus sylvestris*) forests by Grunert (1964), in mediterranean evergreen oak forests by Rapp (1971), in eastern United States deciduous and pine forests (Henderson *et al.*, 1971), and in Scots pine stands in Finland by Malkonen (1974).

Annual variations are also very commonly observed depending upon climatic conditions, occurrence of fructification, or changes in the population of overstory consumer fauna. Since different fractions of litterfall often show considerable differences in their element content, all processes influencing litter composition are also controlling the element input to soil. Independent of seasonal and annual variations, considerable differences in the element content of litter and, consequently, in the element flux reaching the forest floor, were found for the same tree species depending on site conditions like soil chemical and hydrological conditions, climatic conditions (*see* Rodin & Bazilevich, 1967; Duvigneaud & Denaeyer-de Smet, 1969). One common feature of many tree species is the high calcium level in the litter of forests growing on limestone.

In general, restitution of elements to soil by litterfall increases with increase in the productivity of a forest ecosystem. Understory shrubs and ground vegetation may be very important in the total turnover by litterfall. Comparing the two major ways of element input to the forest floor – litterfall and precipitation – we find that litterfall is most important for nitrogen and phosphorus, still very important for potassium, calcium, magnesium, iron and manganese, and of minor importance for sodium, aluminium, sulfur and chlorine (see Chapter 6).

Other pathways of input

There are a number of other ways for an element input to a forest soil which are, though, not of the kind that might be considered essential for a typical forest ecosystem. They may, however, play an important role in some special cases.

Wet sedimentation, sometimes called geological input, may be significant at the base of slopes where eroded material is deposited, or in forests temporarily flooded close to riverbanks, on lowlands or coastal areas. While the rate of deposition in large river basins may be fairly well known from measuring the suspension load of rivers, the estimation of the input will be quite difficult on the basis of individual ecosystems.

Element input with subsurface flow and ground water is even more difficult to account for. The task of estimating these fluxes, when significant for a forest ecosystem, may only be achieved in close cooperation with the hydrologist (*see* Chapter 4).

Fertilization, an important input to some managed forests, is usually done in well defined amounts and may, therefore, easily be taken into account when mineral cycling is investigated. However, the form of the fertilizer has to be considered. Easily soluble fertilizers (e.g. most nitrogen and potassium fertilizers) may be dissolved after a few rainstorms and appear as a high increment in the soil solution compartment. On the other hand, slightly soluble fertilizers (some phosphates, limestone) may act as a moderate source continuously flowing over a long period of time without changing drastically the concentration in soil solution.

Internal processes

Processes controlling fluxes and turnover of elements within the soil are covered by this section. Included are the large-scale transport of elements through soil with percolation water as well as transition of elements from one soil compartment, or binding form, to another, which may take place on a microscopic scale.

Release and mineralization of organic matter, microbial turnover

This subject is treated in detail in the synthesis volume on decomposition and will, therefore, be discussed here only briefly. Processes contributing to the release and mineralization from organic matter form an important step in the internal turnover because only after destruction of biological structures and organic binding forms, and transformation into water soluble forms or gases, are the elements, in general, submissible to large-scale transport processes, to uptake by plant roots, or to sorption on soil particles. There is no generally accepted technique for the experimental assessment of these fluxes, for release of an element and transfer to other compartments (plant root, exchange complex of soil matrix, micro-organisms) often take place within a microscopic scale. The acceptor compartments cannot be seperated from each other without disturbing the whole system.

Three different techniques have been used widely in mineralization studies: litter bags, lysimeters, and isotopic techniques. Froment & Mommaerts-Billiet (1969) and Runge (1971) used litter bags deposited in the very soil layer from which the sample was taken to measure the net mineralization of nitrogen as well as the intensity of the mineralization processes in the different horizons of the organic and mineral soil. In the same way calcium losses from decomposing tree leaves have been measured by Thomas (1970). By use of a similar incubation technique Froment & Tanghe (1972) measured the nitrogen mineralization under laboratory conditions in seven different forest humus types, which showed considerable differences in their characteristics of mineralization. Comparison between in situ methods and incubation methods in nitrogen mineralization studies was performed by Lemée (1967).

Lysimeter techniques (see Chapter 6) have been used for mineralization studies in the laboratory (Rapp, 1971; Baum, 1975) as well as in situ (Mayer, 1972*b*). The latter included the use of funnel-type lysimeters installed at the lower boundary of the humus layer of forest soils after cutting the roots to prevent uptake. From the percolation water collected in the lysimeters the element input by precipitation to the forest floor (throughfall) has to be subtracted to obtain the mineralization flux from the humus layer. A main limitation of this technique is the change in the soil-water conditions induced by the placement of the lysimeters. Therefore, suction lysimeters have been used in recent years by which the continuous soil water flow is not enhanced (Cole, Gessel & Held, 1961; Mayer, 1972, 1974).

Isotopic techniques have been used in decomposition and mineralization studies by Witkamp & Frank (1964), Waller & Olson (1967), Thomas (1969), and Witkamp (1969, 71). These investigators studied the release of isotopes from decomposing litter taken from trees tagged with radioactive cesium, calcium, and other radionuclides.

It has been demonstrated by a great number of investigations that decomposition, and consequently the rate of release of mineral elements from organic matter, is dependent upon the litter species and controlled by environmental factors such as temperature and humidity (Dickinson & Pugh, 1974), which may operate in an indirect way, by influencing primarily the activity of microorganisms and microbial turnover. This explains, in part, the seasonal variations of mineralization as well as the differences between various forest sites. Beyond these effects of environmental factors, the decomposition seems to follow very generally a simple decay law: the rate of mineralization is fairly proportional to the amount not yet minera-lized. This has led many authors to look for a mathematical expression describing decomposition and release of elements with time, under changing environmental conditions during the seasons (Jenny, Gessel & Bingham, 1949; Olson, 1963; Ando, 1970; Mayer, 1972*a*,*b*).

Regarding individual elements, some characteristics common to most forest ecosystems may be stated (cf. Rapp, 1971; Mayer, 1972*a*,*b*). Potassium is set free very quickly from freshly fallen litter; its rate of release is mainly dependent upon the percolation rate of soil water passing through the humus layer. Compared to potassium, the rate of release of phosphorus was found to be about 1/2, and 1/3 in the case of nitrogen. For both elements the rate of release is closely correlated to the temperature and humidity conditions.

These relationships account for seasonal changes in mineralization and release of elements. Irrespective of such short-term variations it is most likely that, within a mature forest ecosystem, the soil organic matter is in steady state, or close to it, with respect to most elements. This means that the rate of mineralization – averaged over a sufficient period of time to exclude seasonal fluctuation – is constant, and equal to the element input by litterfall. It must be stated that the assessment of an increase or decrease of elements stored in the soil organic matter on a significant statistical level would require very extensive sampling over several years. These requirements are not met by most investigations within forest ecosystems.

Physico-chemical equilibria

Any process that leads to a change in the amount of ions in the soil compartment of an ecosystem through the ways of either input or output involves inevitably the ion exchange equilibria, or the adsorption or desorption, or the precipitation and dissolution processes to regulate the system again. This system is composed of two phases, the solid and the liquid phase, which under normal conditions can be expected to be in equilibria with each other. For a heterogeneous system in equilibrium the Phase Rule, $F + P = c + 2$, gives the least number of independent variables (F) required to define the state of the system where the number of phases (P) and the number of components (C) are given. For a two-phase system the number of variables required to define the system equals the number of components involved in the system. The complexity of a forest ecosystem involves a great number of components. For the sake of simplicity, however, usually only one component, e.g. the concentration of an ion, is studied with the expectation that its determination should be sufficient to describe the whole system. Since the most important soil variable that affects and controls a number of chemical and biological processes in soil is the pH of the soil, one has to delineate the pH buffer zones prevailing in soils.

Buffering systems

The natural undisturbed soils show normally two buffer zones, the carbonate and aluminium buffer zones.

In the carbonate buffer zone, the solubility products of carbonates of calcium, magnesium, sodium etc. determine most of the chemical soil characteristics such as the pH values and the amount of exchangeable cations and soil processes like cation exchange, precipitation and dissolution of ions, etc. This zone is characterized by the presence of calcium as the main exchangeable cation and by the absence of aluminium ions in solution and as exchangeable phases. The pH values of soils are usually greater than 6.5 depending upon the CO_2 concentration in the soil air. In

305

calcareous salt-free soils, pH is determined by the system: calcium carbonate–calcium bicarbonate–carbon dioxide–water which is given approximately by the equation

$$pH - 0.5 \; pCa = K + 0.5 \; pCo_{2/(g)}, \qquad (5.19)$$

where $pCO_{2/(g)}$ is the negative logarithm of the partial pressure of carbon dioxide in atmospheres in equilibrium with the solution, pCa is the negative logarithm of the activity of calcium ions, and K is a constant whose value lies between 4.89 (Ulrich, 1961) and 5.2 (*see* Nakayama, 1968, for details) depending upon the value used for the solubility constant of calcium carbonate. Attempts have, thus, been made to show that the term $pH - 0.5$ pCa, also called lime potential, shows constant values under constant carbon dioxide conditions as shown by Schofield & Taylor (1955a). They also suggested the use of 0.01 M calcium chloride solution for the determination of pH of the temperate soils, on the grounds that it approximates to the calcium concentration in the soil solution and reduces the error due to the varying salt concentration in natural soils (for detail refer to Russell, 1973).

The clay minerals and humus in this pH range show an excess of negative charges and all the exchange sites are available to cations so that the total cation exchange capacity (CEC_t), which is normally measured by using N ammonium acetate pH 7.0 or Mehlich's method at pH 8.1, equals the effective cation exchange capacity (CEC_e) of a soil. The total number of negative charges is equal to the sum of permanent charges due to isomorphous substitution in structure of clay and the pH-dependent charges from the clay minerals as well as organic matter. In the aluminium buffer zone the release and fixation of aluminium from and on the clay minerals determines most of the soil chemical processes. The pH of these soils lies between 3.5 and 5.5, and aluminium ion is the main exchangeable cation. Some of the peat soils also show low pH values but these soils, because of the absence of clay minerals and aluminium do not show any buffering capacity against pH changes. When a soil with higher pH value changes on acidification to a low pH value, the change is an irreversible one. This happens because at lower pH values the aluminium released from a clay mineral deposits itself on its surface as polymerized aluminium hydroxide. At low pH values, the soils exhibit lower effective cation exchange capacity which can account for up to 10% of the total, for the following reasons:

1. decrease in the dissociation of acid groups of soil organic matter,
2. decrease in the dissociation of Si-OH groups existing, e.g. at the lateral surfaces of clay minerals,
3. blocking of exchange sites through coating of hydroxyaluminium polymers in the interlayers of 2:1 layer silicates. The hydroxyaluminium polymers themselves exhibit pH dependent charge,

306

4. reduction of net negative charges due to the overlapping of diffuse double layers of the permanent negative charges and the positive charges developed on change in pH.

The amount and type of charges and, thus, the adsorbed ions depend to a great extent on the pH of the soil.

Cation exchange equilibria

The proportions of different exchangeable cations are largely dependent on and are held in equilibrium with their activities in the soil solution. If there are two species of cation involved, any change in the activities of one in the solution will alter the other in such a way that the Ratio Law remains valid (Schofield, 1947) and the equilibrium is maintained. Ratio Law implies:

$$\frac{a_H}{a_K}=\text{constant},\quad \frac{a_H}{\sqrt{a_{Ca}}}=\text{constant and }3\frac{a_H}{\sqrt{a_{Al}}}=\text{constant}, \tag{5.20}$$

where a represents the activities of ions in solution. But this is only valid if all the exchangeable cations are allowed for, the relative concentration of the cations close to the fixed negative charges on the clay surface is high in relation to the external solution (absence of positive charges), and the concentration of a particular species of cation in solution is not sufficiently high to change the cation exchange characteristics of the soil. All these conditions may be fulfilled among natural soils of pH > 6.5 and the validity of this Ratio Law has been shown by Schofield & Taylor (1955b) for the ion pair H–Na, or Ca, or Ca + Mg, or Al; by Taylor (1958) for K–Ca; by Beckett (1964) for K–Ca or Ca + Mg.

The ideal conditions for the application of Ratio Law are not always met in soils, moreover it is of interest to know about relative strengths of cation adsorption and release by soils; therefore, the use of cation exchange equations (for a review see Bolt, 1967) like the most used Gapon equation (the discussion will be restricted to only this equation here), is made to relate the pair of ions in the exchangeable phase to those in the solution phase. For an ion pair A–B the Gapon equation is

$$\frac{A_s}{B_s}=G_{A/B}\cdot\frac{^{zA}\sqrt{a_A}}{^{zB}\sqrt{a_B}}, \tag{5.21}$$

where A_s and B_s are the equivalent amounts of exchangeable cations, $^{zA}\sqrt{a_A}=$ $^{zB}\sqrt{a_B}$ is the activity ratio, also called the reduced activity ratio of cations in solution (usually expressed as mmol/l), z is the valency and $G_{A/B}$ is the Gapon coefficient or constant, because Gapon (1933) considered that it should be a constant independent of the activity ratio. The Gapon coefficient is dimensionless if $z_A = z_B$ but has the following dimensions if $z_A \neq z_B$ and one should be careful when comparing the values of different ion pairs:

$$\text{mol}\left(\frac{1}{z_B} - \frac{1}{z_A}\right) \cdot 1\left(\frac{1}{z_A} - \frac{1}{z_B}\right) \qquad (5.22)$$

if the concentrations are expressed in mol/l.

The Gapon equation has been successfully used by the US Salinity Laboratory for the Na–Ca exchange in alkali soils of arid and semi-arid regions. Bolt (1967) observed that the use of the Gapon equation in the range above 50% exchangeable Na-ions does not appear to be warranted. The Gapon equation has also been used to differentiate among various binding positions on the exchange complex say, for example, for potassium (Schouwenberg & Schuffelen, 1963), on an illite clay or on soils (Ehlers, 1966), where the potassium present on planar, edge and intermicellar sites can be calculated. For other ion pairs its use has been reported by Salmon (1964) for Ca–Mg, by Coulter & Talibudeen (1968) for Ca–Al, and Ulrich, Ahrens & Ulrich (1971) for all possible combinations.

Soils always contain, in fact, more than two species of cations. The Gapon equation, like many other equations based on an ion pair, can be used successfully when one of the cations used for exchange occupies most of the exchangeable sites and is involved in the cation exchange. In cases where this does not happen and cations other than those considered in the exchange equation are involved in the exchange, it is not justified to use equations where only two species of cations are observed. For such cases, Khanna & Ulrich (1973) suggested the use of the chemical potential (μ) of sodium in the solution phase where in addition potassium, calcium, magnesium and aluminium are present

$$\mu_{Na} = RT \ln \frac{a_{Na}^4}{a_K \cdot \sqrt{a_{Ca}} \cdot \sqrt{a_{Mg}} \cdot \sqrt[3]{a_{Al}}}, \qquad (5.23)$$

where R and T are the gas constant and the absolute temperature, respectively. It can then be related empirically to the changes in exchangeable cation. It should also be possible to calculate a multiple Gapon coefficient (Khanna & Ulrich, 1974) which has the following general form for the above given case for Na

$$K_{Na}^G = \frac{Na_s^4}{K_s \cdot Ca_s \cdot Mg_s \cdot Al_s} \cdot \frac{a_K \cdot \sqrt{a_{Ca}} \cdot \sqrt{a_{Mg}} \cdot \sqrt[3]{a_{Al}}}{a_{Na}^4}. \qquad (5.24)$$

For an acid soil this coefficient does provide constant values on changes in the saturation of exchange sites by a single cation species.

Anion adsorption

Adsorption of anions on the surfaces of soil clay particles and the oxides or hydroxides of aluminium and iron occurs mainly in two ways.

1. The anions are required to balance the positive charges. The number of these charges and of the anions balancing them varies with the changes in pH or salt concentration. Different anions behave in their bonding strength as

$$Cl \leqq NO_3 < SO_4 \ll PO_4. \tag{5.25}$$

2. The anions are held by ligand adsorption where the oxygen ions on the crystal surface are replaced under suitable conditions by any other ion that can enter into six coordination with the aluminium or ferric ion (Hingston, Atkinson, Posner & Quirk, 1967). It is always accompanied by an increase in the negative charge or a decrease in the positive charge on the surface with the consequence that the pH of the solution rises. It can take place equally well on surfaces carrying a net negative or a net positive charge and is, thus, quite different from the non-specific anion adsorption. It normally takes place with weak acids and is at a maximum when the pH of the solution is equal to or a little lower than the pK of the acid. A number of anions such as phosphate, silicate, etc. which form weak acids have been shown to undergo this type of adsorption. Although the pK values of sulfuric acid do not indicate the possibility of ligand exchange, Gebhardt & Coleman (1974) view the adsorption of sulfate in allophanic tropical soils as ligand exchange phenomenon.

Dissolution and desorption processes (phosphate equilibria in soils)

Determination of total phosphorus in soils, except in the organic rich horizons and organic soils, has little significance from the ecological point of view because a major part of it may be present in ecologically un-important forms such as the occluded form, the amount of which depends upon a number of soil factors (Ulrich & Khanna, 1968). In an ecological system, the various phosphate forms strive towards a steady state equilibrium through the transformation of phosphates among various chemical and biological forms like aluminium phosphates, iron phosphates, calcium phosphates, organic phosphates, and occluded phosphates and through the redistribution of phosphates in different soil depths mainly by biological activities (Walker, 1965; Ulrich & Khanna, 1969). In the context of uptake of phosphorus by plants, the fraction of phosphorus in the liquid phase and its relationship with the solid phase need special emphasis. The possibility of using the concept of nutrient potential as an analogy to pF of soil water has been suggested by Schofield (1955) and has been discussed in detail by Ulrich (1961). For defining the solubility product of various phosphates in soils where the phosphoric acid potential ($pH + pH_2PO_4$) has been related either to lime potential ($pH - 1/2$ pCa) or to aluminium hydroxide potential ($pH - 1/3$ pAl) the following type of relationship can be used:

$$pH + pH_2PO_4 = a(pH - 1/2\ pCa) - b, \tag{5.26}$$

where a and b values for various calcium phosphates are:

Dicalcium phosphate ($CaHPO_4$) $a = 2$, $b = 0.53$; Octocalcium phosphate ($Ca_4H_2(PO_4)_3OH$) $a = 2.67$, $b = 3.26$; Hydroxy-apatite ($Ca_5(PO_4)_3OH$) $a = 3.33$, $b = 4.70$. For Aluminium phosphates the relationship is:

$$pH + pH_2PO_4 = 3(pH - 1/3\ pAl) + b, \tag{5.27}$$

where the values of b are 2.5 and 0.5 for variscite and amorphous aluminium phosphates, respectively.

Ulrich & Khanna (1968) have plotted the lime or aluminium hydroxide potential against the phosphoric acid potential for 230 German forest soils with varying soil characteristics of pH, $CaCO_3$ content, organic matter content, soil depth etc. and have observed that for soils with a pH above 6.8, the $H_2PO_4^-$ concentration in the soil solution is controlled by the solubility of calcium phosphates more soluble than hydroxy-apatite and less soluble than octocalcium phosphate, unless the soil contains more than 2% $CaCO_3$ when it lies on or above the octophosphate line. The fractionation of soil phosphates according to the modified method of Chang & Jackson (1957) showed calcium phosphates forming the main soil fraction in these soils (pH > 6.8). For the pH range 4.5–6.8, where no buffering system exists and under natural conditions the soils pass through rather quickly, only eight samples were observed. In this range the calcium phosphates change predominantly to aluminium phosphates. The absence of any defined phosphate which controls the phosphate concentration in soil solution for this transitional pH range shows the presence of sorbed phosphates on the surface of clay particles (Russell, 1973). In the soil pH – buffer range between 3.5 and 4.4, the $H_2PO_4^-$ concentration is controlled by aluminium phosphates more soluble than variscite, but less soluble than amorphous aluminium phosphate.

Although the information obtained from the solubility product diagrams and the chemical fractionation method helps in postulating the possible form of a compound involved in controlling the concentration of a given ion in the solution, one requires the so-called Quantity/Intensity (Q/I) relationships to be able to follow the changes in the solution as well as in the solid phases. Simple forms of such Q/I relationships are the adsorption as well as desorption isotherms, where the amount of an ion adsorbed or desorbed is plotted against the equilibrium concentration of the same in solution. Such isotherms may follow a definite adsorption equation like the Langmuir equation for a limited phosphorus concentration in the soil solution (Olsen & Watanabe, 1957), or may follow an empirical relationship like the one suggested by Gunary (1970). In addition to the concentration of phosphates in solution, pH of the equilibrium solution plays an important role for undergoing the ligand type of adsorption or specific adsorption sites as shown for iron and aluminium oxides by Hingston *et al.* (1967).

This approach of using adsorption isotherms for describing the equilibria between the solid and liquid phases of phosphates in soils has one disadvantage in that it does not differentiate the forms of phosphorus present in soils. Each of the soil phosphate forms exists in equilibrium with phosphorus concentration in solution and thus are affected by soil chemical processes (like the processes input or output). Ulrich, Mayer, Khanna & Prenzel (1973) presented another approach where the concept of kinetics of dissolution processes is used in the following manner:

$$R_n = R_n^* \cdot a_n, \tag{5.28}$$

where R_n is the constant rate of exchange at equilibrium, R_n^* the exchange constant of the reaction (precipitation or dissolution), and a_n the isotopically exchangeable phosphorus involved in the reaction; n denotes the type of crystallization (originally it meant the position of localization of phosphorus at calcium phosphate crystal surfaces (Benecke, 1959)) or the form of phosphates involved. A constant value of 0.05 min^{-1} was suggested for R_n^*. The values of a_n for different forms were obtained by assuming that 30, 40 and 10% of the total isotopically exchangeable fraction of aluminium phosphates, iron phosphates and occluded phosphates respectively were involved in the precipitation or dissolution processes. The isotopically exchangeable amounts of different phosphate forms were obtained by the method suggested by Khanna & Ulrich (1967). The above given rate constant showed satisfactory results in a P-model (Ulrich *et al.*, 1973).

Redox equilibria

Biotic activities under conditions of restricted oxygen supply due to impeded diffusion of air may cause the development of reducing conditions in soils. The most important of all biotic activities in soils is the decomposition of organic matter. The consequences of reducing or partially reducing conditions on soil processes can be summarized as below (for details refer to Schlichting & Schwertmann, 1973).

1. On acceptance of electrons, certain compounds such as nitrates and sulfates lose their oxygen and are reduced (nitrates to nitrites and finally to nitrogen gas or ammonia, and sulfates to sulfites and finally to sulfides).

311

This causes, on the one hand, a loss of plant-utilizable nitrogen and sulfur from soils. On the other hand, the presence of highly poisonous nitrites and sulfites in root zones causes damage to the roots and, thus, affects the uptake of water and nutrients.

2. Certain high valence compounds will accept electrons and become reduced to a lower valence state. Trivalent iron and tetravalent manganese may be reduced to divalent ferrous and manganous ions. Ferrous in the presence of sulfide precipitates an insoluble sulfide, and the soil acquires a black color. In the absence of sulfides (and phosphates), ferrous ions may be transported to deeper horizons and be oxidized and deposited there. Because of the difference in the redox potential of iron and manganese, separation during the course of their deposition occurs where iron is leached to deeper layers.

3. The hydrogen ions can accept electrons to become hydrogen gas so that the redox potential (E_h) of a system depends on its hydrogen ion concentration (pH). Such $E_h - $ pH stability diagrams for various chemical systems are available in the literature (Garrels & Christ, 1965).

4. The incomplete oxidation of organic substances during biological activities causes the production of substances such as simple fatty acids, alcohols and ketones, various hydrocarbons, carboxylic acids, etc. Some of these compounds can form soluble chelates with various cations in soils, which can then be leached out of the soil. It has been observed, for example, of that the hydrocarbon ethylene and acetic acid, an organic acid may be produced in soils in quantities toxic to the development and functioning of roots of some plants.

Weathering

Weathering of the mineral components of the soil may be regarded for most elements as an element flux from a very large pool with very little mobility into the soil solution. Speaking in terms of ecosystem modelling we deal with extremely small transfer coefficients. The weatherable minerals form one, more or less effective, buffer controlling the element concentration in the soil solution, and, consequently, affecting element uptake and output by leaching. Weathering, as defined here, is controlled by chemical processes such as dissolution, hydrolysis, desorption, ion exchange, oxidation and reduction – some of which have been discussed previously. Chemical weathering of soil minerals is usually preceded and accompanied by physical (mechanical) weathering.

The presence of weatherable minerals in soil may be a decisive factor determining the productivity of forests (cf. Wittich, 1942). Knowledge of the actual amounts of elements released by weathering is however very limited. The chemistry of weathering and breakdown of mineral structures has been studied by geologists, mineralogists and soil scientists (cf. Henin & Pedro, 1965; Loughnan, 1969). Soil genetics and mineralogy have contributed much to the present knowledge of the rate of weathering by balancing the mineral components in soil profiles of known age (Kundler, 1959; Bosse, 1964; Blume & Schlichting, 1965).

While these methods are useful in the estimation of the flux dynamics, they are hardly suitable to furnish the element flux on a ha/yr basis for a defined area. This may be achieved by other methods, the two most important of which are watershed balances (see Chapters 4 and 6) and the lysimeter technique. Both methods are based on the assessment of the element input to an ecosystem, or watershed, and its element output. The element release by weathering is taken from the input/output difference, after accounting for changes in other compartments of the system (e.g. vegetation and soil organic matter). A great number of watershed investigations have been performed in forested areas. Small watersheds are suitable for the study of single forest ecosystems as defined by Bormann & Likens (1967). Current projects meeting these requirements are described by Johnson & Swank (1973), Elwood & Henderson (1975), and Likens & Bormann (1975).

The information gained from watershed investigations with respect to weathering is limited by the fact that the whole geological and soil stratum from the soil surface to the point where the surface and subsurface outflow of the watershed is captured by a weir, is treated as a 'black box'. All data on net gains or losses have to be related to this black box, which may well include subcompartments irrelevant for the forest ecosystem where gains or losses take place. This limitation would apply, for example, on areas with two-layered soils, where the chemistry within the rooting zone is not determined by the bedrock; yet the element load of the ground water may be definitely changed by the bedrock. Watersheds also require areas with a well defined topographic and phreatic divide, without deep seepage. Level terrain with permeable bedrock is not suitable.

These experimental drawbacks are avoided in part by the lysimeter technique, when suction lysimeters of the type described by Cole *et al.* (1961) are used. These lysimeters are small enough to leave the soil profile fairly undisturbed so that even small soil horizons may be investigated. Therefore, in a detailed study of soil processes within a forest ecosystem that is, at present, the preferred technique provided that a suitable lysimeter material is used which does not interfere with the element concentration of the percolation water.

The lysimeter technique, too, has limitations which do not allow its use in some cases. Installation of lysimeter plates is difficult in rocky or shallow soils. Lysimeters may furnish inaccurate results when used on slopes. The most severe restriction is that they cannot measure the water flux accurately

313

because of technical difficulties in the adjustment of suction and the compensation of the water entry resistance of the lysimeter plate (Koenigs, 1973; van der Ploeg & Beese, 1976). Therefore, the water flux has to be determined by an independent method (*see* Chapter 4).

From investigations as cited above some general conclusions on the dynamics of element release by weathering may be drawn. The rate of release depends on the physical and chemical state of the minerals present in soil, the temperature and humidity conditions, and the chemical quality of the percolation water. In this respect, the acidity of the water plays an important role. In many forest soils a very low pH is found in the top soil layer where weathering consequently is most intensive. From the knowledge of dissolution equilibria effective in the soil system, Ulrich (1975) was able to calculate the amount of silicate weathering within the upper 50 cm of a loess soil under a beech forest.

All quantitative data on weathering based on input-output relationships can be regarded as characteristic for a single ecosystem only when averaged over a large number of years, for annual variations may be considerable owing mainly to changing meteorological (input) conditions.

Transport within soil

Elements may be transported within soil by several transport mechanisms: convective transport (mass flow) in the liquid phase (the element may be present as a solid compound or dissolved) or in the gaseous phase; diffusive transport (diffusion under a thermal and/or a concentration gradient); transport by biological activity (mainly by activity of the soil fauna). Under most conditions, the latter transport mechanism plays a role only in the upper soil. Quantitative data have not been reported.

Transport in the gaseous phase (convective or diffusive) is only important for the elements prevailing in the soil atmosphere (carbon, nitrogen, oxygen). Diffusive transport in the liquid solute phase and in the solid phase is significant mainly in the short-distance transport through non-moving liquid films at the surface of mineral particles and roots, and within mineral structures. Convective transport of solids suspended in the liquid phase may be responsible for the downward movement of clay minerals in some soils (clay migration). The rate of transport is slow. All these element fluxes are either treated in a different chapter (Chapter 3) or are insignificant from in an ecological context, i.e. the amounts transported on a large scale within a short time are negligible.

The remaining transport mechanism, convective transport of elements dissolved in the soil water, is of primary importance for all large-scale element transfers within soils. Investigation of element transport within the soil may be reduced to the task of measuring the soil water flux and analyzing the element concentration in the soil water within certain time intervals (or continuously if highly accurate data are desired). There are considerable difficulties in measuring the 'true' element concentration of that fraction of the soil water which actually is moving in the soil profile. Each isolation of soil solution from the soil matrix – necessary for the chemical analysis – may lead to deviations from the 'true' element concentration (Mayer, 1974).

Under these conditions the suction lysimeter technique, as first used by Cole *et al.* (1961) proves to be most useful. Ceramic plates are installed in a certain depth within the soil without disturbing the overlying profile. By application of suction at the plates adjusted to the soil water suction in the ambient soil (a requirement which is not always met in lysimeter studies), the solution is collected with the aid of a tubing system connected with the plates. The solution collected corresponds in quantity and in chemical quality to the soil solution passing the same level during that period of time. By this technique it has been possible to measure the element transport within the soil as well as the element output with the percolation water in forest ecosystems (Cole & Gessel, 1965; Mayer, 1971; Mayer, 1974). Generalizations about the dynamics of element transport in forest soils can be drawn from these investigations. For a given element concentration, the element flux is dependent on the soil water flow and is consequently governed by precipitation, evapotranspiration, and the soil water flux controlled by hydraulic parameters (permeability, water content, suction gradient) of the soil.

The element concentration itself is dependent on the inputs into, and the outputs from, the solution compartments, e.g. precipitation, release from organic matter, exchange, dissolution, weathering, and root uptake. The information gathered in the single project are, so far, insufficient to establish a complete model of the element transport within soil strictly based on physical and chemical causalities. There are many processes involved which are still not well enough understood, or which cannot be measured directly.

Yet these investigations serve as a basis for the establishment of simulation models which are useful in the interpretation of experimental data and allow reasonable prediction of the behavior of element stores and fluxes (see, e.g. de Wit & van Keulen, 1972; Ulrich *et al.*, 1973; Frissel & Reiniger, 1974). McColl (1973) has expressed the element concentration in the soil solution under a Douglas fir forest on the basis of statistical relationships. He found that the total volume of flow through a specific horizon as well as the duration and temperature of the dry period preceding the flow accounted for most of the variations in soil solution concentration.

Ways of output

Root uptake

A large number of investigations has been undertaken to study uptake of mineral elements by plant roots. Discussing the results of these investigations on the level of recent research progress in biochemistry, membrane physics and cell physiology is beyond the scope of this chapter. The contribution by individual cells, plant tissues or isolated roots is fairly well understood, and hypothetical models explaining uptake have been found to agree with experimental results (Luttge & Pitman, 1976). Yet these models are far from being applicable to trees, or to forest vegetation as a whole. The estimation of root uptake is only possible so far from calculations of the total element budget. Therefore, root uptake in forest ecosystems has only been quantified in a few woodland ecosystem studies which included a fairly complete assessment of all the remaining fluxes (see Chapter 6).

Methods of study

Net root uptake by forest vegetation may be calculated as the sum of the element increment in the vegetation cover ('retention') plus litterfall and leaching ('restitution'). The latter two processes have been discussed in two foregoing sections. The element increment is usually assessed in connection with productivity measurements by multiplying dry matter increments in the different vegetation compartments (bole, branches, roots, bark, etc.) with their respective element contents. Two severe limitations are involved with the use of this calculation procedure.

1. The amounts of elements leached from the canopy by precipitation cannot be measured directly. Therefore, root uptake is either overestimated when the total element increase in the precipitation flux during canopy passage is attached to leaching, or underestimated when the increase is attached to filtering of atmospheric substances (impaction). The error is probably smaller, for most elements in the latter case, since leaching seems to be rather insignificant with the exception of highly mobile elements like potassium and possibly manganese. Leaching is dependent on forest type, soil chemical and meteorological conditions. Therefore, an accurate estimate of root uptake requires definitely the assessment of leaching – otherwise the presentation of element cycling remains incomplete. Methods for the assessment of leaching have been discussed.

2. The increment in dry organic matter of a forest stand is usually measured on an annual basis. The techniques available do not allow the assessment of increments within shorter intervals, or they are too time and money consuming to be feasible. Furthermore, in many investigations the seasonal mass

development of the leaves/needles and, moreover, of the roots are unknown. Finally, there is evidence that root excretion may occur in trees (cf. Thomas, 1969; Witherspoon, 1964). This element transfer has not, until now, been measured quantitatively in a forest ecosystem. As a consequence of these drawbacks in the experimental approach, root uptake may be calculated only on an annual basis. Conclusions on the seasonal characteristics of root uptake cannot be drawn from these data.

Dynamics

Many facts furnished by research are of little value to the interpretation of data on root uptake as long as only annual uptake rates are available. Yet some of these general facts should be mentioned, because they may be helpful in the development of root uptake models aimed at a better understanding of uptake in spite of an inadequate data basis. There is an interrelationship between soil chemical conditions and the plant root. The plant seems to be able to influence the chemical environment of the roots (rhizosphere) and, thus, to control uptake. The same is achieved by root growth, which may open remote pools of nutrient elements. Processes of this kind will be effective in accordance with the physiological activity of the plants, and we would expect annual, seasonal, and daily changes in uptake.

Root uptake is also controlled by environmental (soil) factors in the root environment: chemical and water conditions. The different parameters characterizing these conditions (such as element concentrations in soil solution, pH, soil water content, composition of soil atmosphere) may change quickly, especially within the rhizosphere, so that a complete experimental assessment of these parameters may become impossible. Prediction of uptake from soil parameters may be limited by this fact. There are, however, controlling steps in the root uptake of essential nutrient elements which are dependent on soil parameters. Such steps may be diffusion and mass flow to the roots. These transport processes are fairly well understood (*see* Barber, 1966; Baldwin & Nye, 1974).

Gaseous losses

It was previously stated that only the gas exchange of nitrogen is important in forest ecosystems, if carbon and oxygen are not considered. Losses of nitrogen from soils may occur by denitrification, i.e. the chemical or biologically catalyzed reduction of nitrates to nitrous oxide (N_2O) and free nitrogen gas (N_2). Denitrification is achieved by some microbes (e.g. *Pseudomonas*, *Achromobacter*) under anaerobic conditions in soils especially above pH 5.0. Therefore, denitrification may be important during wet periods in neutral or slightly acid, poorly drained, soils with an adequate supply of decomposable organic matter, when temperatures are not too low. Losses of gaseous ammonia, NH_3, occur only in soils of $pH > 7$, especially at periods of rapid nitrogen mineralization. As in the case of gaseous adsorption by soils, the gaseous losses are very difficult to measure. From the conditions described under which losses may occur it becomes obvious that they are probably negligible for a great number of forest soils in steady state.

Leaching from root zone

The flux of elements in percolation water at a soil depth below the rooting zone from which no upward water fluxes take place is considered as an output from the forest ecosystem. Considerable changes in the element load of the drainage water may occur before it appears as a surface water body. Therefore, the output data from a watershed must not be identical with leaching from the root zone.

Other pathways of output

Erosion output of elements by subsurface and ground water flow, as opposed to sedimentation, is governed by the same principles which applied in the previous discussion on other ways of input. Export is of main importance in managed forests. Since in most cases the amount of the product taken out of a forest are exactly determined, the assessment of the element output connected with the export should pose no difficulties.

Root dynamics

Knowledge of the dynamics of forest ecosystems is extensive. Our knowledge base has progressed rapidly, especially during the period 1968–78, to the point where we can begin to describe the forest ecosystem with some success as a system of mathematical equations representing underlying physiological processes, structural relationships and their interactions with external driving variables of climate. The one exception to this understanding is the below-ground ecosystem, especially roots. Roots comprise the primary interface between the plant and the soil for uptake of water and nutrients. A great deal is known about the biochemistry, cell physiology and membrane physics associated with these important processes (see, for example, Devlin (1966), Larcher (1975) and Pitman (1976)), but it is neither possible nor within the scope of this work to review this important area of physiology.

The question addressed here is the role of the below-ground ecosystem, especially the autotrophic root component, in the structure and function of

318

the forest ecosystem. Beyond the role of anchoring the terrestrial plants and uptake of water and nutrients, this component of the forest ecosystem has been largely neglected.

Forest biomass below-ground is large. Bazilevich & Rodin (1968), in summarizing the reserve of below-ground organic matter in terrestrial ecosystems, indicated that the broad-leaved and subtropical forest types are characterized by a maximum of from 70 000 to 100 000 kg/ha dry weight of root organic matter (here defined as the total of living and recently dead root structure) which constitutes 15% to 33% of the total accumulated biomass. This below-ground organic matter is surpassed only by tropical forests; however, the proportion of the total biomass represented by roots is lowest for forests and ranges up to 90% for tundra and certain steppe vegetation types. Bazilevich & Rodin (1968) conclude from their review that, in general, the root biomass is proportional to the *total* accumulated biomass, and as 'site quality' decreases the proportion of total biomass as roots increases. Estimates of below-ground biomass have been obtained by soil monolith analysis (Karizumi, 1968), allometric analysis (Kira & Ogawa, 1968), and only rarely by whole tree excavation. Besides the extensive reviews by Rodin & Bazilevich (1967) and Bazilevich & Rodin (1968), extant data on root biomass have been summarized by Ovington (1967), Whittaker (1962), Bray (1963) and Santantonio, Hermann & Overton (1977).

Such a large amount of organic matter, accumulated at considerable expense of carbohydrates, clearly could serve several purposes, such as storage of plant sugar and essential nutrients. The following discussion reviews briefly some of the recent findings on the behavior of the below-ground ecosystem and suggests some additional questions yet to be re-solved. In particular, the seasonal production and turnover of root biomass, the role of root processes in nutrient turnover and the significance of below-ground dynamics to the energy balance of the forest ecosystem are considered.

Before proceeding further, some discussion of why understanding the below-ground ecosystem is in order. Our concept of forest root dynamics incorporates several reasonable but generally untested assumptions. These assumptions have arisen primarily because of the extreme methodological problems and labor intensive requirements for research on forest tree roots (Newbould, 1967; Lieth, 1968). Recent progress in this area of study is related to two factors. First, much of the work emanates from the large integrated studies of forest ecosystems such as those initiated as part of the IBP. These studies supported the pool of skilled and dedicated technicians necessary to obtain the requisite data. Second, as more became known about the metabolism of forest ecosystems, the potential role of roots in ecosystem function and their energy demands associated with the accumu-lation and turnover of carbon and other essential elements, surfaced as a

319

central link coupling physiological processes and their environmental constraints with ecosystem behavior. As a result some interesting findings are now available. By no means do we know enough to generalize, but these initial findings are exciting and counter to what had been the common wisdom. The next several years should see considerable addition to the few examples which follow.

Seasonal accumulation and turnover of root organic matter

The root biomass of forests is not static. It changes annually and represents a varying proportion of the total biomass during stand development. By far the most dynamic component of root biomass is the fraction defined as 'fine roots'. No standard definition exists for fine roots; the distinction between fine and large roots is usually based on an arbitrarily chosen diameter limit which has ranged mainly from 0.2 to 1.0 cm. It has long been known that a seasonal periodicity of root growth is common in woody plants (see reviews by Lyr & Hoffmann, (1967) and Kozlowski (1971)). For example, radial growth of woody roots (with secondary xylem thickening) can closely follow the pattern of radial increment growth above-ground (Fayle, 1968). Studies concerned with seasonal periodicity of root elongation, initiation of laterals and subsequent elongation have not clarified whether periods of inactivity reflect physiological or environmentally-mediated dormancy (Santantonio, unpublished observations). Sutton (1969) concluded that evidence for autonomous control is not convincing and that primary growth of roots is probably dominated by environmental conditions. In fact, the real basis for control probably lies with the interaction of endogenous and environmental mechanisms, but this remains to be satisfactorily demonstrated.

While there is considerable information on the phenology of root growth, there is an insufficient basis for making estimates of root production and turnover. The earliest studies of root production are probably those of Heikurainen (1957) and Kalela (1957). Both studies involved Scots pine (*Pinus sylvestris* L.) and both reported a modal pattern of rapid growth to peak root biomass in the spring and a gradual decline during the summer to a low of about 50% of the peak level. A bimodal peak in root biomass has been reported for an oak woodland in central Minnesota, USA (Ovington, Heitkamp & Lawrence, 1963) and a stand of European beech (*Fagus sylvatica* L.) in the Solling area of West Germany (Göttsche, 1972). In both instances, biomass peaked in spring with a second but lower peak in the fall. Investigations by Harris, Kinerson & Edwards (1978) in a yellow poplar (*Liriodendron tulipifera* L.) stand in east Tennessee also revealed a spring–fall bimodal peak, while for loblolly pine (*Pinus taeda* L.) in North Carolina the modality was less clear with peaks observed in late fall, late winter and

320

possibly late spring. As with root elongation, the correlations of periods of peak biomass with environmental patterns are inconclusive. The year-to-year consistency observed by Harris *et al.* (1978) for yellow poplar suggest a strong measure of endogenous control.

The surprising result of recent studies on the seasonal dynamics of fine roots is the large flux of organic matter which is involved – large in an absolute sense as well as relative to the other organic matter fluxes of the forest ecosystem (Harris *et al.*, 1978). Using a coring device consisting of a masonry hole saw blade attached to a steel cylinder with the shaft adapted to a gasoline powered auger assembly, Harris *et al.* (1978) sampled a yellow poplar forest stand intensively over a two-year period. The lateral root biomass of yellow poplar showed considerable variation in the smaller root size classes (Fig. 5.19). Small roots within this forest were characterized by a peak in late winter (1 March), a minimum in mid-May, a second peak in

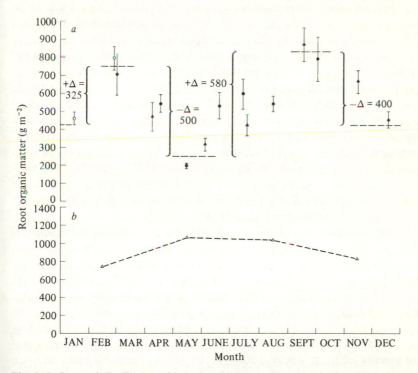

Fig. 5.19. Seasonal distribution of lateral root biomass in a *Liriodendron* forest for (*a*) roots less than 0.5 cm diameter and (*b*) roots of 0.5 cm diameter or greater ($\bar{X} \pm 1$ SE). Net biomass production and turnover were calculated from differences in pool size through the year. Based on monthly summary of core data, no consistent pattern of biomass dynamics could be detected for roots greater than 0.5 cm in diameter. ▲, 1971; ●, 1972; ○, 1973; △, pooled data.

321

mid-September, and a minimum in early winter (December to January). This pattern appears to be consistent among successive years. Based on summation of positive seasonal differences between minimum and subsequent peak biomass estimates, net root biomass production was 9000 kg/ha, with a net annual turnover (translocation and sloughing) of equal magnitude. This value of net annual small root production is 2.8 times larger than mean annual above-ground wood production determined for the study area from allometric equations and periodic (1965 to 1970) dbh inventory (Sollins *et al.* 1973).

Other experimental data on ecosystem carbon metabolism for the *Liriodendron* forest study area corroborate the existence of a large, annual below-ground allocation of carbon. Estimated net photosynthetic influx and soil-litter carbon efflux yield an amount of unaccounted carbon input to soil equivalent to 7500 kg organic matter per ha (Harris *et al.*, 1975; Edwards & Harris, 1977). For the dominant species in this study site, *Liriodendron tulipifera*, ^{14}C-sucrose tracer field studies indicate a vernal allocation of root-associated labile carbon above-ground of approximately 1500 kg/ha organic matter (H. H. Shugart & W. F. Harris, unpublished data). For temperate deciduous forests, the assumption that below-ground primary production is a fraction of above-ground primary production proportional to biomass pool size as is commonly assumed would lead to an underestimate of total annual root production. The results from yellow poplar are not an extreme example. While the number of studies is limited, a sufficient range of forest ecosystem types is represented to suggest that the large flux of organic matter below-ground is a general property of forest ecosystems. For example, McGinty (1976) found primary production of fine roots in an oak–hickory forest to be 6000 kg/ha. Persson (1978) reports root production of 3500 kg/ha in a young Scots pine stand in central Sweden, while Santantonio (unpublished data) has found root production in Douglas fir (*Pseudotsuga menziesii* Franco) of 6000–9000 kg/ha.

The large accumulation of root organic matter is a seasonal phenomenon. The net annual accumulation of root organic matter is much smaller and can best be described as a ratio of total above- to below-ground biomass multiplied by the net annual aboveground production. What then is the fate of the seasonal fluxes of organic matter below-ground? Most of this material is promptly metabolized by the soil heterotrophs (Edwards & Harris, 1977). Analysis of the total and proportional efflux of carbon dioxide from the soil surface (Edwards & Harris, 1977) and experimental measurements of root decay rate (W. F. Harris, unpublished data) all corroborate the prompt metabolism of root detritus. Again, similar findings by McGinty (1976), Persson (1978) and Santantonio (unpublished data) all point to the prompt disappearance of fine root organic matter. We are thus left with the following generalization: the soil of temperate coniferous and deciduous forest ecosystems annually receives an input of root organic matter equal to or much greater than from any other source; this organic matter is promptly metabolized by heterotrophs.

Before leaving the subject of root organic matter dynamics, some mention should be made about the dynamics of larger structural roots ($\geqq 0.5$ cm diameter). Generally, this organic matter component is much more stable, certainly not exhibiting significant seasonal or even annual dynamics. However, it would be a mistaken impression to consider this a static pool. Again, the evidence is sparse, but Kolesnikov (1968) has reported the interesting observation that during the development of a forest stand there is a cyclic renewal of large structural roots. One cannot escape making the analogy of such renewal to the individual forest tree adjusting this main structural component to exploit with its fine root structure 'new' areas of its soil habitat. Whether this generalization will withstand closer observation, and what the controlling factors and mechanisms might be, remain to be answered.

Root organic accumulation during stand development

During forest stand development, the amount of root organic matter increases on an absolute basis, but the proportion of the total biomass as root organic matter decreases. As reviewed by Rodin & Bazilevich (1967), the patterns of root/shoot ratio between deciduous and coniferous forests vary. Generally, coniferous forests reach an equilibrium root/shoot ratio earlier in stand development (i.e., at a lower total biomass) than is the case for deciduous forests. The trend for root organic accumulation in a mixed deciduous forest in east Tennessee (Walker Branch Watershed) is shown in Fig. 5.20.

Total below-ground biomass has ranged from 32 000 to 47 000 kg/ha for the deciduous and mixed coniferous–deciduous forest stands examined which has total above-ground biomass that ranged from 110 000 to 185 000 kg/ha (Fig. 5.20). Above-ground biomass have been estimated allometrically from dbh data and allometric equations reported elsewhere (Harris, Goldstein & Henderson, 1973). The more mesic *Liriodendron* forest followed a pattern of biomass distribution similar to upland hardwood stands with a total above-ground biomass of 130 000 kg/ha and a below-ground biomass of 36 000 kg/ha. A loblolly pine forest in North Carolina deviated considerably from the pattern observed in deciduous forests. The total above-ground estimate was 90 000 kg/ha while below-ground biomass amounted to 21 500 kg/ha (C. W. Ralston, Duke University, unpublished data).

For both plantation and deciduous forest ecosystems, the ratio of below- to above-ground biomass (Fig. 5.19) followed trends similar to those

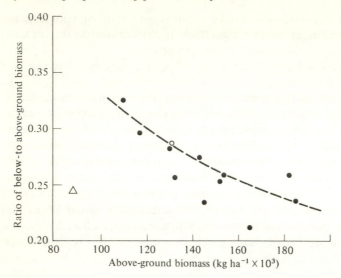

Fig. 5.20. Distribution of below- to above-ground biomass ratios over a range of above-ground biomass pool sizes on Walker Branch Watershed (●). The dashed line is a hand-fitted approximation of the trend noted by Rodin & Bazilevich (1967) for broad-leaved deciduous forests (○, *Liriodendron*). The displacement of the loblolly pine ratio (△, Saxapahaw area) appears to follow the trend for coniferous forests where ratios typically are lower.

reported by Rodin & Bazilevich (1967). They report that the percentage contribution of roots to total forest biomass of coniferous forest ecosystems approaches a constant value of 22.5% as total biomass exceeds 75 000 kg/ha. Roots comprised 19% of the total loblolly pine biomass. The proportion of biomass contributed by roots in deciduous forests approaches a constant value more slowly (Rodin & Bazilevich, 1967). The proportion tends to stabilize at about 20% as total biomass approaches 300 000 kg/ha. The root:shoot ratios of hardwood forests on Walker Branch Watershed follow the trend (dashed line, Fig. 5.19) suggested by the geographically more extensive data summary of Rodin & Bazilevich (1967). The pattern of below- to above-ground biomass distribution, which appears to be consistent over a wide range of forest types, describe (1) the size of the below-ground biomass pool and (2) the net annual biomass production over some interval when estimates of above-ground biomass pool and accumulation are known. While useful as a means of indirectly estimating below-ground biomass pool size and net annual production, this approach cannot be used to determine total annual root production.

On an annual basis, root core analysis and other harvest techniques lack the precision to detect readily net annual below-ground accumulation of biomass. However, applying the rather consistent values of below- to above-ground biomass ratios to stand data on biomass of tops would

324

suggest that total below-ground biomass pool should increase about 3% to 7% each year over the range of stand development examined.

The significance of root dynamics to element inputs to soil and element cycling

If our knowledge of root organic matter dynamics is limited, our knowledge of the role of root production/turnover (sloughing) to element cycling is vanishingly small. Of course, some idea or assumptions about root element content can be used to estimate the flux of elements to the soil. McGinty's work (1976) begins to place the contribution of roots in perspective. In the oak–hickory and eastern white pine forests of Coweeta, North Carolina, roots comprised 28% of the forest biomass but contain 40% of the plant nutrients in hardwoods and 65% of the plant nutrients in pines. Thus, this dynamic root component is a nutrient-rich substrate.

Roots can return nutrients to the soil in several ways, by death and decay, by exudation and leaching and indirectly after consumption. There are no known studies which consider total herbivory on roots. Certainly root feeding nematodes and various larval stages (e.g., cicada) might be principal consumers as described by Ausmus *et al.* (1978).

Studies of leaching and exudation from roots are likewise limited. In a northern hardwood forest ecosystem, Smith (1970) has found root exudation (growing season) to account for 4 kg/ha of carbon, 8 kg/ha of potassium and 34.2 kg/ha of sodium for three principal tree species (*Betula alleghaniensis*, *Fagus grandifolia* and *Acer saccharum*). Although the techniques employed (modified axenic culture) risk introducing artifacts, there is nonetheless a considerable potential for contribution of elements to the soil via exudation.

Radiotracer studies with ^{134}Cs have shown that over 50% of the ^{134}Cs in roots of tagged yellow poplar seedlings was transferred to culture solutions in less than seven days (Cox, 1972). Sandberg, Olson & Clebsch (1969) estimated that during one growing season 75% of ^{137}Cs transfers by yellow poplar seedlings grown in sand was due to exudation-leaching. In their analysis of a cesium-tagged yellow poplar forest, Waller & Olson (1967) considered root exudation leaching processes as important pathways of cesium transfer to the soil based on concentration of ^{137}Cs in soil at various depths. If *in situ* processes of cesium are comparable to those of the chemical analog, potassium, large quantities of potassium could be transferred to the soil annually by leaching exudation processes.

There are very few data which compare the return of elements to the soil via above-ground and below-ground processes. Table 5.7 summarizes a comparison for an extensively studied yellow poplar forest at Oak Ridge, Tennessee (Cox *et al.*, 1978). In this analysis, consumption was assumed to be 10% of the root detritus. Root detritus turnover was estimated by the

Table 5.7. *Annual above- and below-ground organic matter and element returns (kg/ha) to soil in a yellow poplar (Liriodendron tulipifera L.) stand. From Cox et al. 1978*

	Biomass	Nitrogen	Potassium
Above-ground[a]			
Dryfall/wetfall	—	7.2	3.6
Canopy leaching	—	2.3	29.4
Litterfall	3310	42.2	10.0
Total Above-ground return	3310	51.7	43.0
Below-ground[b]			
Root transfer processes			
Death and decay	6750	76	128
Consumption	750	9	14
Exudation-leaching[e]	—	—	128
Total Below-ground return	7500	85.0	270
Total return to soil	10810	136.7	313

[a] Above-ground data from Edwards & Shanks (unpublished data); chemical determinations were made on downed (fallen) material.

[b] Root biomass estimates from a *Liriodendron* stand (>80% *Liriodendron*).

[c] Exudation–leaching data are extrapolated from seedling studies of Cox (1975) assuming equivalent behavior of cesium and potassium. Element losses via root sloughing are based on (amount of sloughed biomass) times (mean root nitrogen content); element flux via consumption assumed to 10% of total below-ground return.

large residual carbon dioxide efflux from the soil unaccounted for after litter decomposition and autotrophic root respiration were subtracted from the total soil carbon dioxide efflux.

In the example illustrated in Table 5.7 annual return of elements to the soil by root processes was three times the combined above-ground inputs (including atmospheric) for potassium and at least 1.5 times above-ground inputs for nitrogen. Of the total (above- and below-ground) return to the soil, root processes accounted for turnover of 70% of the organic matter, 62% of the nitrogen and 86% of potassium. Some additional fraction (here assumed about 10% of the total estimated detrital flux) was transferred to soil consumer pools.

Only with caution can these results be generalized to other forests. However, given the large production/turnover of root material in diverse forests (as discussed above), there is some indirect basis for viewing these results more generally. As this volume is written, there are new studies underway (e.g., by D. Santantonio, Oregon, and H. Persson, Sweden) which should add to these results. However, despite the limited comparative data, the notion that element return to the forest soil occurs principally above-ground is short-lived.

The significance of root sloughing to the forest energy balance

Odum (1969) suggested a strategy of increasing conservation of nutrients in the element cycle during forest ecosystem development. In temperate forest systems, the mechanism leading to a closed cycle for nitrogen could be root sloughing and subsequent microbial mineralization. Biomass and nitrogen accumulate in roots during periods favorable for growth (summer) or just preceding growth (late winter); the winter growth occurs largely at the expense of stored photosynthate. During periods unfavorable for growth (fall–winter) or when above-ground energy demands are high (spring–early summer canopy development), root biomass is sloughed, thus reducing the total energy demand on the temperate forest system at a time when reserves are seasonally depleted and/or plant requirements for carbon are high elsewhere (e.g., canopy development). Nitrogen contained in the sloughed organic matter is conserved as part of the soil detritus by those microbial processes which immobilize it.

The cyclic pattern of photosynthate accumulation in a deciduous forest and the sustained productivity, which is in part dependent on available nitrogen, are closely coupled through the activities of soil microbes on a large systematically replenished substrate rich in nitrogen. In this respect, continuous maintenance of living roots in temperate forests would impose energy limitations on detrivores by reducing the periodic influx of nitrogen-rich root organic matter, because this flux represents 70% of the total detrital input. While data from other deciduous systems are scarce, we hypothesize that evolution of temperate forest species has favored mechanisms which contribute to the systematic return of elements and organic matter through below-ground sloughing. Sloughing, therefore, stabilizes biogeochemical cycles of potentially limiting elements in an environment typified by seasonally limited photosynthate availability.

The energy expenditure to forest ecosystems represented by root sloughing is high. In the yellow poplar forest at Oak Ridge (Harris *et al.*, 1975; Edwards & Harris, 1977), Edwards (in Auerbach, Nelson & Struxness, 1974) estimated that lateral root growth, sloughing and maintenance respiration accounts for 44.8% of the total energy fixed annually in photosynthesis (1.88×10^4 kcal/m^2). As discussed above, root sloughing with subsequent microbial immobilization offers a particularly attractive mechanism to explain retention of essential elements in the uptake zone. In another context, however, it can be argued that maintenance and development of a fertile soil require significant input of energy stored in soil organic matter. For the yellow poplar forest, the energy requirements for maintenance of soil are on the order of 66% of the energy fixed annually by photosynthesis (roots $+21\%$ allocation to leaves). It is thus easily seen that many of man's activities which reduce both the energy fixed photosynthetically and/or the energy input to the soil are most severely manifested by

Table 5.8. *Comparison of turnover times for carbon, nitrogen and calcium in temperate deciduous forests (Tennessee). From O'Neill et al. 1975*

Component	Turnover time (yr)		
	Carbon[a]	Nitrogen[b]	Calcium[c]
Soil	107	109	32[d]
Forest Biomass[e]	155	88	8
Litter (01+02)	1.12	<5	<5
Total[f]	54	1815	445
Decomposers	0.01	0.02	?

[a] Data based on carbon metabolism of yellow poplar forest (Reichle *et al*, 1973a; Harris *et al.*, 1975).

[b] Data based on nitrogen budget for mixed deciduous forest (Henderson & Harris, 1975).

[c] Data based on calcium budget from a *Liriodendron tulipifera* forest (Shugart *et al.*, in press).

[d] Turnover time based on available calcium and assumes all losses of calcium from soil are from the pool of available calcium.

[e] Considers above-ground biomass pool. Cyclic renewal of structural roots (Kolesnikov, 1968) would lower turnover time. Tree mortality estimated from permanent plot resurvey (three-year interval) and probably underestimates the mortality rate over the duration of a forest generation.

[f] Total calculated as *sum* of elements in living and dead components of the ecosystem; element loss based on *sum* of all losses from ecosystem.

degradation of the soil with respect to humified soil organic matter and fertility. The role of energy inputs to the soil (dominated in temperate forests by root sloughing) is to maintain the fertility of the soil component. As illustrated in Table 5.8, carbon (energy) is rapidly metabolized within the ecosystem and lost (as carbon dioxide). Essential elements, on the other hand, because of the interactions of autotrophs and decomposers are retained effectively.

References

Allison, F. E. (1965). Evaluation of incoming and outgoing processes that affect soil nitrogen. In *Soil nitrogen*, ed. W. V. Bartholomew & F. E. Clark. Agronomy No. 10, Madison.

Ando, M. (1970). Litter fall and decomposition in some evergreen coniferous forests. *Japan. J. Ecol.* **20**, 171–81.

Auerbach, S. I., Nelson, D. J. & Struxness, E. G. (1974). Environmental Sciences Division Annual Progress Report. Period ending 30 September 1973. Oak Ridge National Laboratory Report ORNL–4935. Oak Ridge, Tennessee.

Ausmus, B. S., Ferris, J. M., Reichle, D. E. & Williams, E. C. (1978). The role of belowground herbivores in mesic forest root dynamics. *Pedobiologia*, **18**, 289–95.

Baldwin, J. P. & Nye, P. H. (1974). A model to calculate the uptake by a developing root system or root hair system of solutes with concentration variable diffusion coefficients. *Plant and Soil*, **40**, 703–6.

Barber, S. A. (1966). The role of root interception, mass flow and diffusion in regulating the uptake of ions by plants from soil, pp. 45–56. In *Limiting steps in ion uptake by plants from soil*. Technical Report Series No. 65, FAO/IAEA, Vienna.

Baum, U. (1975). Stickstoff-Minerlisation und Stickstoff-Fraktionen von Humusformen unterschiedlicher Wald-Ökosysteme. *Göttinger Bodenkundliche Berichte*, **38**, 1–96.

Bazilevich, N. I. & Rodin, L. E. (1968). Reserves of organic matter in the underground sphere of terrestrial phytocoenoses, pp. 4–8. In *International Symposium, USSR, Methods of Productivity Studies in Root Systems and Rhizosphere Organisms. Reprinted for the International Biological Programme by Biddles, Ltd., Guildford, U.K.*

Beckett, P. (1964). Studies on soil potassium. I. Confirmation of the ratio law: Measurement of potassium potential. *J. soil Sci.* **15**, 1–8.

Beese, F. & van der Ploeg, R. R. (1976). Influence of hysteresis on moisture flow in an undisturbed soil monolith. *Proc. Soil. Sci. Soc. America*, **40**, 480–4.

Benecke, P. (1959). Kinetik des Ionaustausches bei Phosphaten. *Dissertation der Landwirtschaftlichen Fakultat, Universitat Gottingen.*

Benecke, P. (1976). Soil water relations and water exchange of forest ecosystems. In *Ecological studies*, **19**, ed. O. C. Lange, L. Kappen & E. D. Schulze, Springer-Verlag, Berlin-Heidelberg-New York.

Benecke, P. & Mayer, R. (1971). Aspects of soil water behaviour as related to beech and spruce stands – some results of the water balance investigations, pp. 153–63. In *Integrated Experimental Ecology, Ecological Studies*, **2**, ed. H. Ellenberg. Springer-Verlag, Berlin-Heidelberg-New York.

Benecke, P. & van der Ploeg, R. R. (1976). Quantifizierung des zeitlichen Verhaltens der Wasserhaushaltskomponenten eines Buchen-und eines Fichtenaltholzbestandes im Solling mit bodenhydrologischen Methoden. *Verhandlungen der Gesellschaft fur Ökologie*, Göttingen 1976. Dr W. Junk B. V. – The Hague (in Vorbereitung).

Bernhard-Reversat, F. (1975). Nutrients in throughfall and their quantitative importance in rain forest mineral cycles, pp. 153–9. In *Tropical Ecological Systems*, Ecological Studies, **11**, ed. F. B. Golley & E. Medina. Springer-Verlag, Berlin-Heidelberg-New York.

Blume, H. P., Munnich, K. -O. & Zimmermann, U. (1966). Das Verhalten des Wassers in einer Loess-Parabraunerde unter Laubwald. *Zeitschrift fuer Pflanzenernahrung, Dungung, Bodenkunde*, **112**, 156–68.

Blume, H. P. & Schlichting, E. (1965). The relationships between historical and experimental pedology, pp. 340–53. In *Experimental Pedology*, ed. E. G. Hallsworth & D. V. Crawford. Butterworths, London.

Bolin, B. & Persson, Ch. (1975). Regional dispersion of atmospheric pollutants with particular application to sulfur pollution over Western Europe. *Tellus* **27**, 281–310.

Bolt, G. H. (1967). Cation-exchange equations used in soil science – a review. *Netherlands J. agric. Sci.* **15**, 81–103.

Bormann, F. H. & Likens, G. E. (1967). Nutrient cycling. *Science*, Washington, **155**, 424–9.

Bosse, I. (1964). Verwitterungsbilanzen von charakteristischen Bodentypen aus Flugsanden der nordwestdeutschen Geest (Mittelweser-Gebiet). *Dissertation Universität Göttingen.*

Bray, J. R. (1963). Root production and the estimation of net productivity. *Can. J. Bot.* **41**, 65–72.

Brechtel, H. M. (1970). Schneeansammlung und Schneeschmelze im Wald und ihre wasserwirtschaftliche Bedeutung. *'gwf'-wasser und abwasser*, **111**, Jg., H. 7, S. 377–9.

Brechtel, H. M. (1971), Einfluss des Waldes auf Hochwasserabflusse bei Schneeschmelzen. *Wasser und Boden*, 23. Jg., H. 3, S. 60–3.

Brechtel, H. M. & Zahorka, H. (1971). Beeinträchtigt die Umwandlung von Buchen- und Fichtenbestanden die wasserwirtschaftliche Funktion des Waldes. *Allgemeine Forstzeitschrift*, **8**.

Campbell, N. E. R. & Lees, H. (1967). The nitrogen cycle. In *Soil biochemistry*, ed. A. D. McLaren & G. H. Paterson, New York.

Carlisle, A., Brown, A. H. F. & White, E. J. (1966). The organic matter and nutrient elements in the precipitation beneath a sessile oak (*Quercus petrea*) canopy. *J. Ecol.* **54**, 87–98.

Chamberlain, A. C. (1960). Aspects of the deposition of radioactive an other gases and particles. *Int. J. Air Pollution* **3**, 63–88.

Chang, S. C. & Jackson, M. L. (1957). Fractionation of soil phosphorus. *Soil Sci.*, **84**, 133–44.

Cole, D. W. & Gessel, S. P. (1965). Movement of elements through a forest soil as influenced by tree removal and fertilizer additions, pp. 95–104. In *Forest–Soil Relationship in North America.* Ed. C. T. Youngberg.

Cole, D. W., Gessel, S. P. & Held, E. E. (1961). Tension lysimetere studies of ion and moisture movements in glacial till and coral atoll soils. *Proc. Soil Sci. Soc. America*, **25**, 321–5.

Cox, T. L. (1972). Production, mortality and nutrient cycling in root systems of *Liriodendron* seedlings. Ph.D. thesis, University of Tennessee, Knoxville.

Cox, T. L. (1975). Accumulation and mobility of cesium in roots of tulip popular seedlings, pp. 482–488. In *Mineral cycling in southeastern ecosystems*, ed. F. G. Howell, J. B. Gentry & M. H. Smith. ERDA Symposium Series (CONF–740513).

Cox, T. L., Harris, W. F., Ausmus, B. S. & Edwards, N. T. (1978). The role of roots in biogeochemical cycles in an eastern deciduous forest. *Pedobiologia*, **18**, 264–71.

Coulter, B. C. & Talibudeen, O. (1968). Calcium: aluminum exchange equilibria in clay minerals and acid soils. *J. soil Sci.* **19**, 237–50.

Delfs, J., Friedrich, W., Kiesekamp, H. & Wagenhoff, A. (1958). Der Einflu des Waldes und des Kahlschlages auf den Abflussvorgang, den Wasserhaushalt und den Bodenabtrag. – Ergebnisse der ersten 5 Jahre der forstlich-hydrologischen Untersuchungen im Oberharz (1948–53). H. Hannover.

Denaeyer-de Smet, S. (1962). Contribution à l'étude du pluviolessivage du couvert forestier. *Bulletin de la Société Royale de Botanique de Belgique*, **94**, 285–308.

Denaeyer-de Smet, S. (1969). Apports d'éléments minéraux par les eaux de pre-cipitation, d'égouttement sous couvert forestier et d'écoulement le long des troncs (1965, 1966, 1967). *Bulletin de la Société Royale de Botanique de Belgique*, **102**, 355–72.

Devlin, R. M. (1966). *Plant Physiology.* Reinhold Publishing Corp. New York.

Dickinson, C. H. & Pugh, G. J. F. (1974). *Biology of Plant Litter Decomposition.* Academic Press, London and New York.

330

Duvigneaud, P. & Denaeyer-de Smet (1969). Cycle des éléments biogènes dans les écosystèmes forestiers d'Europe (principalement forêts caducifoliées). *Unesco* 1971, *Productivité des écosystèmes forestiers, Actes Coll. Bruxelles* **4**, 527–42.

Duvigneaud, P. & Denaeyer-de Smet (1970). Biological cycling of minerals in temperate deciduous forests, pp. 198–225. In *Analysis of Temperate Forest Ecosystems*, Ecological studies, **1**, ed. D. E. Reichle. Springer-verlag, Berlin-Heidelberg-New York.

Duvigneaud, P., Denaeyer-de Smet, S. & Marbaise, J. L. (1969). Litière totale annuelle et restitution au sol des polyéléments biogènes. *Bulletin de la Société Royale de Botanique de Belgique*, **102**, 339–54.

Edwards, N. T. & Harris, W. F. (1977). Carbon cycling in a mixed deciduous forest floor. *Ecology*, **58**; 431–7.

Ehlers, W. (1966). Beiträge zum Kaliumaustausch des Bodens. Dissertation der Universität Göttingen, Göttingen, Germany.

Elwood, J. W. & Henderson, G. S. (1975). Hydrologic and chemical budgets at Oak Ridge, Tennessee, pp. 30–51. In *Coupling of Land and Water Systems*, Ecological Studies, **10**, ed. A. D. Hasler. Springer-Verlag, Berlin-Heidelberg-New York.

Eriksson, E. (1952). Composition of atmospheric precipitation. I. Nitrogen compounds. *Tellus*, **4**, 215–32.

Eriksson, E. (1963). The yearly circulation of sulfur in nature. *J. geophys. Res.* **68**, 4001–8.

Ernst, W. (1972). Zink- und Cadmium-Immissionen auf Böden und Pflanzen in der Umgebung einer Zinkhütte. *Berichte der Deutschen Botanischen Gesellschaft*, **85**, 295–300.

Eschner, A. R. (1967). Interception and soil moisture distribution. In *International Symposium on Forest Hydrology*, ed. W. E. Sopper & H. W. Lull. Pergamon Press Ltd., Oxford.

Fayle, D. C. F. (1968). *Radial growth in tree roots*. Univ. of Toronto, Fac. For. Tech. Report No. 9.

Feddes, R. A. & Rijtema, P. E. (1972). Water withdrawal by plant roots. *Technical Bull.* 83. Institute of Land and Water Management Research, Wageningen, The Netherlands.

Friedrich, W., Liebscher, H., Rudolph, R. & Wagenhoff, A. (1968). Forstlich-hydrologische Untersuchungen in bewaldeten Versuchsgebieten im Oberharz. – Ergebnisse aus den Abflussjahren 1951–65. *Aus dem Walde* H. 7, Hannover.

Frissel, M. J. & Reiniger, P. (1974). Simulation of accumulation and leaching in soils. *Wageningen*.

Froment, A. & Mommaerts-Billiet, F. (1969). La respiration du sol, l'azote minéral et la décomposition des feuilles de chêne et de hêtre en relation avec les facteurs de l'environnement. *Bulletin de la Société Royale de Botanique de Belgique*, **102**, 317–410.

Froment, A. & Tanghe, M. (1972). L'utilisation du Facteur Azote dans la Caractérisation des Groupes Ecologiques le long d'un Transect de la Région d'Eprave-Rochefort. *Bulletin de la Société Royale de Botanique de Belgique*, **105**, 157–68.

Gapon, E. N. (1933). On the theory of exchange adsorption in soils. *J. Gen. Chemistry U.S.S.R.* **3**, 144.

Gardner, W. R. (1960). Dynamic aspects of water availability to plants. *Soil Sci.* **89**, 63–73.

Garg, R. K. & Vyas, L. N. (1975). Litter production in deciduous forest near Udaipur (South Rajasthan), India, pp. 131–5. In *Tropical ecological systems*, Ecological Studies, **11**, ed. F. B. Golley & E. Medina, Springer-Verlag, Berlin-Heidelberg-New York.

331

Garrels, R. M. & Christ, C. L. (1965). *Solutions, Minerals and Equilibria.* Harper & Row Publishers, New York.

Garstka, W. U. (1964). Snow and snow survey, pp. 10–1 to 10–57. In *Handbook of Applied Hydrology*, ed. Ven Te Chow. McGraw-Hill, Inc., New York.

Gebhardt, H. & Coleman, N. T. (1974), Anion adsorption by allophanic tropical soils: II. Sulfate adsorption. *Proc. Soil Sci. Soc. America*, **38**, 259–62.

Giesel, W., Renger, M. & Strebel, O. (1971). Berechnung des kapillaren Aufstiegs aus dem Grundwasser in den Wurzelraum unter stationären Bedingungen. *Zeitschrift fuer Pflanzenernährung und Bodenkunde* **132**, 17–29.

Giesel, W., Renger, M. & Strebel, O. (1973). Numerical treatment of the unsaturated water flow equation: comparison of experimental and computed results. *Water Resources Res.* **9**, 174–7.

Gloaguen, J. C. & Touffet, J. (1974). Production de litière et apport au sol d'éléments minéraux dans un hêtraie atlantique. *Oecologia Plantarum*, **9**, 11–28.

Göttsche, D. (1972). *Verteilung von Feinwurzeln und Mykorrhizen in Bodenprofileines Buchen-und Fichtenbestandes in Solling.* Kommissionsverlag Buchhandlung Max Wiedebusch, Hamburg.

Granat, L. (1972). On the relation between pH and the chemical composition in atmospheric precipitation. *Tellus* **24**, 550–60.

Granat, L. (1974). On the deposition of chemical substances by precipitation (as observed with the aid of the atmospheric chemistry network in Scandinavia). *Department of Meteorology, University of Stockholm, International Meteorological Institute*, Report AC 27.

Granat, L. (1975). On the variability of rain water composition and errors in estimates of areal wet deposition. *Institute of Meteorology, University of Stockholm, International Institute of Meteorology*, Report AC 30.

Granat, L. & Söderlund, R. (1975). Atmospheric deposition due to long and short distance sources – with special reference to wet and dry deposition of sulfur compounds around an oil-fired power plant. *Department of Meteorology, University of Stockholm, International Meteorological Institute*, Report AC 32.

Grunert, F. (1964). Der biologische Stoffkreislauf in Kiefern-Buchen-Mischbeständen und Kiefernbestäden. *Albrecht-Thaer-Archiv*, **8**, 435–51.

Gunary, D. (1970). A new adsorption isotherm for phosphate in soils. *J. Soil Sci.* **21**, 72–7.

Harris, W. F., Goldstein, R. A. & Henderson, G. S. (1973). Analysis of forest biomass pools, annual primary production and turnover of biomass for a mixed deciduous forest watershed, pp. 41–64. In *Proceedings IUFRO conference on forest biomass studies*, ed. H. Young. Univ. Maine Press, Orono.

Harris, W. F., Sollins, P., Edwards, N. T., Dinger, B. E. & Shugart, H. H. (1975). Analysis of carbon flow and productivity, pp. 116–122. In *Productivity of World Ecosystems, Proc. IBP V General Assembly.* Ed. D. E. Reichle & J. Franklin. National Academy of Science, Washington, 166 p.

Harris, W. F., Kinerson, R. S. & Edwards, N. T. (1978). Comparisons of belowground biomass of natural deciduous forest and loblolly pine plantations. *Pedobiologia*, **17**, 369–81.

Heikurainen, L. (1957). Über Veränderungen in der Wurzelverhältnissen der Kiefernbestände auf Moorböden in Laufe des Jahres. *Act. for. fenn.* **62**, 1–54.

Heiseke, D. (1974). Schneedecke und Wasserhaushalt der Steilen Bramke im Oberharz 1962/63. *Aus dem Walde* H. 22, S. 140–71.

Henderson, G. S. & Harris, W. F. (1975). An ecosystem approach to characterization of the nitrogen cycle in a deciduous forest watershed, pp. 179–193. In

Forest Soils and Forest Land Management, ed. B. Bernier & C. H. Winget, pp. 179–93. Laval University Press, Quebec, Canada.

Henderson, G. S. *et al.* (1971). Walker Branch Watershed: a study of terrestrial and aquatic system interaction, pp. 30–48. In *Ecological Sciences Division Annual Progress Report*, ORNL-4759, Oak Ridge National Laboratory, Oak Ridge, Tennessee.

Henin, S. & Pedro, G. (1965). The laboratory weathering of rocks. In *Experimental pedology*, ed E. G. Hallsworth & D. V. Crawford, Butterworth, London.

Hillel, D. (1971). *Soil and water.* Academic Press, Inc., New York & London.

Hingston, F. J., Atkinson, R. J., Posner, A. M. & Quirk, J. P. (1967). Specific adsorption of anions. *Nature, Lond.* **215**, 1459–61.

Holl, W. & Hampp, R. (1975). Lead and plants. *Residue Rev.* **54**, 97–111.

Jenny, H., Gessel, S. & Bingham, F. T. (1949). Comparative study of decomposition rates of organic matter in temperate and tropical regions. *Soil Sci.* **68**, 419–32.

Jensen, H. L. (1965). Nonsymbiotic nitrogen fixation. In *Soil Nitrogen*, ed. W. V. Bartholomew & F. E. Clark, Agronomy No. 10, Madison, Wisconsin.

Johnson, P. L. & Swank, W. T. (1973). Studies of cation budgets in the southern Appalachians on four experimental watersheds with contrasting vegetation. *Ecology*, **54**, 70–80.

Jost, D. (1969). Survey of the distribution of trace substances in pure and polluted atmospheres, pp. 643–53. In *Air pollution*, International Symposium, Cortina d'Ampezzo.

Junge, Ch.E. (1963). *Air chemistry and radioactivity*, International Geophysics Series 4, Academic Press, New York & London.

Kalela, E. K. (1957). Manniköiden ja Kunsikoiden juurisuhteista. I. *Acta for. fem.* **57**, 1–79 (English summary).

Karizumi, N. (1968). Estimation of root biomass in forests by soil block sampling, pp. 79–86. In: *International Symposium, USSR Methods of Productivity Studies in Root Systems and Rhizosphere Organisms.* Reprinted for the International Biological Programme by Biddles, Ltd., Guildford, U.K.

Khanna, P. K. (1975). Saisonalität der Flüsse von Stickstofformen in einem Buchen- und einem Fichtenwald-Ökosystem. *Mitteilungen der Deutschen Bodenkundlichen Gesellschaft*, **22**, 359–64.

Khanna, P. K. & Ulrich, B. (1967). Phosphatfraktionierung im Boden und isotopisch austauschbares Phosphat verschiedener Phosphatfraktionen. *Zeitschrift für Pflanzenernährung und Bodenkunde*, **117**, 53–65.

Khanna, P. K. & Ulrich, B. (1973). Ion exchange equilibria in an acid soil. *Göttinger Bodenkundliche Berichte*, **29**, 211–30.

Khanna, P. K. & Ulrich B. (1974). Gaponkoeffizienten als Massfür die Kationenselektivität der Bodenaustauscher. *Mitteilungen der Deutschen Bodenkundlichen Gesellschaft*, **18**, 218–78.

Kirkham, D. & Powers, W. L. (1972). *Advanced soil physics.* John Wiley & Sons, Inc.

Kittredge, J. (1948). *Forest influences.* McGraw-Hill Book Company, Inc., New York.

Kira, T. & Ogawa, H. (1968). Indirect estimation of root biomass increment in trees, pp. 96–101. In *International Symposium, USSR, Methods of Productivity Studies in Root Systems and Rhizosphere Organisms.* Reprinted for the International Biological Programme by Biddles, Ltd., Guildford, UK.

Koenigs, F. F. R. (1973). Bedingungen für die Verwendung der Tensiometerplatte als Lysimeter. *Zeitschrift für Pflanzenernährung und Bodenkunde*, **133**, 1–4.

Kolesnikov, V. A. (1968). Cyclic renewal of roots in fruit plants, pp. 102–106. In *International Symposium, USSR, Methods of Productivity Studies in Root*

333

Systems and Rhizosphere Organisms. Reprinted for the International Biological Programme by Biddles, Ltd., Guildford, UK.

Kozlowski, T. T. (1971). *Growth and development of trees. Vol. II.* Academic Press, New York.

Kundler, P. (1959). Zur Methodik der Bilanzierung der Ergebnisse von Boden-bildungsprozessen (Profilbilanzierung), dargestellt am Beispiel eines Textur-profiles auf Geschiebemergel in Norddeutschland. *Zeitschrift für Pflanzenernährung, Düngung und Bodenkunde,* **86**, 215–22.

Lagerwerff, J. V. (1972). Lead, mercury and cadmium as environmental contaminants, pp. 593–636. In *Micronutrients in agriculture.* Madison, Wisconsin, USA.

Larcher, W. (1975). (As translated by M. A. Brederman-Thorson). *Physio-logical Plant Ecology.* Springer-Verlag. New York.

Lemée, G. (1967). Investigations sur la minéralisation de l'azote et son evolution annuelle dans des humus forestiers in situ. *Oecologia Plantarum,* **2**, 285–324.

Lieth, H. (1968). The determination of plant dry-matter production with special emphasis on the underground parts, pp. 179–86. In *Functioning of terrestrial ecosystems at the primary production level. Proc. of the Copenhagen Symposium. Vol. V,* ed. F. E. Eckhardt. UNESCO, Paris.

Likens, G. E. & Bormann, F. H. (1975). An experimental approach to New England landscapes, pp. 7–29. In *Coupling of land and water systems,* Ecological Studies, **10**, ed. A. D. Hasler, Springer-Verlag, Berlin-Heidelberg-New York.

Little, P. (1973). A study of heavy metal contamination of leaf surfaces. *Environmental Pollution,* **5**, 159–72.

Loughnan, F. C. (1969). *Chemical weathering of the silicate minerals.* American Elsevier Publishing Company, New York.

Luttge, U. & Pitman, M. G., ed. (1976). Transport in plants. II, Part B, Tissues and organs. In *Encyclopedia of plant physiology, new series,* ed. A. Pirson & M. H. Zimmermann, Springer-Verlag, Berlin–Heidelberg–New York.

Lyr, H. & Hoffmann, G. (1967). Growth rates and growth periodicity of tree roots, pp. 181–236. In *International review of Forestry Research. Vol.* 2, ed. X. Y. Romberger & X. Y. Mikola. Academic Press, New York.

McColl, J. C. (1973). Environmental factors influencing ion transport in a Douglas-fir forest soil in western Washington. *J. Ecol.* **61**, 71–83.

McGinty, D. T. (1976). Comparative root and soil dynamics on a white pine watershed and in the hardwood forest in the Coweeta basin. Unpublished Ph.D. Dissertation, Univ. of Georgia, Athens.

Madgwick, H. A. I. & Ovington, J. D. (1959). The chemical composition in adjacent forest and open plots. *Forestry* **32**, 1–22.

Malaisse, F., Freson, R., Goffinet, G. & Malaisse-Mousset, M. (1975). Litter fall and litter breakdown in Miombe, pp. 137–52. In *Tropical ecological systems,* Ecological Studies, **11**, ed. F. B. Golley & E. Medina, Springer-Verlag, Berlin-Heidelberg-New York.

Malkonen, E. (1974). Annual primary production and nutrient cycle in some Scots pine stands. *Communcationes Instituti Forestalis Fennial,* 84.5, 87 pp.

Mayer, R. (1971). Bioelement-Transport im Niederschlagswasser und in der Bodenlosung eines Wald-Ökosystems, *Göttinger Bodenkundliche Berichte,* **19**, 1–119.

Mayer, R. (1972a). Untersuchungen uber die Freisetzung der Bioelemente aus der organischen Substanz der Humusauflage in einem Buchenbestand. *Zeitschrift für Pflanzenernährung und Bodenkunde,* **131**, 261–73.

Mayer, R. (1972b). Bioelementflüsse im Wurzelraum saurer Waldböden, *Mitteilungen der Deutschen Bodenkundlichen Gesellschaft,* **16**, 136–45.

334

Mayer, R. (1974). Ermittlung des Stoffaustrags aus Boden mit dem Versickerungs-wasser. *Mitteilungen der Deutschen Bodenkundlichen Gesellschaft*, **20**, 292–9.

Mayer, R. & Ulrich, B. (1974). Conclusion on the filtering action of forests from ecosystem analysis. *Oecologia Plantarum*, **9**, 157–68.

Mecklenburg, R. A., Tukey, H. B., Jr. & Morgan, J. V. (1966). A mechanism for the leaching of calcium from foliage. *Plant Physiol.* **41**, 610–13.

Miller, R. B. (1963). Plant nutrients in hard beech. *New Zealand J. Sci.* **6**, 365–413.

Mitscherlich, G. (1971). *Wald, Wachstum und Umwelt. Zweiter Band: Waldklima und Wasserhaushalt.* J. P. Sauerlanders Verlag, Frankfurt am Main.

Munn, R. E. & Bolin, B. (1971). Global air pollution – meteorological aspects – a survey. *Atmospheric Environment*, **5**, 363–402.

Nakayama, F. S. (1968). Calcium activity, complex and ion-pair in saturated $CaCO_3$ solutions. *Soil Sci.* **106**, 429–34.

Neuberger, H., Hosler, C. L. & Kocmond, W. C. (1967). Vegetation as aerosol filter. *Biometeorology*, **2**, 693–702.

Newbould, D. J. (1967). *Methods for estimating the primary production of forests. IBP Handbook No.* 2. Blackwell Scientific Pub., Oxford.

Nielsen, D. R., Biggar, J. W. & Erh, K. T. (1973). Spatial variability of field-measured soil-water properties. *Hilgardia*, **42**, 215–60.

Nihlgard, B. (1970). Precipitation, its chemical composition and effect on soil water in a beech and a spruce forest in south Sweden. *Oikos, Copenhagen*, **21**, 208–17.

Nimah, M. N. & Hanks, R. J. (1973*a*). Model for estimating soil water, plant, and atmospheric interrelations: I. Description and sensitivity. *Proc. Soil Sci. Soc. Amer.* **37**, 522–7.

Nimah, M. N. & Hanks, R. J. (1973*b*). Model for estimating soil water, plant, and atmospheric interrelations: II. Field test of model. *Proc. Soil Sci. Soc. Amer.* **37**, 528–32.

Nutman, P. S. (1965). Symbiontic nitrogen fixation, pp. 00–00. In *Soil nitrogen*, ed. W. V. Bartholomew & F. E. Clark, Agronomy No. 10, Madison.

Odum, E. P. 1969. The strategy of ecosystem development. *Science*, Washington. **164**, 262–70.

Olsen, S. R. & Watanabe, F. S. (1957). A method to determine a phosphate adsorption maximum of soils as measured by the Langmuir isotherm. *Proc. Soil Sci. Soc. Amer.* **21**, 144–9.

Olson, J. S. (1963). Energy storage and the balance of producers and decomposers in ecological systems. *Ecology*, **44**, 322–31.

O'Neill, R. V., Harris, W. F., Ausmus, B. S. & Reichle, D. E. (1975). A theoretical basis for ecosystem analysis with particular reference to element cycling, pp. 28–40. In *Mineral Cycling in Southeastern Ecosystems*, ed. F. G. Howell, J. B. Gentry & M. H. Smith. ERDA Symposium Series (CONF–740513).

Ovington, J. D. (1962). Quantitative ecology and woodland ecosystem concept, pp. 103–92. In *Advances in ecological research*, **1**, ed. J. B. Cragg. Academic Press, New York.

Ovington, J. D. (1967). Quantitative ecology and the woodland ecosystem concept. *Adv. Ecol. Res.* **1**, 103–92.

Ovington, J. D., Heitkamp, D. & Lawrence, D. B. (1963). Plant biomass of prairie, savanna, oakwood and maize field ecosystems in central Minnesota. *Ecology*, **44**, 52–63.

Pavlov, M. (1972). Bioelement-Inventur von Buchen- und Fichtenbestanden im Solling. *Göttinger Bodenkundliche Berichte*, **25**, 1–174.

Penman, H. L. (1963). Vegetation and hydrology. *Technical Column.* 53. Commonwealth Agriculture Bureaux, Harpenden, England.

Persson, H. (1978). Root dynamics in a young Scots pine stand in Central Sweden. *Oikos*, **30**, 508–19.

Pitman, M. G. (1976). Nutrient uptake by roots and transport to the xylem, pp. 85–100. In *Transport and Transfer Processes in plants*, ed. I. F. Wardlaw & J. B. Passioura. Academic Press, New York.

Poulovassilis, A. (1973). The hysteresis of pore water in presence of non-independent water elements, pp. 161–79. In *Physics of soil water and salts, Ecological studies*, **4**, ed. A. Hadas, D. Schwartzendruber, P. E. Rijtema, M. Fuchs & B. Yaron, Springer–Verlag, Berlin–Heidelberg–New York.

Prather, R. J., Miyamoto, S. & Bohn, H. L. (1973). Nitric oxide sorption by calcareous soils. *Proc. Soil Sci. Soc. Amer.* **37**, 877–9.

Prince, R. & Ross, F. F. (1972). Sulphur in air and soil. *Water, Air and Soil Pollution*, **1**, 286–302.

Quervain, M. R. de & Gand, H. R. (1967). Distribution of snow deposit in a test area for alpine reforestation, pp. 233–9. In *Forest Hydrology*, ed. W. E. Sopper & H. W. Lull, Pergamon Press Ltd., Oxford.

Rapp, M. (1971). Cycle de la matière organique et des éléments minéraux dans quelques écosystèmes méditerranéens. *Editions du Centre Nationale de la Recherche Scientifique, Paris*, Recherche Coopérative sur programme du CNRS, No. 40, 184 pp.

Reichle, D. E., Dinger, B. E., Edwards, N. T., Harris, W. F. & Sollins, P. (1973). Carbon flow and storage in a Forest Ecosystem, pp. 345–365. In *Carbon and the Biosphere*, ed. G. M. Woodwell & E. V. Pecan, pp. 345–65. Proceedings of the 24th Brookhaven Symposium in Biology. AEC–CONF–720510. NTIS, Springfield, Virginia.

Rijtema, P. E. (1965). An analysis of actual evapotranspiration. *Agricultural Research Reports No 659*. Centre for Agricultural Publications and Documentation, Wageningen, Netherlands.

Roberts, B. R. (1974). Foliar sorption of atmospheric sulphur dioxide by woody plants. *Environmental Pollution*, **7**, 133–40.

Rodhe, H. (1972). A study of the sulfur budget for the atmosphere over Northern Europe. *Tellus*, **34**, 128–38.

Rodin, L. E. & Bazilevich, N. I. (1967). *Production and Mineral Cycling in Terrestrial Vegetation*. Oliver & Boyd, Edinburgh & London.

Rose, C. W. (1966). *Agricultural Physics*. Pergamon Press Ltd., Oxford.

Ruhling, A. & Tyler, G. (1973). Heavy metal deposition in Scandinavia. *Water, Air and Soil Pollution*, **2**, 445–55.

Runge, M. (1971). Investigation of the content and the production of mineral nitrogen in soils, pp. 191–202. In *Integrated Experimental Ecology, Ecological Studies*, **2**, ed. H. Ellenberg, Springer-Verlag, Berlin-Heidelberg-New York.

Russell, E. W. (1973). *Soil conditions and plant growth*. Longman, London & New York.

Saito, H. & Shidei, T. (1972). Studies on estimation of leaf fall under model canopy. *Bulletin of the Kyoto University Forests*, No 43, 162–85.

Salmon, R. C. (1964). Cation exchange reactions. *J. Soil Sci.* **15**, 273–83.

Sandberg, G. R., Olson, J. S. & Clebsch, E. E. C. (1969). *Internal distribution and loss from roots by leaching of cesium-137 inoculater into* Liriodendron tulipifera *L. seedlings grown in sand culture*. Oak Ridge National Laboratory Report No. ORNL/TM-2660, Oak Ridge, TN. 68 p.

Santantonio, D., Hermann, R. K. & Overton, W. S. (1977). Root biomass studies in forest ecosystems. *Pedobiologia*, **17**, 1–31.

336

Schlesinger, W. H., Reiners, W. A. & Knopman, D. S. (1974). Heavy metal concentrations and deposition in bulk precipitation in montane ecosystems of New Hampshire, USA. *Environmental Pollution*, **6**, 39–47.

Schlichting, E. & Schwertmann, U. (1973). *Pseudogley and Gley*, Transactions of Commissions V and VI of the International Society of Soil Science, Verlag Chemie GmbH, Weinheim/Bergstr.

Schofield, R. K. (1947). A ratio law governing the equilibrium of cations in the soil solution. *Proceedings of XIth International Congress on Pure and Applied Chemistry*, **3**, 257–61.

Schofield, R. K. (1955). Can a precise meaning be given to 'available' soils phosphorus? *Soils and Fertilizers*, **18**, 373–5.

Schofield, R. K. & Taylor, A. W. (1955a). The measurement of soil pH. *Proc. Soil Sci. Soc. Amer.* **19**, 164–7.

Schofield, R. K. & Taylor, A. W. (1955b). Measurements of activities of bases in soils. *J. Soil Sci.* **6**, 137–46.

Schouwenberg, J. Ch. von & Schuffelen, A. C. (1963). Potassium exchange behaviour of an illite. *Netherlands J. Agric. Sci.* **11**, 13–22.

Singer, S. F., ed. (1970). Global effects of environmental pollution. *Dordrecht*.

Slatyer, R. O. (1967). *Plant–Water Relationships*. Academic Press, London & New York.

Smith, W. H. (1970). Technique for collection of root exudation from mature trees. *Plant and Soil*, **32**, 238–41.

Sollins, P., Reichle, D. E. & Olson, J. S. (1973). Organic matter budget and model for a southern Appalachian *Liriodendron* forest. EDPB–IBP–73–2. Oak Ridge National Laboratory, Oak Ridge, Tennessee.

Stafford, R. G. & Ettinger, H. J. (1971). Comparison of filter media against liquid and solid aerosols. *American Industrial Hygiene Association J.* **31**, 319–26.

Stenlid, G. (1958). Salt losses and restitution of salts in higher plants, pp. 615–37. In *Encyclopedia of plant physiology*, ed. W. Ruhland, **4**, Springer-Verlag, Berlin–Göttingen–Heidelberg.

Sutton, R. F. (1969). *Form and development of conifer root systems*. Tech. Commun. 7. Commonwealth For. Bull., Oxford, England.

Taylor, A. W. (1958). Some equilibrium solution studies on Rothamsted soils. *Proc. Soil Sci. Soc. Amer.* **22**, 511–13.

Thomas, W. A. (1969). Accumulation and cycling of calcium by dogwood trees. *Ecological Monographs*, **39**, 101–20.

Thomas, W. A. (1970). Weight and calcium losses from decomposing tree leaves on land and in water. *J. appl. Ecol.* **7**, 237–41.

Tukey, H. B. Jr. (1970). The leaching of substances from plants. *Annu. Rev. Plant Physiol.* **21**, 305–24.

Ulrich, B. (1961). Die Wechselbeziehungen von Boden und Pflanze in physikalisch-chemischer Betrachtung. *Ferdinand Enke Verlag*, Stuttgart, 114 pp.

Ulrich, B. (1969). Investigations on cycling of bioelements in forests of central Europe, pp. 501–7. In *Unesco 1971, Productivity of Forest Ecosystems, Proceedings of the Brussels Symposium* 1969.

Ulrich, B. (1975). Die Umweltbeeinflussung des Nährstoffhaushaltes eines bodensauren Buchenwaldes. *Forstwissenschaftliches Centralblatt*, 94 Jg., H.6, 280–7.

Ulrich, B. & Khanna, P. K. (1968). Schofield'sche Potentiale und Phosphatformen in Boden. *Geoderma*, **2**, 65–77.

Ulrich, B. & Khanna, P. K. (1969). Ökologisch bedingte Phosphatumlagerung und Phosphatformenwandel bei der Pedogenese. *Flora, Abteilung B*, **158**, 594–622.

Ulrich, B., Ahrens, E. & Ulrich, M. (1971). Soil chemical differences between beech and spruce sites – an example of the methods used, pp. 171–90. In *Integrated Experimental Ecology, Ecological Studies*, **2**, ed. H. Ellenberg, Springer–Verlag, Berlin–Heidelberg–New York.

Ulrich, B., Mayer, R., Khanna, P. K. & Prenzel, J. (1973). Modelling of bioelement cycling in a beech forest of Solling district. *Göttinger Bodenkundliche Berichte*, **29**, 1–54.

Ulrich, B., Mayer, R., Khanna, P. K., Seekamp, G. & Fassbender, H. W. (1976). Input, Output und interner Umsatz von chemischen Elementen in einem Buchen- und einem Fichtenbestand. 6. *Jahresversammlung der* Gesellschaft für Ökologie im Sept. 1976 in Göttingen (Congress Proceedings).

Urone, P. & Schroeder, W. H. (1969). SO_2 in the atmosphere: A wealth of monitoring data, but few reaction rate studies. *Environmental Science and Technology*, **3**, 437–45.

Vachaud, G. & Thony, J. L. (1971). Hysteresis during infiltration and redistribution in a soil column at different initial water contents. *Water Resources Res.* **7**, 111–27.

Van der Ploeg, R. R. (1974). Simulation of moisture transfer in soils: One-dimensional infiltration. *Soil Sci.* **118**, 349–57.

Van der Ploeg, R. R. & Beese, F. (1976). Model calculations for the extraction of soil water by ceramic cups and plates. *Proc. Soil Soc. Amer.*

Van der Ploeg, R. R. & Benecke, P. (1974a). Simulation of one dimensional moisture transfer in unsaturated, layered field soils. *Gottinger Bodenk. Ber.* **30**, 150–69.

Van der Ploeg, R. R. & Benecke, P. (1974b). Unsteady, unsaturated, n-dimensional moisture flow in soil: A computer simulation program. *Proc. Soil Sci. Soc. Amer.* **38**, 881–5.

Van der Ploeg, R. R., Ulrich, B., Prenzel, J. & Benecke, P. (1975). Modeling the mass balance of forest ecosystems, pp. 793–802. In *Proceedings of the 1975 Summer Computer Simulation Conference*, Simulation Councils Inc., LaJolla, California.

Walker, T. W. (1965). The significance of phosphorus in pedogenesis. In *Experimental pedology*, ed. E. G. Hallsworth & D. V. Crawford, Butterworth, London.

Waller, H. D. & Olson, J. S. (1967). Prompt transfers of cesium-137 to the soils of a tagged *Liriodendron* forest. *Ecology*, **48**, 15–25.

Weihe, J. (1968). Zuruckhaltung von Regenniederschlagen durch Buchen und Fichten. AFZ 86–90.

White, E. J. & Turner, F. (1970). A method of estimating income of nutrient in a catch of airborne particles by a woodland canopy. *J. appl. Ecol.* **7**, 441–61.

Whittaker, R. H. (1962). Net production relations of shrubs in the Great Smoky Mountains. *Ecology*, **43**, 357–77.

Wit, C. T. de & Keulen, H. van (1972). Simulation of transport processes in soils, *Wageningen*, 108 pp.

Wit, de C. T. & Arnold, G. W. (1976). Some speculation on simulation, pp. 00–00. In *Critical Evaluation of Systems Analysis in Ecosystems Research and Management*, ed. G. W. Arnold & C. T. de Wit, PUDOC, Wageningen, Centre for Agricultural Publishing and Documentation.

Witherspoon, J. P. (1964). Cycling of cesium-134 in white oak trees. *Ecological Monographs*, **34**, 403–20.

Witkamp, M. (1969). Environmental effects on microbial turnover of some mineral elements. *Soil Biology and Biochemistry*, **1**, 167–84.

Witkamp, M. (1971). Forest soil microflora and mineral cycling, pp. 413–24. In *Productivity of Forest Ecosystems*, Proceedings of the Brussels Symposium, Vol. 4, UNESCO.

Witkamp, M. & Frank, M. L. (1964). First year of movement, distribution and availability of Cs^{137} in the forest floor under tagged tulip poplars. *Radiation Botany*, **4**, 485–95.

Witter, S. H. & Bukovac, M. J. (1969). The uptake of nutrients through leaf surfaces, pp. 235–300. In *Handbuch der Pflanzenernährung und Dungung*, I, 1, ed. Hg. H. Linser, Wien, New York.

Witter, S. H. & Teubner, F. G. (1959). Foliar absorption of mineral nutrients. *Annu. Rev. Plant Physiol.* **10**, 13–27.

Wittich, W. (1942). Natur und Ertragsfahigkeit der Sandboden im Gebiet des norddeutschen Diluviums. *Zeitschrift für das Forst- und Jagdwesen*, **74**, 1–42.

Yamada, Y., Rasmussen, H. P., Bukovac, M. J. & Witter, S. H. (1966). Binding sites for inorganic ions and urea on isolated cuticular membrane surfaces. *Amer. J. Bot.* **53**, 170–2.

Ziemer, R. R. (1968). Soil moisture depletion patterns around scattered trees. *US Forest Service Research Note PSW*–166, US Department of Agriculture.

6. Elemental cycling in forest ecosystems

D. W. COLE & M. RAPP

Contents

Introduction

Among the functions regulating activity and evolution of forest ecosystems, mineral cycling is of critical importance. The study of cycling of elements in forest ecosystems has given us insights into the dynamics of these systems, including nutrient needs, rates of nutrient turnover, loss of nutrients by leaching and addition by weathering, fixation, and the atmosphere. The study of mineral cycling has led also to an understanding of the processes regulating the pathways and rates of flow for the various elements within these systems, information essential for the rational management of our forest lands.

Cycling of elements in forest ecosystems is an integrating process that brings together most other functions of the system. These elements are obviously essential in the plant growth processes of photosynthesis, assimilation, and respiration. The dynamics of nutrient uptake and transport within the soil are considered also within the context of cycling studies. Decomposition of organic materials and weathering of mineral materials are vital aspects of the overall cycle, providing the means by which elements

341

become available to higher plants. Thus, the study of mineral cycling involves study of the ecosystem as a whole. Since one process functions as the precursor to another (i.e., mineralization precedes nitrogen uptake), the flow of elements in a forest ecosystem is inseparably linked in a set of specific interconnected steps that ultimately lead to a set of cyclic pathways.

It is reasonable to expect that, although the processes will remain, their rates will change drastically from one forest type to another, from one climatic zone to another, and from one soil type to another.

Such patterns in cycling have, in all probability, slowly and systematically evolved as a part of the genetic evolution of species. Adaptability of a particular plant to its environment, thus, depends on entire sets of cycling processes that collectively allow the plant to compete within this environment. There are obviously several ways in which these cycling processes have been collectively arranged to allow a plant to adapt to its ecological niche. For example, the strategy of meeting annual nutrient requirements for growth between low and high elevation forests has been handled by different forests' species in quite different ways. Retention of foliage for a number of years (25 yr in the case of black spruce) allows a coniferous forest to occupy sites where only a marginal nutrient supply is available from the soil. Other species, such as alder, have the specific ability through a symbiotic nitrogen-fixing process of providing their own supply of nitrogen and, thus, effectively can occupy disturbed sites where nitrogen levels are low.

To provide clearer insight into the role that mineral cycling has played in the evolution of forest ecosystems, we have collected and compared IBP data sets concerning production and mineral cycling. We here analyze these data sets and make comparisons among them. Our principal contribution has been to assemble and summarize this information. Many of the ideas that we discuss are not new; they were originally presented by such pioneers in the field as Ehwald, Remezov, and Ovington in articles and books on the subject. In recent years Duvigneaud and Ellenberg have added to this literature with much information and many new ideas. It is due to their influence that part of the IBP effort was focused on this subject.

Data base used for this synthesis

One of the primary purposes of IBP forest studies was to provide a general overview of biological productivity in the forest ecosystems of the world and to establish the basic factors that regulate this productivity. As mentioned above, mineral cycling plays a critical role and provides an integrated view of this important process for a forest ecosystem.

Although IBP studies were conducted around the world, under a wide array of forest and environmental conditions located at 117 different sites, unfortunately for the program and this synthesis effort mineral cycling studies were conducted on a far more limited scale. Even at those sites where mineral cycling was studied, the vast bulk of the research was limited to some narrow aspect of the subject. In a detailed review of the available information provided by the Oak Ridge international synthesis group, reasonably complete data adequate for mineral cycling comparisons were available from only 20 forest stands. Contact with the investigators of the other sites expanded the data base to include 32 forest stands from 14 different sites. Workers at the remaining sites either did not have a sufficient data base or failed to respond to the inquiries. It became evident in reviewing the data base supplied to Oak Ridge that there were many obvious errors and other inconsistencies with published results. We hope these differences have been resolved and the data base used here adequately represents these forest stands.

Table 6.1 summarizes the sites from which our final synthesis was completed. These sites are arranged into climatic zones, vegetative types, species, and site location. The name of the scientist(s) responsible for the specific site and through whom the data were obtained is also listed. A more complete description of each site is included in the Appendix to this chapter. Table 6.1 shows that this program did not result in uniform or systematic coverage of areas or geographic regions. Rather, most sites are located within a relatively small geographic area covering North America and Europe. Important ecological areas such as the Mediterranean zone are poorly represented. There is no comprehensive data set for either tropical or equatorial forest types.

Purpose and scope of this synthesis effort

Our purpose in attempting this synthesis of mineral cycling information was twofold. First, we wanted to provide within one document a summary of the cycling information collected as a result of IBP efforts. Second, we wanted to see if a comparison of such information would add a new dimension to our understanding of ecosystem behavior, an understanding that had not or could not emerge without the more extensive data base that this assemblage provided.

In spite of the limitations of the data sets, we believe these IBP studies represent a wide spectrum of conditions and include a variety of regions, life forms, and environmental settings, as indeed is evident from Table 6.1. However, since the diversity of conditions associated with these sites was not selected as a part of a comprehensive experimental design, some of our initial expectations for the synthesis of these data on elemental cycling could not be met.

The difficulties were further magnified by the many problems and caveats

Table 6.1. *IBP study sites and forest types used in this mineral cycling synthesis*

Stand no.	Forest region and species	Stand age (yr)	Country and site	Investigator
	Boreal Coniferous			
1	*Picea mariana*	51	USA; Alaska	Van Cleve
2	*Picea mariana*	55	USA; Alaska	Van Cleve
3	*Picea mariana*	130	USA; Alaska	Van Cleve
	Boreal Deciduous			
4	*Betula papyrifera*	50	USA; Alaska	Van Cleve
	Temperate Coniferous			
5	*Pseudotsuga menziesii*	42	USA; Cedar River	Cole, Turner
6	*Pseudotsuga menziesii*	73	USA; Cedar River	Turner
7	*Pseudotsuga menziesii*	450	USA; Andrews	Grier
8	*Picea abies*	45	USSR	Kazimirov
9	*Picea abies*	60	Sweden; Kongalund	Nihlgard
10	*Picea abies*	34	W. Germany; Solling	Ulrich
11	*Picea abies*	87	W. Germany; Solling	Ulrich
12	*Picea abies*	115	W. Germany; Solling	Ulrich
13	*Pinus echinata*	30	USA; Oak Ridge	Harris
14	*Pinus strobus*	15	USA; Coweeta	Swank
15	*Abies firma*	97–145	Japan	Ando
16	*Tsuga sieboldii*	120–443	Japan	Ando
17	*Tsuga heterophylla*	121	USA; Cascade Head	Grier
	Temperate Deciduous			
22	*Liriodendron tulipifera*	50	USA; Oak Ridge	Reichle *et al.*
23	*Liriodendron-Quercus*	30–80	USA; Oak Ridge	Harris
24	*Quercus-Carya*	30–80	USA; Oak Ridge	Harris
25	*Quercus prinus*	30–80	USA; Oak Ridge	Harris
26	*Quercus-Carya*	60–200	USA; Coweeta	Swank
27	*Acer-Betula-Fagus*	110	USA; Hubbard Brook	Whittaker, Likens
28	*Quercus*-mixed	mixed	Belgium	Duvigneaud
29	*Quercus*-mixed	80	Belgium	Duvigneaud
30	*Fagus sylvatica*	59	W. Germany; Solling	Ulrich
31	*Fagus sylvatica*	80	W. Germany; Solling	Ulrich
32	*Fagus sylvatica*	122	W. Germany; Solling	Ulrich
33	*Quercus-Betula*	mixed	Great Britain; Merlewood	Satchell
34	*Fagus silvatica*	45–130	Sweden; Kongalund	Nihlgard
35	*Alnus rubra*	30	USA; Cedar River	Turner
	Mediterranean			
36	*Quercus ilex*	150	France; Rouquet	Lossaint, Rapp

associated with comparing data collected by different people in different ways at different times. Because of these limitations our primary emphasis is directed toward more general considerations of elemental cycling and productivity in natural ecosystems.

This discussion is presented in five parts:

1. Systematic changes in elemental cycling associated with stand maturity.
2. Biomass and elemental accumulation and elemental cycling for forest regions and ecosystem types.
3. Differences between deciduous and coniferous species relative to accumulation and cycling.
4. Effect of elemental cycling on forest productivity.
5. Addition and losses of nutrient elements from forest ecosystems.

Methods used in this synthesis

Selection of information

The data used in this synthesis were derived from several sources, IBP data files at Oak Ridge (corrected or modified), correspondence with investigators at the IBP field sites, IBP publications, and unpublished data obtained directly from investigators involved in IBP mineral cycling studies. A basic set of data comprising elemental accumulations, increments, and fluxes was tabulated for each study site. This information included both organic matter and elemental nitrogen, potassium, calcium, magnesium and phosphorus.

Unfortunately, a complete data set comprising all the above parameters was never available from any site. In particular, few sites had information on soil and forest floor leaching rates. Elemental addition by way of atmospheric particulate input was seldom measured. Only one study examined rates of nitrogen fixation. The available information was tabulated on a standard entry form in units of kg/ha or kg/ha/yr. These data sheets are included in the Appendix.

Use of information

This information was used in several comparative ways as the chapter outline makes evident. A part of this comparison involved calculating annual elemental uptake, requirement, and interval recycling from the information base. Since these terms have been defined and calculated in many ways, we had to decide on a standardized means of deriving these values. Although it can be argued that these values can be established in a more sophisticated way than in the procedure we used, it should be remembered that this was not possible with the limited data available from most of the study sites. We defined and calculated these terms as follows:

345

Uptake – annual elemental increment associated with bole and branch wood plus annual loss through litterfall, leaf wash, and stemflow.

This calculation assumes a steady-state condition for tree foliage and excludes any consideration of root and bark increment.

Requirement – annual elemental increment associated with bole and branch wood plus the current foliage production.

This calculation is dependent on the same assumptions as stated for uptake. An indication of the extent of the internal recycling of elements can be derived by subtracting uptake from requirement.

Recycling – Requirement minus Uptake

This calculation is obviously based on the same assumption associated with uptake and requirement calculations. In addition it also includes the assumption that litterfall is primarily foliage fall and elements derived from throughfall and stemflow were derived from foliage.

There are many assumptions, some of them perhaps unacceptable, built into these calculations, but the major difficulty derives from lack of information regarding below-ground processes. Very few studies included any root information, and none had annual elemental increment information for the root system. Consequently it was not possible to include roots into any flux calculations.

Most of the studies did include organic matter and the elements nitrogen, phosphorus, potassium, calcium and magnesium. Consequently, we have included all of these elements on the stand summary sheets prepared for each site and included in the Appendix to this chapter. However, for our purposes much of the discussion is focused on organic matter (biomass accumulation and production), nitrogen, potassium and calcium. These elements were most extensively studied at the IBP sites. In addition, they provide an excellent contrast in cycling.

Nitrogen is closely tied to the carbon cycle. Consequently transfer of nitrogen from one compartment to another within a forest (Fig. 6.1) is dependent on transfer of carbon or release of nitrogen from carbon through the process of mineralization. In this way nitrogen behaves in a manner similar to phosphorus and sulfur. Nitrogen also is involved in fixation, nitrification, and denitrification, all critical processes in the mineral cycle.

By contrast, potassium is not associated with organic structures and is transferred from compartment to compartment through the ecosystem and is basically independent of the carbon cycle. Consequently, quantities of potassium found in throughfall are proportionally far higher than those of nitrogen, since transfer of potassium is not dependent on mineralization. The cycling rate of potassium is rapid compared to other elements.

The third element selected was calcium. This element is not a component of protein and, thus, is not dependent on mineralization for its mobility as is nitrogen. However, it is still found as a part of the organic structure of the

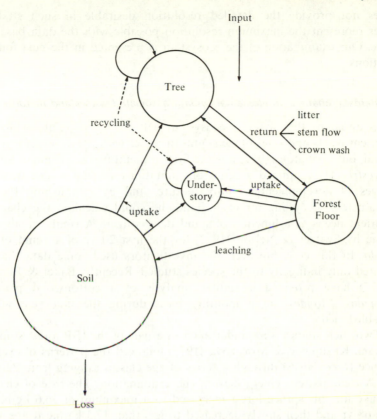

Fig. 6.1. A generalized diagram of nutrient cycling in forest ecosystems. After Turner, 1975.

tree or precipitated as an oxalate or sulfate salt in the plant tissue. Cycling of calcium is, thus, relatively slow within an ecosystem, and recycling within a tree is nearly nonexistant. In this region it behaves in a way similar to magnesium and manganese.

Comparison of these three elements involves in a general way comparison of nearly all of the critical elements associated with mineral cycling of the forest system. We have also included magnesium and phosphorus in many of our tables and some of our discussions when there is an adequate information base for doing so.

Rates of cycling, increments of addition and accumulation of elements within and between ecosystem compartments

The general format in which cycling, increments, and accumulations of elements have been delineated is shown in Fig. 6.1. While this scheme does not provide the detailed resolution desirable in such studies, it does represent the maximum resolution possible with the data base available. Our examination of the ecosystem is presented in the next four subsections.

Systematic changes in elemental cycling associated with stand maturity

It is apparent that a forest ecosystem will change in relationship to its nutrient needs and rates of elemental turnover with the development of the stand, but Ovington (1959) was the first to quantify these changes for *Pinus sylvestris*. He examined a series of plantations varying in age from early stages of development to fully mature, and by establishing the basic parameters of cycling for each stand, he could calculate the changes in nutrient needs as a function of stand development. A similar analysis was done by Switzer & Nelson (1972) for the first 20 yr of a stand of *Pinus taeda*. In this case, however, the investigators used some data that were related only indirectly to the species studied. Recently, Baker & Blackmon (1977) have provided a similar analysis of a cottonwood plantation (*Populus deltoides*) on a monthly basis during the first year after its establishment.

Two such studies were undertaken as a part of the IBP forest studies. In Russia, Kazimirov & Morozova (1973) followed the patterns of cycling in spruce (*Picea abies*) through a series of age classes ranging from 22 to 138 yr. A clear pattern emerged from the examination. The rate of elemental uptake for this spruce forest increased to a maximum of 36.6 kg/ha/yr at age 68 yr and then slowly decreased to less than 17.3 kg/ha at age 138 yr. This decrease in tree uptake corresponded to an increase in the uptake by ground cover vegetation; these uptakes became nearly equal at age 138 yr. Apparently, this reflects changes in cycling associated with the opening of the forest canopy. Up to an age of nearly 70 yr, spruce dominates the site, minimizing the role of the understory flora. Between 70 and 138 yr the understory vegetation assumes a far larger role in the nutrient dynamics of the site until the two are nearly equally important. This relationship for nitrogen is shown in Fig. 6.2.

A similar analysis was undertaken by Turner (1975) for Douglas fir (*Pseudotsuga menziesii*) in Washington State, USA. Turner studied a series of Douglas-fir stands ranging from 9 to 95 yr from which he calculated change in nutrient accumulation and measured transfer with changes in the structure of the forest ecosystem. His analysis makes clear that closure of the forest canopy at about 25 yr played a critical role in the nutrient cycle. Before stand closure, the understory vegetation assumed a prominent position in the structure of this ecosystem. For example, at nine years there was an approximately equal distribution of nitrogen, potassium and cal-

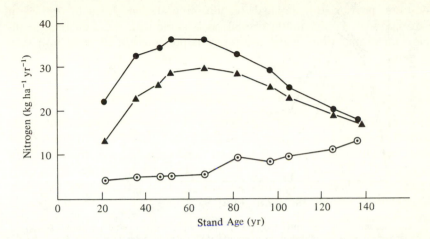

Fig. 6.2. Cycling of nitrogen in spruce (*Picea abies*) from age 22 yr to 138 yr. After Kazimirov & Morozova, 1973. ●—●, tree uptake; ○—○, understory vegetation uptake; ▲—▲, return in litter.

cium between the forest and the understory. This relationship rapidly changed, however, as the crown began to close at age 22 yr. By the age of 30 yr, nutrient accumulation in the understory vegetation declined rapidly (Fig. 6.3).

With closure of the forest canopy at about age 30 yr, there was very little increase in elemental accumulation in the forest foliage (Fig. 6.4). As one would expect, litterfall also remained relatively stable after this period as did a number of other above-ground elemental relationships, such as annual uptake and nutrient requirement by the forest species (Table 6.2). From the summary of stand data in Table 6.2 it is clear that the uptake of both nitrogen and potassium are closely coupled with annual requirements for these two elements. However, in the case of calcium, uptake far exceeded annual requirement for growth, as reflected in the amount of calcium in the current year's growth increment. Excess calcium is stored in older foliage and the woody biomass of the system. It would be speculative, however, to conclude from this that excess calcium is unnecessary to the tree. Its specific role is, however, unclear.

Because litterfall exceeds decomposition in this forest type, the forest floor accumulates biomass with stand age (Fig. 6.5) resulting in a buildup of nutrients in this ecosystem compartment. This accumulation is not uniform between elements. Nitrogen and calcium accumulation closely parallels the accumulation of organic matter. By contrast, potassium is accumulating at a far slower rate because it is not dependent on mineralization for its release (Fig. 6.6).

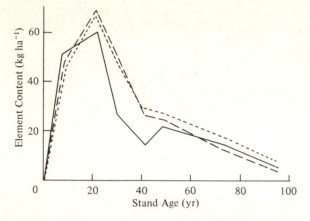

Fig. 6.3. Accumulation of nitrogen (----), potassium (——), and calcium (–––) in the understory vegetation of various aged stands of Douglas fir. After Turner, 1975.

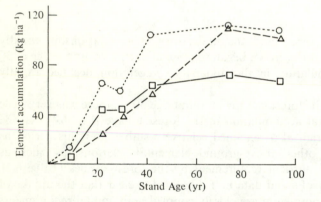

Fig. 6.4. Accumulation of nitrogen (○ . . . .○), potassium (□—□), and calcium (△- - -△) in the foliage of various aged stands of Douglas fir. After Turner, 1975.

Biomass and elemental accumulation and elemental cycling for forest regions and ecosystem types

One of our initial objectives was to make a comparison between the major forest regions for various aspects of accumulation and mineral cycling. The data base developed from the IBP studies was not extensive enough nor adequately distributed between forest regions for regional comparison of types. Significant characteristics were evident, however, in the element accumulation, turnover, and recycling patterns for different forests. Twenty seven of the 32 IBP studies we compared were conducted within the

Table 6.2. *Annual uptake and nutrient requirements of nitrogen, potassium and calcium for Douglas fir between the ages of 9 and 450 yr (after Turner, 1975; Grier et al., 1974)*

Age (yr)	Nitrogen		Potassium		Calcium	
	Uptake	Requirement	Uptake	Requirement	Uptake	Requirement
9	3.7	5.8	3.6	4.5	4.8	2.2
22	33.7	41.8	26.3	31.0	34.4	13.3
30	32.1	32.8	29.5	27.7	55.7	14.4
42	32.8	36.0	27.4	26.1	40.9	12.9
73	32.5	34.7	21.4	24.5	43.2	14.1
95	37.3	28.7	25.5	25.9	51.4	11.9
450	23.7	34.8	21.2	26.7	53.3	17.9

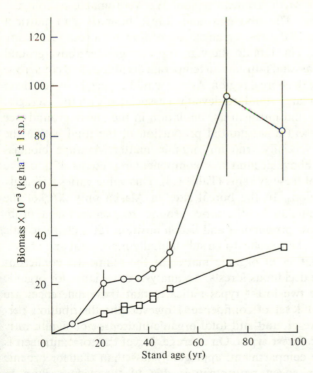

Fig. 6.5. Accumulation of forest floor biomass (O—O, total; □—□, humus) under various aged stands of Douglas fir. After Turner, 1975.

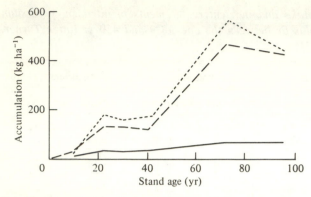

Fig. 6.6. Accumulation of nitrogen (---), calcium (— —), and potassium (—) in the forest floor of various aged stands of Douglas fir. After Turner, 1975.

temperate forest region. Four of the sites were located in the boreal forest region and one in the Mediterranean region. We were unable to compare any from the tropics. The above-ground forest biomass and nutrient accumulation for the IBP sites, arranged according to forest region are tabulated in Table 6.3. The data do show an apparent greater above-ground accumulation of biomass and nutrients in temperate deciduous and coniferous forests than in those of the boreal region. As one would expect, however, these regional values have a large standard deviation associated with the averages.

The organic matter and nitrogen accumulation in the above-ground tree component can represent a substantial proportion of the total within the ecosystem. This is especially true for organic matter. Average biomass accumulation in the above-ground tree component represents 45% of the total organic matter of these systems (Table 6.4). This value varies strikingly however between regions. In the boreal sites in Alaska only 20% of the organic matter is accumulated in the above-ground tree compartment, probably because of the low production and decomposition rates characteristic of northern latitudes. There seems to be substantially more total and above-ground tree accumulation of organic matter in the temperate coniferous than in the temperate deciduous forests. An analysis of variance for organic matter between these two forest types indicates that these differences are significant at the 0.01 level of confidence. However, the distribution percentage between above-ground and total organic matter is not significantly different for these two forest types. On average, 7% of the total nitrogen is found within the tree compartment, appreciably less than that for organic matter. This lower nitrogen percentage is due to the wide carbon to nitrogen ratio for woody biomass. Again, boreal forests have a far smaller

352

Table 6.3. *Above-ground organic matter and nutrient accumulation for the IBP study sites arranged according to forest types and regions*

Forest region	Stand no.	Accumulation (kg/ha)					
		Biomass	Nitrogen	Potassium	Calcium	Magnesium	Phosphorus
Boreal coniferous	1	16597	40	24	30	7	6
	2	24018	95	31	120	10	14
	3	113189	215	76	625	60	29
	Average	51268	116	44	258	26	16
	S.D.	±53753	±89	±28	±321	±30	±12
Boreal deciduous	4	97343	221	104	164	38	20
Temperate coniferous	5	126495	227	132	202	29	112
	6	129400	316	205	358	53	54
	7	802310	566	189	687	105	86
	8	78200	282	105	241	31	28
	9	305000	720	406	453	67	80
	10	142700	449	349	291	62	59
	11	244490	729	401	413	56	39
	12	232990	628	342	379	88	65
	13	120600	215	125	307	40	18
	14	69599	224	109	98	—	—
	15	379562	573	1426	1402	130	82
	16	448601	582	441	934	63	25
	17	915480	721	189	478	60	167
	Average	307341	479	340	480	65	68
	S.D.	±271547	±203	±349	±351	±30	±42
Temperate deciduous	22	124700	304	173	456	—	47
	23	109100	267	172	537	59	21
	24	121600	369	220	856	67	24
	25	137300	397	242	852	45	26
	26	137496	405	227	543	—	—
	27	134000	367	154	402	38	33
	28	291850	728	398	1142	92	46
	29	115085	368	200	830	77	28
	30	154658	407	130	324	26	52
	31	158764	404	202	303	26	36
	32	273958	581	273	303	67	20
	33	113065	278	213	480	26	17
	34	104000	1071	456	606	105	85
	35	151020	240	74	170	60	23
	Average	151900	442	224	557	57	35
	S.D.	±58130	±221	±100	±273	±26	±19
Mediterranean	36	269000	745	626	3853	151	224
Overall averages, all stands		207526	429	263	589	60	52
	S.D.	±196483	±229	±253	±672	±34	±47

Table 6.4. *Total and above-ground tree accumulation of organic matter and nitrogen for the various forest regions*

Forest region	No. of sites	Organic matter (kg/ha)			Nitrogen (kg/ha)		
		Tree	Total	%[a]	Tree	Total	%[a]
Boreal coniferous	3	51000	226000	19	116	3250	4
Boreal deciduous	1	97000	491000	20	221	3780	6
Temperate coniferous	13	307000	618000	54	479	7300	7
Temperate deciduous	14	152000	389000	40	442	5619	8
Mediterranean	1	269000	326000	83	745	1025	73
Average	32	208000	468000	45	429	5893	7

[a] Percent of total contained in above-ground component.

percentage of nitrogen in the above-ground tree compartment than that found for the other forest regions.

Accumulation of organic matter and the five nutrient elements within the forest floor of the 32 study sites is listed in Table 6.5. In general, the largest forest floor organic accumulations are found in the boreal coniferous forests followed by temperate coniferous, boreal deciduous, temperate deciduous, and Mediterranean in that order. The accumulation of nutrient elements within the forest floor generally follows the same distribution pattern between forest regions noted above for organic matter. These differences between forest regions cannot be statistically substantiated, however, because we lack an adequate number of sites within several of these regions and because of the large standard deviation associated with the data sets.

The rate of organic matter and nutrient return by forest regions is summarized in Table 6.6. As expected, the largest return of organic matter is in the temperate deciduous forest, probably due to the high productivity and deciduous character of this forest type. The high annual rate of organic return is reflected in the return of most of the nutrient elements. On average, over 60 kg/ha of nitrogen is returned each year under a temperate deciduous forest. Only 36 kg/ha of nitrogen is returned annually under a temperate coniferous forest. This contrast is obviously due to the extended period in which coniferous species retain foliage. The duration of foliage retention varied greatly between temperate coniferous forests, ranging from six years for older Douglas-fir stands (stands 6 and 7) to 1.2 years for short leaf pine (stand 13).

It should be recognized that return of the nutrient elements is only in part carried by litterfall. Averaging of all 32 forest stands considered here showed that 83% of nitrogen return was by way of litterfall. Similarly 85% of phosphorus was so returned. By contrast, only 41% of the potassium

Table 6.5. *Forest-floor organic matter and nutrient accumulation for the IBP study sites arranged according to forest types and regions*

Forest type	Stand no.	Accumulation (kg/ha)					
		Biomass	Nitrogen	Potassium	Calcium	Magnesium	Phosphorus
Boreal coniferous	1	88664	486	60	396	96	106
	2	133260	656	145	231	224	152
	3	119235	710	122	452	99	87
	Average	113720	617	109	360	140	115
	S.D.	±22804	±117	±44	±115	±73	±33
Boreal deciduous	4	68772	548	99	489	139	79
Temperate coniferous	5	20540	178	38	114	54	26
	6	26670	223	44	130	54	30
	7	218520	445	80	619	160	62
	8	19200	—	—	—	—	—
	9	18500	245	15	48	8	15
	10	52000	1430	105	72	44	199
	11	49000	960	41	83	20	53
	12	111000	2260	115	258	46	98
	13	27000	290	21	256	23	18
	14	—	—	—	—	—	—
	15	4279	484	113	182	67	32
	16	50780	507	76	159	30	34
	17	301080	474	118	343	83	98
	Average	74881	681	70	206	53	60
	S.D.	±92299	±640	±39	±165	±42	±54
Temperate deciduous	22	6000	78	9	100	—	5
	23	1500	187	14	294	22	11
	24	27000	334	26	517	32	22
	25	25000	298	26	318	22	18
	26	9500	110	29	130	—	—
	27	48016	126	66	372	38	8
	28	5600	—	—	—	—	—
	29	4762	44	17	107	5	2
	30	29000	815	85	86	44	62
	31	39000	1050	123	118	44	72
	32	29700	810	83	91	34	52
	33	6118	74	12	101	8	3
	34	5200	86	104	34	5	6
	35	66350	887	91	391	57	45
	Average	21625	377	53	205	28	25
	S.D.	±19624	±391	±40	±153	±17	±25
Mediterranean	36	11400	125	10	361	20	4
Overall averages, all stands		52343	515	65	236	55	50
	S.D.	±66410	±482	±43	±161	±51	±48

Table 6.6. *Average litterfall and nutrient return by forest regions as determined at the IBP study sites*

Forest region	No. of sites	Litterfall (kg/ha/yr)	Nutrient return (kg/ha/yr)				
			Nitrogen	Potassium	Calcium	Magnesium	Phosphorus
Boreal coniferous	3	322	2.9	1.1	3.8	0.3	0.7
Boreal deciduous	1	2645	20.2	9.8	35.5	9.7	5.2
Temperate coniferous	13	4377	36.6	26.1	37.3	5.6	4.4
Temperate deciduous	14	5399	61.4	41.6	67.7	11.0	4.0
Mediterranean	1	3842	34.5	44.0	95.0	9.0	4.7
All stands	32	4373	43.3	27.9	47.3	7.4	3.7

returned to the forest floor was found in litterfall, the majority returned via throughfall and stemflow, indicating the highly mobile nature of this element. Return of calcium and magnesium seems to be intermediate between that of nitrogen and phosphorus and that of potassium, with 71% of calcium and 60% of magnesium returned via litterfall.

By comparing the rate of nutrient return to total accumulation within the forest floor we can calculate mean residence or turnover times of the forest floor. Such a calculation is based on the assumption that the forest floor and rate of return are both in steady state. Although there is little reason to believe that this is the case, calculation still provides a relative index useful for comparative purposes. Turnover periods have been calculated for each of the forest regions; the influence of latitude and forest type on decomposition is clearly shown in Table 6.7. The more northerly boreal forests have an exceedingly long mean residence time for both organic matter and all five elements. On average, coniferous boreal forests retain nitrogen nearly 100 times longer than the Mediterranean forest, 13 times longer than the temperate coniferous forest, and 42 times longer than the temperate deciduous forest. Similar extremes in turnover periods can be seen for other elements.

Annual elemental uptake and requirement was calculated for each site. These values, including above-ground tree production, are listed in the Appendix to this chapter, and are summarized by forest regions in Table 6.8. We recognize the limitations of the data, but nevertheless, our calculations illustrate some important cycling characteristics of forest ecosystems. It is clear, for example, that forests are relatively modest in removing

356

Table 6.7. *Mean residence (turnover) time in years for the forest floor and its mineral elements by forest regions as determined at the IBP sites. Values calculated by dividing annual return into total forest floor accumulation. A steady state condition is assumed*

Forest region	No. of sites	Mean residence time (yr)					
		Organic matter	N	K	Ca	Mg	P
Boreal coniferous	3	353	230	94	149	455	324
Boreal deciduous	1	26	27.1	10.0	13.8	14.2	15.2
Temperate coniferous	13	17	17.9	2.2	5.9	12.9	15.3
Temperate deciduous	14	4.0	5.5	1.3	3.0	3.4	5.8
Mediterranean	1	3.0	3.6	0.2	3.8	2.2	0.9
All stands	32	12.0	34.1	13.0	21.8	61.4	46.0

nutrients from the soil. Annual uptake of nitrogen for the 32 sites averaged only 55 kg/ha. Maximum average uptake of 75 kg/ha was found in the temperate deciduous region. The amount of nitrogen required for the growth increment of bole and stem wood and current foliage production is also quite low with the highest value noted again in the temperate deciduous forest.

These values demonstrate that uptake by the temperate deciduous forest for nitrogen is 63% greater than the coniferous forest of the same region and 15 times greater than the spruce in the boreal region. Annual requirement for nitrogen by the temperate deciduous forests is more than twice that of the temperate coniferous forests and 20 times that of the boreal spruce. These differences between sample means are all significant at the 99% confidence level.

Differences in uptake and requirement between forest regions were similar for potassium. Again these differences are significant at the 99% confidence level.

Except for the single evergreen oak site in the Mediterranean region, the temperate deciduous forest also had the highest uptake and requirement for calcium, exceeding the temperate coniferous forest by a factor of 1.9 and 2.4 times, and the boreal coniferous forest by a factor of 14 and 19 times in these two cycling categories. These differences between sample means are significant at the 99% confidence level. The high calcium uptake by the Mediterranean oak is probably related to the calcareous soil from which the forest grows. The uptake and requirement of magnesium and phosphorus follows no consistent pattern between regions except for the consistently low values noted for the black spruce in the boreal region.

Table 6.8. *Annual production and nutrient cycling (uptake and requirement) for the IBP study sites arranged by forest region*

Forest type	Stand no.	Total above-ground production biomass (kg/ha/yr)	Nitrogen (kg/ha/yr)		Potassium (kg/ha/yr)		Calcium (kg/ha/yr)		Phosphorus (kg/ha/yr)		Magnesium (kg/ha/yr)	
			Uptake	Require-ment	Uptake	Require-ment	Uptake	Require-ment	Uptake	Require-ment	Uptake	Require-ment
Boreal coniferous	1	627	2.6	2.8	1.3	1.7	2.7	2.0	1.6	0.3	0.4	0.3
	2	1402	6.3	8.2	1.9	3.8	8.0	4.3	0.8	1.2	0.6	0.9
	3	1590	6.6	3.3	3.1	1.8	7.8	3.0	0.8	0.4	0.9	0.6
Average S.D.		1206 ±510	5.1 ±2.2	4.7 ±3.0	2.1 ±0.9	2.4 ±1.2	6.1 ±3.0	3.1 ±1.1	1.1 ±0.5	0.6 ±0.5	0.6 ±0.3	0.6 ±0.3
Boreal deciduous	4	5164	25.0	55.9	12.8	25.7	38.9	18.3	5.7	5.7	10.6	9.0
Temperate coniferous	5	8770	33.7	41.8	26.3	31.0	34.4	13.3	6.3	5.9	9.8	6.8
	6	6570	35.9	36.0	29.7	26.1	49.6	12.9	4.9	5.0	9.2	5.9
	7	6080	23.7	34.8	21.2	26.7	53.3	17.9	5.8	7.1	5.6	3.7
	8	5150	34.4	38.9	15.3	16.6	38.1	25.9	3.7	3.5	5.1	3.5
	9	13800	88.0	67.0	44.4	43.0	45.7	12.7	6.8	9.1	10.2	4.8
	10	8501	53.3	47.1	—	—	—	—	4.8	5.7	3.6	4.5
	11	8864	63.0	49.2	42.9	25.9	44.8	20.2	5.6	4.5	2.8	1.6
	12	6517	55.6	35.6	—	—	—	—	4.9	3.2	2.8	1.6
	13	7878	48.6	49.7	35.0	31.2	73.8	40.2	2.9	3.1	10.8	9.4

14	13530	50.0	84.4	48.1	30.0	32.8	25.6	—	—	—	—
15	8365	43.6	44.9	—	—	—	—	—	—	—	—
16	5069	27.6	34.0	—	—	—	—	—	—	—	—
17	9510	58.7	41.4	30.2	20.2	28.8	8.6	10.3	6.9	11.4	3.8
Average	8354	47.4	46.5	32.6	27.9	44.6	19.7	5.6	5.5	7.1	4.6
S.D.	±2758	±17.3	±14.4	±11.0	±7.5	±13.6	±9.7	±2.0	±1.9	±3.5	±2.4
Temperate deciduous											
22	6029	47.9	61.0	55.0	55.6	105.3	44.6	4.2	7.3	—	—
23	8650	58.1	87.9	40.0	47.5	87.8	82.6	3.4	6.3	12.4	21.7
24	10400	69.1	87.2	48.2	57.5	97.2	91.0	3.7	6.4	13.8	18.4
25	10700	68.8	98.1	40.3	58.2	88.8	80.3	3.3	5.5	11.5	13.1
26	7966	43.0	92.1	48.3	47.1	59.0	50.5	—	—	—	—
27	7889	74.3	90.0	53.4	37.8	62.3	37.7	6.2	7.4	8.9	6.6
28	7980	86.4	108.9	51.0	52.0	91.0	59.0	5.4	8.0	19.7	10.0
29	14775	80.0	96.1	48.8	47.0	169.4	106.3	4.6	6.3	16.4	9.2
30	11813	91.3	113.9	45.2	32.7	47.7	18.4	7.2	9.1	7.3	4.4
31	9648	87.6	106.7	46.7	38.1	55.0	28.4	5.7	6.7	6.4	3.4
32	10352	75.6	88.5	45.0	31.3	50.2	18.8	6.5	6.7	7.0	3.9
33	6615	78.9	91.9	59.0	43.8	124.1	54.7	3.6	4.9	25.0	10.5
34	15100	78.3	129.3	42.2	43.3	63.1	49.3	8.6	10.5	10.2	10.9
35	12780	114.9	119.0	86.1	77.5	91.2	56.0	10.3	8.1	19.5	12.8
Average	10050	75.4	97.9	50.7	47.8	85.0	55.6	5.6	7.2	13.2	10.4
S.D.	±2807	±18.2	±16.7	±11.6	±11.9	±33.1	±26.3	±2.1	±1.5	±5.9	±5.6
Mediterranean overall 36	7100	47.7	77.2	52.9	36.9	120.7	63.7	7.3	8.6	11.2	8.2
Averages all stands	8287	55.0	66.3	38.4	35.4	63.3	37.4	5.2	5.9	9.4	7.0
S.D.	±3589	±27.2	±34.6	±19.2	±17.6	±38.2	±28.5	±2.4	±2.6	±6.0	±5.3

Differences between deciduous and coniferous species relative to accumulation and cycling

Because the foliage of deciduous species is replaced each year, unlike that of coniferous species, it is to be expected that many aspects of mineral cycling between these major taxonomic groups will also be different. Recognizing that such differences should exist, we summarized the IBP data sets in a way that would provide direct comparison between two systems relative to mineral cycling parameters. We included all deciduous and coniferous sites in our analysis.

Our comparison clearly shows that the rate of cycling is far more rapid in deciduous than coniferous species for all five elements (Table 6.9). For example, the nitrogen uptake rate of coniferous species is only 56% that of deciduous stands. The potassium, calcium and magnesium uptake rates of coniferous species approximate 50% that of deciduous sites. Only in the case of phosphorus is the difference relatively small (within 80%). Annual requirement and return of these elements showed similar differences when we compared these two forest types.

Deciduous species translocate significantly more nitrogen from old to new tissue than do conifers (Table 9). Average nitrogen uptake for deciduous species is 70.5 ± 21 kg/ha. However, 94 ± 19 kg/ha is needed to meet annual requirements for nitrogen in new tissue. Thus, about 1/3 of annual requirement for nitrogen is met through translocation from older to new tissue. In the case of coniferous species, the data suggest little if any translocation. Essentially identical quantities of nitrogen are taken up $(39 \pm 22$ kg/ha/yr) as are required to produce new tissue $(39 \pm 21$ kg/ha/yr). There is very little apparent translocation of potassium for either deciduous or coniferous species. Calcium uptake, by way of contrast, greatly exceeds annual requirement for both deciduous and coniferous species. In the case of the coniferous species, calcium uptake is about double the requirement. Much of the calcium taken up each year accumulates in older tissue and is not directly incorporated in the growth of new material.

The pattern for magnesium is similar to that for calcium. Uptake greatly exceeds annual requirement for both the deciduous and coniferous species, suggesting, as for calcium, that magnesium accumulates in the older tissue of the tree.

Deciduous and coniferous species were significantly different at the 99% confidence level in the uptake, requirement, and return of all four of these elements. Only for phosphorus did we fail to find a significant difference in the amount of uptake and return. The annual requirement was different at the 99% confidence level.

In contrast, the distribution of these elements within the ecosystems of these same deciduous and coniferous forests was typically not significantly different (Table 6.10). This comparison was made for elemental accumu-

Table 6.9. *Comparison of deciduous and coniferous species relative to element uptake, requirement and return*

Element	Uptake (kg/ha/yr)			Requirement (kg/ha/yr)			Return (kg/ha/yr)		
	Deciduous	Coniferous	%[a] Sign.	Deciduous	Coniferous	%[a] Sign.	Deciduous	Coniferous	%[a] Sign.
Nitrogen	70	39	99	94	39	99	57	30	99
Potassium	48	25	99	46	22	99	40	20	99
Calcium	84	35	99	54	16	99	67	29	99
Magnesium	13	6	99	10	4	99	11	4	99
Phosphorus	6	5	79	7	4	99	4	4	60

[a]%significance between means.

361

Table 6.10. *Comparison between deciduous and coniferous species relative to accumulation of nitrogen, potassium and calcium within certain ecosystem compartments*

Element	Above-ground accumulation (kg/ha)			Forest floor accumulation (kg/ha)			Soil (kg/ha)		
	Deciduous	Coniferous	%[a] Sign.	Deciduous	Coniferous	%[a] Sign.	Deciduous	Coniferous	%[a] Sign.
Nitrogen	447	411	32	368	668	90	5658	6701	30
Potassium	242	284	36	51	78	90	b	b	b
Calcium	737	439	78	223	239	21	b	b	b
Magnesium	63	57	31	35	72	93	b	b	b
Phosphorus	47	38	46	27	72	98	b	b	b

[a] % significance between means.
[b] Soils data for potassium, calcium, magnesium and phosphorus were not compared statistically because of differences in analyses.

Table 6.11. *Comparison of nitrogen accumulation and cycling for deciduous and coniferous species at three IBP study sites*

Site	Species	Accumulation (kg/ha)			Cycling (kg/ha/yr)		
		Above-ground	Forest floor	Soil	Uptake	Requirement	Return
Solling[a]	Beech	464	892	7708	66	106	53
	Spruce	601	1550	6937	57	44	47
Kongalund	Beech	1071	86	7800	78	129	70
	Spruce	720	245	6900	80	67	66
Cedar River	Alder	405	887	5450	115	119	96
	Douglas fir	316	223	2476	36	36	26

[a] Average of three forest stands.

lation above-ground, in the forest floor, and within the soil. Unfortunately, we could compare only nitrogen in the soil compartment because of the many different ways in which the other elements had been analyzed. Only phosphorus accumulation within the forest floor shows a significant difference between a deciduous and a coniferous ecosystem. Since these deciduous and coniferous ecosystems were significantly different in elemental cycling but not in elemental distribution, we inferred that the rate of cycling (Tables 6.9, 6.10) is an inherent property of deciduous and coniferous species and is not significantly regulated by differences in elemental accumulation. However, a better comparison between the elemental composition of the soils of these sites is needed before such a conclusion can be safely drawn.

In several IBP studies, deciduous and coniferous species were directly compared in adjacent plots. The advantage of such a comparison is that soil and climatic conditions should be reasonably similar. The beech (*Fagus silvatica*) and spruce (*Picea abies*) stands of the Kongalund site in southern Sweden (stands No. 9 and 34), the beech and spruce stands at Solling (stands No. 10, 11, 12, 30, 31, 32) and Douglas fir (*Pseudotsuga menziesii*) and red alder (*Alnus rubra*) at the Cedar River site in the USA (stands No. 6 and 35) are examples of this form of comparison. Results form these comparisons have been reported (Nihlgard, 1972; Heinrichs & Mayer, 1977; Cole, Gessel & Turner, 1978). Although we cannot provide a statistical analysis of these studies, as we could for the comparison of all of the plots (as discussed above), there were many of the same differences and similarities (Table 6.11). Except for alder at Cedar River, there is no clear relationship between occurrence of a coniferous or deciduous species and accumulation of nitrogen by the ecosystems. The greater nitrogen accumulation in the alder ecosystem compared to Douglas fir is due to nitrogen fixation. This relationship and the comparison of these two stands relative to nitrogen fixation is discussed later. The spruce forest seems to be accumulating more nitrogen in the forest floor than beech at both the Solling and Kongalund sites. This is probably due to the lower decomposition rates associated with coniferous litterfall.

In general, we can conclude from both this comparison and the one made earlier that the rate of cycling is higher in deciduous stands. Except at the Kongalund site, nitrogen uptake is greater for deciduous species. At all of the deciduous sites nitrogen requirement is substantially higher as is nitrogen return. Productivity of these forests will certainly affect these cycling rates. This interaction is discussed separately.

Effect of elemental cycling on forest productivity

The relationship between elemental cycling and productivity has been addressed in the extensive reviews on this subject by Ovington (1962) and Rodin & Bazilevich (1967). Specific studies have been made comparing elemental cycling, elemental distribution, and forest productivity for individual forest ecosystems and ecosystems that have received fertilization (Heilman & Gessel, 1963; Madgwick, White, Xydias & Leaf, 1970; Fagerstrom & Lohm, 1977). The synthesis of the IBP studies provided another opportunity to examine the general relationship between elemental uptake, requirement, and return as it affects or is affected by forest production.

The coefficients of determination between productivity and elemental cycling in deciduous and coniferous forests are given in Table 6.12. Clearly, production is far more strongly correlated to cycling of nitrogen and potassium than it is to the other three elements. In addition it is equally apparent that production is more strongly correlated to elemental cycling in coniferous ecosystems than in deciduous ones. The highest correlation, $r^2 = 0.92$, was found between coniferous production and nitrogen requirement associated with annual growth of the forest. In contrast, there was little if any correlation between production and cycling of calcium in either coniferous or deciduous forests.

These data suggest several roles for cycling in the production of forest biomass. Uptake and requirement for nitrogen and potassium are far more strongly correlated with production of coniferous than with deciduous species (Table 6.12). The difference in correlation between these two forest types is probably due to the greater amount of nitrogen and potassium that deciduous species translocate before leaffall, thus providing a certain degree of independence from the soil nitrogen supply and a need to supply the annual requirement from uptake. Earlier data (Table 6.3) lent strength to this suggestion. Coniferous species appear to meet all of their annual

364

Table 6.12. *Coefficients of determination (r²) of above-ground biomass production and cycling of nitrogen, potassium and calcium in deciduous and coniferous ecosystems*

		Coefficients of determination (r^2)	
Element	Process	Deciduous	Coniferous
Nitrogen	Uptake	0.42	0.78
	Requirement	0.58	0.92
	Return	0.33	0.62
Potassium	Uptake	0.07	0.90
	Requirement	0.07	0.81
	Return	0.00	0.80
Calcium	Uptake	0.02	0.32
	Requirement	0.09	0.22
	Return	0.00	0.21
Magnesium	Uptake	0.01	0.55
	Requirement	0.00	0.34
	Return	0.09	0.48
Phosphorus	Uptake	0.16	0.63
	Requirement	0.17	0.77
	Return	0.04	0.63

requirement for nitrogen through the uptake process (39 kg/ha out of 39 kg/ha), while deciduous species are meeting only 67% of their needs through the uptake process (64 kg/ha out of 95 kg/ha). The rest apparently is derived through translocation from older tissue.

We tested the hypothesis that the lower correlation between production and uptake of nutrients by deciduous species is due to higher soil fertility in areas where deciduous sites are located. Obviously if this was indeed true, a lower correlation would be expected because other factors such as soil moisture conditions or climatic factors could also be involved in regulating production. To some extent, we tested and rejected this hypothesis in the statistical analysis presented in Table 6.10. Deciduous forests of the IBP sites were found to be growing on soils that did not differ statistically in total nitrogen from those supporting the coniferous forest.

Lack of any meaningful correlation between production and cycling of calcium and magnesium for either coniferous or deciduous forests can be attributed to several possible reasons: (1) there are no calcium and magnesium deficiencies at the study sites, (2) calcium and magnesium are accumulated in older as well as new tissue, (3) there is a large variation in soil calcium and magnesium levels between sites, depending on whether they are located on calcareous deposits and the age of the soil. These

Table 6.13. *Above-ground biomass production for deciduous and coniferous forest normalized to the annual uptake, requirement, and return of nitrogen, potassium, calcium, magnesium and phosphorus*

Element	Process	Forest type	Normalized Production (kg/ha/yr)	%[a] Sign.
Nitrogen	Uptake	Deciduous	143 ± 36	99
		Coniferous	194 ± 48	
	Requirement	Deciduous	102 ± 20	99
		Coniferous	201 ± 81	
	Return	Deciduous	180 ± 51	99
		Coniferous	283 ± 126	
Potassium	Uptake	Deciduous	216 ± 85	99
		Coniferous	354 ± 154	
	Requirement	Deciduous	220 ± 78	99
		Coniferous	380 ± 182	
	Return	Deciduous	271 ± 132	98
		Coniferous	521 ± 333	
Calcium	Uptake	Deciduous	130 ± 61	99
		Coniferous	217 ± 95	
	Requirement	Deciduous	232 ± 161	99
		Coniferous	520 ± 303	
	Return	Deciduous	167 ± 86	98
		Coniferous	302 ± 160	
Magnesium	Uptake	Deciduous	915 ± 445	98
		Coniferous	1559 ± 783	
	Requirement	Deciduous	1292 ± 846	96
		Coniferous	2257 ± 1313	
	Return	Deciduous	1151 ± 642	99
		Coniferous	2404 ± 1465	
Phosphorus	Uptake	Deciduous	1859 ± 745	80
		Coniferous	1519 ± 582	
	Requirement	Deciduous	1374 ± 417	90
		Coniferous	1759 ± 732	
	Return	Deciduous	2496 ± 965	37
		Coniferous	2263 ± 1524	

[a] Percent significance between means.

possibilities, either individually or collectively, are undoubtedly influencing calcium and magnesium cycling at the IBP sites.

To compare directly cycling effects between deciduous and coniferous forests we normalized the above-ground production rates to rates of mineral cycling. This comparison (Table 6.13) illustrates that the coniferous forest has a marked advantage over the deciduous in respect to amount of nitrogen needed to produce the same quantity of biomass. Similarly, the

Table 6.14. *Above-ground biomass production for the various forest regions and types normalized to the uptake, requirement, and return of nitrogen*

Forest region	No. of sites	Normalized production (kg/ha/yr)		
		Uptake	Requirement	Return
Boreal coniferous	3	236 ± 12	295 ± 169	425 ± 139
Boreal deciduous	1	207	92	256
Temperate coniferous	13	184 ± 49	179 ± 26	250 ± 102
Temperate deciduous	14	138 ± 34	103 ± 21	173 ± 50
Mediterranean	1	149	92	206
All stands	32	168 ± 49	151 ± 77	232 ± 108

nitrogen requirement for a coniferous forest is only 50% as large as that of a deciduous forest having the same level of production. This apparently greater nitrogen efficiency of a coniferous forest is partly due to needle retention. Since a coniferous forest does not have to replace annually its total foliage, it should not need as much nitrogen to maintain its canopy as does a deciduous forest. The differences between means are all significant at the 99% confidence level.

The relationship between nitrogen uptake (kg/ha/yr) and production (above-ground increment, kg/ha/yr) can be expressed by the regressions:

Deciduous forest production $= 4242 + 85 \times$ uptake,
Coniferous forest production $= 1201 + 146 \times$ uptake.

Similarly the relationship between nitrogen requirement and production can be expressed by the regressions:

Deciduous forest production $= 1069 + 114 \times$ requirement,
Coniferous forest production $= 406 + 170 \times$ requirement.

Similar differences between deciduous and coniferous forests can also be seen in potassium, calcium and magnesium cycling. These differences between means are nearly all significant at the 98 or 99% confidence level. The deciduous and coniferous forests were statistically similar only for the uptake and return of phosphorus (Table 6.13).

Due to unequal distribution of the number of sites between forest regions, as discussed earlier, we cannot provide a statistical comparison between production and cycling by region. The general relationship that exists between production and cycling of nitrogen is summarized in Table 6.14. As in Table 6.13, production values have been normalized to the various parameters of cycling (uptake, requirement, return). Sufficient information to evaluate the data statistically was available only for the boreal coniferous, temperate coniferous, and temperate deciduous forests. Within this limitation of data, several patterns of production and cycling still emerge. Coniferous forests are consistently more efficient than the deciduous in producing biomass with the same amount of nitrogen. The northern species seem to be more efficient than the more southerly species in producing biomass for a given amount of nitrogen. This could be caused by a greater deficiency of available nitrogen in these boreal sites, resulting in a higher conversion ratio between production and the uptake and utilization of nitrogen.

The requirement of nitrogen shows a similar trend. The coniferous forests within a given region require less nitrogen per unit of production than the deciduous forests. The more northern forests seem to need less than those to the south. Collectively, these data strongly suggest that the production of forest ecosystems is nitrogen-limited, with coniferous forests, especially those in northern latitudes, more limited than are deciduous forests.

Additions and losses of nutrient elements from forest ecosystems

Accumulation of nutrient elements in either an ionic form or incorporated within the organic component of an ecosystem is dependent on the net difference between those processes regulating inputs and those controlling losses. The additions and losses of elements from forested plots and small watersheds have been extensively reported in the literature (Cole, Gessel & Dice, 1967; Duvigneaud & Denaeyer-DeSmet, 1970; Likens *et al.*, 1977; Henderson, Swank, Waide & Grier, 1978). A comprehensive analysis of these processes where the input–output fluxes were isolated and individually assessed as not undertaken at any of the IBP sites. However, some measurements of input–output were made at most of the sites, typically representing the summation or interaction of several processes. For example, few studies isolated precipitation inputs from dry fallout – natural inputs derived through atmospheric processes from inputs associated with atmospheric pollutants. Nutrients losses by soil leaching were not compared to inputs through soil weathering processes. Leaching losses measured from small watersheds were not separated into losses associated with the rooting zone and losses due to mineral weathering below the rooting zone. In addition, only two studies (stands No. 7 and 35) examined the potential input of nitrogen through the process of biological fixation.

We have here limited our discussion of input–output fluxes and processes to those additions resulting from atmospheric precipitation and biological fixation and losses from leaching below the rooting zone. Data are insufficient at the IBP sites to consider comprehensively the other aspects of nutrient additions and losses mentioned above.

Atmospheric nitrogen additions

The amount of elemental addition to forest ecosystems by way of atmospheric precipitation and dryfall varies greatly between study sites. Typically, additions of nutrients by these means are relatively small compared to the amount taken up by forest vegetation. As discussed above, average nitrogen uptake for all sites was 55 ± 27 kg/ha/yr. Average atmospheric addition of nitrogen was 9.8 kg/ha/yr, or about 18% of average annual uptake. However, this value varied greatly between sites ranging from 1.1 kg/ha/yr at the spruce site in the USSR to 22.8 kg/ha/yr at the Solling site in West Germany. This variability in input appears to be highly dependent on the proximity to sources of input such as industrial or population centers, a common observation (e.g., Junge, 1963; Likens & Bormann, 1974). At the Cedar River site in Washington, northwest USA, less than 2 kg/ha of nitrogen is added annually by precipitation, or less than 5% of the annual uptake of second growth Douglas-fir trees at that site. At the Coweeta, Walker Branch and Hubbard Brook sites in the eastern USA, annual nitrogen input is about 8 kg/ha, evidently due to the larger populations and increased industrial activity in that area of the United States. Such an input rate could potentially supply 14% of the uptake needs at these sites. In contrast, at the Solling site in West Germany, atmospheric input is nearly 23 kg/ha/yr, or 32% of the average annual needs.

High rates of nitrogen input should have a positive influence on productivity of such areas, decreasing the potential of a nitrogen deficiency. The possibility that this occurs at the Solling site is shown by our earlier calculations that compared biomass production to nitrogen uptake. Both deciduous and coniferous forests at this site show substantially less production per unit of uptake than that noted for the average deciduous and coniferous stands in the temperate region, indicative of a nitrogen surplus. For example, the spruce stands at Solling produce 135 kg/ha of biomass per kg of nitrogen uptake, while the average for the temperate coniferous forest is 184 kg/ha. The beech stands produce 124 kg/ha of biomass per kg of nitrogen uptake. Average production for temperate deciduous forests is 138 kg/ha. In contrast, nitrogen-deficient Douglas-fir stands on the Cedar River site produce 214 kg/ha of biomass for each kg of nitrogen uptake. The Cedar River site, as we indicated earlier receives only 1.7 kg/ha/yr of nitrogen from the atmosphere.

These results suggest that atmospheric nitrogen inputs can play an important role in minimizing nitrogen deficiencies that occur in many forest stands. In a nitrogen-deficient forest, an increase in atmospheric nitrogen should result in a decrease in the amount of biomass produced per unit of nitrogen uptake. From these data sets, a correlation coefficient (r) of -0.35 exists between production values normalized to nitrogen uptake and the amount of nitrogen added from the atmosphere. Considering only temperate coniferous forests, this correlation coefficient is -0.54. In the temperate deciduous forests where our earlier analysis indicated less likelihood of a nitrogen deficiency (Table 6.12), atmospheric addition of nitrogen has had less effect on changing productivity. In this case the correlation coefficient between normalized production value and atmospheric nitrogen additions is only -0.13.

The atmospheric addition of elements other than nitrogen does not seem to have as important a role in the productivity of these stands. As we discussed earlier, the only other element closely correlated to productivity was potassium (Table 6.12). However atmospheric addition of potassium was seldom reported to be high (maximum of 5 kg/ha/yr) and varied by only 2.3 ± 1.3 kg/ha/yr between sites.

Addition of nitrogen by fixation

In many forest ecosystems nitrogen is also added through the process of biological fixation. The amount of fixation varies, depending on the specific ecosystem, fixation process, and the organism involved. The amount of fixation reported in the literature varies widely, ranging from as little as 1 to 2 kg/ha/yr for free-living fixers to over 200 kg/ha/yr for symbiotic fixation associated with legumes, alder and other higher plants.

In the IBP forest ecosystem program, nitrogen fixation rate and its impact on the elemental cycle was followed at only one site, a red alder ecosystem at the Cedar River site (stand No. 35). This study was designed directly to compare alder and Douglas fir relative to elemental accumulation and cycling rates. The two stands are on comparable soils and located adjacently. The Douglas-fir site was established as a plantation in 1933 after the area was logged. The red alder site was established as natural regeneration adjacent to the Douglas-fir plantation boundaries. We have assumed in comparing these two ecosystems that the differences currently found in elemental accumulation and cycling are caused by the differences in the vegetation of these two sites.

It is evident by comparing distribution of nitrogen within the two ecosystems that alder has yielded an apparent increase of total nitrogen of 3240 kg/ha over a 38-yr period. Although most of this increase can be found in the soil (2180 kg/ha), a substantial amount is also in the forest floor and above-ground vegetation (Table 6.15). These values provide estimates (Cole *et al.*, 1978) that nitrogen accumulates in this alder ecosystem at a rate of 85.3 kg/ha/yr. This value is not markedly different from that reported by other researchers in this field (Tarrant & Miller, 1963; Newton, el Hassan & Zavitkovski, 1968).

These two ecosystems are also very different in nearly every aspect of

370

Table 6.15. *Estimated nitrogen accumulation and average annual accumulation by 38-yr-old red alder (Alnus rubra)*

	Estimated nitrogen accumulation (kg/ha)		Increase in red alder over Douglas fir (kg/ha)	Average annual accumulation (kg/ha/yr)
	Douglas fir	Red alder		
Overstory	320	590	270	7.1
Understory	10	100	90	2.4
Forest floor	180	880	700	18.4
Soil	3270	5450	2180	57.4
Total	3780	7020	3240	85.3

elemental cycling (Table 6.16). Annual nitrogen uptake by alder is 115 kg/ha, three times higher than the Douglas fir stand. Annual return of nitrogen is 94 kg/ha, six times greater than Douglas fir. In the Douglas fir ecosystem, about 19% of the annual nitrogen requirement is met through recycling from older foliage. In the case of alder only 3% of nitrogen is recycled, apparently due to the high nitrogen levels available to the tree from fixation. Because of this low recycling in alder, the mean residence time of nitrogen in the foliage is only 1.1 yr. By contrast, the mean residence time of nitrogen in the foliage of Douglas fir is 6.5 yr.

These results clearly show that nitrogen fixation by alder results in major nitrogen accretion to these ecosystems. It is also evident that nitrogen cycling is quite different in alder than in the adjacent Douglas fir ecosystem partly because of this added nitrogen. This cycling difference could be attributed in part to differences between a coniferous and a deciduous species, as we have already mentioned. To contrast differences between two deciduous species, of which only one fixes nitrogen, we compared the accumulation and cycling in alder (stand No. 35) with that of a 59-yr-old beech ecosystem (stand No. 30) at Solling, West Germany. The two stands are quite similar in productivity, above-ground nitrogen accumulation, nitrogen content in the forest floor, and annual nitrogen requirement for biomass production (Table 6.16). However, the two ecosystems are strikingly different in nitrogen recycling at the time of senescence. Only 4.1 kg/ha or 3% of the nitrogen required for growth is derived by recycling nitrogen from senescing alder foliage, while 22.6 kg/ha or 20% of the nitrogen is derived from recycling in the beech foliage. As previously indicated (Table 6.8), deciduous stands in the temperate region, on average, recycle 22.9 kg/ha or 23% of the nitrogen for growth.

This lack of nitrogen recycling in alder could explain the longer period of

Table 6.16. *Effect of nitrogen fixation on accumulation and cycling. A comparison between fixing (alder) and non-fixing (beech) deciduous species*

Process	Red alder Stand no. 35	Douglas fir Stand no. 5	Beech Stand no. 30
Accumulation (kg/ha)			
Total tree	416	227	407
Foliage	100	65	97
Forest floor	887	178	815
Cycling (kg/ha/yr)			
Uptake	115	34	91
Requirement	119	42	114
Returned	94	22	75
Recycled	4	8	23
% Recycled	3	19	20

autumnal foliage retention. Since this species does not need to conserve nitrogen through a recycling process it can afford to retain its leaves well into the winter months when they finally freeze. Alder thus has a significantly longer season for photosynthesis than is typically available to deciduous species.

Elemental losses by leaching

Elemental losses through the leaching process were studied at a limited number of IBP sites. Two quite different procedures were used: (1) Soil leachates were collected with lysimeters placed below the rooting zone (stands No. 6, 8, 11, 22, 32, 35), and (2) stream drainage waters were collected from a defined small watershed unit in which the stand is located (stands No. 7, 13, 14, 23, 24, 25). These two techniques for assessing losses are not directly comparable. The lysimeter technique will collect only soil solutions that reach the assigned depth in which the lysimeters have been placed. Stream collections from the small watershed will collect not only solutions leached beyond the rooting zone, but also those elements added to these solutions by bedrock weathering and surface erosional processes. Consequently, stream collections will typically have higher elemental concentrations than those collected by lysimeters (Table 6.17). This is especially true for most of the base elements that are involved in soil and geologic weathering such as calcium, magnesium, iron, manganese and potassium. Since nitrogen is not readily involved in mineralogical weathering processes, concentrations found below the rooting zone are not appreciably different

Table 6.17. *Elemental losses through leaching and stream runoff at the IBP sites*

Sites	Stand no.	Elements (kg/ha/yr)				
		Nitrogen	Potassium	Calcium	Phosphorus	Magnesium
		Deep leaching				
Douglas fir (100 cm)[a]	6	0.6	1.0	4.5	0.02	—
Picea abies (100 cm)	8	0.9	2.2	2.3	0.06	0.45
Spruce (50 cm)	11	14.9	2.1	13.5	0.02	3.7
Yellow poplar (60 cm)	22	3.5	8.9	44.5	0.05	—
Beech (50 cm)	32	6.0	2.9	12.7	0.10	3.7
Oak–Birch	33	12.6	8.3	59.8	0.2	6.0
Red alder (100 cm)	35	1.7		2.2		
		Watershed runoff				
H. J. Andrews	7	1.7	9.7	121.8	0.6	10.7
Coweeta	14	0.2	4.5	5.9	—	—
Hubbard Brook	27	3.9	1.9	13.7	0.01	3.1
Walker Branch	13, 23, 24, 25	1.8	6.8	147.5	0.02	77.1

[a] Depth of leachate collection.

from those in a drainage stream. If anything, there seems to be a tendency for nitrogen concentrations to be lower in runoff waters because of nitrogen immobilization by aquatic biological processes.

These results indicate that most ecosystems receive more nitrogen from atmospheric addition (average of 8.8 kg/ha/yr) than is lost through soil leaching (4.3 kg/ha/yr). This net balance of nitrogen (4.5 kg/ha/yr) is perhaps misleading due to the high nitrogen inputs experienced at a few of the sites. However, only one site (stand No. 33, Merlewood in Great Britain) showed a net deficit. At all other sites where both input and losses were measured, there was a positive net balance of nitrogen. This was true even at those sites where only very small increments of nitrogen were being added each year, such as the USSR site, and the Cedar River and H. J. Andrews sites in the USA. Losses were correspondingly smaller, leaving a net positive accumulation.

Mechanisms by which nitrogen and other ions are conserved in forest ecosystems have been discussed frequently. The important role played by vegetative uptake cannot be minimized (Vitousek *et al.*, 1979). In addition, the soil also has a critical role in retaining this element as long as it remains in the form of ammonium ions (Cole, Crane & Grier, 1975). Transformation of ammonia to nitrate will, however, release nitrogen from the soil exchange sites and increase its potential for leaching. Without active uptake, nitrate will readily leach through the soil such as was reported for the Hubbard Brook watershed (Bormann *et al.*, 1974).

Leaching losses of the base elements occur as one would expect (Table 6.17). At those sites where the bedrock material is high in calcium, leaching of calcium is also high. This is seen in both the Coweeta and Walker Branch watershed studies. The leaching losses of phosphorus are nearly always low, both within the soil and drainage waters. This is probably due to the high phosphorus sorption capacity associated with nearly all soils.

Conclusions

We have attempted here to tabulate, compare, and summarize the elemental studies conducted under IBP sponsorship. Thirty-two forest stands from 14 sites are included. Since the sites are not uniformly distributed, they do not cover the major regions and forest ecosystems in a systematic manner. This has, of course, limited the possibilities of synthesizing the data collected.

At the beginning of this chapter we stated that the most significant aspect of our contribution would be the collection and summarization of the information on elemental cycling that resulted from the IBP. In addition, we also analyzed this information to see whether new insights would emerge to help us better to understand ecosystem behavior. From the synthesis of these data we were also able to test, modify, clarify, and expand some ideas previously developed and discussed in the literature. Some of the more significant observations of our synthesis follow.

1. The cycling of elements is not stable over the life of an ecosystem, rather it changes dramatically during this period. Thus, it is critical to know the development stage of an ecosystem before reaching conclusions about its cycling properties.
2. The cycling of different elements within an ecosystem is markedly different. Apparently those elements present in deficient amounts are cycled far more efficiently than those present in excess. Ecosystems seem to have evolved strategies for efficiently recycling or using those elements that are in short supply. In most of the sites studied, nitrogen appears to be present at deficiency levels.
3. On average, 45% of organic matter and 7% of nitrogen is held in the above-ground tree components of an ecosystem. These percentages tend to decrease at higher latitudes and increase at lower latitudes.

4. The mean residence time of elements in the forest floor varies widely between regions. Turnover periods are longer in the boreal region and significantly shorter in the temperate and Mediterranean regions. Coniferous forest floors have longer turnover periods than deciduous forest floors.

5. The rate of uptake and requirement is significantly higher in deciduous forests than in coniferous forests for all elements except phosphorus.

6. Deciduous species translocate significantly more nitrogen from older foliage before litterfall than do conifers. Very little potassium is translocated for either coniferous or deciduous species. Neither calcium nor magnesium is translocated, rather the uptake of these two elements greatly exceeds annual requirement for both coniferous and deciduous species.

7. In coniferous species, uptake and requirement of nitrogen and potassium are strongly correlated with biomass production. The correlation is somewhat weaker for deciduous species in regards to nitrogen uptake and requirement. There is little, if any, correlation between production, cycling, and the other elements studied for either deciduous or coniferous species.

8. Coniferous species are consistently more efficient than deciduous species in producing biomass with equal amounts of nitrogen uptake. Similarly, species in the boreal sites are more efficient than those in the temperate region. Apparently, the efficiency of production per unit of nitrogen uptake increases as nitrogen becomes more limiting.

9. The atmospheric addition of nitrogen to forest ecosystems varies greatly between sites. In many areas it represents as little as 5% of the average annual uptake by the vegetation. However, at sites adjacent to industrial and population centers this value may be as much as 32% of the average annual uptake. Nitrogen deficiencies seem to be minimized in those areas that receive large quantities of atmospheric nitrogen.

10. The fixation of nitrogen by alder causes a significant increase in nitrogen accumulation. It also results in other changes in the elemental cycle, including an increase in nitrogen uptake and a decrease in amount of nitrogen recycled by foliage.

11. In general, more nitrogen is added annually by way of precipitation than is lost in drainage waters from the forest ecosystems considered in our synthesis. On average, atmospheric additions are twice as large as losses.

Appendix. Elemental cycling data for 32 IBP stands

Stand No. 1
Site. Black Spruce Muskeg, Site 1, Alaska, USA
Investigator(s): Van Cleve
Forest type: Black spruce forest (*Picea mariana*) Muskeg
Geology: Parent material is loess: alkaline in reaction
Soil type: Pergelic cryaquept
Stand age: 51 years
Annual precipitation: 268.8 mm
Mean annual temperature: $-3.4°C$
Length growing season: 60–80 days
Altitude: 166.7 m
Latitude: 64° N
Institution: Univ. Alaska, USA

		Elements				
	Org. Mat.	N	K	Ca	P	Mg
Amounts (kg/ha)						
Overstory foliage total	5012.0	21.0	9.8	3.5	2.8	2.9
Overstory branches	3590.0	7.4	3.9	1.8	1.1	1.8
Overstory boles	7995.0	11.5	10.3	24.2	1.6	2.2
Overstory roots	12457.0	18.5	13.7	21.4	2.9	3.1
Understory vegetation	8639.0	66.7	20.0	31.7	7.6	11.3
Forest litter layer	88664.0	486.0	59.7	395.6	105.6	95.7
Soil-rooting zone	3951.0	689.0	124.0	1691.6	2.0	853.8
Increments (kg/ha/yr)						
Overstory foliage	233	2.1	1.2	0.6	0.2	0.2
Overstory branches	109	0.22	0.11	0.58	0.03	0.05
Overstory boles	285	0.45	0.38	0.83	0.05	0.08
Overstory roots						
Understory vegetation						
Fluxes (kg/ha/yr)						
Atmosphere precipitation						
Atmosphere particulates						
Overstory litterfall	143	1.1	0.03	0.11	1.4	0.09
Leaf wash (approx)		0.8	0.83	1.14	0.1	0.14
Stem flow						
Leaching–forest floor						
Leaching–rooting depth						
Leaching–watershed						

Stand No. 2
Site: Black Spruce Muskeg, Site 2, Alaska, USA
Investigator(s): Van Cleve
Forest type: Black spruce forest (*Picea mariana*) Muskeg
Geology: Bedrock: Birch Creek schist: Loessic soil
Soil type: Pergelic cryaquept
Stand age: 55 yr
Annual precipitation: 286.8 mm
Mean annual temperature: −3.4°C
Length growing season: 60–80 days
Altitude: 469.7 m
Latitude: 64°N
Institution: Univ. Alaska, USA

		Elements				
	Org. Mat.	N	K	Ca	P	Mg
Amounts (kg/ha)						
Overstory foliage total	5310.0	26.4	13.7	53.3	3.6	2.6
Overstory branches	4691.0	19.2	5.4	17.9	1.7	3.1
Overstory boles	14017.0	49.0	12.2	48.7	8.9	4.8
Overstory roots	10401.0	39.5	41.2	88.7	5.1	10.4
Understory vegetation	6283.0	50.8	15.3	22.0	5.2	6.4
Forest litter layer	133260.0	656.5	145.4	231.3	152.1	224.5
Soil-rooting zone	35350.0	2198.0	169.0	609.0	5.4	121.0
Increments (kg/ha/yr)						
Overstory foliage	484	4.3	3.0	1.4	0.6	0.6
Overstory branches	220	0.6	0.2	0.7	0.1	0.1
Overstory boles	698	3.3	0.6	2.2	0.5	0.2
Overstory roots						
Understory vegetation						
Fluxes (kg/ha/yr)						
Atmosphere precipitation						
Atmosphere particulates						
Overstory litterfall	290	1.6	0.26	4.0	0.12	0.17
Leaf wash (approx)		0.8	0.83	1.1	0.09	0.14
Stem flow						
Leaching–forest floor						
Leaching–rooting depth						
Leaching–watershed						

377

Dynamic properties of forest ecosystems

Stand No. 3
Site: Black Spruce–Feather Moss Site, Alaska, USA
Investigator(s): Van Cleve
Forest type: Black spruce forest, (*Picea mariana*) Feather moss
Geology: Bedrock: Birth Creek schist, basic in reaction
Soil type: Pergelic cryaquept
Stand Age: 130 yr
Annual precipitation: 268 mm
Mean annual temperature: −3.4°C
Length growing season: 60–80 days
Altitude:
Latitude: 64° N
Institution: Univ. Alaska, USA

	Org. Mat.	N	K	Ca	P	Mg
			Elements			
Amounts (kg/ha)						
Overstory foliage total	14196.0	70.8	17.6	163.3	8.5	15.9
Overstory branches	12947.0	49.0	6.2	71.0	6.1	10.7
Overstory boles	86046.0	95.1	51.8	390.7	14.3	33.4
Overstory roots	51697.0	70.8	24.5	97.5	7.6	11.4
Understory vegetation	7598.4	46.7	24.4	28.4	6.9	7.4
Forest litter layer	119235.0	709.7	122.1	452.2	87.4	99.3
Soil-rooting zone	47490.0	2362.0	286.0	5052.0	4.4	912.0
Increments (kg/ha/yr)						
Overstory foliage	147.0	1.10	0.24	0.44	0.17	0.13
Overstory branches	259.0	0.64	0.80	0.85	0.09	0.12
Overstory boles	1184.0	1.52	0.72	1.69	0.18	0.35
Overstory roots						
Understory vegetation						
Fluxes (kg/ha/yr)						
Atmosphere precipitation						
Atmosphere particulates						
Overstory litterfall	534.3	3.6	0.7	4.1	0.4	0.3
Leaf wash		0.8	0.83	1.14	0.09	0.1
Stem flow						
Leaching–forest floor						
Leaching–rooting depth						
Leaching–watershed						

378

Stand No. 4
Site: Birch Site, Alaska, USA
Investigator(s): Van Cleve
Forest type: Paper birch (*Betula papyrifera*)
Geology: Bedrock: Birch Creek schist
Soil type: Pergelic cryaquept, silt loam
Stand Age: 50 yr
Annual precipitation: 268 mm
Mean annual temperature: −3.4°C
Length growing season: 60–80 days
Altitude:
Latitude: 64° N
Institution: Univ. Alaska, USA

		Elements				
	Org. Mat.	N	K	Ca	P	Mg
Amounts (kg/ha)						
Overstory foliage total	2362.8	51.1	22.8	14.9	5.2	8.2
Overstory branches	11025.2	39.8	17.1	37.9	4.9	5.7
Overstory boles	83954.9	130.0	64.3	110.9	9.4	24.2
Overstory roots	44297.0	131.1	43.3	99.3	15.4	20.6
Understory vegetation	95.0	1.0	44.8	134.2	15.5	23.5
Forest litter layer	68772.0	548.0	99.0	489.0	79.0	139.0
Soil-rooting zone	280100.0	2879.0	186.0	7545.7	26.2	2317.1
Increments (kg/ha/yr)						
Overstory foliage	2362.8	51.1	22.8	14.9	5.2	8.2
Overstory branches	386.9	1.3	0.7	1.2	0.2	0.2
Overstory boles	2414.0	3.5	2.2	2.2	0.3	0.6
Overstory roots	2375.3	25.7	11.8	8.4	2.6	4.1
Understory vegetation						
Fluxes (kg/ha/yr)						
Atmosphere precipitation	37.21	2.1	0.16	0.98	0.09	0.06
Atmosphere particulates						
Overstory litterfall	2645.7	18.0	8.3	34.5	5.0	9.2
Leaf wash	30.33	2.1	1.46	1.01	0.20	0.55
Stem flow	3.42	0.1	0.11	0.03	0.006	0.01
Leaching–forest floor						
Leaching–rooting depth						
Leaching–watershed						

Dynamic properties of forest ecosystems

Stand No. 5
Site: Thompson Research Center, Washington, USA
Investigator(s): Turner
Forest type: Douglas-fir plantation (*Pseudotsuga menziesii*)
Geology: River terraces, glacial outwash
Soil type: Typic haplorthod, Everett gravelly sandy loam
Stand age: 22 yr
Annual precipitation: 1360 mm
Mean annual temperature: 9.8°C
Length growing season: 214 days
Altitude: 210 m
Latitude: 47° 23′N
Institution: Univ. Washington, USA

		Elements				
	Org. Mat.	N	K	Ca	P	Mg
Amounts (kg/ha)						
Overstory foliage total	4995	65.1	42.6	27.0	12.5	5.8
Overstory branches	8162	17.6	40.3	34.7	7.1	7.9
Overstory boles	113338	144.6	49.0	140.6	92.6	15.3
Overstory roots						
Understory vegetation	7638	66.6	59.4	68.7	8.7	18.5
Forest litter layer	20540	177.7	37.8	114.0	25.6	54.0
Soil-rooting zone						
Increments (kg/ha/yr)						
Overstory foliage	2100	29.8	21.4	7.0	4.6	2.9
Overstory branches	540	2.2	3.0	3.4	0.5	0.6
Overstory boles	6130	9.8	6.6	2.9	0.8	3.3
Overstory roots						
Understory vegetation						
Fluxes (kg/ha/yr)						
Atmosphere precipitation		1.7	2.2	2.2	2.3	0.5
Atmosphere particulates						
Overstory litterfall	2836	18.8	9.7	22.9	3.2	4.4
Leaf wash	Data included in stem flow					
Stem flow		2.9	7.0	5.2	1.8	1.5
Leaching–forest floor						
Leaching–rooting depth						
Leaching–watershed						

Stand No. 6
Size: Thompson Research Center, Washington, USA
Investigator(s): Turner, Cole
Forest type: Douglas-fir plantation (*Pseudotsuga menziesii*)
Geology: River terraces, glacial outwash
Soil type: Typic haplorthod, Everett gravelly sandy loam
Stand age: 42 yr
Annual precipitation: 1360 mm
Mean annual temperature: 9.8°C
Length growing season: 214 days
Altitude: 210 m
Latitude: 47° 23'N
Institution: Univ. Washington, USA

	Org. Mat.	Elements				
		N	K	Ca	P	Mg
Amounts (kg/ha)						
Overstory foliage total	9440	98	66.7	73.1	21.1	14.6
Overstory branches	13720	49	28.7	70.5	9.3	7.4
Overstory boles	106240	169	110.0	214.5	24.1	31.1
Overstory roots						
Understory vegetation	3390	21	24.1	25.1	2.7	5.7
Forest litter layer	26670	223	44.1	129.5	30.3	53.7
Soil-rooting zone		2476	209.5	661.0		97.8
Increments (kg/ha/yr)						
Overstory foliage	2440	26	19.4	9.2	5.2	3.0
Overstory branches	480	4	2.7	1.9	0.3	0.7
Overstory boles	3650	6	4.0	1.8	0.5	2.2
Overstory roots						
Understory vegetation						
Fluxes (kg/ha/yr)						
Atmosphere precipitation		1.7	2.5	0.3	2.3	0.5
Atmosphere particulates						
Overstory litterfall	5607	25.4	12.5	40.5	3.3	6.0
Leaf wash		0.5	10.3	5.2	0.8	0.3
Stem flow		0.03	0.25	0.18	0.01	0.03
Leaching–forest floor		7.3	14.6	24.0	2.2	3.6
Leaching–rooting depth (100 cm)		0.6	1.0	4.5	0.02	
Leaching–watershed						

Dynamic properties of forest ecosystems

Stand No. 7
Site: Andrews Experimental Forest, Watershed 10, Oregon, USA
Investigator(s): Grier
Forest type: Douglas-fir forest (*Pseudotsuga menziesii*)
Geology: Miocene tuffs and breccias
Soil type: Inceptosol
Stand age: 450 yr
Annual precipitation: 2250 mm
Mean annual temperature: 8.5°C
Length growing season: 150 days
Altitude: 430–670 m
Latitude: 44° 15'N
Institution: Univ. Oregon, USA

	Org. Mat.	Elements				
		N	K	Ca	P	Mg
Amounts (kg/ha)						
Overstory foliage total	14120	147	81	102	33	16
Overstory branches	54220	70	60	184	17	15
Overstory boles	733970	349	48	401	36	74
Overstory roots	172800	140	50	225	12	52
Understory vegetation	5600	14	10	2	2	
Forest litter layer	218520	445	80	619	62	160
Soil-rooting zone	120000	4560	660	2040	34	560
Increments (kg/ha/yr)						
Overstory foliage	3090	33.1	26	12	6.8	3.4
Overstory branches	410	0.5	0.5	1.4	0.1	0.1
Overstory boles	2580	1.2	0.2	4.5	0.2	0.2
Overstory roots						
Understory vegetation						
Fluxes (kg/ha/yr)						
Atmosphere precipitation		2.0	1.2	3.1	0.3	1.2
Atmosphere particulates						
Overstory litterfall	6138	18.8	7.3	40.4	4.5	3.3
Leaf wash		3.2	13.2	7.0	1.0	2.0
Stem flow			Negligible			
Leaching–forest floor						
Leaching–rooting depth						
Leaching–watershed		1.7	9.7	121.8	0.6	10.7

Stand No. 8
Site: Southern Karelian Spruce, Site 10, Karelia, USSR
Investigator(s): Kazimirov, Morozova
Forest type: Spruce forest (*Picea abies*)
Geology: Sand moraine
Soil type: Humus iron podsol
Stand age: 45 yr
Annual precipitation: 650 mm
Mean annual temperature: 2.2°C
Length growing season: 150 days
Altitude: 140 m
Latitude: 62° N
Institution: Karelian Branch of the
Academy of Sciences
Petrozavodsk USSR

		Elements				
	Org. Mat.	N	K	Ca	P	Mg
Amounts (kg/ha)						
Overstory foliage total	9800	101.9	50.0	95.1	11.8	13.7
Overstory branches	12100	55.7	15.7	44.8	4.8	6.0
Overstory boles	56300	123.9	39.4	101.3	11.2	11.2
Overstory roots	15800	63.2	23.7	44.2	7.9	6.3
Understory vegetation	1600	22.1	12.0	10.5	3.1	2.0
Forest litter layer	19200					
Soil-rooting zone						
Increments (kg/ha/yr)						
Overstory foliage	2900	29.9	13.8	18.6	2.7	2.6
Overstory branches	190	2.7	0.8	2.2	0.2	0.3
Overstory boles	2060	6.3	2.0	5.1	0.6	0.6
Overstory roots	520	4.2	1.6	2.9	0.5	0.4
Understory vegetation	20	5.4	3.0	2.6	0.7	0.5
Fluxes (kg/ha/yr)						
Atmosphere precipitation		1.1	1.0	2.3		0.9
Atmosphere particulates						
Overstory litterfall	3700	23.8	10.6	28.3	2.6	4.2
Leaf wash		Data included in stem flow				
Stem flow		1.6	1.9	2.5	0.3	
Leaching–forest floor						
Leaching–rooting depth		0.87	2.2	2.3	0.06	0.45
Leaching–watershed						

Stand No. 9
Site: Kongalund Spruce Site, Sweden
Investigator(s): Nihlgard
Forest type: Spruce plantation (*Picea abies*)
Geology: Cambria shales and sandstone, stony-sandy moraine
Soil type: Brown forest soil (acid)
Stand age: 60 yr
Annual precipitation: 800 mm
Mean annual temperature: 7°C
Length growing season: 230 days
Altitude: 120 m
Latitude: 55° 59′N
Institution: Univ. Lund
Sweden

		Elements				
	Org. Mat.	N	K	Ca	P	Mg
Amounts (kg/ha)						
Overstory foliage total	1800	220	122	84	21.8	9.6
Overstory branches	25000	230	112	86	30.2	18.2
Overstory boles	262000	270	172	283	28.5	38.9
Overstory roots	59000	90	65	45	5.7	10.5
Understory vegetation	10	0.001	0.02	0.1	0.01	0.02
Forest litter layer	18500	245	15	47.9	15.4	7.5
Soil-rooting zone	207000	6900	65	150	146	31
Increments (kg/ha/yr)						
Overstory foliage	3900	53	34.5	0.7	7.5	2.9
Overstory branches						
Overstory boles	9900	14	8.5	12.0	1.6	1.9
Overstory roots						
Understory vegetation	10					
Fluxes (kg/ha/yr)						
Atmosphere precipitation		· 8.2	1.9	3.5	0.1	0.9
Atmosphere particulates						
Overstory litterfall	5720	58	10.7	19.8	4.8	3.1
Leaf wash		Data included in stem flow				
Stem flow		16	25.2	13.9	0.4	5.2
Leaching–forest floor						
Leaching–rooting depth						
Leaching–watershed						

Stand No. 10
Site: Solling Project, Site F 3, Federal Republic of Germany
Investigator(s): Ulrich, Ellenberg
Forest type: Spruce plantation (*Picea abies*)
Geology: Buntsandstein
Soil type: Brown forest soil (acid)
Stand age: 34 yr
Annual precipitation: 1063 mm
Mean annual temperature: 5.9°C
Length growing season: 132 days
Altitude: 390 m
Latitude: 51° 45′N
Institution: Univ. Göttingen, Fed. Repbl
         Germany

	Org. Mat.	N	K	Ca	P	Mg
			Elements			
Amounts (kg/ha)						
Overstory foliage total	18870	247.8	132.2	60.5	29.6	17.9
Overstory branches	18730	99.4	111.9	35.3	22.8	21.2
Overstory boles	105100	102.2	105.1	195.4	6.3	22.5
Overstory roots	34560					
Understory vegetation						
Forest litter layer	52000	1430	105	72	199	44
Soil-rooting zone	190000	6650	340	170	2660	36
Increments (kg/ha/yr)						
Overstory foliage	2979	39.0	19.4	9.5	4.6	2.8
Overstory branches	627	3.3	3.8	1.2	0.8	0.7
Overstory boles	4895	4.8	4.9	9.1	0.3	1.0
Overstory roots	1593					
Understory vegetation						
Fluxes (kg/ha/yr)						
Atmosphere precipitation		21.8	3.7	12.6	0.5	2.6
Atmosphere particulates						
Overstory litterfall	2924	41.5	1.2	14.9	3.5	1.7
Leaf wash	Data included in stem flow					
Stem flow		3.7			0.2	
Leaching–forest floor						
Leaching–rooting depth						
Leaching–watershed						

Dynamic properties of forest ecosystems

Stand No. 11
Site: Solling Project, Site F 1, Federal Republic of Germany
Investigator(s): Ulrich, Ellenberg
Forest type: Spruce plantation (*Picea abies*)
Geology: Buntsandstein
Soil type: Brown forest soil (acid)
Stand age: 87 yr
Annual precipitation: 1063 mm
Mean annual temperature: 5.9°C
Length growing season: 132 days
Altitude: 505 m
Latitude: 51° 49′ N
Institution: Univ. Göttingen, Fed. Repbl. Germany

		Elements				
	Org. Mat.	N	K	Ca	P	Mg
Amounts (kg/ha)						
Overstory foliage total	17880	228.0	118.8	70.3	13.8	5.6
Overstory branches	28210	242.7	183.1	74.8	17.3	19.7
Overstory boles	198400	258.4	99.3	267.6	7.9	30.7
Overstory roots	71720					
Understory vegetation						
Forest litter layer	49000	960	41	83.0	53.0	20.0
Soil-rooting zone	190000	7100	546	283	2630	35
Increments (kg/ha/yr)						
Overstory foliage	2901	37.0	19.3	11.4	3.1	0.9
Overstory branches	603	5.2	3.9	1.6	0.4	0.4
Overstory boles	5360	7.0	2.7	7.2	1.0	0.3
Overstory roots						
Understory vegetation						
Fluxes (kg/ha/yr)						
Atmosphere precipitation		21.8	3.7	12.6	0.51	2.6
Atmosphere particulates						
Overstory litterfall	3393	47.1	12.7	17.0	4.0	1.9
Leaf wash		3.7	23.6	19.0	0.2	0.2
Stem flow	Data included in leaf wash					
Leaching–forest floor						
Leaching–rooting depth (50 cm)		14.9	2.1	13.5	0.02	3.7
Leaching–watershed						

Stand No. 12
Site: Solling Project, Site F 2, Federal Republic of Germany
Investigator(s): Ulrich, Ellenberg
Forest type: Spruce plantation (*Picea abies*)
Geology: Buntsandstein
Soil type: Brown forest soil (acid)
Stand age: 115 yr
Annual precipitation: 1063 mm
Mean annual temperature: 5.9°C
Length growing season: 132 days
Altitude: 440 m
Latitude: 51° 44′ N
Institution: Univ. Göttingen, Fed. Repbl. Germany

		Elements				
	Org. Mat.	N	K	Ca	P	Mg
Amounts (kg/ha)						
Overstory foliage total	12660	161.4	84.1	49.7	13.3	40.0
Overstory branches	24630	212	160.0	65.3	15.1	17.2
Overstory boles	195700	255	98.0	264.1	36.9	30.3
Overstory roots	74930					
Understory vegetation						
Forest litter layer	111000	2260	115	258	98	46
Soil-rooting zone	251000	7060	342	160	1530	38
Increments (kg/ha/yr)						
Overstory foliage	2123	27.0	14.1	8.3	2.2	0.7
Overstory branches	390	3.4	2.5	1.0	0.2	0.3
Overstory boles	4004	5.2	2.0	5.4	0.8	0.6
Overstory roots	853					
Understory vegetation						
Fluxes (kg/ha/yr)						
Atmosphere precipitation		21.8	3.7	12.6	0.5	2.6
Atmosphere particulates						
Overstory litterfall	3076	43.3	1.2	15.6	3.7	1.7
Leaf wash		3.7			0.2	
Stem flow	Data included in leaf wash					
Leaching–forest floor						
Leaching–rooting depth						
Leaching–watershed						

Dynamic properties of forest ecosystems

Stand No. 13
Site: Walker Branch Site 2, Oak Ridge, Tennessee, USA
Investigator(s): Harris
Forest type: Short leaf pine forest (*Pinus echinata*)
Geology: Knox dolomite
Soil type: Typic paleudults derived from dolomitic residuum
Stand age: 30 yr
Annual precipitation: 1400 mm
Mean annual temperature: 13.3°C
Length growing season: 180 days
Altitude: 265–360 m
Latitude: 35° 58′ N
Institution: Oak Ridge National Laboratory, USA

		Elements				
	Org. Mat.	N	K	Ca	P	Mg
Amounts (kg/ha)						
Overstory foliage total	4600	51	36	43	4	12
Overstory branches	27000	64	29	100	7	11
Overstory boles	89000	100	60	164	7	17
Overstory roots	34000	117	128	187	21	17
Understory vegetation						
Forest litter layer	27000	290	21	256	18	23
Soil-rooting zone	116000	4100	38000	3600	1200	10900
Increments (kg/ha/yr)						
Overstory foliage	3728	41.5	29.3	34.8	3	9
Overstory branches						
Overstory boles	4150	8.2	1.9	5.4	0.1	0.4
Overstory roots						
Understory vegetation						
Fluxes (kg/ha/yr)						
Atmosphere precipitation		8.7	1.0	9.1	0.06	1.0
Atmosphere particulates			2.2	5.3	0.48	1.1
Overstory litterfall	4130	37.5	14.4	51.0	2.5	7.6
Leaf wash		2.9	18.7	17.4	0.3	2.8
Stem flow	Data included in leaf wash					
Leaching–forest floor						
Leaching–rooting depth						
Leaching–watershed		1.8	6.8	147.5	0.02	77.1

Stand No. 14
Site: Watershed 1, Coweeta, North Carolina, USA
Investigator(s): Swank
Forest type: White pine plantation (*Pinus strobus*)
Geology: Granite, mica schists, and gneisses (bedrock)
Soil type: Saluda stony loam (typic hapludult, loamy, mixed, mesic, shallow family)
Stand age: 15 yr
Annual precipitation: 1628 mm
Mean annual temperature: 13.6°C
Length growing season: 150 days
Altitude: 706–988 m
Latitude: 35° 04′ N
Institution: Coweeta Hydrologic Laboratory,
        USA

		Elements				
	Org. Mat.	N	K	Ca	P	Mg
Amounts (kg/ha)						
Overstory foliage total	4664	72.80	22.7	17.6		
Overstory branches	22825	78.80	45.4	47.5		
Overstory boles	42110	72.40	40.9	32.6		
Overstory roots	60300	422.10	163.2	203.6		
Understory vegetation						
Forest litter layer						
Soil-rooting zone						
Increments (kg/ha/yr)						
Overstory foliage	3630	60.90	15.4	14.0		
Overstory branches	3050	10.80	6.6	6.1		
Overstory boles	6850	12.70	8.0	5.5		
Overstory roots	6900	48.30	18.6	23.5		
Understory vegetation						
Fluxes (kg/ha/yr)						
Atmosphere precipitation		5.50	2.0	4.1		
Atmosphere particulates		3.30				
Overstory litterfall	3184	26.50	5.5	19.2		
Leaf wash			28.0	2.0		
Stem flow	Data included in leaf wash					
Leaching–forest floor			41.9	31.2		
Leaching–rooting depth						
Leaching–watershed		0.19	4.5	5.9		

Dynamic properties of forest ecosystems

Stand No. 15
Site: JPTF–71 Yusuhara Takatoriyama, Japan
Investigator(s): Ando
Forest type: True fir forest (*Abies firma*)
Geology: Cretaceous/Mudstone
Soil type: Brown forest soil (moderately wet)
Stand age: 97–145 yr
Annual precipitation: 2748 mm
Mean annual temperature: 13.6°C
Length growing season:
Altitude: 420 m
Latitude: 33° 20′ N
Institution: Shikoku Branch, Government Forest
　　　　　Experiment Station, Kochi, Japan

		Elements				
	Org. Mat.	N	K	Ca	P	Mg
Amounts (kg/ha)						
Overstory foliage total	15134	145.7	157.4	125.6	12.3	14.4
Overstory branches	57760	173.7	255.8	315.5	22.5	33.8
Overstory boles	306668	254.0	1012.3	961.0	46.9	81.7
Overstory roots	145647					
Understory vegetation	121957	363.4	436.9	851.6	40.0	52.4
Forest litter layer	4279	484	113	182	32	67
Soil-rooting zone		3389	368	278	5	143
Increments (kg/ha/yr)						
Overstory foliage	3156	32.9	55.5	20.4	3.4	4.1
Overstory branches	1300	9.4	20.3	9.6	1.6	2.7
Overstory boles	3909	2.6	12.4	10.9	0.6	1.0
Overstory roots						
Understory vegetation	3309	2.9	32.6	30.3	2.4	4.4
Fluxes (kg/ha/yr)						
Atmosphere precipitation						
Atmosphere particulates						
Overstory litterfall	5376	31.6	11.4	37.0	1.9	4.7
Leaf wash						
Stem flow						
Leaching–forest floor						
Leaching–rooting depth						
Leaching–watershed						

Stand No. 16
Site: JPTF–70 Yusuhara Kubotaniyama, Japan
Investigator(s): Ando
Forest type: Hemlock forest (*Tsuga sieboldii*)
Geology: Cretaceous/Sandstone
Soil type: Brown forest soil (moderately wet)
Stand age: 120–443 yr
Annual precipitation: 2748 mm
Mean annual temperature: 13.6°C
Length growing season:
Altitude: 720 m
Latitude: 33° 20′ N
Institution: Shikoku Branch, Government Forest
　　　　　　Experiment Station, Kochi, Japan

		Elements				
	Org. Mat.	N	K	Ca	P	Mg
Amounts (kg/ha)						
Overstory foliage total	7846	84.7	46.7	39.7	5.3	11.7
Overstory branches	92345	205.7	105.4	430.7	13.4	5.1
Overstory boles	348410	291.8	289.1	463.4	6.0	45.8
Overstory roots	165565					
Understory vegetation	118318	281.1	220.5	283.1	14.7	46.6
Forest litter layer	50780	506.8	76.3	159.0	34.1	30.0
Soil-rooting zone		2732	62	49	1.3	21
Increments (kg/ha/yr)						
Overstory foliage	2390	26.6	14.9	8.7	1.8	3.5
Overstory branches	767	6.0	5.6	3.3	0.6	0.7
Overstory boles	1912	1.4	1.5	1.5	0.03	0.2
Overstory roots						
Understory vegetation	2968	25.0	18.8	15.6	1.5	5.1
Fluxes (kg/ha/yr)						
Atmosphere precipitation						
Atmosphere particulates						
Overstory litterfall	4686	20.2	6.8	24.8	1.5	3.6
Leaf wash						
Stem flow						
Leaching–forest floor						
Leaching–rooting depth						
Leaching–watershed						

Dynamic properties of forest ecosystems

Stand No. 17
Site: Cascade Head, Oregon, USA
Investigator(s): Grier
Forest type: Western hemlock forest (*Tsuga heterophylla*)
Geology: Tyee formation marine siltstone
Soil type: Utisol
Stand age: 121 yr
Annual precipitation: 2500 mm
Mean annual temperature: 10.1°C
Length growing season: 130–140 days
Altitude: 200 m
Latitude: 45° N
Institution: Cascade Head Experimental Forest, US Forest Service
             Oregon, USA

	Org. Mat.	Elements				
		N	K	Ca	P	Mg
Amounts (kg/ha)						
Overstory foliage total	8030	85	31	19	11	8
Overstory branches	50580	52	20	26	7	5
Overstory boles	856870	584	138	433	149	47
Overstory roots	189200	179	61	153	44	34
Understory vegetation	4320	16	10	8	3	2
Forest litter layer	301080	474	118	347	98	83
Soil-rooting zone	776000	34900	720	1600	70	1600
Increments (kg/ha/yr)						
Overstory foliage	2870	28.9	13.5	4.7	3.9	2.4
Overstory branches	1880	9.4	6.2	2.4	2.0	1.1
Overstory boles	4760	3.1	0.5	1.5	1.0	0.3
Overstory roots						
Understory vegetation						
Fluxes (kg/ha/yr)						
Atmosphere precipitation		5.7	2.7	5.3	0.9	13.0
Atmosphere particulates						
Overstory litterfall	6140	44.0	12.9	21.3	6.9	7.8
Leaf wash		2.0	8.0	3.3	0.3	2.1
Stem flow		0.2	2.6	0.3	0.1	0.1
Leaching–forest floor						
Leaching–rooting depth						
Leaching–watershed						

Stand No. 22
Site: Liriodendron Site, Oak Ridge, Tennessee, USA
Investigators(s): Reichle, Harris, Edwards
Forest type: Yellow popular forest, mesic (*Liriodendron tulipifera*)
Geology: Knox dolomite limestone
Soil type: Deep alluvial emory silt loam
Stand age: 50 yr
Annual precipitation: 1400 mm
Mean annual temperature: 13.3°C
Length growing season: 180 days
Altitude: 225 m
Latitude: 35° 55′ N
Institution: Oak Ridge National Laboratory,
        USA

			Elements			
	Org. Mat.	N	K	Ca	P	Mg
Amounts (kg/ha)						
Overstory foliage total	3200	53.9	52.3	35.2	6.3	
Overstory branches	27100	78.6	46.1	143.7	30.6	
Overstory boles	94400	172.0	74.8	276.8	9.7	
Overstory roots	36000	122.7	111.5	150.8	18.5	
Understory vegetation	8800	23.8	32.4	115.2	5.7	
Forest litter layer	6000	77.9	9.2	99.6	5.3	
Soil-rooting zone	159000	7650	38960	8130	2840	
Increments (kg/ha/yr)						
Overstory foliage	3200	53.9	52.3	35.2	6.3	
Overstory branches	575	1.7	1.0	3.1	0.7	
Overstory boles	2254	5.4	2.3	6.3	0.3	
Overstory roots	513	1.8	1.2	2.2	0.2	
Understory vegetation	81	0.3	0.2	0.6	0.1	
Fluxes (kg/ha/yr)						
Atmosphere precipitation		7.7	0.73	6.1	0.06	
Atmosphere particulates			2.56	4.5	0.7	
Overstory litterfall	4290	31.3	19.0	77.7	2.8	
Leaf wash		9.4	31.7	17.8	0.4	
Stem flow		0.1	1.0	0.4	0.002	
Leaching–forest floor						
Leaching–rooting depth (60 cm)		3.5	8.9	44.5	0.05	
Leaching–watershed						

Stand No. 23
Site: Walker Branch Site 4, Oak Ridge, Tennessee, USA
Investigator(s): Harris, Henderson
Forest type: Yellow poplar – mixed hardwoods forest (*Liriodendron tulipifera, Quercus*)
Geology: Knox dolomite
Soil type: Typic paleudults derived from dolomitic residuum
Stand age: 30–80 yr
Annual precipitation: 1400 mm
Mean annual temperature: 13.3°C
Length growing season: 180 days
Altitude: 265–360 m
Latitude: 35° 58′ N
Institution: Oak Ridge National Laboratory,
            USA

		Elements				
	Org. Mat.	N	K	Ca	P	Mg
Amounts (kg/ha)						
Overstory foliage total	5100	78	45	75	6	21
Overstory branches	21000	54	37	155	7	13
Overstory boles	83000	135	90	307	8	25
Overstory roots	31000	122	132	211	22	19
Understory vegetation						
Forest litter layer	1500	187	14	294	11	22
Soil-rooting zone	189000	7300	36000	6300	1400	8700
Increments (kg/ha/yr)						
Overstory foliage	5100	78	45	75	6	21
Overstory branches						
Overstory boles	3550	9.9	2.5	7.6	0.3	0.7
Overstory roots						
Understory vegetation						
Fluxes (kg/ha/yr)						
Atmosphere precipitation		8.7	1.0	9.1	0.06	1.0
Atmosphere particulates			2.2	5.3	0.48	1.1
Overstory litterfall	4330	36.2	19.1	58.3	2.7	8.3
Leaf wash		12.0	18.4	21.9	0.4	3.4
Stem flow	Data included in leaf wash					
Leaching–forest floor						
Leaching–rooting depth						
Leaching–watershed		1.8	6.8	147.5	0.02	77.1

Stand No. 24
Site: Walker Branch Site 1, Oak Ridge, Tennessee, USA
Investigator(s): Harris, Henderson
Forest type: Oak–Hickory forest (*Quercus–Carya*)
Geology: Knox dolomite
Soil type: Typic paleudults derived from dolomitic residuum
Stand age: 30–80 yr
Annual precipitation: 1400 mm
Mean annual temperature: 13.3°C
Length growing season: 180 days
Altitude: 265–360 m
Latitude: 35° 58′ N
Institution: Oak Ridge National Laboratory,
　　　　　USA

		Elements				
	Org. Mat.	N	K	Ca	P	Mg
Amounts (kg/ha)						
Overstory foliage total	5600	67	53	70	6	17
Overstory branches	26000	131	53	288	9	19
Overstory boles	90000	171	114	498	9	31
Overstory roots	33000	128	136	244	22	19
Understory vegetation						
Forest litter layer	27000	334	26	517	22	32
Soil-rooting zone	116000	4500	38000	3600	1200	10900
Increments (kg/ha/yr)						
Overstory foliage	5600	67	53	70	6	17
Overstory branches						
Overstory boles	4800	20.2	4.5	21.0	0.4	1.4
Overstory roots						
Understory vegetation						
Fluxes (kg/ha/yr)						
Atmosphere precipitation		8.7	1.0	9.1	0.06	1.0
Atmosphere particulates			2.2	5.3	0.48	1.1
Overstory litterfall	4800	36.5	19.8	49.1	2.7	8.7
Leaf wash		12.4	23.9	24.1	0.6	3.7
Stem flow	Data included in leaf wash					
Leaching–forest floor						
Leaching–rooting depth						
Leaching–watershed		1.8	6.8	147.5	0.02	77.1

Dynamic properties of forest ecosystems

Stand No. 25
Site: Walker Branch Site 3, Oak Ridge, Tennessee, USA
Investigator(s): Harris, Henderson
Forest type: Chestnut oak forest (*Quercus prinus*)
Geology: Knox dolomite
Soil type: Typic paleudults derived from dolomitic residuum
Stand age: 30–80 yr
Annual precipitation: 1400 mm
Mean annual temperature: 13.3°C
Length growing season: 180 days
Altitude: 265–360 m
Latitude: 35° 58′ N
Institution: Oak Ridge National Laboratory,
USA

	Org. Mat.	Elements				
		N	K	Ca	P	Mg
Amounts (kg/ha)						
Overstory foliage total	5300	75	52	58	5	12
Overstory branches	30000	139	61	294	10	13
Overstory boles	102000	183	129	500	11	20
Overstory roots	36000	132	140	249	22	18
Understory vegetation						
Forest litter layer	25000	298	26	318	18	22
Soil-rooting zone	116000	4700	38000	3600	1200	10900
Increments (kg/ha/yr)						
Overstory foliage	5300	75	52	58	5	12
Overstory branches						
Overstory boles	5400	23.1	6.2	22.3	0.5	1.1
Overstory roots						
Understory vegetation						
Fluxes (kg/ha/yr)						
Atmosphere precipitation		8.7	1.0	9.1	0.06	1.0
Atmosphere particulates			2.2	5.3	0.48	1.1
Overstory litterfall	4450	34.1	16.4	45.0	2.4	7.5
Leaf wash		11.6	17.7	21.5	0.4	2.9
Stem flow	Data included in leaf wash					
Leaching–forest floor						
Leaching–rooting depth						
Leaching–watershed		1.8	6.8	147.5	0.02	77.1

Stand No. 26
Site: Watershed 1, Coweeta, North Carolina, USA
Investigator(s): Swank
Forest type: Oak–Hickory forest (*Quercus-Carya*)
Geology: Granite, mica schist, and gneisses (bedrock)
Soil type: Saluda stony loam (typic hapludult, loamy, mixed, mesic, shallow family)
Stand age: 60–200 yr
Annual precipitation: 1628 mm
Mean annual temperature: 13.6°C
Length growing season: 150 days
Altitude: 706–988 m
Latitude: 35° 04′ N
Institution: Coweeta Hydrologic Laboratory,
　　　　　USA

		Elements				
	Org. Mat.	N	K	Ca	P	Mg
Amounts (kg/ha)						
Overstory foliage total	5584	95.00	46.0	49.0		
Overstory branches	25599	116.00	46.0	133.0		
Overstory boles	106313	194.00	135.0	361.0		
Overstory roots	52525	434.00	167.0	278.0		
Understory vegetation						
Forest litter layer	9500	110.00	29.0	130.0		
Soil-rooting zone						
Increments (kg/ha/yr)						
Overstory foliage	4195	83.00	41.0	36.0		
Overstory branches	901	3.40	2.1	4.7		
Overstory boles	2870	5.70	4.0	9.8		
Overstory roots	6000	43.20	16.8	27.6		
Understory vegetation						
Fluxes (kg/ha/yr)						
Atmosphere precipitation		4.90	2.1	4.8		
Atmosphere particulates		3.30				
Overstory litterfall	4369	33.90	18.1	44.5		
Leaf wash			24.1			
Stem flow	Data included in leaf wash					
Leaching–forest floor			31.7	23.4		
Leaching–rooting depth			2.1	1.9		
Leaching–watershed		3.17	5.6	7.7		

397

Stand No. 27
Site: Hubbard Brook, New Hampshire, USA
Investigator(s): Whittaker, Likens
Forest type: Northern hardwoods forest (*Acer, Betula, Fagus*)
Geology: Littleton gneiss
Soil type: Boulders, glacial till, podzolic-haplorthod
Stand age: 60 yr
Annual precipitation: 1250 mm
Mean annual temperature:
Length growing season: 110 days
Altitude: 550–710 m
Latitude: 44° 00′ N
Institution: Cornell University,
        USA

		Elements				
	Org. Mat.	N	K	Ca	P	Mg
Amounts (kg/ha)						
Overstory foliage total	3180	70.1	30.0	19.5	5.5	4.8
Overstory branches	38980	162.5	52.9	189.3	16.2	13.8
Overstory boles	91840	134.6	70.8	193.2	11.0	18.9
Overstory roots (+crowns)	28560	181	63.2	101	52.7	13.5
Understory vegetation	54	9.0	2.2	3.4	0.8	1.6
Forest litter layer	48016	125.6	66	372	7.8	38.0
Soil-rooting zone	174900					
Increments (kg/ha/yr)						
Overstory foliage	3211	74.8	32.2	21.5	5.9	5.2
Overstory branches	2249	11.3	3.8	11.0	1.2	0.9
Overstory boles	2429	3.6	1.8	5.2	0.3	0.5
Overstory roots	1644	10.5	3.7	6.0	3.1	0.8
Understory vegetation	232	4.7	3.5	1.3	0.4	0.8
Fluxes (kg/ha/yr)						
Atmosphere precipitation[a]		6.5	0.9	2.2	0.04	0.6
Atmosphere particulates						
Overstory litterfall	5860	54.2	18.3	40.7	4.0	5.9
Leaf wash	104	4.7	26.1	5.0	0.6	1.5
Stem flow	11.5	0.5	3.4	0.4	0.1	0.1
Leaching–forest floor						
Leaching–rooting depth						
Leaching–watershed[a]		3.9	1.9	13.7	0.01	3.1

[a] Long-term averages 1963–1974

Stand No. 28
Site: Virelles, Belgium
Investigator(s): Duvigneaud, Denaeyer-DeSmet
Forest type: Mixed oak forest (*Quercus*-mixed)
Geology: Devon chalk
Soil type: Calcareous brown soil
Stand age: 115–160 yr
Annual precipitation: 951.6 mm
Mean annual temperature: 8.5°C
Length growing season: 155 days
Altitude: 245 m
Latitude: 50° 4′ N
Institution: Univ. Libre de Brussels,
        Belgium

	Org. Mat.	Elements				
		N	K	Ca	P	Mg
Overstory foliage total	3500	83	40	34	6	7
Overstory branches	78300	259	139	339	23	27
Overstory boles	210050	386	219	769	17	58
Overstory roots	35300	297	112	301.7	29.3	27.7
Understory vegetation	30580	166	112	172	14	32
Forest litter layer	5600					
Soil-rooting zone	300000	13800	767	13865	2100	1007
ncrements (kg/ha/yr)						
Overstory foliage	3500	83	40	34	6	7
Overstory branches	4480	25	12	25	2	3
Overstory boles						
Overstory roots	1000	5	6	9	1	1
Understory vegetation	1100	10	3	8	1	1
`luxes (kg/ha/yr)						
Atmosphere precipitation		8.7	4	15		
Atmosphere particulates						
Overstory litterfall	5600	59	28	66	3.4	9.7
Leaf wash		2.2	9			7
Stem flow		0.2	2			
Leaching–forest floor						
Leaching–rooting depth						
Leaching–watershed						

Stand No. 29
Site: Virelles, Belgium
Investigator(s): Duvigneaud, Denaeyer-De Smet
Forest type: Mixed oak forest (*Quercus*-mixed)
Geology: Devon chalk
Soil type: Calcareous brown soil
Stand age: 80 yr
Annual precipitation: 951.6 mm
Mean annual temperature: 8.5°C
Length growing season: 155 days
Altitude: 245 m
Latitude: 50° 4' N
Institution: Univ. Brussels,
          Belgium

		Elements				
	Org. Mat.	N	K	Ca	P	Mg
Amounts (kg/ha)						
Overstory foliage total	3458	73	36	54	4.7	4.6
Overstory branches	36425	295	164	776	23	72
Overstory boles	75202					
Overstory roots	34600	121	92	372	10	19
Understory vegetation	4700	46	46	70	4.6	6
Forest litter layer	4762	44	17	107	2	5
Soil-rooting zone	220000	4480	157	13600	920	151
Increments (kg/ha/yr)						
Overstory foliage	3458	73	36	54	4.7	4.6
Overstory branches	5263	23.1	11	52.3	1.6	4.6
Overstory boles	6054					
Overstory roots	2000	7	5.4	22	0.6	1.1
Understory vegetation	874					
Fluxes (kg/ha/yr)						
Atmosphere precipitation	1	13	5	19		5.8
Atmosphere particulates						
Overstory litterfall	5287	50	21	110	2.4	5.6
Leaf wash		0.9	16	6.2	0.6	5.6
Stem flow		6	0.8	0.9	0	0.6
Leaching–forest floor						
Leaching–rooting depth						
Leaching–watershed						

Stand No. 30
Site: Solling Project, Site B4, Federal Republic of Germany
Investigator(s): Ulrich, Ellenberg
Forest type: Beech forest (*Fagus silvatica*)
Geology: Buntsandstein
Soil type: Brown forest soil (acid)
Stand age: 59 yr
Annual precipitation: 1063 mm
Mean annual temperature: 6.3°C
Length growing season: 145 days
Altitude: 430 m
Latitude: 51° 45' N
Institution: Univ. Göttingen, Federal Repbl. Germany

		Elements				
	Org. Mat.	N	K	Ca	P	Mg
Amounts (kg/ha)						
Overstory foliage total	3158	97.2	26.9	14.3	6.5	2.5
Overstory branches	41500	105	30.5	118.1	13.7	5.7
Overstory boles	110000	205	72.7	191.1	31.8	17.9
Overstory roots	23950	62.4	34.6	27.8	11.9	5.9
Understory vegetation						
Forest litter layer	29000	815	85.0	86.0	62	44
Soil-rooting zone	186000	6332	500	245	3150	45
Increments (kg/ha/yr)						
Overstory foliage	3158	97.2	26.9	14.3	6.5	2.5
Overstory branches	989	2.5	0.7	28	0.3	0.2
Overstory boles	7666	14.2	5.1	1.3	2.3	1.7
Overstory roots	1263	3.3	1.8	1.5	0.6	0.3
Understory vegetation						
Fluxes (kg/ha/yr)						
Atmosphere precipitation		21.8	3.7	12.6	0.5	2.6
Atmosphere particulates						
Overstory litterfall	3418	52.5	16.9	17.3	4.2	1.5
Leaf wash		19.6	15.9	22.0	0.4	3.3
Stem flow		2.5	6.6	4.3		0.6
Leaching–forest floor						
Leaching–rooting depth						
Leaching–watershed						

Dynamic properties of forest ecosystems

Stand No. 31
Site: Solling Project, Site B3, Federal Republic of Germany
Investigator(s): Ulrich, Ellenberg
Forest type: Beech forest (*Fagus silvatica*)
Geology: Buntsandstein
Soil type: Brown forest soil (acid)
Stand age: 80 yr
Annual precipitation: 1063 mm
Mean annual temperature: 6.1°C
Length growing season: 144 days
Altitude: 470 m
Latitude: 51° 45′ N
Institution: Univ. Göttingen, Federal Repbl. Germany

		Elements				
	Org. Mat.	N	K	Ca	P	Mg
Amounts (kg/ha)						
Overstory foliage total	3294	95.5	31.4	17.5	5.7	2.5
Overstory branches	25870	97.1	37.4	68.9	11.6	5.7
Overstory boles	129600	211.5	133.7	216.3	18.7	17.9
Overstory roots	22080	57.5	31.9	25.7	10.9	5.9
Understory vegetation						
Forest litter layer	39000	1050	123	118	72	44
Soil-rooting zone	212000	9452	550	280	2310	45
Increments (kg/ha/yr)						
Overstory foliage	3294	95.5	31.4	17.5	5.7	2.5
Overstory branches	447	1.6	0.6	1.2	0.2	0.1
Overstory boles	5907	9.6	6.1	9.7	0.8	0.8
Overstory roots	633	1.6	0.9	0.7	0.3	0.2
Understory vegetation						
Fluxes (kg/ha/yr)						
Atmosphere precipitation		21.8	3.7	12.6	6.5	2.6
Atmosphere particulates						
Overstory litterfall	4046	54.3	17.5	17.8	4.3	1.6
Leaf wash		19.6	15.9	22.0	0.4	3.3
Stem flow		2.5	6.6	4.3		0.6
Leaching–forest floor						
Leaching–rooting depth						
Leaching–watershed						

Stand No. 32
Site: Solling Project, Site B2, Federal Republic of Germany
Investigator(s): Ulrich, Ellenberg
Forest type: Beech forest (*Fagus silvatica*)
Geology: Buntsandstein
Soil type: Brown forest soil (acid)
Stand age: 122 yr
Annual precipitation: 1063 mm
Mean annual temperature: 6.1°C
Length growing season: 144 days
Altitude: 470 m
Latitude: 51° 45′ N
Institution: Univ. Göttingen, Federal Repbl. Germany

		Elements				
	Org. Mat.	N	K	Ca	P	Mg
Amounts (kg/ha)						
Overstory foliage total	3078	84.4	24.6	11.1	4.6	2.2
Overstory branches	32480	128.7	42.7	76.7	9.8	8.0
Overstory boles	238400	367.9	205.5	215.2	5.5	56.7
Overstory roots	30040	107	55.5	46	19.0	11.6
Understory vegetation	13					
Forest litter layer	29700	810	83	91	52	34
Soil-rooting zone	190000	7340	387	268	2870	40
Increments (kg/ha/yr)						
Overstory foliage	3078	84.4	24.6	11.1	4.6	2.2
Overstory branches	784	3.1	1.0	1.9	0.6	0.2
Overstory boles	6490	1.0	5.6	5.8	1.5	1.5
Overstory roots	660	2.3	1.3	1.0	0.4	0.2
Understory vegetation	13					
Fluxes (kg/ha/yr)						
Atmosphere precipitation		21.8	3.7	12.6	0.5	2.6
Atmosphere particulates						
Overstory litterfall	3783	49.4	15.9	16.2	4.0	1.4
Leaf wash		19.6	15.9	22	0.4	3.3
Stem flow		2.5	6.6	4.3		0.6
Leaching–forest floor						
Leaching–rooting depth (50 cm)		4.4	2.6	8.4	0.01	2.1
Leaching–watershed						

Dynamic properties of forest ecosystems

Stand No. 33
Site: Meathop Wood, United Kingdom
Investigator(s): Satchell
Forest type: Mixed deciduous (*Quercus–Betula*)
Geology: Carboniferous limestone
Soil type: Glacial drift and brown earths
Stand age: 80 yr
Annual precipitation: 1115 mm
Mean annual temperature: 7.8°C
Length growing season: 244 days
Altitude: 45 m
Latitude: 54° 12.5′ N
Institution: The Nature Conservancy, Merlewood, Research Station
      Great Britain

	Org. Mat.	Elements				
		N	K	Ca	P	Mg
Amounts (kg/ha)						
Overstory foliage total	3505	85.7	38.9	41.9	4.6	9.5
Overstory branches	33640	58.8	61.1	137.4	3.7	8.0
Overstory boles	75920	133.6	113.1	301	8.4	8.5
Overstory roots	242181	223.0	112.8	235.3	11.9	30.8
Understory vegetation	15529	8.1	42.9	66.9	5.4	9.7
Forest litter layer	6118	74.4	12	101	3.3	8.1
Soil-rooting zone	138000		1543	1766.4	1512	81.5
Increments (kg/ha/yr)						
Overstory foliage	3505	85.7	38.9	41.9	4.6	9.5
Overstory branches	948	2.1	1.5	3.9	0.1	0.3
Overstory boles	2162	4.1	3.4	8.9	0.2	0.7
Overstory roots	1978	13.2	6.6	14.3	0.6	1.7
Understory vegetation	343	3.3	2.4	2.2	0.2	0.5
Fluxes (kg/ha/yr)						
Atmosphere precipitation		5.8	3.3	6.9	0.2	5.4
Atmosphere particulates						
Overstory litterfall	3697	63.5	19	83.3	2.6	9.7
Leaf wash		8.8	30.3	22.8	0.7	11.9
Stem flow		0.4	4.8	5.2	0.03	2.4
Leaching–forest floor						
Leaching–rooting depth		12.6	8.3	59.8	0.2	6.0
Leaching–watershed						

Stand No. 34
Site: Kongalund Beech Site, Sweden
Investigator(s): Nihlgard
Forest type: Beech forest (*Fagus silvatica*)
Geology: Cambrian shales and sandstone, stony-sandy moraine
Soil type: Brown forest soil (acid)
Stand age: 45–130 yr
Annual precipitation: 880 mm
Mean annual temperature: 7°C
Length growing season: 230 days
Altitude: 120 m
Latitude: 55° 59′ N
Institution: Univ. Lund,
  Sweden

		Elements				
	Org. Mat.	N	K	Ca	P	Mg
Amounts (kg/ha)						
Overstory foliage total	5000	121	26.7	24.5	7.0	7.5
Overstory branches	99000	660	228	381	55.2	40.9
Overstory boles		290	201	201	22.4	57
Overstory roots	51000	150	79	42	15.6	13.5
Understory vegetation	200	10	8	0.9	0.7	0.1
Forest litter layer	5200	86	104	34.2	5.8	4.8
Soil-rooting zone	207000	7800	56	175	38	84
Increments (kg/ha/yr)						
Overstory foliage	5000	121	26.7	24.5	7.0	7.5
Overstory branches						
Overstory boles	10100	8.3	16.6	24.8	3.5	3.4
Overstory roots						
Understory vegetation						
Fluxes kg/ha/yr						
Atmosphere precipitation		8.2	1.9	3.5	0.1	0.9
Atmosphere particulates						
Overstory litterfall	5690	69	14.4	31.7	5	4.3
Leaf wash	Data included in stem flow					
Stem flow		1.0	11.2	6.6	0.1	2.5
Leaching–forest floor						
Leaching–rooting depth						
Leaching–watershed						

Stand No. 35
Site: Thompson Research Center, Washington, USA
Investigator(s): Turner
Forest type: Red alder (*Alnus rubra*)
Geology: Glacial till
Soil type: Typic haplorthod, alderwood gravelly sandy loam
Stand age: 30 yr
Annual precipitation: 1360 mm
Mean annual temperature: 9.8°C
Length growing season: 214 days
Altitude: 210 m
Latitude: 47° 23' N
Institution: Univ. Washington,
　　　　　USA

		Elements				
	Org. Mat.	N	K	Ca	P	Mg
Amounts (kg/ha)						
Overstory foliage total	4060	100	43	42	5	8.4
Overstory branches	19380	20	4	77	2	6.8
Overstory boles	127580	120	27	51	16	45.1
Overstory roots	35230	176	7	123	4	12.4
Understory vegetation	9530	103	132	95	7	16
Forest litter layer	66350	877	91	391	45	57.
Soil-rooting zone	158520	5450				
Increments (kg/ha/yr)						
Overstory foliage	4060	100	43	42	5.0	8.4
Overstory branches	1150	3	4.5	2	0.4	0.6
Overstory boles	7570	16	30.0	12	2.7	3.8
Overstory roots						
Understory vegetation						
Fluxes (kg/ha/yr)						
Atmosphere precipitation		1.7	2.2	2.2	0.3	0.5
Atmosphere particulates						
Overstory litterfall	15972	87	39	67	6.2	10
Leaf wash		8.8	12.2	10	1	5
Stem flow		0.1	0.4	0.2		0.08
Leaching–forest floor		15	42.5	77	10.1	14.7
Leaching–rooting depth						
Leaching–watershed						

Stand No. 36
Site: Rouquet, France
Investigator(s): Lossaint, Rapp
Forest type: Mediterranean evergreen oak forest (*Quercus ilex*)
Geology: Portlandien–Kimmeridgien limestone
Soil type: Brunified Mediterranean red soil
Stand age: 150 yr
Annual precipitation: 987 mm
Mean annual temperature: 12.4°C
Length growing season: 365 days
Altitude: 180 m
Latitude: 43° 42′ N
Institution: CNRS, Montpellier, France

		Elements				
	Org. Mat.	N	K	Ca	P	Mg
Amounts (kg/ha)						
Overstory foliage total	7000	93	43	70	10	9
Overstory branches	27000	135	90	493	40	25
Overstory boles	235000	517	493	3290	174	117
Overstory roots	45000	153	85	1147	81	7.2
Understory vegetation	340	2.3	17	4.1	0.1	1.1
Forest litter layer	11400	124.7	10.2	361.2	4	19.7
Soil-rooting zone						
Increments (kg/ha/yr)						
Overstory foliage	4500	64	28	38	6	6
Overstory branches	1100	9.9	5.8	21.7	1.6	1.5
Overstory boles	1500	3.3	3.1	4.0	1.0	0.7
Overstory roots						
Understory vegetation						
Fluxes (kg/ha/yr)						
Atmosphere precipitation		15.6	2.4	11.7	1.0	2.1
Atmosphere particulates						
Overstory litterfall	3842	32.8	16.2	63.9	2.8	4.6
Leaf wash		0.5	21.3	23.3	1.6	3.6
Stem flow		1.2	6.5	7.8	0.3	0.8
Leaching–forest floor						
Leaching–rooting depth						
Leaching–watershed						

References

Baker, J. B. & Blackmon, B. G. (1977). Biomass and nutrient accumulation in a cottonwood plantation – the first growing season. *Soil Sci. Soc. Am. J.* **41**, 632–6.

Bormann, F. H., Likens G. E., Siccama, T. G., Pierce, R. S. & Eaton, J. S. (1974). The export of nutrients and recovery of stable conditions following deforestation at Hubbard Brook. *Ecol. Monog.* **44**, 255–77.

Cole, D. W., Gessel, S. P. & Dice, S. F. (1967). Distribution and cycling of nitrogen, phosphorus, potassium, and calcium in a second-growth Douglas fir ecosystem. In: *Symposium, Primary Productivity and Mineral Cycling in Natural Ecosystems.* pp. 197–232. *Am. Assoc. Adv. Sci. 13th Annual Meeting, New York, December* 1967. Univ. Maine Press, Orono.

Cole, D. W., Crane, W. J. B. & Grier, C. C. (1975). The effect of forest management practices on water chemistry in a second-growth Douglas-fir ecosystem. In: *Forest Soils and Forest Land Management,* ed. B. Bernier & C. H. Winget, pp. 195–207. Laval Univ. Press, Quebec.

Cole, D. W., Gessel, S. P. & Turner, J. (1978). Comparative mineral cycling in red alder and Douglas-fir. In *Utilization and Management of Alder,* ed. D. G. Briggs, D. S. DeBell & W. A. Atkinson, pp. 327–36. Pac. NW For. and Rang Exp. Stn., Portland, Oregon.

Duvigneaud, P. & Denaeyer-DeSmet, S. (1970). Biological cycling of minerals in temperature deciduous forests. In *Analysis of Temperate Forest Ecosystems,* ed. D. E. Reichle, pp. 199–225. Springer-Verlag.

Fagerstrom T. & Lohm, U., (1977). Growth in Scots pine (*Pinus silvestris* L.). Mechanism of response to nitrogen. *Oecologia* (Berl.), **26**, 305–15.

Grier, C. C., Cole, D. W., Dyrness, C. T. & Fredriksen, R. L. (1974). Nutrient cycling in 37- and 450-year-old Douglas-fir ecosystems. In *Integrated research in the Coniferous Forest Biome,* pp. 21–34, Coniferous Forest Biome Bulletin No. 5, University of Washington, Seattle, WA.

Heilman, P. E. & Gessel, S. P. (1963). Nitrogen requirement and the biological cycling of nitrogen in Douglas-fir stands in relationship to the effects of nitrogen fertilization. *Plant and Soil,* **18**, 386–402.

Heinrichs, H. & Mayer, R. (1977). Distribution and cycling of major and trace elements in two central European forest ecosystems. *J. Environ. Qual.* **6**, 402–47.

Henderson, G. S., Swank, W. T., Waide, J. B. & Grier, C. C. (1978). Nutrient budgets of Appalachian and Cascade region watersheds: a comparison. *For. Sci.* **24**, 385–97.

Junge, C. E. (1963). *Air chemistry and radioactivity.* Academic Press, New York.

Kazimirov, N. I. & Morozova, R. N. (1973). *Biological cycling of matter in spruce forests of Karelia.* Nauka Publ. House, Leningrad.

Likens, G. E. & Bormann, F. H. (1974). Acid rain: a serious regional environmental problem. *Science,* **184**, 1176–9.

Likens, G. E., Bormann, F. H., Pierce, R. S., Eaton, J. S. & Johnson, N. M. (1977). *Biogeochemistry of a forested watershed.* Springer-Verlag, New York.

Madgwick, H. A. I., White, E. H., Xydias, G. K. & Leaf, A. L. (1970). Biomass of *Pinus resinosa* in relation to potassium nutrition. *For. Sci.* **16**, 154–9.

Newton, M., el-Hassan, B. A., & Zavitkovski, J. (1968). Role of red alder in western Oregon forest succession. In *Biology of Alder,* ed. J. M. Trappe, J. F. Franklin, R. F. Tarrant & G. M. Hansen, pp. 73–84. Pac. Northwest For. and Range Exp. Stn., Portland, Oregon.

Nihlgard, B. (1972). Plant biomass, primary productivity and distribution of chemical elements in a beech and planted spruce forest in South Sweden. *Oikos,* **23**, 69–81.

Ovington, J. D. (1959). The circulation of minerals in plantations of *Pinus silvestris* L. *Annals Bot.*, **23**, 229–39.

Ovington, J. D. (1962). Quantitative ecology and the woodland ecosystem concept. *Advanc. Ecol. Res.* **1**, 103–92.

Rodin, L. E. & Bazilevich, N. I. (1967). *Production and mineral cycling in terrestrial vegetation.* Oliver & Boyd, Ltd., London.

Switzer, G. L. & Nelson, L. E. (1972). Nutrient accumulation and cycling in Loblolly pine (*Pinus taeda* L.) plantation ecosystems: the first twenty years. *Proc. Soil Sci. Soc. Am.* **36**, 143–7.

Tarrant, R. F. & Miller, R. F. (1963). Accumulation of organic matter and soil nitrogen beneath a plantation of red alder and Douglas-fir. *Proc. Soil Sci. Soc. Am.* **27**, 231–4.

Turner, J. (1975). Nutrient cycling in a Douglas-fir ecosystem with respect to age and nutrient status. *Ph.D. Thesis*, University of Washington, Seattle.

Vitousek, P. M., Gosz, J. R., Grier, C. C., Melillo, J. M., Reiners, W. A. & Todd, R. L. (1979). Nitrate losses from disturbed ecosystems. *Science*, **204**, 469–74.

7. Comparative productivity and biomass relations of forest ecosystems*

R. V. O'NEILL & D. L. DeANGELIS

Contents

Introduction

The Woodlands Data Set (WDS) (Chapter 11) provides a unique resource for the analysis of forest ecosystems. Previous attempts to compare the forests of the world have been severely limited by the small number of ecosystems which had been documented with methods and parameters similar enough to permit comparison. With the completion of the International Biological Programme, we now have available one of the most extensive bodies of data on forest ecosystems that has yet been assembled.

The great range of forest sites included in the data set provides both an advantage and a disadvantage. The advantage is that a relatively complete sample of forest types is available, in principle permitting generalizations on a global scale. Within some forest types, such as temperate deciduous forests, a sufficient number of sites is available to characterize a wide range of independent variables such as soil, moisture, and fertility conditions. The disadvantage is that with so many independent variables, it is difficult to locate any consistent pattern.

The emphasis in this chapter is on a search for general characteristics of

*Research sponsored by the Eastern Deciduous Forest Bione, US–IBP, funded by the National Science Foundation under Interagency Agreement AG–199, BMS76–00761 with the US Department of Energy. Eastern Deciduous Forest Biome Contribution Number 286. Publication No. 1038, Environmental Sciences Division, Oak Ridge National Laboratory. Operated by Union Carbide Corporation under contract with the US Department of Energy.

forests, seeking those patterns which emerge from large arrays of sites. One can expect that relatively few characteristics can be ascribed to forests as a general class. But once consistent patterns are discovered, it may be possible to make generalizations valid for all forests.

The diversity of site conditions represented in the data precludes the approaches of classical forestry or plant ecology, such as the establishment of allometric relationships, or the development of functional relationships between productivity and climate, as these have been based on forest sites with well-defined constraints. Thus, productivity as a function of soil parameters has been studied by comparing forests of the same type, in the same climatic region, and of a similar age. In this way the differences in productivity due to soil factors, for example, have been isolated and accurately analyzed. The WDS is of a very different nature, with almost every possible combination of site conditions represented.

The approach in this chapter, therefore, is somewhat different. We begin with a discussion of the constraints associated with a comparative analysis of the data sets and provide some specific examples illustrating the difficulties. Then we examine a number of published studies which have generalized about world forests and test their conclusions with this new data set. Next we present relationships which appear to be valid across all site conditions. Finally, we explore recent hypotheses in ecosystem theory and draw upon the data set to provide tests for various theoretical predictions.

Constraints associated with analyzing the woodlands data set

Perhaps the most persistent problem experienced in attempting a comparative analysis is the great range of site conditions represented in the data set. The generalizations which form a large part of the theory of the structure and function of forest ecosystems are based on the analysis of controlled experiments in which only a few factors are permitted to vary. The WDS represents a factorial experiment with literally dozens of independent variables such as age, species, soils, temperatures, etc. The result is that straightforward attempts at multiple correlation analysis (e.g. correlation of annual net primary production with all possible independent variables) produced only insignificant correlation coefficients.

In addition to the variability among sites, the differences in measurement techniques compound the analyses. Particularly problematic is the characterization of abiotic variables. In some sites, temperature, moisture, and other parameters were measured on the site and averaged over the period of the study. However, in other cases, abiotic parameters were taken from nearby meteorological stations and represent average values which may not be characteristic of the seasons of measurement. Under such circumstances, correlations of forest characteristics with climatic variables are difficult to establish.

412

In most cases, site measurements were taken to test specific hypotheses, which influenced the methodology. As a specific example, in many cases the effect of heterotrophs was considered inconsequential to the experimental design. Consumption was either negligible or approximately equal across sites within an experiment. In other cases, the investigators specifically included the effects of treatments on heterotrophic populations. Therefore, variable detail exists with respect to heterotrophic biomass and consumption data.

Effects of stand age

The effects of stand age on forest characteristics frequently necessitated the omission of very young or very old stands from comparisons across forest types. The effect is difficult to account for because aging effects appear at different times for different forest types.

The effect can be seen clearly by comparing production characteristics across different aged stands. Fig. 7.1 shows the ratio of branch and bole production to leaf production (leaf litterfall) for five forest types. The ratio of bole to leaf production is an indicator of wood produced per unit of leaf tissue production. We will utilize this 'wood production efficiency' as an indicator of stand vigor. The effects of age on wood production efficiency appear in each of five age series. The greatest variation is expressed at intermediate ages (sapling to pole transition) when other site characteristics such as fertility, moisture, and crowding have their greatest effect. Following the intermediate years, stands show a consistent decrease in the efficiency of wood production with increasing age.

The effects of aging are most striking in the series of boreal needle-leaved stands from the USSR (Fig. 7.1e). Stands in this series are subject to similar climatic conditions and are approximately even-aged. One would expect the aging effect to be most conspicuous in even-aged stands where all trees are in equivalent stages of their life cycle. In many of the other forests in Fig. 7.1, such as the temperate deciduous series Fig. 7.1d, the canopy is composed of several taxa of varying ages so that mortality and replacement may have become more nearly equilibrium phenomena. As a result, the deciduous series shows considerably more variability in the wood efficiency values. However, the consistency of the pattern of aging across all series indicates that reduction in wood production efficiency is a generalizable characteristic of forest stand age, even for natural stands.

Many of these aging effects have been pointed out by Rodin & Bazilevich (1965). However, it is interesting that the age at which efficiency begins to decrease appears to be related to the rate of productivity and the environment to which the stand is subjected. Aging effects begin to appear in conifer plantations (Fig. 7.1b) and highly productive beech stands (Fig. 7.1c) between 50 and 100 years old. The same effects begin in natural conifer (Fig.

413

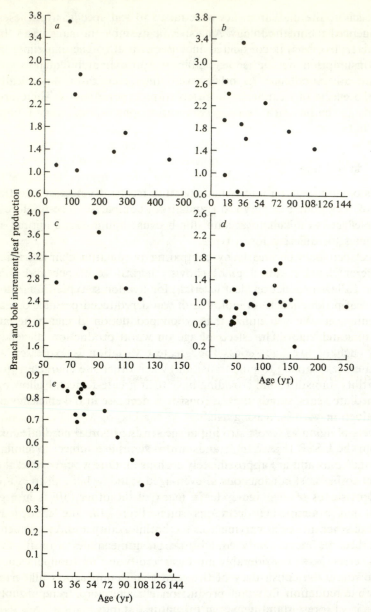

Fig. 7.1. Wood production efficiency (branch and bole production/leaf production) as a function of stand age for five forest types. *a*, natural conifer stands; *b*, conifer plantations; *c*, temperate beech stands; *d*, temperate deciduous stands; *e*, Russian boreal conifer series.

414

7.1*a*) and deciduous stands (Fig. 7.1*d*) between 150 and 200 years old. The rapid aging of the boreal conifers can be explained by the suggestion of Sukachev & Dylis (1964) that a cold climate decreases the duration of vigor in a stand by shortening the period over which it can maintain positive biomass and energy balance. In the case of conifer plantations and highly productive beech stands, biomass accumulation approaches its limit relatively early in the lifetime of trees, leading to correspondingly early aging effects in the forest.

While it is interesting to examine the differences in aging patterns of wood production efficiency, these differences are quite problematic in analyzing the total data set. Because the aging effects appear at different times in different forest types, it is not a simple matter to choose a single cut-off age to eliminate older stands. In many of the analyses which follow, 130 years of age was used to eliminate stands even though this criterion is not appropriate for all stands.

Another difficulty in finding consistent productivity patterns in the data develops when plantations are combined with natural stands. Because the plantations are managed specifically to increase wood production, productivity rates in these stands are quite high compared to natural stands. It was frequently necessary to eliminate the managed stands to find patterns that were consistent across forest types.

Comparisons with other analyses of forest ecosystems

The variability of sites in the WDS provides a range of stand conditions which has not previously been available. A number of attempts to find general patterns in world forests have appeared in the literature but are for the most part based on limited data. It is of considerable interest, therefore, to extract some of these proposed relationships and test their validity against this new and expanded data set.

Over a period of several years, Lieth (Lieth & Radford, 1971; Lieth, 1972*a,b*, 3) has attempted to relate productivity to climatic variables. Having established a consistent relationship from literature values, Lieth utilizes worldwide climate patterns to estimate geographic productivity patterns. The relationships proposed by Lieth are based on limited available data, without benefit of the extensive effort and cooperation that has gone into the Woodlands Data Set. It is interesting, therefore, to reproduce Lieth's relationships graphically and superimpose the information from the present data set.

Fig. 7.2*a* shows the proposed relationship between productivity and annual average temperature. Lieth's original data points were used to fit the curve on the figure (Whittaker & Likens, 1973) and are shown as open circles. The solid dots represent data from the WDS. In general, it appears

Dynamic properties of forest ecosystems

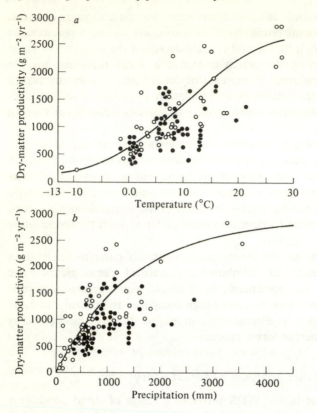

Fig. 7.2. Relationships proposed by Lieth (Lieth & Radford, 1971; Leith, 1972a,b, 3) between net primary production and temperature (a) or precipitation (b). The line represents the relationship fitted by Lieth and open circles show data points used in his analysis. Data from WDS (Chapter 11) are represented as solid dots. Redrawn from Whittaker & Likens, 1973.

that the relationship proposed by Lieth is valid, although the WDS lacks the same temperature extremes. The majority of the WDS sites fall reasonably close to the projected line and lend confirmation. On the other hand, about a dozen sites fall well below the line, indicating unexpectedly low productivity for a given temperature. In many cases, but not all, these sites represent old, senescent stands. The low productivity of these stands does not seem to be correlated with xeric conditions as might be expected.

Fig. 7.2b shows the relationship between annual precipitation and productivity proposed by Lieth. The projected line is slightly high relative to the WDS data, but the majority of the sites fall reasonably close to Lieth's proposed relationship. Once again, the WDS lacks extreme values of precipitation. About a dozen stands show unexpectedly low productivity. In general, the data points which fall well below the line in both Figs. 7.2a and b are

416

the same stands. This reinforces our observation that these sites show neither temperature nor moisture stress but are displaying lowered productivity associated with age.

From this comparison we conclude that the WDS (omitting the old stands) would motivate little change in the relationship proposed, though perhaps the precipitation curve should be lowered slightly. A recent paper (Lieth & Box, 1972) indicates that a stronger correlation can be established between productivity and evapotranspiration which combines both temperature and precipitation information. These regressions against evapotranspiration are not reproduced here because the monthly temperature and precipitation data from which evapotranspiration is calculated were available for only a few sites. However, combining temperature and moisture would not eliminate the problem associated with the old stands, since they lie below the expected line in both the precipitation and temperature curves.

The intriguing question which emerges from this comparison is whether the original data sets used to establish the above relationships were systematically biased on age. The inclusion of old stands tends to obscure the productivity relationships, and the inclusion of a relatively large number of older stands would have altered the regressions. If only younger, more vigorous stands are used in obtaining these relationships, it could lead to a serious overestimate of worldwide productivity of forest systems. The answer to this question lies beyond the scope of the present analysis but does raise the possibility of systematic bias in the present methods used to estimate global productivity.

Another general relationship has been proposed by Whittaker & Likens (1973) between aboveground biomass and annual net primary production. Fig. 7.3 shows the expected correlation with the original data (open circles) and data from the WDS (solid dots). Many of the WDS sites fall within the solid lines. Some points fall slightly outside the boundaries as might be expected, since the present data represent such a wide range of site conditions. A set of eight sites falls well outside the boundaries, showing significantly greater productivity than would be expected for their standing crops. These eight sites represent evergreen plantations and managed beech sites which have been specifically manipulated for maximum productivity. It would seem, therefore, that the relationship proposed by Whittaker and Likens is verified by the WDS, at least for natural stands. The only modifications necessary are slightly broader boundaries to include a greater range of site conditions.

Bray & Gorham (1964) have provided a very extensive analysis of litterfall patterns in the forests of the world. While it is not possible to use the WDS sites in all of their analyses, in many cases equivalent data are available. Table 7.1 compares their analysis of litter composition (leaves,

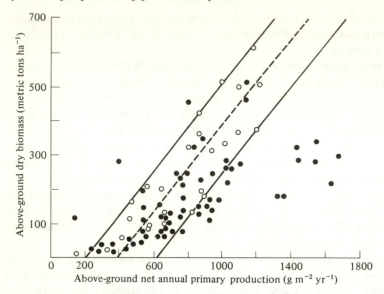

Fig. 7.3. Proposed relationship between above-ground biomass and above-ground primary production. Lines show the central tendency and extremes represented in the original data (open circles). The solid dots are data from WDS (Chapter 11). Redrawn from Whittaker & Likens, 1973.

fruit, and woody tissue) with the WDS. Unfortunately, Bray & Gorham did not express error terms for mean values and only subjective judgments are possible. It is immediately evident from the range of percent composition for leaf and fruit litter that the WDS contains a greater diversity of sites. In both cases, the range of values in the WDS exceeds the published range, but the central tendency is similar. That is, the mean values calculated from the WDS fall toward the center of the range given by Bray & Gorham. Mean values of percent woody tissue for angiosperms and gymnosperms are similar, though the more inclusive coverage of the present data set drops the mean value for angiosperm percentages significantly.

In comparing the percent woody tissue by type and climatic region, most values do not appear to be significantly different. The possible exception is the gymnosperm forests in warm temperate North America where the WDS indicates a value of 18%. In most other cases, the values stated by Bray & Gorham fall within or reasonably close to the error bounds of the WDS.

Bray & Gorham also analyzed the quantity of leaf and total litterfall by type and region (Table 7.2). Values from the WDS for evergreen forests are remarkably similar, but deciduous data are much higher due to the inclusion of tropical deciduous forests from India. It appears that the data analyzed by Bray & Gorham were biased toward more northern deciduous

418

Table 7.1. *Comparison of litter composition data from Bray & Gorham (1964) with equivalent information from the Woodlands Data Set. All values represent percentage $\pm$ one standard error. Values in parentheses are estimates based on a single value*

	Bray & Gorham		Woodlands data set	
% of total litterfall				
Leaves, range	60–76		29–95	
mean $\pm$ S.E.	—		69 ± 2	
Fruit, range	1–17		1–22	
mean $\pm$ S.E.	—		8 ± 1	
Wood, mean	30		25 ± 2	
Angiosperms	30		21 ± 2	
Gymnosperms	29		30 ± 3	
% woody litter by region and type				
	Gymnosperm	Angiosperm	Gymnosperm	Angiosperm
Equatorial	—	(33)	—	27 ± 8
Warm temperate				
Australia	(39)	42	—	(42)
North America	37	23	18 ± 11	19 ± 3
Cold temperate	23	21	28 ± 9	24 ± 3
Arctic	(39)	21	34 ± 1	15 ± 3

stands which lowered litterfall values. Because of this bias, they concluded that evergreen stands as a class were slightly more productive than deciduous stands. This generalization appears to be reversed in the present analysis. Obviously, the conclusion from such an analysis would be strongly biased by the number of tropical evergreen and deciduous stands which are included in the means.

The analysis of litterfall by region shows a striking increase in both leaf litter and total litterfall as one proceeds to lower latitudes. The WDS bears this out. The major discrepancy appears in the arctic and cold temperate values where mean values in the WDS data are higher and closer together.

Bray & Gorham continued their analysis of litterfall by region with a graphic representation which is redrawn in Fig. 7.4. Their data are represented by open circles and the line shows their proposed relationship. The closed circles represent values from the WDS. Once again, the WDS data show a greater variance but seem to confirm the relationship. Careful examination shows that the proposed regression is perhaps too steep. At lower latitudes, the present data tend to fall below the line and at higher latitudes a significant number of points fall well above the line. Thus, while a general tendency toward lowered litterfall at higher latitudes is confirmed, the slope of the line is probably less than originally proposed.

419

Table 7.2. *Leaf litter and total litterfall compared across forest types and major climatic zones. Values represent g m^{-2} yr^{-1}* $\pm$ *S.E.*

	Bray & Gorham		Woodlands data set	
	Leaf litter	Total litter	Leaf litter	Total litter
Evergreen	260	370	270±21	375±24
Deciduous	240	320	446±37	522±49
Arctic	70	100	217±17	329±25
Cold temperate	250	350	301±13	457±46
Warm temperate	360	550	350±31	473±28
Equatorial	680	1090	659±75	934±266

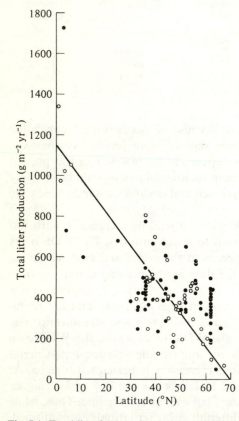

Fig. 7.4. Total litterfall as a function of latitude (redrawn from Bray & Gorham, 1964). WDS (Chapter 11; solid dots) are superimposed on the original data (open circles) and line fitted by Bray & Gorham.

420

In discussing the effects of age on production in general and on litterfall in particular, Bray & Gorham cite several early workers (Ebermayer, 1876; Danckelmann, 1887*a,b*; Sonn, 1960) whose studies confirm the aging phenomenon discussed in an earlier section. Litterfall increases during early years, peaks, and then decreases with increasing age. Bray & Gorham attempted to confirm this relationship by combining all forest types in their data but were unable to show the same relationship. Their conclusion was probably biased by the lumping procedure, as different forest types exhibit aging phenomenon at different times. The WDS confirms the results of the earlier workers and contradicts the conclusions of Bray & Gorham. The phenomenon is particularly evident in the more carefully controlled site series in USSR and India, where the aging phenomenon is clearly indicated in the litterfall data.

Bray & Gorham also cite several workers (Bonnevie-Svendsen & Gjems, 1957; Grosby 1961) as finding significant correlations between litterfall and basal area of stands but this was not strongly indicated in their data. Once again the relationship may be obscured by their lumping procedure and by the lack of stands with very high basal area in their data set. Fig. 7.5 shows

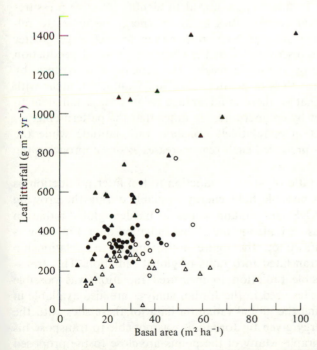

Fig. 7.5. Leaf litterfall as a function of total basal area. WDS (Chapter 11) show that relationships exist but are complicated by forest type and management practices. ●, temperate deciduous; ○, temperate evergreen plantations; ▲, tropical; △, temperate and boreal evergreen.

that a correlation may indeed exist, but that it is complicated by forest type and management practices. All forest types appear similar at lower basal areas. The tropical stands with their higher productivity show a reasonable correlation, with increased basal area associated with increased litterfall. Temperate deciduous and temperate evergreen plantations are not well represented at higher basal areas and are not clearly distinguishable from each other. These stands display weak correlation that lies toward the middle of the graph. Boreal and temperate evergreen stands follow the general trend at lower basal areas but approach an asymptote somewhere between basal areas of 20 to 30 m^2 ha^{-1}. In this case, many of the stands with basal areas over 30 m^2 ha^{-1} are also old stands which are showing lowered productivity. Overall, there appears to be a relationship in the WDS data even though it is complicated by climate, management practices, and age effects.

Jordan (1971) examined world patterns of plant energetics, seeking explanations for productivity relationships. Basically, he sought a basis for calculating productive potential by identifying patterns that minimized local variations in conditions and emphasized global patterns. Therefore, he de-emphasized productivity itself and sought patterns in the way plants allocated fixed energy. In this way he hoped to identify selective pressures which would cause plant communities to divert energy resources to relatively permanent tissues such as boles rather than to encourage greater leaf production. Jordan discovered that the efficiency of wood production increases as solar energy input decreases. He explains this pattern by postulating that fast growth is of greater selective advantage in areas with relatively little light. That is, there is advantage to growing a tall stem to avoid being shaded out by competitors. He notes that the pattern emerges because wood production is relatively constant with latitude while the amount of leaf biomass produced each year increases as one approaches the equator (*see* Fig. 7.4).

Jordan graphed the ratio of wood production to leaf litter production as a function of the total possible light energy available during the growing season. A number of WDS sites contain values for incident global radiation during the growing season. Utilizing the global radiation data, Fig. 7.6 was produced. The line drawn on the figure approximates the relationship proposed by Jordan, translated into units of global radiation. This translation from total possible radiation to measured radiation was possible because several of the sites used in the Jordan analysis are also available in the WDS. By comparing the global radiation data for these sites with the total possible light energy given by Jordan, it was possible to transpose his line onto the present graph. Many of the points are close to the proposed line, but several fall significantly below it. Jordan notes that some sites can be expected to fall below the line because of moisture stress. Jordan also

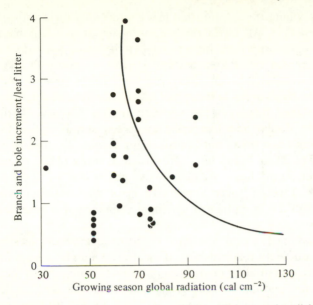

Fig. 7.6. Relationship between wood production efficiency and available light energy proposed by Jordan (1971). The line approximates to the relationship given by Jordan. ● WDS.

discovered that very young or very old stands had uncharacteristically low ratios and he omitted them from his analysis. To provide a consistent comparison only stands between 50 and 130 years of age have been graphed so the age factor does not account for the low values. Examination of precipitation data does not indicate that these sites would be expected to show moisture stress to any great extent.

It is unfortunate that growing season global radiation data are not available for more sites, particularly those at higher energy inputs, since the available data do not span the entire range of values presented by Jordan. One can try to explain our results in Fig. 7.6 by studying Figs. 7.4 and 7.7. It would appear from Fig. 7.4 that litter production does indeed increase toward the equator. Fig 7.7 shows that wood production data do not show a very consistent pattern with latitude. At the same time, it is difficult to maintain that wood production is a constant. There appears to be a definite increase in wood production closer to the equator and a definite decrease at higher latitudes. It is probable that there is no consistent trend at intermediate latitudes.

Jordan assumed that wood production showed no distinct pattern with latitude (i.e. it was a constant) and leaf production decreased with increasing latitude. If these premises were true then his expected relationship (Fig. 7.6) would hold. It appears, however, that his assumption about constant wood production is not exhibited in the WDS data at very high and very

423

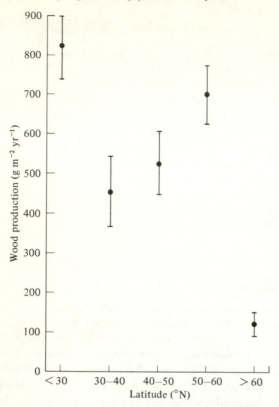

Fig. 7.7. Wood production (branch and bole increment) as a function of latitude from the WDS. Points represent means; bars indicate one standard error.

low latitudes, and therefore Jordan's correlation is not seen. As a result of the lack of information on radiation, only a limited number of points can be compared in Fig. 7.6. Because many of the points fall close to the line proposed, it is difficult to reject Jordan's hypothesis. The outliers are not explained by age or moisture stress as Jordan supposed, but are boreal sites at high latitudes where his assumption of constant wood production breaks down.

Structural characteristics of world forests

One of the goals of the present analysis is the discovery of forest character-istics which are relatively independent of site and species differences. Because the differences among the WDS sites are quite large, it is not surprising that relatively few general properties are easily discerned. Nevertheless, some patterns are consistent across all forests, while other characteristics hold true for major forest types, such as evergreen and deciduous or gymnosperms and angiosperms.

Fig. 7.8 shows total above-ground biomass versus stand height. There is a strong correlation between these quantities, but with considerable remaining variance. It should be noted that the graph is drawn on logarithmic scales which tend to minimize the spread of points at higher values. Some of the variability can be explained by the fact that total above-ground biomass is represented, including understory, herbaceous strata, and epiphytes. For some sites with high biomass values, stand height is below the expected value because extensive development of understory vegetation has contributed biomass without adding to stand height. In other cases relatively low stem density apparently has reduced competition, resulting in greater biomass increment diverted to diameter growth or to other plant parts. Nevertheless, the general relationship appears to hold across all forest types represented.

A consistent relationship also exists between diameter classes (mean diameter of trees on the stand) and the density of trees (Fig. 7.9). Such figures are frequently produced to show changes in stands of different age, and it should be noted that we are representing the mean diameter for the total stand, including all species. Decreasing density with increasing diameter is intuitively expected as competition between individuals eliminates

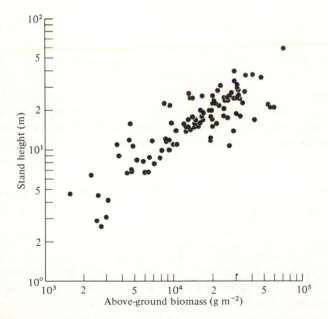

Fig. 7.8. Stand height as a function of total above-ground biomass, including understory, herbaceous layer, and epiphytes.

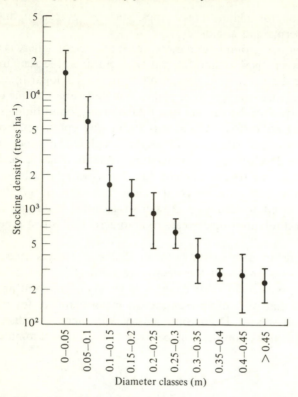

Fig. 7.9. Relationship between number of trees per hectare and average tree diameter. Points represent means for diameters classes and error bars show one standard error.

some trees. The interesting feature is the concavity of the relationship. West & Johnson (1980) have recently noted that a convex relationship is to be expected for pioneer species and a concave relationship for climax vegetation. The present analysis verifies their analysis, since the majority of the sites represented in Fig. 7.9 are mature stands and more likely to resemble climax than pioneer stands.

An aspect of forest structure related to the topics discussed above involves the formula,

$$y = c_o p^{-\alpha}, \tag{7.1}$$

where $\alpha = 0.05$, derived empirically by Yoda, Kira & Hozumi (1963), relating the mean weight of biomass per unit area, y, and plant stocking density, p, in pure stands. This formula should be applicable to cases where self-thinning is occurring. The WDS (Chapter 11) includes a large number of pure stands, many of the same species, and hence affords an opportunity for testing eqn. 7.1. Perhaps the best series of data for this purpose are

426

twelve adjacent stands of the *Picea abies–Myrtillosum* association. Eleven of the twelve points are easily fitted by hand (Fig. 7.10*a*) to give a value of $\alpha = 0.44$. Similarly the pure and almost pure beech stands in the WDS can be plotted to yield $\alpha = 0.31$ (Fig. 7.10*b*). In both of these types of stands the self-thinning is greater than that predicted by Yoda's formula. Quite

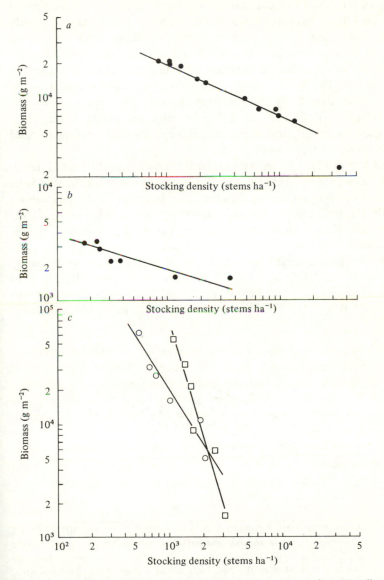

Fig. 7.10. Biomass per unit area versus stocking density. *a, Picea abies–Myrtillosum* stands; *b*, beech stands; *c*, teak stands, ○, and sal stands, □.

different results are observed in the case of tropical deciduous teak and sal plantations (Fig. 7.10c). In these cases α is 1.59 and 3.5, respectively, which implies that the stands have not undergone much self-thinning, despite the fact that they have reached biomass levels which in the preceding cases were accompanied by strong self-thinning. Harper (1967) mentions attempts to apply Yoda's formula to mixed stands. We found that a large number of temperate deciduous natural stands (other than the beech stands already mentioned) fell along a line with α = 0.55, but there were also some significant deviations from that line which are not easy to account for.

Fig. 7.11 shows a correlation between the basal area of the average tree and stand height. The basal area of the average tree was calculated as the total basal area divided by the stocking density. The figure shows the expected relationship of increasing height with increasing basal, area. Although there is considerable variation, the relationship appears to hold true across all forest types and species.

The solid line of Fig. 7.11 represents the theoretical buckling relationship for trees as calculated by McMahon (1973). As in his study, all the stands lie to the right of this line, indicating that there is a basic physical limitation on tree growth. He averaged the heights reached by record individuals of over 500 species and found them to be on the order of one-fourth of this maximum value.

Although the lowest height for a given basal area appears to be highly variable, the greatest height achieved appears to be more tightly constrained, with all points lying below the dotted line representing half the buckling height (Fig. 7.11). The lower boundary of the ensemble of points is

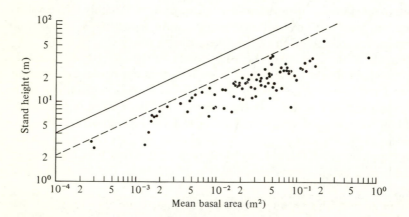

Fig. 7.11. Stand height as a function of the basal area of the average tree (total basal area/ stocking density) for the WDS. Solid line indicates the height at which wood tissue of a given basal area would buckle or collapse under stress. All stands lie below the interrupted line which is one half of the buckling height.

irregular, but the top is almost 'flat'. Doubtless this relationship is partially caused by the basic elastic strength of wood fibers which requires a certain basal area to support a given height. At greater heights, there is the possibility of 'buckling' or having the top collapse under its own weight when subjected to stress such as wind.

Utilizing the physical relationships given by McMahon, the equation of the dotted line can be given as:

$$\text{height} = \frac{0.792}{2} \left| \frac{E}{\rho} \right|^{1/3} d^{2/3}, \tag{7.2}$$

where E is the elastic modulus of the wood (given by McMahon as 1.05×10^5 kg m^{-2}), ρ is density (given as 6.18×10^2 kg m^{-3}), and d is diameter at breast height (*dbh*). The equation as given by McMahon is based on *dbh*, which in turn is obviously related to basal area of the average tree, $BA(\text{m}^2)$, by

$$d = 2 \sqrt{\frac{BA}{\pi}}. \tag{7.3}$$

It should be carefully noted that the relationship suggested by Fig. 7.11 is a property of forest stands, not individual trees. Obviously, the height achieved by a tree growing in a forest canopy can be greater than that achieved by a tree growing in isolation, without the environmental modifying effects of surrounding vegetation. One would expect, therefore, that the minimal basal area relationship is related to the effects of surrounding vegetation as well as the elastic strength of wood. Because many of the points fall well below the interrupted line in the figure, factors other than buckling strength must also be limiting height growth. Possibly, cohesion of the capillary water column within the conductive tissue of the plants and the anatomical configuration of the xylem play a role.

Examination of the points which lie closest to the interrupted line (i.e. maximum stand height for a given basal area) does not reveal any specific pattern. They represent a variety of systems and species including both boreal and temperate systems, both evergreen and deciduous, both needle-leaved and broad-leaved. About half are boreal and half temperate; 69% are needle-leaved stands, and 31% are broad-leaved; 30% are managed stands and 70% are natural. If there is any pattern at all, it is that boreal evergreen and temperate beech stands generally occur closer to the buckling limit than any other types.

Because litterfall forms an important component of above-ground net primary production, one would expect to find a strong positive correlation between these two variables. Fig. 7.12 shows a relatively linear correlation at lower values, but many points lie below the line at intermediate and high values. For these sites a larger portion of the net primary production is

429

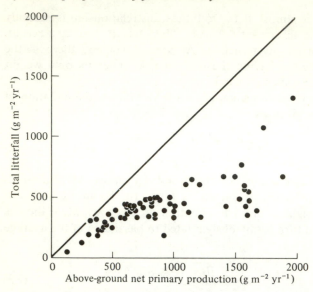

Fig. 7.12. Total annual litterfall as a function of above-ground net primary production for the WDS. The solid line represents the hypothetical situation where there is no wood production and litterfall is 100% of the productivity.

going into the production of wood tissue. Stands which show higher litterfall at greater productivity levels appear to be older stands where mature trees are showing loss of branch or bole tissue into the litter. Thus, the low points are systems in which the greatest net amount of bole tissue is being produced.

The line in Fig. 7.12 represents the situation in which net primary production is completely accounted for by litter production; i.e. no wood tissue is being accumulated. The data points diverge from the line, indicating that the more productive the system, the greater the photosynthate available for growth and the greater the amount of wood tissue produced. Therefore, the relationship shown in Fig. 7.13, between increased productivity and greater net wood production, is expected. Five stands over 130 years old with very low or negative net wood production, have been omitted from this graph. In these old stands, productivity continues, but sloughing of wood tissue exceeds new wood production. We can hypothesize that these stands are allocating increasingly large fractions of their net photosynthate to repair and maintenance.

In general, one would expect the annual production of wood (branch increment plus bole increment) to increase with the size of the photosynthesizing base (leaf biomass). In Fig. 7.14, needle-leaved stands (excluding plantations) are shown as open circles, while broad-leaved stands are shown as filled circles. It is quite obvious that wood increment is positively

430

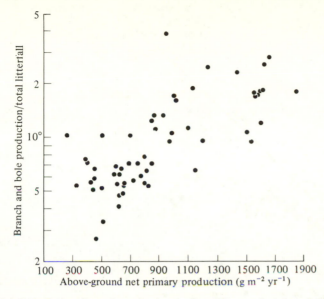

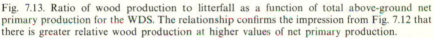

Fig. 7.13. Ratio of wood production to litterfall as a function of total above-ground net primary production for the WDS. The relationship confirms the impression from Fig. 7.12 that there is greater relative wood production at higher values of net primary production.

correlated with increasing leaf biomass in gymnosperm systems. In contrast, the wood increment appears to be relatively independent of the leaf biomass in broad-leaved (mostly temperate) stands. The majority of the needle-leaved systems shown on the graph are boreal and the majority of the broad-leaved are temperate. It may be hypothesized that radiation limitations play a more important role in the boreal systems, and additional leaf biomass in the cone-shaped geometry of the canopy provides greater energy-capturing tissue and results in greater wood production by the stand as a whole. On the other hand, temperate systems may not be limited by radiation in the same manner, and other factors may dominate the production of wood tissue.

Fig. 7.15 shows the general structure of tissue components for angiosperm and gymnosperm communities in the WDS. In both community types, woody tissue makes up the majority of the biomass and the straight line simply indicates that almost all of the biomass is composed of woody tissue. The line is almost exactly the 45° angle which would be achieved if 100% of the tissue were branches and boles. . The line tends to become more rigorously defined at higher biomass levels where leaf tissue makes up an insignificant percentage of the total. The woody tissue tends to show slightly greater variability at lower values where variations in the leaf tissue contribute a more significant fraction to the total.

By displaying the data on logarithmic paper, an impression of the

431

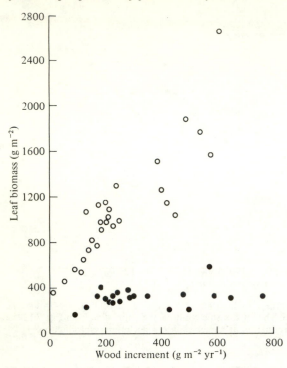

Fig. 7.14. Relationship between leaf biomass and wood production (branch and bole increment) for the WDS. Needle-leaved stands are shown as open circles and broad-leaved stands as filled ones.

variability of the root and leaf tissue is obtained, because each tends to occupy a different cycle on the paper (the values differ by an order of magnitude). The amount of root tissue supporting the above-ground biomass appears to be quite similar for both types of community. The data fall in a nearly straight band down the center of the figure. This could be partly tautological because the root biomass was not directly measured from some sites, but merely estimated as some fraction of above-ground biomass.

The leaf data are the most variable, as might be expected, and above-ground biomass alone is a poor indicator of the standing crop of leaf tissue. As indicated in Fig. 7.14, the amount of gymnosperm leaf tissue is higher for a given biomass value and might be slightly less variable in comparison with the broad-leaved systems.

Forest characteristics and radiation summation

During our attempt to find consistent relationships across the entire WDS, numerous abiotic parameters were examined. During the numerous at-

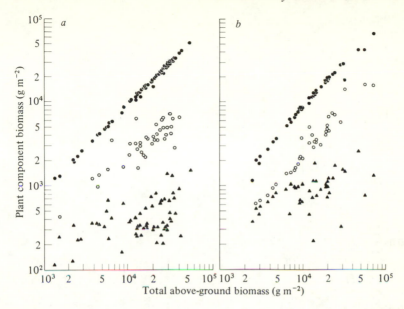

Fig. 7.15. Biomass of various plant components as a function of total above-ground biomass for forests from the WDS, *a*, angiosperms; *b*, gymnosperms. The tight fit indicated for branch and biomass simply reflects the fact that the majority of the biomass is composed of these tissues. ●, bole and branch; ○, roots; △, leaves.

tempts to characterize the forests, using climatic variables, the index which continually emerged as most useful was R, characterizing the radiation received by the stand during its existence. This value was calculated by taking the average temperature in °C during the growing season, G, as an indicator of radiation input. This temperature was multiplied by the age in years of the stand, Y, and by the fraction of the year which represented the average length of the growing season, L:

$$R = G \cdot Y \cdot \frac{L}{365}. \tag{7.4}$$

The units are degree-years and R represents the temperature summation over the growing seasons of the stand. The index combines three determiners of forest growth: age, length of growing season, and temperature. By combining these parameters into one index, it was possible to detect some clear patterns.

Fig. 7.16 shows the relationship between the temperature summation, R, and the volume of the average tree in each stand. The average volume was calculated as the basal area times the average stand height and divided by stocking density. There is a marked positive correlation between R and

433

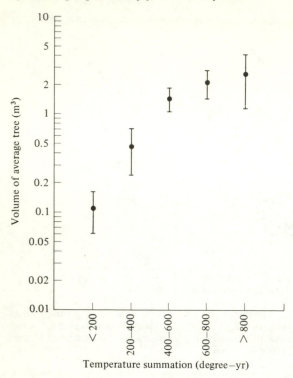

Fig. 7.16. Volume of average tree on a stand (basal area times stand height divided by stocking density) as a function of the temperature summation for the WDS. The asymptote is caused by reduced productivity of senescent stands.

average tree volume. The logarithm of tree volume seems to approach an asymptote at higher values of *R*.

The asymptote at large *R* may be related to aging phenomena because lowered productivity of these older stands tends to decrease the accumulated volume of wood. Another important factor which leads to increased variability is the inclusion of plantations which have relatively large wood volumes. As a result, the needle-leaved plantations tend to lie consistently above the mean values. In the comparisons which follow, the effects of age and management practices were minimized by eliminating stands over 130 years of age and managed plantations from the analysis.

Fig. 7.17 shows that a weak but perceptible relationship exists between the temperature summation, *R*, and the ratio of standing crop to productivity. The ratio increases with increasing age and radiation input, and the amount of biomass per unit productivity increases with older stands growing under more favorable temperature conditions. It is quite evident that the variance in the data increases with increasing values of *R* because

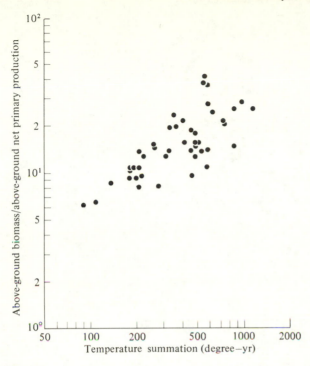

Fig. 7.17. Ratio of above-ground biomass to primary production as a function of temperature summation for the WDS. A relationship appears to exist though variance increases greatly at higher temperature summations.

additional site factors begin to play a role in mature stands and in stands where favorable temperature causes other environmental conditions to become limiting. The correlation is best in young sites with unfavorable temperature regimes (largely boreal forests) where temperature is a fair indicator of radiation limitations and a more extreme environment.

One might expect an index based on average temperature to show a tighter fit to data on standing crop, which is the result of average growing conditions over the age of the stand. Fig. 7.18 illustrates that a correlation does indeed exist between aboveground biomass and R. That is, stands which have grown for a considerable period of time under favorable climatic conditions have accumulated the greatest biomass, while younger stands under less favorable conditions have been unable to accumulate biomass to the same degree.

When old and managed stands are eliminated from the analysis, a correlation is found between the temperature summation and the annual net primary production (Fig. 7.19). The figure shows that mature stands under favorable conditions have higher productivity than young stands

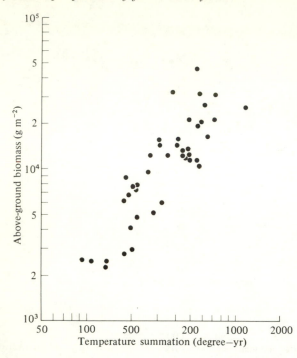

Fig. 7.18. Above-ground biomass as a function of temperature summation for the WDS. The relationship confirms our intuition that mature stands with favorable temperatures have accumulated greater biomass.

under less favorable conditions. The variance remaining in Fig. 7.19 (i.e. deviations presumably due to moisture, soil and other site conditions) is very difficult to analyze. Soil characteristics and moisture parameters derivable from the existing data sets do not seem to account for it. Thus, it was not possible to relate the deviations in the figure with other abiotic variables in the data set.

Other approaches to analyzing forest ecosystems

The difficulties of characterizing the WDS are revealed in the attempt at a site index approach. By this method, relations between biological character-istics of the forest are assumed. Deviations from the values predicted by these relations are taken to indicate site conditions not directly accounted for in the original assumption. Attempts are then made to explain why particular sites have higher or lower values than predicted.

One type of site index has been proposed by Czarnowski (1964). The analysis begins with a hypothesized relationship between stocking density

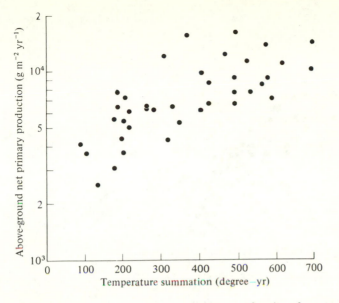

Fig. 7.19. Above-ground net primary production as a function of temperature summation for the WDS. Stands over 130 years old and managed systems have been omitted.

(stems ha^{-1}), D, and the inverse of the square of stand height, H in meters,

$$D = N\frac{P}{H^2}, \qquad (7.5)$$

where P is the surface area considered (10 000 m^2 in the present case) and N is a proportionality constant.

Fig. 7.20 graphs the stocking density versus the square of the stand height for the WDS. Analysis yields a mean value for N of 41.65 ± 0.29 S.E. for the 95 sites included. The value for individual sites is then larger than this average if growing conditions are more favorable than the average and less if site conditions are more extreme. One can then calculate the expected values of N for individual sites from eqn 7.5 and seek soil or moisture conditions which might explain unusually high or low values. The value of N was calculated for each of the sites shown individually in Fig. 7.19, but no significant relationships could be found with precipitation or the temperature summation.

Carrying his analysis further, Czarnowski theorized that the site index would be related to the leaf biomass, M (g m^{-2}), according to the relationship

$$M = \alpha\sqrt{1 + \beta N} \qquad (7.6)$$

where α and β are constants. But again no significant relationship of this

437

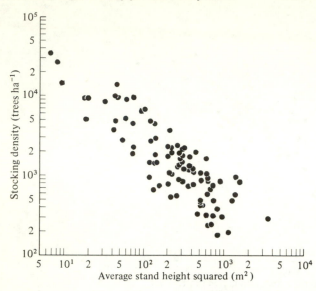

Fig. 7.20. Relationship between tree density and the square of stand height as proposed by Czarnowski (1964) calculated for the WDS. On the basis of this relationship, Czarnowski derived his site index.

form could be discovered in the WDS data set. As a result of these analyses, the site index N was rejected as a useful indicator of growth conditions for the WDS sites. This type of index appears more useful for discerning relationships among even-aged, single-species stands.

To circumvent the difficulties experienced with the index, N, an attempt was made to develop another index defined as the proportionality constant between temperature summation, R, and the wood volume (basal area multiplied by stand height). Following an approach similar to that used with the index, N, the average proportionality constant between wood volume and temperature summation is 1.26 ± 0.12 s.e. based on the 64 sites (Fig. 7.21) for which the temperature summation can be calculated. Sites with values above 1.25 have produced greater wood volume than can be explained on the basis of the temperature summation and it is hypothesized that this greater production is due to species differences and other site factors such as soil fertility or moisture.

The new site index was then calculated for each site and the deviations from the average value investigated. But once again no significant relationships could be found with other parameters of site conditions such as soil depth, precipitation, or radiation balance. The only general pattern which emerged was that managed stands tended to have higher values for the index (mean of 2.11 ± 0.23 s.e.) compared to the other stands (mean of

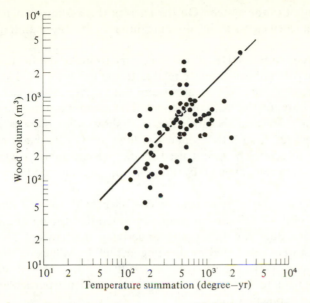

Fig. 7.21. Wood volume (basal area × stand height) as a function of temperature summation for the WDS.

0.97 ± 0.11 s.e.). In addition, it was possible again to discern the effects of aging with stands over 130 years of age showing a lower site index (mean of 0.62 ± 0.15 s.e.) compared to the rest of the stands (mean of 1.35 ± 0.13 s.e.). In general, it would appear that the extremes of variation represented in the WDS and the variety of species and site conditions are of sufficient magnitude that any simple approach to a site index method that would be valid across all forests is not possible.

Testing ecosystem hypotheses with the woodlands data set

Ecosystem analysis is a relatively new field which has, as yet, generated little theory which can be tested with existing data. Nevertheless, some current hypotheses can be tested with the WDS.

Olson (1963) analyzed the quantity of litter and rates of decomposition for various forest types. He predicted differences in the dynamics of evergreen versus deciduous forests and substantiated these predictions with model results and literature data. The basis for his comparisons was the relationship between organic matter ·on the forest floor and the rate at which this organic matter is turned over. To calculate a turnover coefficient, he assumed that the rate of loss of litter was at least approximated by the rate of litterfall. Thus, even though the amount of litter on the forest floor might be gradually increasing, the decomposition rate would be closely

439

related to the litter input to the system. On the basis of this assumption, he calculated a turnover coefficient, k, by dividing litterfall by the standing crop of litter.

To apply this analysis to the WDS, it was necessary to make the further assumption that total litterfall was linearly related to the leaf litterfall. This assumption was necessary to include many sites in the analysis which gave litterfall information only for the leaf component. The assumption appears to be a reasonable one, however, because Fig. 7.22 shows that there is a linear relationship between total litterfall and leaf litterfall. On the basis of this relationship, the turnover coefficient used in the present analysis was calculated as the leaf litterfall divided by the soil top organic matter as given in Chapter 11.

Fig. 7.23 compares the value of the turnover coefficient with the soil top organic matter for 60 of the WDS sites. It should be noted that the tightness of fit to a straight line does not imply any underlying mechanism. The values on the ordinate were calculated by dividing leaf fall by values on the abscissa (i.e. soil top organic matter). As a result of this method of calculation, the data tend to fall along a straight line and no particular significance should be applied to the fit.

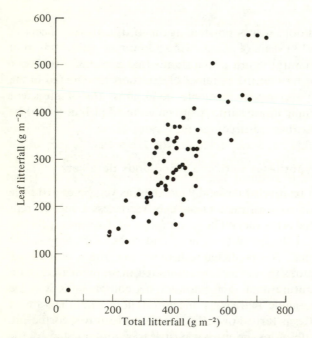

Fig. 7.22. Relationship between leaf litterfall and total litterfall for the WDS. On the basis of this relationship, it appears reasonable to assume a linear proportionality between the two variables.

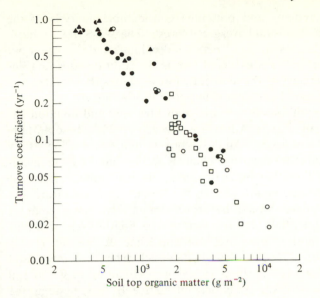

Fig. 7.23. Litter turnover coefficient as a function of the amount of soil top organic matter for the WDS (Chapter 11). The tightness of fit is an artifact of the method for calculating the coefficient and does not imply any underlying mechanism. The important feature is the arrangement of the forest types along the line. ▲, tropical broad-leaved type; □, boreal needle-leaved type; ●, temperate broad-leaved type; ○, temperate needle-leaved type.

The real information in the figure is contained in the patterns of the forest types as they lie along this line. First, tropical systems tend to lie in the upper left and boreal systems in the lower right. This distribution was predicted by Olson (1963). Rapid turnover rates in the tropics lead to small amounts of litter, while reduced decomposition rates cause the build-up of organic matter in boreal climates. However, the temperate system in the figure spans the entire range of conditions. This relationship was not predicted by Olson, who assumed that intermediate temperatures would cause these systems to lie toward the middle of the line. It appears that the range of moisture, soil, and nutrient conditions represented by the temperate forest sites in the WDS include a broad range of decomposition and litter conditions.

A second pattern which can be distinguished in the figure is that broad-leaved systems generally lie in the upper portions of the graph, while needle-leaved systems are concentrated in the lower portion. There is a broad overlap, but the distinction is visually evident. The pattern may be a consequence of the recalcitrant nature of the needles which retards the rate of decomposition and leads to more litter on the forest floor.

The patterns which were predicted by Olson (1963) seem to be confirmed by the present analysis. Olson correctly predicted the relationship between tropical and boreal systems and presented considerable analysis of the difference between deciduous and evergreen forests. The WDS data indicate that the significant relationship is not between deciduous and evergreen but between broad-leaved and needle-leaved forests. In addition, this analysis suggests that site variability in the temperate forests causes these systems to span a very broad range of decomposition/litter conditions.

A different set of hypotheses about ecosystem structure and function has been offered by O'Neill, Harris, Ausmus & Reichle (1975). These authors consider ecosystems with particular reference to mineral cycling. They observe that systems in more favorable climates are not limited by temperature or moisture and, therefore, nutrient conserving mechanisms should be best developed under these conditions. In the absence of temperature and moisture stress, ecosystems would be expected to evolve more complex mechanisms for the retention and recycling of the limiting resource nutrients. They cite several studies (e.g. Jordan, Kline & Sasscer (1972); Shugart, Harris' Edwards & Ausmus (1978) which infer that mesic and tropical forests have evolved nutrient recycling mechanisms in which soil organic matter and wood tissue are closely interactive. In these systems, the combined wood and soil organic matter form a large reservoir in which nutrients are stored and very slowly released through mortality and decomposition. If the soil organic matter and wood components do indeed serve as an important nutrient reservoir, then we would expect this reservoir to be larger and more important as climatic conditions for growth become more favorable. Stated more precisely, we hypothesize that the amount of soil organic matter plus wood tissue per gram of active leaf tissue should increase as climatic factors become more favorable.

To test this hypothesis, it is necessary to derive some index of climate that combines both temperature and moisture conditions. As an indicator of moisture stress, potential evapotranspiration, PE, is proportional to the summation of daily mean temperatures over $10°C$. Since daily mean temperatures are not available in the data sets, we estimated the summation by taking the average growing season temperature, G, multiplied by the length of the growing season, L. Because the growing season is most likely to have temperatures in excess of $10°C$, the approximation should be close. The exact relationship is given by

$$PE = 0.18 \, G \cdot L. \qquad (7.7)$$

By dividing the potential evapotranspiration into the actual precipitation during the growing season, P, we arrive at an index of moisture stress. When this value exceeds 1.0, precipitation is greater than potential evapotranspiration and, as the index falls below unity, moisture stress becomes more likely. We have then combined this index of moisture stress with the average annual temperature, T, to derive a measure of climatic conditions, π,

$$\pi = T\frac{P}{PE}. \tag{7.8}$$

To test the hypothesis that forest ecosystems in more favorable climatic regions maintain greater nutrient retention capability, we can compare the index, π, with the ratio of active leaf tissue to reservoir material (bole and branch biomass plus soil top organic matter). Several additional factors need to be considered. Management practices on many forest stands are specifically designed to increase bole biomass and could be expected seriously to distort the present analysis. Therefore, it was necessary to concentrate on natural stands and eliminate managed systems from the comparison. It is also obvious that the age of the stand will affect the amount of bole material present. To minimize the error introduced by age effects, stands between 50 and 130 years of age were selected for the comparison.

Unfortunately, the stringent requirements for data permit the inclusion of only 16 stands. Increasing values of the climatic index, indicating increasingly favorable temperature and moisture regimes, are associated with increasing size of the hypothesized nutrient reservoir (Fig. 7.24). Although it may be an artifact of the small number of sites involved, the relationship approximates a straight line. The boreal forests tend to lie toward the origin and tropical systems in the far upper right, with temperate and broad-leaved evergreen forests occupying intermediate positions. We conclude that the information from the WDS confirms the prediction of O'Neill *et al.* (1975) that systems in more favorable conditions will tend to accumulate organic matter to store and retain nutrients.

A third set of hypotheses about ecosystem function has been analyzed by O'Neill (1976). In this study, careful attention was paid to the role that heterotrophic organisms play in ecosystem function. Consumers and decomposers have often been considered simply as consuming excess production and playing a minor role in the maintenance and persistence of the ecosystem. However, recent studies, (Golley, 1973; Chew, 1974; Lee & Inman, 1975; Mattson & Addy, 1975; McNaughton, 1976) have emphasized that heterotrophs play important regulation roles in the system, providing feedback loops that help to stabilize the ecosystem.

O'Neill (1976) pointed out that systems differ in the functional nature of their energy-capturing base (i.e. photosynthesizing populations). Some ecosystems have a number of species, each with a different spectrum of environmental responses and each capable of rapid reproduction (e.g. phytoplankton communities). Other systems, such as forests, have large

443

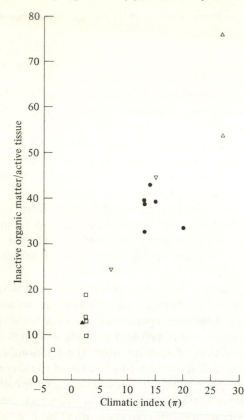

Fig. 7.24. The ratio of inactive organic matter (branch and bole biomass plus soil top organic) to active leaf tissue as a function of the climatic index, π. The index is calculated as average annual temperature multiplied by precipitation during the growing season divided by potential evapotranspiration calculated according to the approximation of Budyko (1956). The linear relationship implies that systems under more favorable conditions maintain a greater nutrient retention reservoir (inactive tissue) to support a unit of active photosynthesizing tissue. $\triangle$, tropical evergreen forests; $\blacktriangle$, tropical deciduous forests; $\triangledown$, broad-leaved evergreen forests; $\bullet$, temperate deciduous forests; $\square$, boreal evergreen forests.

structural components to maintain the plants through unfavorable periods. Without some form of rate regulation, the potential exists for autotrophic populations to exhaust available water and nutrient resources, resulting in collapse of the energy base with disastrous consequences for the ecosystem. The potential for exhausting the available resources is greatest in those systems with the most rapid turnover of autotrophic populations and therefore it is not surprising to find that planktonic systems maintain an order of magnitude greater heterotrophic biomass (e.g. 21–32 g dry matter m^{-2} (Harvey, 1950; Riley, 1956)) than do forests, (e.g. 0.31–0.25 g m^{-2} (Satchell, 1971; Reichle *et al.*, 1973)).

444

Heterotrophic populations may exert more control than would be indicated by the small amount of organic matter actually consumed. If the ecosystem is conceived as merely the chance coincidence of populations with overlapping environmental requirements, one might expect the heterotrophic standing crop to be proportional to net primary production. With a larger supply of available energy, one might expect more heterotrophs to be supported without disrupting the system. The data analyzed by O'Neill (1976) did not show any relationship between net primary production and heterotrophic biomass, and analysis of the WDS also produced negative results. Therefore, the data do not seem to support the contention that heterotrophs are simply 'parasitic' on the net primary production of the system.

On the other hand, if the heterotrophs are actively participating in the functional dynamics of the system and are playing a role of rate regulation, we would expect that greater heterotrophic biomass would be supported per unit of autotrophic biomass as the need for this regulation increased. That is, as the potential for rapid population increases and the potential for rapid exhaustion of limited nutrient resources increases, the need for rate regulation would increase. The hypothesis then develops that the more rapid the turnover of the total ecosystem, the greater the need for heterotrophs. That is, one would expect greater heterotrophic standing crop per unit autotrophic standing crop as the production turnover of the total ecosystem increases (i.e. as the potential for rapid change increases). O'Neill (1976) indeed found this relationship by comparing different ecosystems ranging from tundra through forests to pond ecosystems.

It is possible to test this hypothesis with recently published data summarized by Whittaker & Likens (1973). These authors construct the biotic sections of a global carbon budget and present data on a wide range of ecosystems types. Fig. 7.25 compares the heterotroph/autotroph ratio for 10 systems with increasing turnover times. The turnover time was calculated as the total organic matter above-ground (plants, animals, and litter) divided by the net primary production. The result is the number of years required to turn over the standing crop of organic matter. The graph clearly shows that as turnover times increase, the heterotroph/autotroph ratio decreases. As only above-ground components are listed and no aquatic systems are included in the comparison, the relationship between the systems is slightly altered from that presented by O'Neill (1976). In the present analysis, grassland and savanna systems have the most rapid turnover times and the greatest heterotroph ratios. The grasslands are followed by marshes, woodlands, and deserts, then temperate deciduous and tropical forests, and finally temperate evergreen and boreal forests. The tundra systems appear to fall off the main line and this is probably due to the omission of the large and slow soil organic matter pool in these systems which would tend to

445

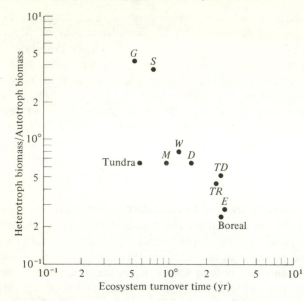

Fig. 7.25. Ratio of heterotroph to autotroph biomass as a function of total ecosystem turnover time. Data taken from Whittaker & Likens (1973). System turnover time is calculated as total above-ground biomass plus litter divided by above-ground net primary production (O'Neill, 1976). The relationship indicates that greater relative biomass of heterotrophs is maintained in systems with more rapid turnover and greater potential to exhaust limited resources. This seems to verify the rate regulatory role of heterotrophs in the ecosystem. *G*, grasslands; *S*, Savannas; *M* marshes; *W*, woodlands; *D*, deserts; *TD*, temperate deciduous forests; *TR*; tropical forests; *E*, temperate evergreen forests; *B*, boreal forests.

show a much longer turnover time and imply a closer correlation with the rest of the systems.

It is also possible to test this hypothesis with four of the WDS sites, where sufficient information is available on heterotrophic biomass. Again the relationships calculated were based on above-ground biomass of the system. Fig. 7.26 shows that the expected relationship exists, with the lowered heterotroph ratio being associated with systems with longer turn-over times. Both analyses (i.e. Figs. 7.24 and 7.25) reinforce the hypothesis that consumers play an active rate regulation role in ecosystems, and the more rapid the turnover of the system, the greater the heterotroph biomass required to maintain and regulate energy flow through the system. O'Neill (1976) offers considerable analysis of this theoretical concept and shows that effective regulation of the autotrophic components of the ecosystem can be expected even though the required changes in the heterotrophic components needed to effect the regulation are quite small and might be extremely difficult to detect in normal field sampling.

The analyses presented in this section illustrate that forest ecosystems are well regulated functional systems. Viewed at the ecosystem level, beyond the

446

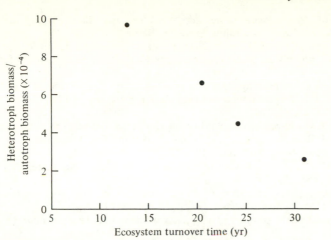

Fig. 7.26. Ratio of heterotroph to autotroph biomass as a function of total ecoystem turnover time with data from WDS. Only four sites have sufficient information to permit inclusion in the figure. Calculations are based on above-ground biomass only. Although it appears possible to draw a straight line through these points, other analyses (Fig. 7.25 and O'Neill, 1976) do not show such a simple relationship.

questions of adaptation and productivity of individual species, patterns emerge indicating that complex functional entities have evolved toward stability and persistence. This concept, which has rapidly emerged in recent years, is reinforced by testing against the Woodlands Data Set.

References

Bonnevie-Svendsen, C. Gjems, O. (1957). Amount and chemical composition of the litter from larch, beech, Norway spruce and Scots pine stands and its effect on the soil. *Meddelelser fra det Norske Skogfasosksvesen* **48**, 111–74.

Bray, J. R. & Gorham, E. (1964). Litter production in forests of the world. *Ad. ecol. Res.* **2**, 101–57.

Budyko, M. I. (1956). Teplovoy balans zemmoy poverkhnosti (*The heat balance of the earth's surface.* Translated by N. I. Stepanova, 1958. US Weather Bureau, Washington). Gidrometeoi, Leningrad.

Chew, R. M. (1974). Consumers as regulators of ecosystems: An alternative to energetics. *Ohio J. Sci.*, **74**, 359–70.

Crosby, J. S. (1961). *Litter and duff fuel in shortleaf pine stands in southeast Missouri.* Technical Paper, Central States Forest Experiment Station 178.

Czarnowski, M. S. (1964). *Productive capacity of locality as a function of soil and climate with particular reference to forest land.* Louisiana State University Press, Baton Rouge.

Danckelmann, B. (1887a). Streuertragstafel fur Kiefern bestands. *Zeitschrift für forst und Jagdwesn*, **19**, 457–66.

Danckelmann, B. (1887b). Streuertragstafel fur Buchen und Fichtenhochwaldungen. *Zeitschrift für Forst und Jagdwesn*, **19**, 577–87.

447

Ebermayer, E. (1876). *Die gesamte Lehre der Waldstreu mit Rucksicht auf die chemische Statik des Waldbaues*. Julius Springer, Berlin.

Golley, F. B. (1973). Impact of small mammals on primary production. In *Ecological energetics of homeotherms*, ed. J. A. Gessaman, pp. 142–7. Utah State University Press, Logan, Utah.

Harper, J. L. (1967). A Darwinian approach to plant ecology. *J. Ecol.* **55**, 247–70.

Harvey, H. W. (1950). On the production of living matter in the sea off Plymouth. *J. mar. biol. Ass. UK* **29**, 97–137.

Jordan, C. F. (1971). A world pattern in plant energetics. *Amer. Scientist*, **59**, 425–33.

Jordan, C. F., Kline, J. R. & Sasscer, D. S. (1972). Relative stability at mineral cycles in forest ecosystems. *Amer. Naturalist*, **106**, 237–53.

Lee, J. J. & Inman, D. L. (1975). The ecological role of consumers – An aggregated systems view. *Ecology*, **56**, 1455–8.

Lieth, H. (1972*a*). Computer mapping of forest data. *Proc. 51st Annu. Mtg, Appalachian Sect. of the Soc. of Amer. Foresters*, pp. 53–79.

Lieth, H. (1972*b*). Modeling the primary productivity of the world. *Tropical Ecol.* **13**, 125–30.

Lieth, H. (1973). Primary production: terrestrial ecosystems. *Human Ecol.* **1**, 303–32.

Lieth, H. & Box, E. (1972). Evapotranspiration and primary productivity. In *Thornthwaite memorial model*, **2**, *Papers on selected topics in climatology*, ed. J. R. Matter, pp. 37–46. *Publications in climatology*, **25**, No. 3. C. W. Thornthwaite associates, Centerton, N. J.

Lieth, H. Radford, J. S. (1971). Phenology, resource management and synagraphic computer mapping. *BioScience*, **21**, 62–70.

McMahon, T. (1973). Size and Shape in Biology. *Science*, Washington **179**, 1201–4.

McNaughton, S. J. (1976). Serengeti migratory wildebeest: facilitation of energy flow by grazing. *Science*, Washington, **191**, 92–4.

Mattson, W. J. & Addy, N. D. (1975). Phytophagous insects as regulators of forest primary production. *Science*, Washington, **190**, 515–22.

Olson, J. S. (1963). Energy storage and the balance of producers and decomposers in ecological systems. *Ecology*, **44**, 322–31.

O'Neill, R. V. (1976). Ecosystem persistence and heterotrophic regulation. *Ecology*, **57**, 1244–53.

O'Neill, R. V. Harris, W. F., Ausmus, B. S. & Reichle, D. E. (1975). A theoretical basis for ecosystem analysis with particular reference to element cycling. In *Mineral cycling in southeastern ecosystems*, ed. F. G. Howell, J. B. Gentry & M. H. Smith. ERDA Symp. Series. CONF–740513. Washington, DC.

Reichle, D. E., Dinger, B. E., Edwards, N. T., Harris, W. F. & Sollins, P. (1973). Carbon flow and storage in a woodland ecosystem. In *Carbon and the biosphere*, ed. G. M. Woodwell & E. Pecan. AEC Symp. Series. CONF–720510. Washington, DC.

Reichle, D. E., O'Neill, R. V. & Harris, W. F. (1975). Principles of energy and material exchange in ecosystems. In *Unifying concepts in ecology*. ed. W. H. Van Dobben & R. H. Lowe-McConnell. Dr W. Junk, The Hague.

Riley, G. A. (1956). Oceanography of Long Island Sound, 1952–4. IX. Production and utilization of organic matter. *Bulletin Binghamton Oceanographic Collection*, **15**, 324–44.

Rodin, L. E. & Bazilevich, N. I. (1965). *Production and mineral cycling in terrestrial vegetation*. Translated by G. E. Fogg. Oliver & Boyd, London.

Satchell, J. E. (1971). Feasibility study of an energy budget for Meathop wood. In *Productivity of forest ecosystems*, ed. P. Duvigneaud, pp. 619–30. UNESCO, Paris.

448

Shugart, H. H., Harris, W. F., Edwards, N. T. & Ausmus, B. S. Modeling of the belowground processes associated with roots in deciduous forest ecosystems. *Pedobiologia*, **17**, 382–8.

Soon, S. W. (1960). *Der Einfluss des Waldes auf die Boden.* Translated by P. Kundln. Gustav Fischer Verlag, Jena.

Sukachev, V. & Dylis, N. (1964). *Fundamentals of forest biogeology.* Translated by J. M. MacLennan. Oliver & Boyd, Edinburgh & London.

West, D. C. & Johnson, W. C. (1980). The density–diameter distribution as a function of tolerance. *Ecology*, (in press).

Whittaker, R. H. & Likens, G. E. (1973). Carbon in the biota. In *Carbon and the biosphere*, ed. G. M. Woodwell & E. Pecan, pp. 281–300. AEC Symp. Series. CONF–720510. Washington, DC.

Yoda, K. Kira, T. & Hozumi, K. (1963). Self-thinning in overcrowded pure stands under cultivated and natural conditions. *J. Biol. Osaka City University*, **14**, 107–29.

8. Analysis of biomass allocation in forest ecosystems of the IBP*

R. H. GARDNER & J. B. MANKIN

Contents

Evidently for science there is neither truth nor reality but only the possibility of rationalization and the hope of reliability, and both of these can often be achieved in many ways. Apart from purpose and inspiration little remains but economy to guide our choice.   J. G. Skellam (1971)

Forested ecosystems contain approximately 90% of the earth's biomass and cover approximately 40% of its land surfaces (Whittaker & Likens, 1973; Olson, 1975). The importance of forests for fuel and fiber has long been evident (Burgess, 1978); new investigations are emphasizing the importance of other aspects of forest systems. For example, the cutting and burning of forests may have played a significant role in the recent elevation of atmospheric carbon dioxide (Baes, Goeller, Olson & Rothy, 1976; 1977; Bolin, 1977; Adams, Mantovani & Lundell, 1977). This problem is particularly acute in developing countries where the per capita consumption of

*Research supported by the Eastern Deciduous Forest Biome, US–IBP, funded by the National Science Foundation under Interagency Agreement AG–199, DEB76–00761 with the US Department of Energy. Publication No. 1496, Environmental Sciences Division, Oak Ridge National Laboratory and Contribution No. 339, Eastern Deciduous Forest Biome, US–IBP.

451

forest products is high and expected to increase. Because of the importance of diminishing forest resources the uncertainties which remain about their productivity and potential (Whittaker & Likens, 1973; Sharpe, 1975) and the effects of large scale changes in forested areas (Boremann *et al.*, 1974) require immediate attention.

The need for accurate and extensive data about forests and man's relationship with these systems was an early goal of the PT (Terrestrial Productivity) Section of IBP (Worthington, 1975). Accurate information on land areas, species composition, biomass pools and fluxes, and forest productivity were recognized as necessary, if an assessment of current resources and confident recommendations for long range management policies were to be made. Biomass studies were recognized as the first step in terrestrial productivity investigations (Worthington, 1975), because these data help define the structure of the system and provide a framework for process studies. It was also recognized that these studies would be greatly facilitated by the methods of systems modeling and analysis (Neuhold, 1975; Worthington, 1975). The woodlands efforts were particularly successful in synthesizing these aspects of ecosystem studies and produced both a variety of forest models (O'Neill, Chapter 7) and a comprehensive data set of forest biomass and productivity values (DeAngelis, Gardner & Shugart, Chapter 11).

The purpose of this chapter is to examine the IBP woodlands data on forest biomass pools and fluxes and to develop a method for comparing sites. The problems of this type of comparison are unique because the intent is to detect differences between terrestrial ecosystems rather than the dynamics of an individual ecosystem. Differences which may arise as artifacts of a particular site, or due to uneven representation of a given ecosystem, confound the problems of comparison. Therefore, in the first part of this chapter a simple generalized model is developed, the above problems are considered, and the suitability of this model as a comparative and analytical tool is examined. In the second portion of the chapter, the model is used to summarize and characterize the IBP woodlands data.

Donor-controlled linear compartment models

Donor-controlled linear compartment models (DCLMs) have been extensively used to display pools, fluxes and annual budgets of elemental and biomass dynamics of ecosystems (Smith, 1970; Caswell, Koenig, Resh & Ross, 1972). We will not review this subject, (for a general treatment see Waide, Krebs, Clarkson & Setzler, 1974; Weigert, 1975; Patten, 1975) but some characteristics of DCLMs need to be briefly discussed.

The concept of mass (or energy) flow between distinct components of an ecosystem is a fundamental ecological paradigm (Lindeman, 1942; Odum,

1957; Golley, 1965). Model conceptualization requires that an ecosystem be divided into structural or functional components with the separation largely dependent on the objectives and perception of the modeler. For example, to model concentrations of elements in a forest one might lump all species of autotrophs together and partition the forest into bole, branch, leaves, etc. If the objective were to display food web dynamics, the lumping might be based on trophic level. In this manner pertinent information concerning pools, fluxes, and annual budgets of ecosystems can be parsimoniously presented (Reichle, 1975). The process of partitioning is particularly suited for the convenient comparison and contrast of similar ecosystems, such as the forest data compiled in the IBP Woodlands Data Set (Ulrich, Mayer & Heller, 1974; DeAngelis *et al.*, Chapter 11).

The description of the mass and/or energy flow by systems of linear, ordinary differential equations presents a potentially powerful tool for analyzing and comparing ecosystem functions and processes. Models based on data sets which describe ecosystems for short periods of time can be extrapolated to longer time spans and manipulations of the model can be quickly considered to test alternative hypotheses. Such an exercise depends strongly on the structure and mathematical representation of the model. The assumption of linearity restricts responses to a relatively small range (e.g. an unperturbed system and relatively stable environmental conditions), and the requirement of stability imposes the donor-controlled restriction. This produces a model with restricted properties (Reeves, 1971) but with desirable mathematical stability (Funderlic & Heath, 1971; Waide *et al.* 1974). If successful, the changes in a compartment with time (transient response) and the amount in a compartment at ecosystem maturity (steady state value) of each state variable can be used to describe the intermediate and final states of the modeled system.

The adequacy of DCLMs to model ecosystem dynamics has been a controversial topic discussed and reviewed by a number of authors (*see* Patten, 1975; Wiegert, 1975). It is clear from these discussions that DCLMs have some serious deficiencies which prevent them from being universally applicable. For instance, the logistic curve, typical of many biological phenomena, can only be represented by nonlinear systems. The assumption of donor-controlled systems is also inadequate for explaining feeding and competition interactions of many herbivores when the amount of plant material eaten is dependent upon factors such as the number and size of consumers, palatability of the food, season, etc. Generally, nonlinear systems allow more detailed and realistic expressions of biological phenomena and interactions (Wiegert, 1975). However, there are many advantages to using linear systems to represent ecosystem dynamics (Patten, 1975), including the fact that linear systems are easily implemented and a great body of theory exists with which to interpret results.

453

Descriptions of forest biomass distributions

The IBP Woodlands Data Set is a collection of information on annual pools and fluxes of biomass. Because these data are not complete for any site and because no causal relationships are described it is unrealistic to consider a nonlinear model as a general representation of woodland sites. The use of a general DCLM to describe and compare sites has some distinct advantages including:

(1) The diversity of ages among forests can be removed.
(2) The sensitivity of certain parameters and model relationships to the behavior of an individual system can be tested and compared within and between forest types.
(3) By introducing variability and statistical dependencies into the parameter estimates, confidence intervals can be placed about the transient and steady-state conditions of a model (Gardner, Mankin & Shugart, 1976). Quantitative comparisons between forest sites can then be made, or, alternatively, the required confidence needed to make assertions and inferences can be determined for a specific model.

With these possibilities in mind and the availability of forest biomass data from the IBP studies (Chapter 11) it is our objective to:

(1) investigate the effects that the model structure (number and size of compartments and the fluxes between them) has on the behavior of the model.
(2) investigate the behavior of models when parameters are allowed to exhibit statistical variability and interdependencies.
(3) use what has been learned from pursuing objectives (1) and (2) to develop a general model to compare the dynamics of biomass allocation in the IBP woodland ecosystems.

The COMEX computer code

General description and rationale

A general Monte Carlo simulation program, COMEX, was specifically developed to meet the objectives listed above. For each DCLM considered, COMEX requires the means and variances for all parameters and then generates sets of solutions by randomly selecting parameter values from the multinormal distribution. Thus, the statistical variabilities of the parameters are combined by COMEX so that their cumulative effect on model results can be quantitatively assessed. Details of the mathematical methods used by COMEX are given in Appendix A. This approach is particularly useful for

454

forest model applications which attempt to extrapolate measurements over a short time span to predict transient conditions over centuries. Even where measurements for many stands of different ages are combined, it is difficult to incorporate the necessary information into a model and adequately to judge the results.

COMEX provides a method of comparing models with minor structural differences so that the 'best' (or the few 'best') model structures are identified. Because 'best' is a subjective criterion, it is necessary to first state the modeling objectives and then use COMEX to provide the necessary quantitative information. For example, if the objective is to find the 'best model' to predict the amount of material in compartment i and all models are equally realistic, then COMEX can combine the variances and statistically characterize the model results. The 'best model', in this case, would be the one with the narrowest confidence interval around the predicted amounts in compartment i.

COMEX can be used to determine, before any measurements have been made, what variability in parameter values is acceptable for prespecified levels of variability in the model results. This is done by reiteratively running a model with gradual increases in the levels of parameter variability. Those results which are less variable or equal in variability to the objective criterion then indicate the acceptable level of parameter variances. Further examination can then be made to identify sensitive parameters (i.e. those requiring smaller variances for a prespecified level of variability in the results), or indicate an overall level of variability for the model parameters. The results can establish some requirements of the data before resources have been committed.

The IBP woodlands general linear model

Model description

The general DCLM illustrated in Fig. 8.1 was parametized by using data from the Oak Ridge *Liriodendron* forest (Reichle & Edwards, 1973; Sollins, Reichle & Olson, 1973). Some model parameters were adjusted so that the predicted values of each compartment at 50 years of age matched the mean values calculated from the direct measurements of the forest. The values for the resulting transfer matrix are given in Table 8.1. The manner in which parameters are calculated and fit to a DCLM can have profound effects on the final steady state values and the transient responses of the model. Initial estimates of biomass amounts, increments, fluxes and losses are subject to sampling errors and natural variability so care must be exercised to prevent the introduction of additional error. We have used the following procedure:

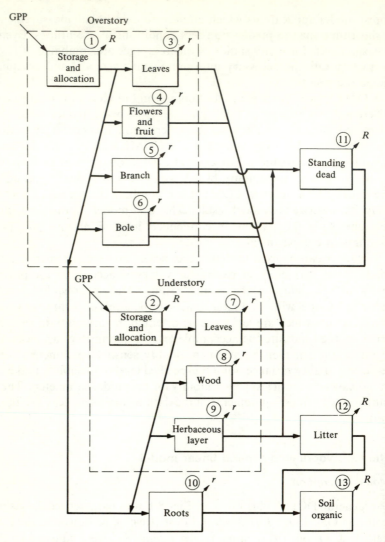

Fig. 8.1. General linear model for the IBP Woodlands Data Set. This generalized 13 compartment model represents the major over- and understory components of a forest. GPP indicates the fraction of gross primary production for overstory and understory components. The model may be parameterized in two ways: (1) Model I – all respirations accounted for by the storage compartments (O, all r=0). (2) Model II – respiration values are estimated for each compartment and R or r indicate the amount of respiration for that compartment only.

(1) Initial parameter estimates were made by considering each compartment as isolated from the system. The initial estimate of the transfer coefficient from compartment *i* to compartment *j* was obtained by using

Table 8.1. *Transfer coefficient matrix and vector of forcing functions for the general linear model*

Ā matrix (upper number = Model I, lower number = Model II)

	1	2	3	4	5	6	7	8	9	10	11	12	13	b vector
Overstory storage — 1	-0.99588 / -0.99754													2592.8
Understory storage — 2		-0.69133 / -0.6931												125.0
Overstory leaves — 3	0.0838 / 0.1947		-0.68 / -1.58											0.
Overstory flowers — 4	0.00608 / 0.01404			-0.69315 / -1.60815										0.
Overstory branch — 5	0.058 / 0.1245				-0.05143 / -0.10993									0.
Overstory bole — 6	0.15 / 0.214					-0.03317 / -0.04747								0.
Understory leaves — 7		0.17 / 0.273					-0.61366 / -0.97426							0. / 0.
Understory wood — 8		0.13 / 0.2						-0.1518 / -0.01598						0. / 0.
Herbaceous — 9		0.1074 / 0.12							-0.69315 / -0.77465					0. / 0.
Roots — 10	0.098 / 0.45	0.09713								-0.165 / -0.72				0. / 0.
Standing dead — 11					0.01741 / 0.01741	0.0295 / 0.29					-0.52303 / -0.64603			0. / 0.
Litter — 12			0.68 / 0.68	0.69315 / 0.69315	0.03402 / 0.03402	0.00367 / 0.00367	0.61366 / 0.61366	0.1518 / 0.1518	0.69315 / 0.69315		0.13103 / 0.13103	-0.76257 / -0.76257		0. / 0.
Soil organic — 13										0.165 / 0.160		0.06257 / 0.06257	-0.013 / -0.0127	0. / 0.

The general linear model is illustrated in Fig. 8.1. The upper number in the Ā matrix represents Model I and the lower number represents Model II. The b vector represents the yearly input from gross primary production (GPP) and is the same for both models.

457

the measured flux from compartment i to compartment j, f_{ij}, and the amount of material in compartment i, q_{ij}, at time t. Since the IBP Woodlands Data Set reports average annual increment, we obtain the following initial estimates of annual flux and biomass

$$f_{ij} = \frac{q_{i(t+1)} - q_{i(t)}}{(t+1) - t},\tag{8.1}$$

and

$$q_{i(t+1)} = e^{a_{ij}} q_{i(t)},\tag{8.2}$$

where a_{in} is the transfer coefficient (element of the A matrix) from compartment i to compartment j. Therefore,

$$f_{ij} = (e^{a_{ij}} - 1) q_{i(t)},\tag{8.3}$$

or

$$a_{ij} = \ln(f_{ij}/q_i + 1).\tag{8.4}$$

(2) These initial estimates were then adjusted by iterative fitting on a digital interactive linear differential equation modeling program (Rust & Mankin, 1976). Since the initial estimates did not take into account interaction between compartment and forcing functions, the parameters were adjusted observing the following constraints:

(a) The measured initial conditions at year Y (the age of the forest) plus the reported yearly increment must equal the amount in each compartment in year $Y+1$ with error less than 1%.
(b) The soil organic matter (compartment 13) was assumed to be at steady state.
(c) Where choices were necessary, the final adjustments were made by altering the respiration terms rather than transfers from one compartment to another.

Care was taken during the initialization process to ensure that the results reflected the measurements actually made. In some instances it was necessary to obtain information from other sources (Sollins, Reichle & Olson, 1973; Edwards & Sollins, 1973; Edwards & Harris, 1977) to completely parameterize a model.

Among the parameters that must be estimated, respiration values are the most difficult to obtain. For this reason, two versions of the general model were formulated (Fig. 8.1). Model I shows all of the respiration parameters on the two storage–biomass allocation compartments (R for 1 and 2), along with R for compartments 11, 12, and 13 (standing dead, ground litter, and soil organic, respectively). In this configuration, losses due to respiration and/or consumption are applied directly to the storage compartments.

Thus, two values replace seven individual values of r for compartments 3–10 in Model II (Fig. 8.1). Model I has the advantage that forest sites in the IBP Woodland Data Set that have not reported respiration values can be compared with a minimum number of assumptions. Model II, although a more realistic formulation, has few data sets to support it and is presented only for purposes of comparison. For this reason we fitted the transients of Model II so that they fell within 1% of Model I at steady state rather than pass directly through the exact values for the Oak Ridge *Liriodendron* site. The matrix of transfer coefficients and the forcing functions for these two models are listed in Table 8.1.

The data for each forest site in the IBP Woodlands Data Set were collected for a variety of reasons by numerous techniques. Differences due to variabilities in the method of calculation or reporting of yearly budgets required many hours of inspection, correspondence, and recalculation before the data could be uniformly tabulated (*see* Chapter 11 for full details). Nevertheless, deficiencies for certain critical values still exist necessitating still further calculations, assumptions, and model modifications before a final model structure (Fig. 8.1) and parametization (Table 8.1) could be completed. The iterative process between model development and data fitting is common to systems ecology and is often justified, as here, on grounds of computational convenience and/or because the purposes of the model require a different level of resolution. However, the cumulative effect of these processes on the dynamic behavior of the final model is poorly understood and rarely quantified. DCLMs are particularly sensitive to this problem because the only 'control' over the system is the amount of material available in the donor compartments(*s*). Therefore, before using COMEX to evaluate the IBP Woodlands Data Set, we must first discuss the effects and interactions of model structure and parameter variability.

Evaluation of model structure

During the early phases of development of COMEX it became apparent that the direct effects of parameter variation are influenced by model structure (i.e. number of compartments in the model, the number of transfers per compartment, the relative magnitude of transfer coefficients, etc.). Investigations with COMEX have also revealed that several additional factors may directly affect the behavior of transient responses, steady state values, and system performance. Some of these factors are:

1. The statistical distributions of steady state values when model parameters are variable.
2. The location of a compartment within a model.
3. The variability of the turnover rate of each compartment.
4. The statistical relationships (covariances) between model parameters.

459

5. The range of differences between parameters.
6. Combining (or splitting) parameters between compartments.

We have attempted to construct examples which isolate and illustrate these factors, but even simple examples are subject to indirect and subtle effects of model manipulation. It is also difficult to specify the effect that the above factors have on the variability of model components and/or the variability of the entire system. For this reason we have chosen to use the variability of the time to steady state (τ) as an index of system sensitivity, and the amounts at steady state and the variability of the steady state values as an indication of the effects on individual model components. This does not exhaust all possible relationships between parameter variability and model structure, nor does the following discussion present a complete analysis of these factors. The intention, however, is to consider those cases that have the most pertinent bearing on the analysis and interpretation of the IBP models and the Woodlands Data Set. Some examples lend themselves to direct analytical solutions and, where appropriate, these solutions are presented. Where the analytical solutions are more difficult, heavier emphasis is placed on the Monte Carlo simulations of COMEX.

Distributions of steady state values

When parameters of an ecosystem model are randomly varied, the distribution of steady state values of compartments are skewed to the right. Table 8.2 compares the deterministic solution for the model illustrated in Fig. 8.1, along with the means, medians, and means of the logs of the steady state values when parameters are varied by 10, 20, or 30%. In all cases, the exact distribution from which COMEX selects parameters is known. This permits the analytical determination of the distribution of compartment steady state values for a few simple examples. Inspection of the terms for linear compartment models indicates that the analytical solutions for each compartment will be determined separately, although their solutions will be dependent upon the previous donor compartments. If no complex feedback loops exist, then these solutions should be straightforward. However, consideration of a variety of cases makes this a complex task.

The results presented in Table 8.2 have important implications for model predictions and for the estimation of model parameters from the original data sets. The distributions of steady state values are skewed to the right with the mean overestimating the central point of the distribution (Fig. 8.2). This clearly biases predictions about the deterministic case or the central case for stochastic systems. Table 8.2 shows that the median is preferred as a more accurate measure of the center of distributions. Parametric confidence limits around the central transients or steady state values should be unequal with the upper confidence limits farther from the median. When the

Table 8.2. *Distribution of steady state values for the model illustrated in Fig. 8.2*

Compartment number	Deterministic solution	Parameter variance								
		10%			20%			30%		
		Mean	Median	LNM	Mean	Median	LNM	Mean	Median	LNM
1	6666.7	6751.2	6700.4	6729.8	6666.3	6544.1	6592.1	7039.4	6645.5	6844.4
2	2666.7	2670.6	2674.3	2662.8	2713.4	2659.4	2692.3	2757.1	2694.3	2677.0
3	1777.8	1782.0	1774.5	1769.1	1851.3	1809.6	1796.5	1941.9	1824.9	1791.9
4	3555.6	3586.5	3516.8	3547.0	3712.9	3566.4	3559.0	3979.4	3518.6	3309.3
5	1777.8	1781.5	1758.6	1769.9	1839.1	1808.9	1781.5	1828.3	1671.3	1700.9
6	3555.6	3559.4	3472.9	3522.0	3792.2	3568.8	3613.2	3923.7	3291.0	3359.3
MSE[a]		36.9	40.7	29.8	123.9	53.8	40.6	286.3	119.8	151.1

[a] $MSE = (\Sigma(s-d)^2/6)^{1/2}$ where s = stochastic result and d = deterministic solution. (Mean and median are for 300 simulations at each level of variability. LNM refers to the exponentiated mean of the natural log of each simulation.)

461

Dynamic properties of forest ecosystems

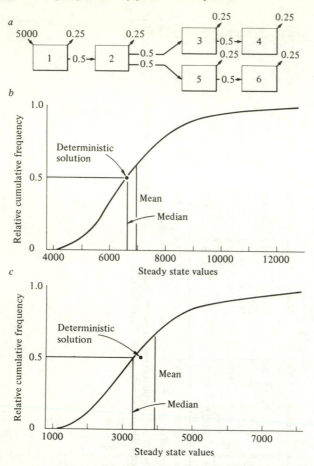

Fig. 8.2. *a*, hypothetical model; *b*, distribution of steady state values for compartment 1 of 300 simulations of the hypothetical model; *c*, distribution of steady state values for compartment 4 from 300 simulations of the hypothetical model. The model forcing (yearly, addition of fixed biomass) is 5000 g. The transfer coefficients are all equal to 0.5 and the respiration coefficients (losses from the system) are all equal to 0.25. All units are arbitrary. The deterministic solution is the steady state value when all parameter variances are 0.0. The mean and medians refer to arithmetic average and middle number of the distributions.

underlying probability distribution is unknown, as is usually the case, it is preferable to substitute upper and lower percentiles for parametric statistics.

If the distribution of parameters in natural systems is skewed, then the mean is a biased estimate and median values for increments and fluxes are preferred for parameterizing models. The results above suggest that, in models with no feedbacks, a normal distribution of steady state values can only result from parameters whose distribution is skewed to the left.

462

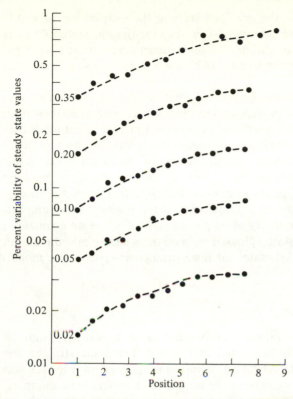

Fig. 8.3. Variability of steady-state values as a function of the *position* of the compartment. Each line is the result of 300 iterations of a 12 compartment cascading model with transfer coefficients equal to 0.5 and respiration coefficients (losses from the system) equal to 0.25. The percent variability of all parameters in the model is indicated for each line. Position of a compartment has been adopted from Kercher & Shugart's (1975) method of calculating trophic position. It indicates for each compartment the actual amount of biomass received per unit of time as a function of the total amount of biomass it could possibly receive.

Position of a compartment within an ecosystem model

The concept of position of a compartment within an ecosystem model has been adapted from Kercher & Shugart's (1975) discussion of trophic position. We use it here to indicate the proportion of material, in this case, biomass, received by a compartment as a fraction of the total amount of material it could possibly receive. By drawing an analogy with food webs passing units of biomass or energy along the trophic levels, the position of a compartment within an ecosystem model indicates the relative extent, or number of times, that a given unit has been passed from one compartment to another. The effect of position on the variability of individual compartments is illustrated in Fig. 8.3. The farther the compartment is from the forcing functions, the higher its position in the model, and the more variable

463

its biomass. As shown in the previous section, the distribution of steady state values becomes more extremely skewed to the right as material passes from one compartment to the next. For compartments located near position 7, the variability of the steady state values is nearly twice that of the individual parameters.

In the absence of feedback loops, the greatest amount of variability in ecosystems models should be expected in those compartments farthest from the forcing functions. Although this is sometimes apparent in nature, it may be desirable in some modeling tasks to reduce this 'whiplash' effect at the far end of the model. Some alternatives are to reduce the length of the model by lumping compartments or by expending greater efforts to measure precisely the parameters of these compartments further from the forcing functions. Fig. 8.3 indicates that knowledge of the position of a compartment and the relative variability of its parameters will allow an estimate of the variability of steady state values. This method is simple and direct, but is only suited to relative estimates for lower triangular systems (i.e. models with no feedback loops).

Variability of turnover rates

The effect of a variable number of parameters in a turnover rate on compartment dynamics is illustrated in Fig. 8.4. The same straight line model with direct transfers between compartments used in Fig. 8.3 was modified by successively increasing the number of transfers from compartments 6 and 7 while maintaining a constant mean turnover rate. With one transfer from each compartment, the mean value of that transfer coefficient is 0.5; with two transfers, each parameter is equal to 0.25; with three transfers, each is equal to 0.1666667; and with four transfers, each is equal to 0.125. The mean respiration of each compartment was maintained at 0.25 and all parameters were varied at 30% of their mean value.

The direct effect of increasing the number of parameters in a turnover rate, while maintaining a constant mean value, is a steady decline in the variability of the turnover rate (Fig. 8.3). This result is easily predicted from the equation for combining variances

$$\sigma^2_{\text{sum}} = \Sigma\sigma^2_i + 2\sum_{i=1}^{n}\sum_{j=1}^{i-1} \rho_{ij}\sigma_i\sigma_j, \tag{8.5}$$

where σ^2_{sum} is the variance of the sum, σ^2_i is the variance of the ith parameter, ρ_{ij} is the correlation between parameters i and j, and $\sigma_i\sigma_j$ is the product of the standard deviations for compartments i and j. In this example the second term disappears because all ρ_{ij} are equal to zero. Thus, the variances of the turnover rates are equal to the sum of the variances of the individual parameters. By taking the square root of σ^2_{sum} and dividing by 0.75 we obtain the expected percent variability of the turnover rate for compart-

464

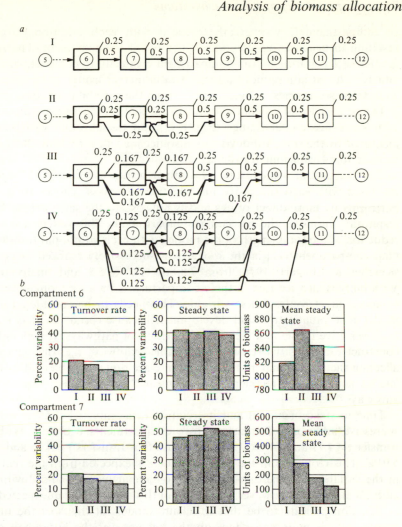

Fig. 8.4. The direct and indirect effects of increasing the number of transfers in a turnover rate. *a*, the successive modifications (I–IV) to compartments 6 and 7 of a 12 compartment model. Each successive modification introduces an additional transfer from compartments 6 and 7 while maintaining constant mean turnover rates for all compartments. *b*, the results of 300 simulations with all parameters (transfers) varying at 10% of their mean value. Turnover rate (a_{ii}) is the sum of transfers from compartment i to all other compartments. The steady state values are the amount of material in the compartment at equilibrium.

ments 6 and 7. For the models considered in Fig. 8.3, these are 22.4%, 17.3% 15.3%, and 14.1% for 1, 2, 3, and 4 transfers per turnover rate.

The biomass at steady state of compartment 6 did not significantly change with successive model modifications, and as a result of the decreasing variability of the turnover rate of the compartment, the relative variability of steady state values steadily declined. However, the mean biomass

465

in compartment 7 was greatly reduced with each additional transfer, resulting in increased relative variability of this compartment. The reason for this is that transfers from compartment 6 directly to compartments 8, 9, and 10 resulted in a reduced transport of biomass through compartment 7. This decreased the mean value and increased the variability of compartment 7. The percent variability of the turnover rates of compartment 7 did decline with successive model change (Fig. 8.4b) and the absolute variability, as measured by the standard deviation, also declined steadily from 263 for one transfer to 74 with four transfers.

The biomass at steady state of compartments 8 through 12 generally increased with each successive modification. Biomass reaches these compartments by more direct routes and is not lost, to the same extent, by the respiration terms of the preceding compartments. This is reflected in the reduction of the total length of the model; with each additional transfer, the final compartment (12) in the model was successively reduced in position (Kercher & Shugart, 1975) from 7.8, to 6.2, to 5.2 and finally to 4.5. Variability is also reduced in these compartments because biomass received by more direct routes is less variable. This needs to be considered during model conceptualization and development because certain models may be quite sensitive to the modification or deletion of pathways. It is possible to construct examples where several 'variable' pathways (i.e. the parameters affecting the flow of material along a route are quite variable) will result in less variance by the recipient compartment than by fewer, 'less variable' pathways.

Time to steady state (τ) and its relative variability (CV_{τ}) decreased with successive additions of transfers from compartments 6 and 7. With one transfer $\tau = 4.77$ and $CV_{\tau} = 113.2\%$ and with four transfers $\tau = 4.16$ and $CV_{\tau} = 6.91\%$. This reduction in the variability of τ is expected from the reduction in the variability of the turnover rates of compartments 6 and 7 with each additional transfer. However, the extent of this reduction is unexpected. The explanation appears to be that in 'chain' models of this sort the time to steady state is most dependent on the behavior of the latter half of the model and, in particular, the behavior of the last compartment (see Fig. 8.3). Additional transfers from compartments 6 and 7 cause this portion of the model to be 'better behaved' because it receives more predictable amounts of biomass. Thus, although the variability of the parameters has remained constant, the model is better behaved when additional transfers between compartments are possible.

Statistical relationships between parameters

Correlations between parameters of DCLMs exist. Their presence would imply causal relationships between parameters. The usefulness of considering these correlations when exploring ecosystem model behavior has not been considered. However, COMEX can make use of this information if these correlations can be theoretically identified or empirically measured. Inspection of eqn 8.5 indicates that considerable change in sensitivity can result from strong correlations between parameters. As an example, consider three transfer coefficients (a_{ij}) of one compartment, all equal to 0.25, that sum to the transfer rate (a_{ii}) of 0.75. If each a_{ij} varies at 30%, then the expected relative variability of the a_{ii} would be 17.32%. If, however, these three a_{ij}s are correlated at 0.8, then the variability of a_{ii} increases to 23.24% but is reduced to 7.75% if the correlation changes sign ($\rho_{ij} = -0.8$).

More complex associations, such as correlations between parameters of different compartments, can be postulated. Such relationships might be used to imply that all transfers of a certain type, e.g. respiration, behave in a similar manner (positive correlation) or transfers to one compartment imply lesser transfers to other compartments (negative correlations). The existence of these relationships modifies the statistical behavior of DCLMs and in some circumstances imitates the results of nonlinear systems. As an example, consider the case where respiration rates of compartment i are positively correlated ($\rho_{ij} = 0.8$) with the transfer coefficient from compartment j to compartment i. This implies, in an ecosystem–biomass transfer model, that as more material is received by a compartment, the rate at which that compartment loses energy through respiration increases. A real world example is that consumers sometimes expend more energy when the rate of intake increases. Applying this hypothetical situation to the previous model (*see* Fig. 8.2), all parameters were varied at 30% and correlations equal to 0.8 between respiration rates and the transfer coefficients from the preceding compartment were specified. The mean variability of each compartment is then reduced from 44.14% without correlations to 36.16% with correlations. This reduction was primarily due to the smaller range of variability observed between compartments, especially compartments in the higher positions (compartments 9 through 12). The observed range of variability with correlations was from 27.1% to 43.6%, and without correlations it was from 25.9% to 59.2%. Time to steady state (τ) of the model also dropped from 113.2% to 75.2% as a result of these changes.

In general, a variance–covariance (correlation) matrix could have an unexpected influence on the behavior of linear systems. The effect that these empirical and/or causal relationships might have needs further examination. Although the estimation of correlations between parameters will be difficult to obtain, this information may provide insight into ecosystem behavior and the models used to represent them.

Table 8.3. *The variability of turnover rates and relationship to τ^**

Compartment 3		Compartment 6		τ	
CV	r	CV	r	$\bar{x}$	CV
13.7	0.092 (0.1)	21.3	0.540 (0.0001)	4.61	19.9
20.8	0.344 (0.0001)	20.2	0.317 (0.0001)	4.72	21.9
27.2	0.733 (0.0001)	22.0	0.147 (0.01)	5.02	38.1
31.9	0.789 (0.0001)	20.6	0.193 (0.001)	5.18	42.6

The results in this table are based on a 6 compartment model. Respiration coefficients for each compartment are 0.25 and transfer coefficients between each compartment $(a_{j+1, j})$ are 0.5. There is an additional transfer between compartments 2 and 4 $(a_{4, 2} = 0.25)$. CV is the coefficient of variation (standard deviation/mean) of the turnover rates of compartments 3 and 6 and the time to steady state (τ). Listed in the column below r is the correlation of the turnover rate with τ (upper number) and the significance level of this correlation are the mean values ($\bar{x}$) of time to steady state and its relative variability (CV).

Relative magnitudes of parameter variances

The relative variability of particular parameters of some models can be important to total system behavior. One such example is the effect that the turnover rate (a_{ii}) of the slowest compartment has on τ and its variability. Table 8.3 shows an example of two compartments (3 and 6) with equal turnover rates (0.25). When the variability of the turnover rates of these compartments are approximately equal then their correlation with τ is approximately the same, as indicated by the correlation coefficient. Increasing the CV of one compartment, in this case compartment 3, results in a stronger correlation τ. A lower CV results in a weaker correlation with τ. The mean and variability of τ also respond to these changes.

If the parameters in a model become inordinantly variable, then the sensitivity of the model can be greatly affected. Table 8.3 demonstrates that as the variability of a_{33} increases, the mean variability of τ rapidly increases. When parameter variation is greater than 50% then model sensitivity is drastically affected. Instances of high parameter variability within a model may be compensated for by the behavior of other parameters (Tiwari & Hobbie, 1976). The existence of such mechanisms implies that methods may be found to make the behavior and variabilities of models more closely resemble the ecosystems they represent. Such relationships may be expressed or simulated by correlations between parameters. Constrained behavior may thus be imposed on systems which show wide fluctuations in individual parameter values. The possible existence of these relationships remains to be demonstrated.

Accounting for respirational losses

Systems with respirational losses deducted from an initial storage/allocation compartment (Model I, Fig. 8.1) demonstrate dynamic behavior different from those that account for respirational losses from the individual compartments (Model II, Fig. 8.1). This can be analytically demonstrated by simplifying these alternate model formulations into a two compartment case (see Appendix for the mathematical details). Although the parameters are calculated such that the steady state conditions for the two models are identical, the transients (amount of material at any point in time until system maturity) for the model with all respirational losses deducted from the first compartment are consistantly less than when respirational losses are taken from the individual compartments. This simplified model (in the sense of fewer parameters) underestimates the predicted amount of material (biomass) by any point in time until system maturity. Because these systems are donor-controlled, changes in the first few compartments of more complex models will also affect the dynamics of other dependent compartments.

Summary of model analysis

COMEX has proven to be a valuable computational and heuristic tool which quickly provides interpretable results about the variability of model behavior and the effects of alterations to model structure. Although it may be possible to solve analytically probabilistic models and describe precisely the variability of DCLMs, the characterization of model behavior before applying it to ecosystem analysis can now be efficiently accomplished by Monte Carlo methods.

Comparison of model predictions with the IBP woodlands data set

The constraints imposed upon the two forest models (Fig. 8.1) during development and formulation have a noticeable effect on their dynamic behavior. Because these models were simultaneously developed from the same assumptions and data base, we assume that the inevitable limitations that were beyond our control have affected both to an equal degree. Deliberate manipulation of the two models allows examination of differ-

469

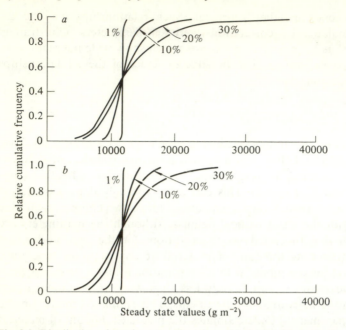

Fig. 8.5. Distribution of steady state values for *a*. Model I (with respiration) and *b*, Model II (without respiration). Relative cumulative frequency (*RCF*) is calculated for overstory boles by rank ordering the observations, assigning them a number, *i*, from 1 to *N*, *RCF* then equals *i/N*.

ences in behavior and predictability. In Model II, all respiration energy is withdrawn from the individual compartments, while in Model I the respiration terms for the compartments representing living plant tissues are consolidated into fluxes from each of the two photosynthate storage/allocation compartments. Although the fluxes from the storage compartments are less in Model I, because respiration has been accounted for, the two models are the same in all other respects. (Recall that Model I was fitted to the data for the Oak Ridge *Liriodendron* forest with the transients passing through the data points at 50 years of age (Fig. 8.6). Model II was intended as a comparison to Model I so its transients were made to be within 1% of Model I at steady state (approximately 250 years). The alternative would have been to fit both models directly to the data with resulting differences in steady state values).

Fig. 8.5 shows the distribution of steady state values for overstory boles for both Model I and Model II at 1, 10, 20, and 30% parameter variation. The cumulative frequency plots show that as parameter variability increases the range of steady state values becomes broader, and the values are increasingly skewed to the right. Although the median values for each model simulation are similar to one another, the means increasingly depart

470

from the central point of their distribution. The range of values for each level of variability is narrower for Model II, with fewer values skewed to the right (Fig. 8.5).

In the previous sections it was shown that the location from which respiration is removed can result in substantial differences in transient responses. Formulations similar to Model II have transients that reach steady state more rapidly than models in which respiration is removed from a central storage compartment (e.g. Model I). The variability of compartment steady states also is affected by the variability of the turnover rate of the compartment (*see* eqn 8.5). If all transfer coefficients from a compartment have equal levels of variability, the addition of another term will decrease the variability of the turnover rate. For instance, if all parameters of the overstory bole compartment (compartment 6 in Fig. 8.1) are varied at 20% with no correlations between parameters, eqn (8.5) shows that the relative variability of the turnover rate for boles will be 13.8% for Model II and 17.9% for model I. At 30% parameter variation, the relative variability for compartment 6 (boles) will be 20.7% and 26.9% for II and I, respectively. In general, Model II has narrower confidence bands around its transient responses and steady state values than Model I.

If confidence bands were our only concern, and all data sets had the required information for estimating respiration parameters, then Model II would clearly be preferred for making predictions and comparisons because, given the errors associated with the data, Model II indicates with greater precision (smaller confidence regions) where the true transient and steady state values lie. However, the data sets we wish to compare are incomplete with little information available for estimating respiration coefficients and uncertain levels of accuracy and precision for the data presented.

Model I satisfactorily meets the objectives of providing estimates of biomass amounts for a variety of forest components with appropriate empirical confidence bounds for these transient and steady state responses. Although the transient responses of Model I are monotonic and do not reflect the rich diversity of behavior that is typical of forest systems, it is an important 'first step' in the construction of more sophisticated comparative models. The steps used in the development of Model I are not dependent on simple linear systems, but can be extended to other models and/or formulations as time, data, and opportunities permit.

Selection of reasonable levels of parameter variability

Respiration values are particularly difficult to obtain and, when available, are probably measured with lower levels of precision than biomass. Table 8.4 shows what would happen to the variability of turnover rates for bole, branch, root, litter and soil organic matter compartments with different

Table 8.4. *Effect of variation of respiration terms on variability of compartment turnover rates*

CV_r^b	Percent variability of turnover rates for compartments of Model II[a]				
	Bole	Branch	Root	Litter	Soil organic
20.0	14.78	12.71	16.10	18.43	20.0
25.0	15.59	15.01	19.84	23.01	25.0
30.0	16.53	17.41	23.62	27.59	30.0
35.0	17.58	19.88	27.41	32.17	35.0
40.0	18.71	22.39	31.23	36.75	40.0
45.0	19.92	24.94	35.06	41.34	45.0
50.0	21.18	27.50	38.89	45.93	50.0

[a] Model II is shown in Fig. 8.1. Respiration terms represent the yearly losses of biomass accounted for by each compartment.
[b] CV_r = Coefficient of variation for respiration (standard deviation/mean). The effect of increasing CV_r on the CV of the turnover rate (sum of transfers from the compartment) is calculated by eqn 8.5 with no correlations between parameters and variability of all non-respiration parameters held constant at 20%.

levels of variability for the respiration term, assuming a constant level of variability for all other parameters of 20%. The bole and branch compartments have low respiration rates with two transfer coefficients determining the turnover rate, while roots and litter have relatively high respiration rates and only one biomass transfer determining the turnover rate. If the objective is to keep the variability of the turnover rates below 20% (the level assumed for the biomass transfers) then the variability of the bole and branch respiration might be as high as 45% and 35%, respectively, but for roots and litter it may be no higher than 25% and 20%, respectively. Although we have little choice but to use Model I for comparisons of the IBP Woodlands Data Set, it is interesting to note that it is probably less desirable to use a model with high levels of variability for certain parameters if a simpler and more accurate alternative exists (O'Neill, 1973). Although Model I produces some bias when compared to Model II (Fig. 8.6), its confidence limits are wider and, with the current data limitations, it is probably more reliable for prediction and comparison.

The level of variability in the general linear model (Fig. 8.1) has been favorably affected by its basic structure and some of the parameterization procedures. The number of transfers from the source (GPP) to the final compartment (soil organic matter) is not great. Furthermore, soil organic matter was assumed to be at steady state (or nearly so) over the life of the

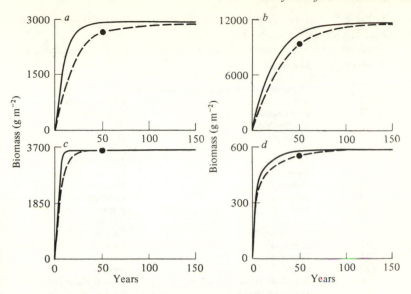

Fig. 8.6. Transients for overstory branches, *a*, and boles, *b*, total roots, *c*, and litter layer, *d*. Solid line refers to Model II and dashed line to Model I (*see* Fig. 8.1). The single point on each graph represents the *Liriodendron* forest.

forest. Therefore, soil organic matter is effectively a constant term and does not enter into the model dynamics. The positional effects illustrated in Fig. 8.3 are not severe in these models.

Data set comparisons

Figs. 8.7 through 8.12 compare data of individual forest sites from the IBP Woodlands Data Set against the transient responses and confidence intervals for the deciduous forest Model I. A list of the sites illustrated in these figures is given in Table 8.5 and the figures in which each is shown are listed in Table 8.6. These confidence intervals were calculated by varying all parameters in the model at the same relative level (i.e., either 10, 20, or 30%) and then plotting the upper 97.5 and lower 2.5 percentiles. For example, Fig. 8.7 shows nine of 24 deciduous forest with standing crops of bole wood within the 95% probability space occupied when all parameters of Model I vary 10%. Meaningful interpretation of these graphs requires a knowledge of the experimental errors, natural ˙variability, and biases involved in measuring and calculating these parameters.

Experimental errors can be estimated by direct inspection of sampling intervals and procedures, however, bias inherent in a particular technique or due to a series of 'unusual' events (i.e. a series of dry years during the

Table 8.5. *Listing of the IBP woodland data set sites illustrated in Fig. 8.7 through Fig. 8.13*

Number	Site name	Country
	Temperate broad-leaved deciduous forests	
1	Bab	Czechoslovakia
2	Hestehaven[b]	Denmark
3	Fontainbleu[b]	France
4	Sikfokut	Hungary
5	Ashu[b]	Japan
6	Meerdink	Netherlands
7	Ispina	Poland
8	Babadag 1	Rumania
9	Babadag 2	Rumania
10	Sinaia 1[b]	Rumania
11	Andersby	Sweden
12	Kongalund[b]	Sweden
13	Langarod[b]	Sweden
14	Linnebjer	Sweden
15	Oved[b]	Sweden
16	Meathop Woods	United Kingdom
17	Hubbard Brook	USA
18	Brookhaven	USA
19	Coweeta 18	USA
20	Liriodendron	USA
21	Walker Branch 1	USA
22	Walker Branch 3	USA
23	Walker Branch 4	USA
24	Nakoma	USA
25	Tallish Oak 1	USSR
26	Vorskla River 8	USSR
27	Solling B1[b]	West Germany
28	Solling B3[b]	West Germany
29	Solling B4[b]	West Germany
	Temperate needle-leaved evergreen forests	
30	Yusuhara Kubotaniyama	Japan
31	Yusuhara Takatoriyama	Japan
32	Shigayama	Japan
33	Kampinos	Poland
34	San Juan	Spain
	Temperate needle-leaved evergreen plantation	
35	Duke Forest	USA
36	Saxapahaw	USA
37	Coweeta 1	USA
38	Thompson	USA
39	Walker Branch 2	USA
	Temperate needle-leaved deciduous forests	
40	Koiwai-JPTF 66	Japan
	Boreal needle-leaved evergreen forests	
41	Black Spruce, Site 1	USA
42	Feather Moss Site	USA
43	Southern Karelian, Site 11	USSR
44	Southern Karelian, Site 12	USSR

Table 8.5 (*continued*)

Number	Site name	Country
45	Southern Karelian, Site 13	USSR
46	Southern Karelian, Site 14	USSR
47	Southern Karelian, Site 15	USSR
48	Central Forest	USSR
49	Koinas	USSR
	Boreal needle-leaved evergreen plantation	
50	Kongalund	Sweden
51	Solling F1	West Germany
52	Solling F2	West Germany
53	Solling F3	West Germany
54	Faeglharting	West Germany
	Mediterranean broad-leaved evergreen forests	
55	Mt Disappointment	Australia
56	Rouquet	France
	Tropical broad-leaved deciduous forests	
57	Chakia-Bandu	India
58	Lubumbashi	Zaire
	Tropical broad-leaved evergreen forests	
59	Manaus	Brazil
60	Pasoh	Malaysia

[b] Indicates Beech (*Fagus sylvatica*) forests.

Table 8.6. *Cross tabulation of IBP Woodland sites and figures in which they appear*

Site number	Figure number						
	8.7	8.8	8.9	8.10	8.11	8.12	8.13
1	X	X	X	—	—	X	—
2	X	X	X	—	—	X	X
3	X	X	X	—	—	X	X
4	X	X	—	—	—	—	—
5	X	X	—	—	—	—	X
6	X	X	X	—	—	X	X
7	X	X	—	—	—	X	X
8	X	X	—	—	—	—	—
9	X	X	X	—	—	—	—
10	X	X	X	—	—	—	—
11	X	X	—	—	—	—	—
12	X	X	—	—	—	X	X
13	X	X	X	—	—	X	X
14	X	X	—	—	—	X	X
15	X	X	X	—	—	X	X
16	X	X	X	—	—	X	X

Table 8.6 (*continued*)

Site number	Figure number						
	8.7	8.8	8.9	8.10	8.11	8.12	8.13
17	X	X	—	—	—	X	X
18	X	X	X	—	—	X	—
19	X	X	X	—	—	X	X
20	X	X	—	—	—	X	X
21	X	X	—	—	—	X	—
22	X	X	—	—	—	X	X
23	X	X	—	—	—	—	—
24	X	X	—	—	—	—	—
25	X	X	X	—	—	—	X
26	X	X	X	—	—	X	—
27	X	X	X	—	—	X	X
28	X	X	X	—	—	X	X
29	X	X	X	—	—	X	X
30	—	—	—	X	X	X	X
31	—	—	—	X	X	X	X
32	—	—	—	X	X	—	X
33	—	—	—	X	X	X	—
34	—	—	—	X	X	X	—
35	—	—	—	X	X	X	—
36	—	—	—	X	X	X	X
37	—	—	—	X	X	—	X
38	—	—	—	X	X	X	X
39	—	—	—	X	X	X	X
40	—	—	—	X	X	X	X
41	—	—	—	X	X	X	X
42	—	—	—	X	X	X	X
43	—	—	—	X	X	X	X
44	—	—	—	X	X	X	X
45	—	—	—	X	X	X	X
46	—	—	—	X	X	X	X
47	—	—	—	X	X	X	X
48	—	—	—	X	X	X	—
49	—	—	—	X	X	X	—
50	—	—	—	X	X	X	X
51	—	—	—	X	X	X	X
52	—	—	—	X	X	X	X
53	—	—	—	X	X	X	X
54	—	—	—	X	X	—	—
55	—	—	—	X	X	X	X
56	—	—	—	X	X	X	X
57	—	—	—	X	X	X	X
58	—	—	—	X	X	X	—
59	—	—	—	X	X	X	X
60	—	—	—	X	X	X	X

course of the study) can cause serious problems because their presence is difficult to detect and/or take into account. Examination of the forest biomass and productivity data listed by Art & Marks (1971) shows that after accounting for differences in age of the forest stands and their

476

composition, significant differences in the techniques used to estimate above-ground biomass and production remain. Where sufficient data were available, the mean tree method estimated lower values than the regression methods which were, in turn, lower than the stratified tree technique ($F = 5.64$, with 2 and 67 d.f., $P < 0.006$). That is, the amounts calculated are somewhat dependent on the methods used to estimate them.

The potential effect of even small biases is evident by inspection of the series of regression equations used to estimate the standing crops for the Oak Ridge *Liriodendron* site (Sollins & Anderson, 1971; Sollins *et al.*, 1973). The percent variance accounted for by the regression of the log of bole weight on the log of tree diameter at breast height exceeds, in some cases, 98%. The variability around the exponentiated mean is approximately 16%. When these variabilities are combined for several trees we see that, if no correlations exist, the relative variability of the sum will decrease approximately as CV/n (eqn 8.5), where n is the number of trees in the sum. Thus, for an estimate of 100 trees the variability of the standing crop of boles would be less than 2%. This result indicates a level of precision that is difficult to accept. Examination of the original equations for bole weight (Sollins *et al.*, 1973) indicates that even a 1% shift (bias) in the slope of log–log regression equations can have an effect on the biomass of trees in the 36 to 54 dbh category by approximately 9%. Bias, in this case, is not reduced by increasing sample size.

Madgwick (1970) has considered the problems of estimating foliage, branch and bole biomass for canopy species and found that error and bias are evident with appropriate confidence limits ranging from 7% for boles to as high as 90% for branch and leaf biomass. Ovington, Forrest & Armstrong (1967) compared several combinations of sampling methods and biomass estimations for a *Pinus radiata* stand and, depending upon the sampling scheme, found the range of confidence for regressions from 1% to 28% for boles and 1% to 22% for branches. Swank & Schreuder (1974) discussed the statistical limitations of regression methods and determined that the percent variability for a *Pinus strobus* stand ranged from 5% to 10% for branch and bole biomass. Satoo (1970) has also compared biomass estimation methods and obtained results within 5% of actual measurements for *Betula* boles, branches, and leaves.

Consistent dimensional relationships (e.g. dbh and biomass) might be expected for trees grown under uniform conditions, but bias is introduced into these techniques whenever competition (Satoo, 1967) or unusual stand history has taken place. Whittaker (1966) has noted several strong relationships between biomass, productivity and volume estimates, but when disturbed or 'unstable' stands were considered, these relationships deteriorated. Woodwell & Botkin (1970) compared three methods of calculating NPP (net primary production), two methods based on dimensional

477

analysis, and one on gas exchange techniques. The final results differed from one another by 20%.

Calculation of errors associated with model parameters are difficult. Assumptions, extrapolations of one kind or another, and the final adjustments of a model provide ample opportunities for adding biases to the natural variability of parameters and the experimental errors involved in their measurement. The danger of these biases has been noted by Goodall (1972), who considers their cumulative effects to be the major source of uncertainty in ecological models. For this reason, and because we are concerned with general patterns of forest biomass dynamics, we have plotted several confidence bands around model transients (Fig. 8.7, 8.8, 8.10–8.13).

Examination of the Oak Ridge *Liriodendron* data set indicates ranges of variabilities for the model compartments. Standing crops for above-ground biomass of individual trees vary from 3 to 16% for boles, from 5 to 30% for branches, and from 130 to 200% for leaves (taken from the data presented by Sollins & Anderson, 1971). Overall levels of litterfall estimated from the original data collections (Edwards, personal communication) indicate that leaffall will vary by approximately 15–18%, reproductive parts by 50–75%, and branches by 120–210%.

Respiration rates are quite dependent on year-to-year differences in temperature patterns. Years with unusually warm winters may produce as much as 4% or more carbon dioxide from the forest floor than years with cooler winters (Edwards, 1975). The predictions of soil respiration from temperature are quite good ($R^2 = 0.94$) with less than 3% difference between monthly estimates and measurements. However, spatial heterogeneity, sampling and measurement errors result in an actual variability of monthly measurements of 86.8% (from Table 1 of Edwards, 1975). Bole and branch respiration rates have been estimated for *Pinus taeda* by logarithmic regression equations with R^2s ranging from 0.91 to 0.99 (Kinerson, 1975). These equations are subject to the same problems that were mentioned for biomass estimation equations; if the biomass or surface of boles and branches have first been estimated by these equations, we find that several parameters for a model will all be related to diameter at breast height. For a more complete discussion of the many aspects of the problem of estimating carbon dioxide production and utilization from forests see Yoda *et al.* (1965) and Woodwell & Botkin (1970).

Taken together, the above results indicate that variations of 20% for model parameters will produce reasonable confidence limits for the prediction of above-ground biomass. Overstory bole and branch biomass are estimated with unusually high precision so 10% parameter variability seems satisfactory for the prediction of amounts and confidence intervals by Model I. Because of the uncertainties and strong possibility of serious bias,

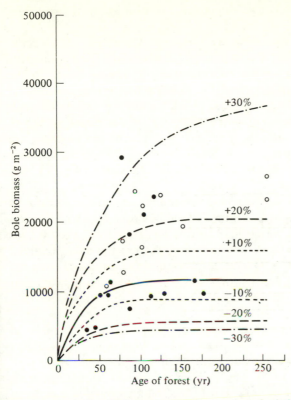

Fig. 8.7. The bold solid line represents the transients for bole biomass values predicted from the generalized model. The dotted lines represent the confidence limits with 10%, 20% and 30% parameter variation. The points on the figure represent deciduous (●) and beech (○) forest sites from the IBP Woodlands Data Set, which are listed in Tables 8.5, 8.6.

the below-ground components probably should be estimated with 30% variability around the model parameters.

Boles and branch biomass comparisons

Figs. 8.7 and 8.8 show the patterns of bole and branch biomass of deciduous forests of the IBP Woodlands Data Set compared to the transients of the Oak Ridge *Liriodendron* site predicted by Model I. These two graphs imply a similar distribution of biomass among generically similar forests of the world by assuming that temperate deciduous forests would be found within similar sets of soil and climatic conditions and would have similar growth patterns and phenodynamics. Although certain aspects of growth and accumulation of biomass are similar across all forest types, the

479

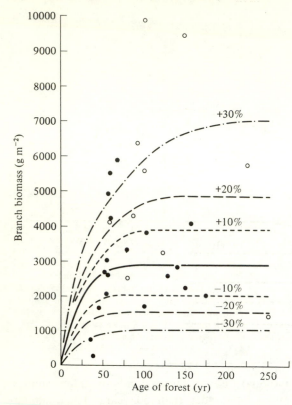

Fig. 8.8. The bold solid line represents the transients for branch biomass values predicted from the generalized model. The dotted lines represent the confidence limits with 10%, 20%, and 30% parameter variation. The points on the figure represent deciduous (●) and beech (○) forest sites from the IBP Woodlands Data Set, which are listed in Tables 8.5, 8.6.

differences within temperate deciduous forests ought to be less than between temperate deciduous and other forest types. Furthermore, because of their importance for fuel and fiber, bole and branch growth are probably the most precisely measured forest biomass compartments. By arranging deciduous sites (Figs. 8.7 and 8.8) in descending order of similarity, first those whose bole and branch biomasses both fall within 10%, then 20%, etc. we can rank order the forests from most to least similar to Model I and the Oak Ridge *Liriodendron* site.

Examination of basal area and stocking density of these stands with age (Fig. 8.9) shows one aspect of the development and appearance of these forests, and indicates that caution is necessary in interpreting the similarity of results from diverse ecosystems. Although forests most similar to Model I (within 20% confidence bounds) show large differences in age and number

480

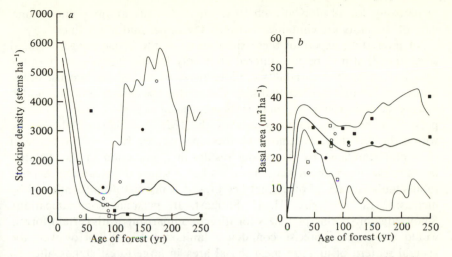

Fig. 8.9. The bold solid line represents the mean value for stocking density (*a*) and basal area (*b*) of 100 computer simulated forest plots with radius of 16.5 m (from the FORET model, Shugart & West, 1977; see text for further details). The lower line represents the 3rd and the upper line the 97th percentile of the rank ordered plots at each 10 yr age interval. The symbols indicate deciduous forests from the IBP Woodlands Data Set that were within the indicated percent confidence bands for both bole and branch biomass (Fig. 8.7 and 8.8); ●, 10%; ○, 20%; ■, 30%; □, >30%. A listing of these sites may be found in Tables·8.5 and 8.6.

of stems per hectare, the basal area of these stands is quite similar (within 20 to 25 m^2 ha^{-1}) for sites older than 50 years. Thus, similar biomass distributions have been produced by a wide range of stems per hectare.

The local history of a forest stand has a significant effect on biomass allocation and productivity, but the nature and extent of this effect is difficult to identify. A computer simulation program, FORET, has been developed by Shugart & West (1977) that provides this information. FORET simulates stand growth within southern Appalachian deciduous forests where *Liriodendron* often dominates. The model provides a lucid picture of the interaction of germination, growth, competition and death processes within a forest by starting with a bare plot of 16.3 m radius (1/12 ha), assuming uniform soil conditions, and, based on the life history patterns of the indigenous tree species (i.e. light preferences, growth rates, germination characteristics, etc.), stochastically simulates the dynamics of individual forest stands.

We have used FORET to simulate 100 plots and obtain information on the changes in basal area and number of stems with time (Fig. 8.9; all results expressed on a per hectare basis). The mean is represented by the broad line and the upper and lower lines represent the 97th and 3rd percentiles,

481

respectively, for all plots at each 10-yr interval. Thus, at any point in time, 94% of the plots are contained between the upper and lower lines.

An interesting aspect of these simulations is the broad range of stand conditions that may be encountered after only 50–75 years of growth of the *same forest type*. The variability of plots increases at this time because of the death of a few mature trees that have dominated the canopy and the sudden emergence and growth of young trees competing to replace them. As noted previously, most deciduous forests of the IBP Woodlands Data Set lie within the broad area bounded by these percentiles. Increasing the size of each plot simulated by the model results in a reduction in the distance between the upper and lower percentiles but, because the mean represents a total simulated area of over eight hectares, it is not significantly affected by increasing the plot size (H. H. Shugart, Jr, personal communication). Knowledge of specific plot sizes for direct comparisons with certain forests would allow more precise confidence limits for Fig. 8.9. However, the general pattern of uniform mean basal area in large forest stands after 50 years, but high variability of the number of stems in local areas, is evident from both the IBP Woodlands Data Set and FORET results.

Site specific climatic conditions may also affect the amount of biomass and productivity of a forest. O'Neill (Chapter 7) has used the information available in the IBP Woodlands Data Set to approximate π, Budyko's (1956) estimation of potential evapotranspiration. Increasing values of π indicate increasingly favorable temperature and moisture regimes. Comparing this index for the 16 sites with sufficient information, the rank ordering of the temperate deciduous forests shows that for those falling within 10% bounds for both branch and bole (Figs. 8.7 and 8.8), the range of π was 9.7 to 20.5; within 20% bounds the range was 5.3 to 14.2; and within 30% bounds the range was 5.1 to 12.2. Beyond this last confidence band, the range decreases still further to 3.8 to 11.9. High values of π indicate favorable conditions of temperature and moisture (the value for the *Liriodendron* stand that Model I is based upon is 13.2). In spite of the limitations of this comparison, the pattern of decreasing divergence from the growing conditions of Model I with increasing divergence in resulting biomass accumulation in bole and branch compartments is evident and may explain some of the differences between deciduous sites.

Bole and branch biomass of other forests are compared in Fig. 8.10 and 8.11. This between-forest comparison must be viewed with the additional cautionary note that, although these forests may occupy the same confidence space as the *Liriodendron* site, there is less reason to suppose that the actual rates of growth biomass accumulation and turnover are actually similar to those predicted by Model I. That is, their rates may be quite different, but the transient responses may follow similar pathways over part of their growth history. Thus there might be similarities among quite

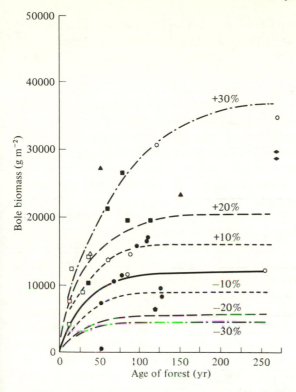

Fig. 8.10. The bold solid line represents the transients for bole biomass values predicted from the generalized model. The dotted lines represent the confidence limits with 10%, 20% and 30% parameter variation. The points represent other forests from the IBP Woodlands Data Set: O, temperate needle leaved evergreen; □, temperate needle leaved evergreen plantation; △, temperate needle leaved deciduous; ●, boreal needle evergreen; ■, boreal needle leaved evergreen plantations; ▲, Mediterranean Broad Leaved Deciduous; ◆ tropical broad leaved deciduous; ◆ tropical broad leaved evergreen. A listing of these sites can be found in Tables 8.5 and 8.6.

divergent forests if they are predominantly young sites, or if they have suffered disturbances or stresses during part of their growth history. These effects are present among all forest sites, but the extent of their effect is difficult to ascertain.

The outer 30% confidence bands for this model extend from 4570 to 37 000 g m^{-2} for boles and from 1000 to 1040 g m^{-2} for branch biomass. Beyond these limits lie fast-growing *Eucalyptus* (boles, Fig. 8.10) and giant evergreen and tropical broad-leaved evergreen sites (branch, Fig. 8.11). *Eucalyptus*, giant evergreens, and tropical broad-leaved evergreens are clearly different from the predicted values for the Oak Ridge *Liriodendron* forest, so their location beyond the 30% confidence band is not suprising. What is difficult to explain is the large variance of deciduous sites expected

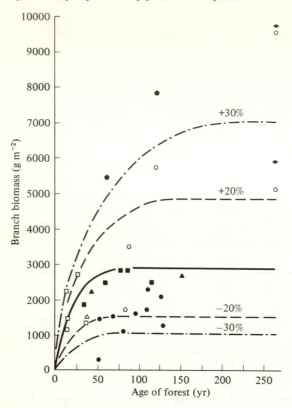

Fig. 8.11. The bold solid line represents the transients for branch biomass values predicted from the generalized model. The dotted lines represent the confidence limits with 10%, 20% and 30% parameter variation. The points represent other forests from the IBP Woodlands Data Set: ○, temperate needle leaved evergreen; □, temperatre needle leaved evergreen plantation; △, temperate needle leaved deciduous; ●, boreal needle leaved evergreen: ■, boreal needle leaved evergreen plantations; ▲, Mediterranian broad leaved deciduous; ⬟, tropical broad leaved deciduous; ◆, tropical broad leaved evergreen. A listing of these sites can be found in Tables 8.5 and 8.6.

to be more similar to the Oak Ridge site. That is, based only on approximate variances for model parameters and the resulting variances of predicted biomass allocation, it is difficult to distinguish all but the most different forests on their mean values of biomass allocation. Although all the reasons for this result are not clear, it is evident that they are not entirely due to the behavior of the model.

Inspection of Figs. 8.10 and 8.11 show that the age of most of these forests fall within the area where we would expect them to be most similar. Another reason for similarities in spite of differences in actual growth rates is that some forest types (e.g. boreal) are growing under much more restricted conditions than temperate forests. Because of uneven representations of many forest types and ages it becomes difficult to make definitive statements about these types. The most interesting aspect is that even after 100 years of growth, several boreal sites are similar to the temperate deciduous forests (i.e. within the 20% confidence bounds for both bole and branch biomass).

The calculation of π (an estimate of potential evapotranspiration) for non-deciduous temperate forests show that only one site, the Oak Ridge needle-leaved evergreen site, has a value of π similar to the 13.2 of the *Liriodendron* site. All other sites fall below this value with some boreal sites being as low as 2.6. This indicates that processes leading to the accumulation of similar levels of biomass in these forests are apparently due to different strategies of allocation. Total dynamics of accumulation rates and standing crops of organic matter need to be considered. O'Neill (Chapter 7) has found separation with total accumulations versus π discernibly less for boreal sites. All these patterns are somewhat difficult to determine, because of the variable amount of information available from each of the sites in the IBP Woodlands Data Set and the few sites that have sufficient data for a large number of comparisons that are inevitably required by *a posteriori* speculations.

Tropical broad-leaved evergreen sites are clearly different from the temperate deciduous sites, but tropical broad-leaved deciduous sites are similar in bole weights (within the 20% band). Giant evergreen sites (stands 37, 38) are easily identified in Fig. 8.8 and 8.10, although the value of π for these sites is below that of the *Liriodendron* stand.

Root and soil organic matter

Figs. 8.12 and 8.13 illustrate the transient, steady state, and confidence bands for root biomass and total soil organic matter. The increased uncertainties around these compartments and the skewed distributions, particularly for soil organic matter, are evident by the upper bounds for the 30% confidence bands. The uncertainties for these two compartments are much higher than for bole and branch (Figs. 8.7 and 8.8) because of the position of the compartment within the model (see Figs. 8.1 and 8.3). It is possible to manipulate the model to obtain 'tighter' confidence bands around the transients and final steady state values for root and soil organic matter, but without specific information about the natural variabilities of these compartments there would be little confidence in such procedures. Most forest sites lie within 20% of the transient response for Model I. However, there was considerable confusion about what constituted litter and soil organic layers, with values for temperate deciduous forests ranging from 11.6 to 20 550.0 g/m^2. We combined litter and soil organic matter compartments into total soil organic matter (Fig. 8.13) and made little

485

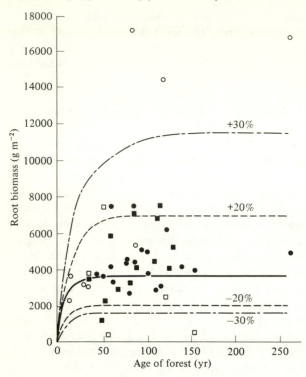

Fig. 8.12. The bold solid line represents the mean transient for root biomass values predicted from the generalized model. The dotted lines represent the confidence limits with 20% and 30% parameter variation. The symbols represent other forests from the IBP Woodlands Data Set: ●, all deciduous forests including beech; ○, all temperate needle leaved evergreen forests including plantations; ■, all boreal forests including plantations; □, other IBP sites not falling into the above three categories. These sites are listed in Tables 8.5 and 8.6.

attempt to screen the data from the IBP Woodlands Data Set. The value of Figs. 8.12 and 8.13 are a reference for specific comparisons where additional information allows the estimation of errors around the transients.

Covariance terms and other considerations

Considerable attention was devoted in the earlier sections to a discussion, with appropriate examples, of the effects of interrelationships between model parameters. Because of the current state-of-the-art of ecosystem investigation and modeling, it is not feasible to estimate the covariance terms for ecosystem parameters. However, there are several reasons for considering their existence and speculating about their effects and about predictions made on ecosystem models.

First, as pointed out in earlier sections, the methods of calculating many

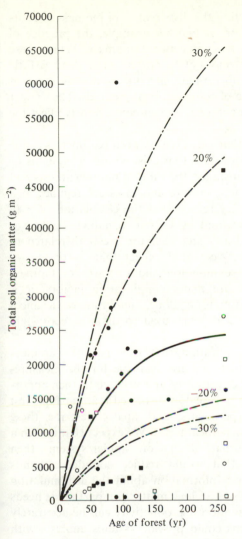

Fig. 8.13. The bold solid line represents the mean transient for total soil organic matter (all surface layers plus the dead organic matter within the soil) predicted from the generalized model. The dotted lines represent the confidence limits with 20% and 30% parameter variation. The symbols represent other forests from the **IBP** Woodlands Data Set: ●, all deciduous forests including beech; ○, all temperate needle leaved evergreen forests including plantations; ■, all boreal forests including plantations; □, other **IBP** sites not falling into the above three categories. These sites are listed in Tables 8.5 and 8.6.

parameter values depend upon certain underlying assumptions. There is a great deal of redundancy and interdependency with several numbers relying on the accurate estimation of a much smaller set of numbers in the final

487

model. This is obvious from inspecting the distribution of biomass of roots in the IBP Woodlands Data Set, or, as another example, the practice of calculating respiration as a ratio of compartment biomass. While these methods and assumptions may or may not be correct (and there is little choice in the matter), this practice unwittingly introduces interrelationships between parameters merely because of computational methods. The danger of this practice is obvious – an error in one number can seriously affect the calculations of several others.

Another interesting relationship that may exist between parameters is due to the biological and physical relationships that exist within ecosystems or between forest compartments. For instance, the rates of photosynthesis and respiration depend strongly upon temperature and moisture regimes in a forest. This is, of course, the primary reason that the calculation of π (a measure of potential evapotranspiration) by O'Neill (Chapter 7) has validity. The availability of carbohydrates and their use are both related to water and temperature. Therefore, we would expect that as these two varied, the processes of growth accumulation and turnover of biomass would also vary. Although there are many complicating factors, it is reasonable to suppose that for this reason alone, significant covariances would exist between the mathematical terms used to describe ecosystem processes.

Time lags may be another complication that will make covariances difficult to determine. Numerical methods are available for the measurement of serial self-correlations and covariances and provide further opportunities to use statistical techniques to reveal unsuspected relationships, test their existence with definitive experiments, and finally describe these relationships in biological and mathematical terms. We expect that if such a course were vigorously pursued, perhaps at first on simpler systems, then relationships could be identified that would reduce confidence bands around ecosystem model predictions. Information about the accumulation of organic matter in litter, root and soil organic matter compartments needs to be refined so that ideas concerning forest productivity can be accurately tested. In addition, covariance terms could provide systems analysts with empirical information that would define unsuspected relationships within ecosystems as well as more precisely define the predictions made from models.

The nature and extent of relationships between parameters that would achieve these results are currently unknown. We can expect strong covariances between compartments to reduce model sensitivity without appreciably affecting the variability of compartments near the source (forcing function) of energy or matter, but the variability of compartments farther from the source may be changed. Strong correlations between compartments might indicate that external forces, such as temperature and moisture, affect the behavior of parameters in a similar manner, while negative correlations might indicate compensatory mechanisms.

The possible existence of strong covariance terms and their locations within a system can only be speculative. Their inclusion within any model analysis provides a straightforward and understandable method of representing the rich nature of information that is necessary to describe adequately the dynamics of ecosystems. Inclusion of these values in realistic models, whether these be linear or nonlinear, process models or yearly averages, will require careful planning and analysis of future experiments and investigations.

We hope that it will soon be possible to examine optimal rates of growth, accumulation and turnover (i.e. ones unaffected by the history of the ecosystem, local perturbation, etc.) for data from the IBP Woodlands Data Set and then examine all interrelationships between forest sites. However, because of the diversity of forest types, local conditions, etc., we can only expect that the data now available and the methods investigated here can help move us closer to the estimation of those relationships which will explain the remarkable diversity of forest ecosystems.

Summary

A Monte Carlo program for the efficient evaluation of ecosystem models and quantitative comparison of the IBP Woodlands Data Set was developed specifically for this synthesis effort. The development and application of this program, called COMEX, resulted in the following conclusions.

1. Knowledge of the variances of model parameters allows quantitative comparisons of biomass allocation between woodland data sets. While certain levels of variability may be expected for the typical woodland site of a given forest type, the level of variance actually measured will be site specific and depend upon the size of the stand, local successional sequences, history of the stand, climatic variability, sampling procedures, etc. Seperate lines of evidence indicate that the partitioning of variances due to the within-site factors can be important in reducing the variance of model predictions. The levels of within-site variance are presently high and obscure many differences between sites. Future application of the Monte Carlo methods developed here will need to consider time lags, serial correlations and terms which represent interactions and dependencies between parameters in order to handle this information.
2. Monte Carlo simulations are of value in understanding model dynamics and providing quantitative bases for model alteration and improvement. This procedure is a multivariate sensitivity analysis which compares the correlations between parameter and model output variations. When Monte

Carlo simulations are used in this manner a precise knowledge of parameter variances is not required.

3. Predictions and comparisons based on current models and available data are limited because many factors (e.g. measurement techniques, sampling frequency, availability of data, etc.) influence the structure and dynamics of the models. Even though the models employed here are simple ones, these factors have a noticeable effect on model behavior and predictions. It may never be possible to control all the variances which affect the modeling process, but it is desirable to identify the effects of alternate assumptions and model formulations before a final model selection is made.

4. The intriguing problems which remain include those concerned with the application of Monte Carlo methods to more complex models, especially models which describe ecological processes in greater detail. The investigation of new methods to partition variances during all phases of a study and the possibility of including covariance terms (correlations) between parameters promises to have a profound effect on the precision of model predictions.

Appendix. Some mathematical considerations

Numerical methods of COMEX

COMEX is a FORTRAN program written in double precision for the IBM 360 system. The program is modularly structured with most information passed between subroutines by common blocks. Complete program documentation is available elsewhere (Gardner *et al.*, 1976).

The model parameters used in COMEX are the $N \times N$ matrix of mean transfer coefficients ($\bar{A}$), the N-vector of forcing functions ($\bar{b}$) and an N-vector of losses from the system ($\bar{r}$). In the case of the IBP model, the loss vector represents the respirational terms of the system. Since mass balance is preserved, the diagonal elements of $\bar{A}$ may be calculated by

$$\bar{a}_{ii} = - \sum_{\substack{j=1 \\ i \neq j}}^{n} \bar{a}_{ji} - \bar{r}_i. \quad i = 1, 2, \ldots, N \tag{A.1}$$

The mean model is then given by the set of ordinary differential equations:

$$\frac{\mathrm{d}\bar{x}}{\mathrm{d}t} = \bar{A}\bar{x} + \bar{b}, \quad \bar{x}(t_o) = x_o. \tag{A.2}$$

The off-diagonal elements of the $\bar{A}$ matrix represent turnover rates between the various compartments. For instance, the ij^{th} element represents the turnover rate of material flowing from the ith compartment to the jth compartment. The diagonal elements represent the turnover rate of the compartments. The elements of $\bar{b}$ are the time rate at which material is input to the compartments from outside the system.

For each Monte Carlo iteration, the elements of a new coefficient matrix (A) are selected randomly from a multivariate–normal distribution around the elements of the mean coefficient matrix ($\bar{A}$), where there are M parameters to be varied. The variance–covariance matrix must now be formed. Since there are M parameters to be varied, this is an $M \times M$ matrix where the diagonal elements are the variance terms and the off-diagonal elements are the covariance terms. If the ij^{th} element at A is the kth variate, we have the variance given by

$$s_{ij}^k = (\bar{a}_{ij}*c_{ij})^2, \quad k=1, 2,\ldots, M; \quad i\varepsilon[1, M]; \quad j\varepsilon[1, M] \tag{A.3}$$

where c_{ij} is the coefficient of variation of the ij^{th} element. The elements of the variance–covariance matrix, Σ, are calculated

$$\sigma_{kl} = \gamma_{kl}*(s_{ij}^k s_{mn}^l)^{1/2}, \quad k, l=1, 2,\ldots, M \tag{A.4}$$

where Σ is an $M \times M$ matrix, and γ_{kl} is a specified correlation between the ij^{th} and the mn^{th} elements. The matrix Σ is then factored in the form

$$\Sigma = LL^T, \tag{A.5}$$

and the matrix Δ is formed by

$$\Delta = DL, \tag{A.6}$$

where D is a matrix of normal deviates, $\Delta_{ij}\ \varepsilon(0, 1)$. As was done for the mean model, the elements of each new coefficient matrix are then formed by

$$a_{ij} = \bar{a}_{ij} + \delta_{ij}, \quad i \neq j, \tag{A.7}$$

$$a_{ii} = -\sum_{\substack{j=1 \\ j\neq i}}^{n} a_{ji} - r_i. \tag{A.8}$$

This sequence of events is performed for each Monte Carlo iteration.

The use of multivariate–normal distribution to generate a randomly perturbed system matrix is a statistical and computational convenience. It does not imply that the elements of the system matrix A or the solutions of the ordinary linear differential equations are necessarily normally distributed. In fact, because randomly selected negative parameter values are discarded decidedly non-normal results may be produced when the level of parameter variation is high (Mankin, Gardner & Shugart, 1976).

We then have the following system of differential equations to be solved

$$\dot{x} = Ax + b, \quad x(t_0) = x_0. \tag{A.9}$$

The determination of the steady state solutions of eqn 8.28 is rapid, but the transient solutions are more time consuming. Due to the structure of the IBP model, the transient responses are relatively smooth. Therefore, the steady state solution contains almost as much information about the model variability as the transient solutions. For this reason, we compute the steady state solution at each iteration and the transient response only at every tenth iteration.

The steady state solution is determined by assuming that all derivatives vanish at steady state. This is true if the A matrix is non-singular. If the matrix A is non-singular, steady state values for each compartment and time to steady state are calculated by solving eqn A.1 for the steady state value of x given by

$$x(\infty) = -A^{-1}b. \tag{A.10}$$

Time to steady state is defined as the time for the slowest transient to achieve 98% of

491

Dynamic properties of forest ecosystems

its final value. This is approximately equal to

$$\tau = 4./\min_i (\text{real part of } \lambda_i), \quad i = 1, 2, \ldots, n \tag{A.11}$$

where λ_i is the eigenvalue.

The transient responses are calculated by solving (eqn A.20) giving

$$x(t) = \phi(t, t_o)x_0 - [1 - \phi(t, t_o)]A^{-1}b, \tag{A.12}$$

where I is the $N \times N$ identity matrix and $\phi(t, t_o)$ is the $N \times N$ transition matrix defined by

$$\phi(t, t_o) = e^{A(t-t_o)} = \sum_{i=0}^{\infty} \frac{A^i t^i}{i!} \tag{A.13}$$

In practice, it is much more efficient and accurate to solve for the transition matrix by computing the eigenvalues and eigenvectors of the system matrix, A. This is done in the following manner:

$$AT = T\Lambda, \tag{A.14}$$

where Λ is an $N \times N$ diagonal matrix whose elements are the eigenvalues of A and where T is a matrix whose columns are the N eigenvectors. Then

$$A = T\Lambda T^{-1}. \tag{A.15}$$

Therefore,

$$\phi(t, t_o) = e^{A(t-t_o)} = e^{T\Lambda T^{-1}(t-t_o)}. \tag{A.16}$$

It can be shown that

$$e^{T\Lambda T^{-1}(t-t_o)} = Te^{\Lambda(t-t_o)}T^{-1}. \tag{A.17}$$

Therefore

$$x(t) = Te^{\Lambda(t-t_o)}T^{-1}x_o + \Lambda[1 - e^{\Lambda(t-t_o)}]\Lambda^{-1}T^{-1}b. \tag{A.18}$$

The inverse Λ^{-1} is trivial to compute, and the inverse of T need not be computed. One merely has to solve the linear systems

$$Ty = x_o, \tag{A.19}$$

and

$$Tg = b, \tag{A.20}$$

for y and g.

Lumping parameters in a two-compartment system

Consider the simplified two-compartment model shown in Fig. A.1. Compartment 1 is analogous to a photosynthate storage compartment, and Compartment 2 is analogous to leaves, branches, or boles. Accounting for all respirations from a single compartment is equivalent to the system of equations:

$$\dot{x}_1^* = -a_1^* x_1^* - r_1^* x_1^* + f^*, \quad x_1^*(0) = k_1^*, \tag{A.21}$$

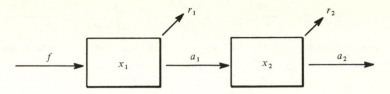

Fig. A.1. Simplified two compartment model. The f represents the forcing function for the model; x_1 and x_2 represent the amount of material, r_1 and r_2 represent respirational losses, and a_1 and a_2 represent the transfer coefficients from compartment 1 and 2 respectively.

$$\dot{x}_2^* = a_1^* x_1^* - a_2^* x_2^*, \quad x_2^*(0) = k_2^*, \tag{A.22}$$

If respiration terms are greater than 0, for all compartments we have a system of equations:

$$\dot{x}_2 = -a_1 x_1 - r_1 x_1 + f, \quad x_1(0) = k_1, \tag{A.23}$$

$$\dot{x}_2 = a_1 x_1 - a_2 x_2 - r_2 x_2, \quad x_2(0) = k_2. \tag{A.24}$$

The systems are related to each other by the following criteria:

1. The initial conditions are the same ($k_i = k^*$),
2. The steady state is the same [$x_i(\infty) = x_i^*(\infty)$].
3. The forcing function is the same ($f = f^*$), and
4. The transfer out of compartment 2 is the same ($a_2 = a_2^*$).

Using these conditions, the solution to the first system (eqns A.20 and A.21)

$$x_1^*(t) = e^{-(a_1^* + r_1^*)t} k_1 + \frac{f}{(a_1^* + r_1^*)} [1 - e^{-(a_1^* + r_1^*)t}], \tag{A.25}$$

$$x_2^*(t) = e^{-a_2 t} k_2 + \frac{(1 - e^{a_2 t}) a_1^* f}{(a_1^* + r_1^*) a_2} +$$

$$\frac{(e^{-a_2 t} - e^{-(a_1^* + r_1^*)t}) a_1^* (k_1 - f_1^{(a_1^* + r_1^*)})}{(a_1^* + r_1^* - a_2)} \tag{A.26}$$

and the solution to the second system (eqns A.24 and A.25) is:

$$x_1(t) = e^{-(a_1 + r_1)t} k_1 + \frac{f}{a_1 + r_1} [1 - e^{-(a_1 + r_1)t}], \tag{A.27}$$

$$x_2(t) = e^{-(a_2 + r_2)t} k_2 + \frac{(1 - e^{-(a_a + r_2)t}) a_1 f}{(a_1 + r_1)(a_2 + r_2)} +$$

$$\frac{(e^{-(a_2 + r_2)t} - e^{-(a_1 + r_1)t}) a_1 (k_1 - f_1^{(a_1 + r_1)})}{(a_1 + r_1 - a_2 - r_2)}. \tag{A.28}$$

These systems have the steady-state solutions given by

$$x_1(\infty) = x_1^*(\infty) = \frac{f}{a_1 + r_1} = \frac{f}{a_1^* + r_1^*}, \tag{A.29}$$

493

$$x_2(\infty) = x_2^*(\infty) = \frac{a_1 f}{(a_2 + r_2)(a_1^* + r_1^*)} = \frac{a_1^* f}{a_2(a_1^* + r_1^*)}. \tag{A.30}$$

From these equations you have the relationships between the parameters given by

$$a_1 = \frac{a_1 a_2}{a_2 + r_2} \tag{A.31}$$

$$r_1^* = r_1 + \frac{a_1 r_2}{a_2 + r_2}. \tag{A.32}$$

If eqns (A.30) and (A.31) are substituted into eqns (A.24) and (A.25) we have response for the first system given by

$$x_1^*(t) = e^{-(a_1 + r_1)t} k_1 + \frac{f}{a_1 + r_1} [1 - e^{-(a_1 + r_1)t}], \tag{A.33}$$

$$x_2^*(t) = e^{-a_2 t} k_2 + \frac{(1 - e^{-a_2 t}) a_1 f}{(a_1 + r_1)} + \frac{(e^{-a_2 t} - e^{-(a_1 + r_1)t}) a_1 a_2 (k_1 - f_1^{(a_1 + r_1)})}{(a_2 + r_2)(a_1 + r_1 - a_2)}. \tag{A.34}$$

If we compare eqns (A.23) and (A.33) to eqns (A.26) and (A.27), we see that the transient response of compartment 1 is the same for both models. However, the transient response of compartment 2 for the two models differs, even though the steady state is the same. This affects any compartment which receives material from compartment 2.

Acknowledgements

We wish to express our thanks to N. T. Edwards, W. F. Harris, and S. B. McLaughlin for supplying critical data and advice for model parameters and to D. L. DeAngelis, R. V. O'Neill, D. E. Reichle, and H. H. Shugart for advice and criticisms throughout various phases of this project. We also wish to thank D. Mielke, D. C. West, and H. H. Shugart for helping with the special runs of the FORET model.

References

Adams, J. A. S., Mantovami, M. S. M. & Lundell, L. L. (1977). Wood versus fossil fuel as a source of excess carbon dioxide in the atmosphere: a preliminary report. *Science*, Washington, **196**, 54–6.

Art, H. W. & Marks, P. L. (1971). A summary table of biomass and net annual production in forest ecosystems of the world, pp. 3–28. In *Forest biomass studies*, ed. H. E. Young, University of Maine Press, Orono.

Baes, C. F., Jr., Goeller, H. E., Olson, J. S. & Rotty, R. M. (1976). *The global carbon dioxide problem.* ORNL–5194. Oak Ridge National Laboratory, Oak Ridge, Tennessee.

Baes, C. F., Jr., Goeller, H. E., Olson, J. S. & Rotty, R. M. (1977). Carbon dioxide and the climate: the uncontrolled experiment. *Amer. Scientist*, **65**, 310–20.

Bolin, B. 1977. Changes in land biota and their importance for the carbon cycle. *Science* Washington, **196**, 613–5.

Boremann, F. H., Likens, G. E., Siccama, T. G., Pierce, R. S. & Eaton, J. S. (1974). The export of nutrients and recovery of stable conditions following defore-station at Hubbard Brook. *Ecol. Monog.* **44**, 255–77.

494

Budyko, M. I. (1956). Teplovoy balans zemmoy poverkhnosti (The heat balance of the earth's surface. Translated by N. I. Stepanova, 1958. US Weather Bureau, Washington). Gidrometeoizdat, Leningrad.

Burgess, R. L. (1978). Potential of forest fuels for producing electrical energy. *J. For.* **76**, 154–7.

Burgess, R. L. & O'Neill, R. V. (eds). (1976). Eastern Deciduous Forest Biome progress report, 1 September, 1974 – 31 August, 1975. EDFB/IBP–76/5. Oak Ridge National Laboratory, Oak Ridge, Tennessee.

Caswell, H., Koenig, H. E., Resh, J. A. & Ross, Q. E. (1972). An introduction to systems science for ecologists. In *Systems analysis and simulation in ecology*, **2**, ed. B. C. Patten, pp. 3–78. Academic Press, New York.

Edwards, N. T. (1975). Effects of temperature and moisture on carbon dioxide evolution in a mixed deciduous forest floor. *Proc. Soil Sci. Soc. America*, **39**, 361–5.

Edwards, N. T. & Harris, W. F. (1977). Carbon cycling in a mixed deciduous forest floor. *Ecology*, **58**, 431–7.

Edwards, N. T. & Sollins, P. (1973). Continuous measurement of carbon dioxide evolution from partitioned forest floor components. *Ecology*, **54**, 406–12.

Funderlic, R. E. & Heath, M. T. (1971). *Linear compartment analysis of ecosystems.* ORNL/IBP–71/4. Oak Ridge National Laboratory, Oak Ridge, Tennessee.

Gardner, R. H., Mankin, J. B. & Shugart, H. H., Jr., (1976). *The COMEX computer code.* EDFB/IBP–76/4. Oak Ridge National Laboratory, Oak Ridge, Tennessee.

Golley, F. B. (1965). Structure and function of an old-field broomsedge community. *Ecol. Monogr.* **35**, 113–31.

Goodall, D. W. (1972). Building and testing ecosystem models. In *Mathematical models in ecology*, ed. J. N. R. Jeffers, pp. 173–94. Blackwell, Oxford.

Kercher, J. R. & Shugart, H. H., Jr. (1975). Trophic structure, effective trophic position, and connectivity in food webs. *Amer. Naturalist*, **109**, 191–206.

Kinerson, R. S. (1975). Relationships between plant surface area and respiration in loblolly pine. *J. App. Ecol.* **12**, 965–71.

Lindeman, R. L. (1942). The trophic–dynamic aspect of ecology. *Ecology* **23**, 499–518.

Madgwick, H. A. I. (1970). Biomass and productivity models of forest canopies. In *Analysis of temperate forest ecosystems*, ed. D. E. Reichle, pp. 47–54. Springer–Verlag, New York.

Mankin, J. B., Gardner, R. H. & Shugart, H. H., Jr., (1976). The COMEX computer code: Monte Carlo analysis of ecosystem attributes. In *Proceedings of the 1976 Summer Computer Simulation Conference*, pp. 433–6. Simulation Councils, Inc., LaJolla, California.

Neuhold, J. M. (1975). Introduction to modeling in the biomes. In *Systems analysis and simulation in ecology*, 3 bold type ed. B. C. Patten, pp. 7–12. Academic Press, New York.

Odum, H. T. (1957). Trophic structure and productivity of Silver Springs, Florida. *Ecol. Monogr.* **27**, 55–112.

Olson, J. S. (1975). Productivity of forest ecosystems. In *Productivity of world ecosystems*, ed. D. E. Reichle, J. F. Franklin & D. W. Goodall, pp. 33–43. National Academy Sciences, Washington, D.C.

O'Neill, R. V. (1973). Error analysis of ecological models. In *Radionuclides in ecosystems, Proceedings of the third national symposium on radioecology*, **1**, ed., D. J. Nelson, pp. 87898–908. CONF–710501. National Technical Information Service, Springfield, Virginia.

495

O'Neill, R. V. (1975). Modeling in the Eastern Deciduous Forest Biome. In *Systems Analysis and Simulation in Ecology*, 3, ed. B. C. Patten, pp. 49–71. Academic Press, New York.

Ovington, J. D., Forrest, W. G. & Armstrong, J. S. (1968). Tree biomass estimation. In *Symposium: Primary Productivity and Mineral Cycling in Natural Ecosystems*, ed. H. E. Young, pp. 4–31. University of Maine Press, Orono.

Patten, B. C. (1975). Ecosystem linearization: an evolutionary design problem. *Amer. Naturalist*, **109**, 529–39.

Reeves, M., III. (1971). *A code for linear modeling of an ecosystem*. ORNL/IBP–71/2. Oak Ridge National Laboratory, Oak Ridge, Tennessee.

Reichle, D. E. (1975). Advances in ecosystem analysis. *BioScience*, **25**, 257–64.

Reichle, D. E. & Edwards, N. T. (1973). IBP–Eastern Deciduous Forest Biome – Oak Ridge site. In *Modeling Forest Ecosystems*, ed. D. E. Reichle, R. V. O'Neill & J. S. Olson, pp. 151–202. EDFB/IBP–73/7. Oak Ridge National Laboratory, Oak Ridge, Tennessee.

Reichle, D. E., O'Neill, R. V. & Olson, J. S. (1973). *Modelling forest ecosystems*. EDFB/IBP–73/7. Oak Ridge National Laboratory, Oak Ridge, Tennessee.

Rust, B. W. & Mankin, J. B. (1976). *An interactive differential equations modeling program*. EDFB/IBP–74/2. Oak Ridge National Laboratory, Oak Ridge, Tennessee.

Satoo, T. (1967). Primary production relations in woodlands of *Pinus densiplura*. In *Symposium: Primary Productivity and Mineral Cycling in Natural Ecosystems*, ed. H. E. Young, pp. 52–80. University of Maine Press Orono.

Satoo, T. (1970). A synthesis of studies by the harvest method: primary production relations in the temperate deciduous forests of Japan. In *Analysis of Temperate Forest Ecosystems*, ed. D. E. Reichle, pp. 55–72. Springer–Verlag, New York.

Sharpe, D. M. (1975). Methods of assessing the primary production of regions. In *Primary Productivity of the Biosphere*, eds. H. Leith & R. H. Whittaker, pp. 147–66. Springer–Verlag, New York.

Shugart, H. H. & West, D. C. (1977). Development of an Appalachian deciduous forest succession model and its application to assessment of the impact of the chestnut–blight. *J. Environmental Management*, **5**, 161–79.

Skellam, J. G. (1971). Some philosophical aspects of mathematical modelling in empirical science with special reference to ecology. In *Mathematical Models in Ecology*, ed. J. N. R. Jeffers, pp. 13–28. Blackwell, Oxford.

Smith, F. E. (1970). Analysis of ecosystems. In *Analysis of Temperate Forest Ecosystems*, ed. D. E. Reichle, pp. 7–18. Springer–Verlag, New York.

Sollins, P. & Anderson, R. M. (1971). *Dry-weight and other data for trees and woody shrubs of the southeastern United States*. ORNL/IBP–71/6. Oak Ridge National Laboratory, Oak Ridge, Tennessee.

Sollins, R., Reichle, D. E. & Olson, J. S. (1973). Organic matter budget and model for a southern Appalachian *Liriodendron* forest. EDFB/IBP–73/2. Oak Ridge National Laboratory, Oak Ridge, Tennessee.

Swank, W. T. & Schreuder, H. T. (1974). Comparison of three methods of estimating surface area and biomass for a forest of young eastern white pine. *Forest Sci.* **20**, 91–100.

Tiwari, J. L. & Hobbie, J. E. (1976). Random differential equations as models of ecosystems: Monte Carlo simulation approach. *Math. Bios.* **38**, 25–44.

Ulrich, B., Mayer, R. Heller, H. (1974). Goettinger Bodenkundliche Berichte 30. Institute of Soil Science and Forest Nutrition of Goettingen University, D-34, Gottingen, Busgenweg 2.

Waide, J. B., Krebs, J. E., Clarkson, S. P. & Setzler, E. M. (1974). A linear analysis of the calcium cycle in a forested watershed ecosystem. *Prog. theor. Biol.* **3**, 261–344.

Weigert, R. G. (1975). Simulation models of ecosystems. In *Annual review of ecology and systematic*, **6**, ed. R. F. Johnson, P. W. Frank, P. W. Frank & C. D. Michener, pp. 311–38. Annual Review, Inc., Palo Alto, California.

Whittaker, R. H. (1966). Forest dimensions and production in the Great Smoky Mountains. *Ecology*, **47**, 103–21.

Whittaker, R. H. & Likens, G. E. (1973). Carbon in the biota. In *Carbon and the Biosphere*, ed. G. M. Woodwell & E. V. Pecan, pp. 281–302. Proc. 24th Brookhaven Symposium in Biology CONF-720510. National Technical Information Service, Springfield, Virginia.

Woodwell, G. M. & Botkin, D. B. (1970). Metabolism of terrestrial ecosystems by gas exchange techniques: the Brookhaven approach. In *Analysis of Temperate Forest Ecosystems*, ed. D. E. Reichle. Springer–Verlag, New York.

Worthington, E. B. (ed.). (1975). *Evolution of the IBP*. Cambridge University Press, London.

Yoda, K., Shinozaki, K., Ogawa, H., Hozumi, K. & Kira, T. (1965). Estimation of the total amount of respiration in woody organs of trees and forest communities. *J. Biol. Osaka City University*, **16**, 15–26.

9. Carbon metabolism in terrestrial ecosystems*

N. T. EDWARDS, H. H. SHUGART, JR., S. B. McLAUGHLIN, W. F. HARRIS, & D. E. REICHLE

Contents

But nobody denies the necessity for investigation of *all* the components of the ecosystem and of the ways they interact to bring about approximation to dynamic equilibrium. That is the prime task of the ecology of the future.   A. G. Tansley (1935).

One of the hallmarks of the recent decade of ecological research was the consideration of ecosystems as objects of study per se. Although many of the ecosystems concepts had been formulated for some time (Tansley, 1935; Lindeman, 1942), the studies of participants in IBP were instrumental in the development of an ecosystem ecology during the late 1960s. To determine the metabolism of a forest ecosystem, one must determine the standing crops of the various pools of matter and the fluxes of material in and out of the system. Compared to many ecosystem types, this is a task of herculean magnitude. The size of trees makes the measurement of the standing crops logistically difficult. The time constants of fluxes associated with soils and trees are long in relation to the lives of humans. Other time constants associated with microbial populations and respiratory losses are very fast and require intensive sampling stratagems. The most natural units for measuring ecosystem metabolism are either biomass, carbon, or energy. Usually the basic data on ecosystem metabolism are determined on biomass measurements of standing crops (or pools of matter) and by biomass and

*Research supported by Eastern Deciduous Forest Biome US–IBP, funded by the National Science Foundation under Interagency Agreement AG–199, DEB76–00761 with the DOE–ORNL. Contribution No. 338 from the Eastern Deciduous Forest Biome, US–IBP. Operated by Union Carbide Corporation under contract W–7405–eng 26 with the U.S. Department of Energy. Environmental Sciences Division Publication No. 1494, Oak Ridge National Laboratory.

499

Reprinted from *Dynamic properties of forest ecosystems* edited by D. E. Reichle. International Biological Programme 23. © Cambridge University Press 1980. Printed in Malta.

carbon dioxide measurements of ecosystem fluxes. Investigators usually convert their data to common units using empirically determined concentration factors. In a given ecosystem the conversions among biomass, carbon and energy do not vary greatly.

If one conceptualizes the terrestrial biosphere as a mosaic of ecosystem types, forest ecosystems are unique in terms of their pools of carbon. Fig. 9.1 (from Baes, Goeller, Olson & Rotty, 1977) shows the dominance of forest ecosystems at a global scale both for slowly exchanging and rapidly exchanging pools of carbon. Forest ecosystems are relatively similar to agricultural ecosystems in terms of the concentration of pools of carbon per unit area, whereas the tundra, desert, savanna, and other non-wooded ecosystems have much lower concentrations. It is the carbon concentration multiplied by a large area that makes forests the singular most important terrestrial ecosystem in terms of global carbon pools.

In this paper, we will discuss the relative magnitudes of the carbon reserves of different ecosystem types in the biosphere. We will compare carbon metabolisms in different ecosystems. Additional comparisons among forest ecosystems will be made. Detailed discussions will treat problems in measurement and ecological patterns of various mechanisms involved in carbon metabolism.

Carbon metabolism in various ecosystems

The global inventories of carbon shown in Fig. 9.1 have associated fluxes of carbon (the exchanges with the atmosphere measured as units of mass time^{-1}) in and out of each ecosystem type. One might consider the matabolism of an ecosystem to be described by the magnitudes and relationships among these fluxes. Useful measures of ecosystem metabolism include (Reichle, 1975):

1. Total standing crop (TSC) – total carbon mass of the ecosystem.
2. Gross primary production (GPP) – the total photosynthetic fixation of carbon by the ecosystem.
3. Net primary production (NPP) – gross primary production minus the autotrophic (plant) respiration.
4. Autotrophic respiration (R_A).
5. Heterotrophic respiration (R_H).
6. Ecosystem respiration (R_E) [$R_A + R_H = R_E$].
7. Net ecosystem production (NEP) [$GPP - R_E = NEP$].

The metabolic parameters can take a wide range of values depending on the age of the ecosystem, the environmental conditions, and the type of ecosystem. Fig. 9.2 shows the range of net primary production associated with various types of ecosystems as an example of the variability of this

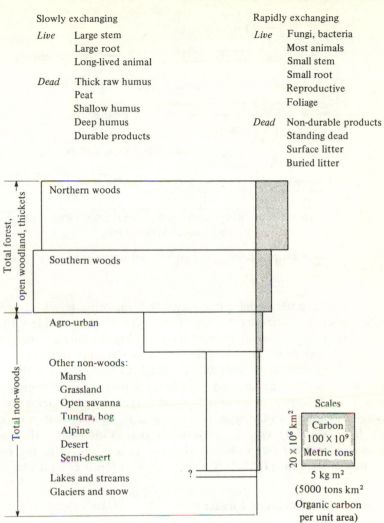

Slowly exchanging			Rapidly exchanging	
Live	Large stem		*Live*	Fungi, bacteria
	Large root			Most animals
	Long-lived animal			Small stem
				Small root
Dead	Thick raw humus			Reproductive
	Peat			Foliage
	Shallow humus			
	Deep humus		*Dead*	Non-durable products
	Durable products			Standing dead
				Surface litter
				Buried litter

Fig. 9.1. The quantities of carbon in various terrestrial ecosystems are shown as concentrations (horizontal direction of the histograms) and as area (vertical direction). The vertical line divides pools of carbon that have slow exchanges of carbon with the atmosphere from those that have a rapid exchange with the atmosphere. From Baes *et al.* 1977.

measure of ecosystem metabolism. Note that the range of net primary production associated with forests increases by almost an order of magnitude between temperate and tropical forest ecosystems. This change is roughly the same as that found between marine ecosystems with a substrate (reefs, estuaries) and the open-ocean ecosystem.

501

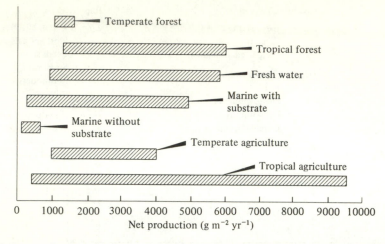

Fig. 9.2. Range of net production associated with different types of ecosystems. Data from Woodwell (1978).

Forests, in general, differ considerably from other terrestial ecosystems with regard to their metabolism (Table 9.1). If one considers the broad groupings of terrestrial ecosystems shown in Fig. 9.1, forest ecosystems have a much higher gross primary production, but the net primary production for the forest ecosystems (Fig. 9.2) is similar to that of agricultural systems. The net ecosystem production (which takes into account heterotrophic as well as autotrophic respiration) among agricultural systems and forest systems are even more similar. It is the large respiratory cost of maintaining the large standing crop of phytomass that distinguishes the temperate forests from other temperate ecosystems. This is shown in the high production costs ratio (Table 9.1) associated with forest ecosystems.

Carbon metabolism of forests

The first step in analysis of a carbon cycle of a forest ecosystem is to develop a budget. The amount of carbon in the various structural components of the ecosystem must be measured as well as the annual increment of these components. Exchanges of carbon among components and between the atmosphere and certain components must be determined. Transfer rates have been measured by a variety of different methods (e.g. gas analysis against harvest/allometric analysis) usually depending on availability of equipment and manpower. Parameters for eight forests located in various parts of the world are compared in Table 9.2. In the discussion that follows, we compare and contrast the carbon cycles of three of these forested ecosystems: a mesic deciduous forest in Tennessee, a xeric oak–pine forest in

502

Table 9.1. *Comparative metabolic parameters of five different ecosystems using indices from Reichle (1975). All values above the dotted line on the table are in grams of carbon $m^{-2}yr^{-1}$; values below the dotted line are dimensionless indices*

Comparative metabolic parameters	Agro-urban		Other non-woods		
	Potato field[a]	Rye field[a]	Shortgrass prairie[b]	Tundra[c]	Temperate forest[d]
Gross primary production (GPP)	1286	1006	641	240	2505
Autotrophic respiration (R_A)	431	342	218	120	1398
Net primary production (NPP)	849	664	423	120	1108
Heterotrophic respiration (R_H)	500	310	294	108	547
Net ecosystem production (NEP)	355	354	129	12	561
Ecosystem respiration (R_E)	931	652	512	228	1945
Production efficiency (R_A/GPP)	0.34	0.34	0.34	0.50	0.56
Effective production (NPP/GPP)	0.66	0.66	0.66	0.50	0.44
Maintenance efficiency (R_A/NPP)	0.52	0.52	0.52	1.00	1.26
Respiration allocation (R_H/RA)	0.86	0.91	1.34	0.90	0.39
Ecosystem productivity (NEP/GPP)	0.28	0.35	0.21	0.005	0.22

[a] From L. Ryszkowski (personal communication).
[b] From Andrews, Coleman, Ellis & Singh (1974).
[c] From Reichle (1975).
[d] Average values from three forest ecosystems (pine in North Carolina, oak–pine in New York and mixed deciduous in Tennessee). Data from Woodwell & Botkin (1970). Harris *et al.* (1975), Kinerson *et al.* (1977).

New York, and a pine plantation in North Carolina. The mesic deciduous forest is discussed in more detail than the other forests as an example of methodologies used in constructing forest carbon cycles.

Carbon budgets of forests

Mesophytic deciduous forest

The *Liriodendron* forest, located on the DOE Reservation at Oak Ridge in East Tennessee, has been the site of intensive nutrient cycling and gas

503

Table 9.2. *Comparative parameters of eight different forests using indices from Reichle (1975). All values above the dotted line are in grams of carbon m^{-2} and grams of carbon $m^{-2}yr^{-1}$ except values in parentheses; values below the dotted line are dimensionless indices. All conversions from author's biomass values assume carbon content of biomass to be 50%*

Parameters	Forest types and locations							
	Mixed deciduous[a] Tenn., USA	Pine plantation[b] N. Carolina, USA	Oak–Pine[c] New York, USA	Subalpine coniferous[d] Japan	Beech[e] Denmark	Rain forest[e] Ivory Coast, Africa	Oak[f] Britain	Oak–Hornbeam[g] Poland
TSC	8760	7062	5096	15905	—	—	7466	(1377)
GPP	2162	4124	1231	1910	1175	2675	2330	(133)
NPP	726	2056	542	535	675	675	1918	(82)
R_A	1436	2068	690	1375	500	2000	1412	(51)
R_H	670	694	282	331	675	675	564	—
R_E	2106	2762	966	1706	1175	2675	1976	—
NEP	56	1362	265	204	—[e]	—[e]	354	—
R_A/GPP	0.66	0.50	0.56	0.72	0.43	0.75	0.61	0.38
NPP/GPP	0.34	0.50	0.44	0.28	0.57	0.25	0.39	0.62
R_A/NPP	1.98	1.00	1.27	2.57	0.74	2.96	1.54	0.62
R_H/R_A	0.47	0.34	0.41	0.24	1.35	0.34	0.40	—
NEP/GPP	0.03	0.33	0.22	0.11	—	—	0.15	—

[a] From Harris *et al.* (1975).
[b] From Kinerson *et al.* (1977).
[c] From Woodwell & Botkin (1970), which report all values as biomass. They assumed that 1g carbon dioxide represents 0.614 g of dry matter. We back-calculated to obtain a value for carbon dioxide, then multiplied by 0.27 to convert carbon dioxide to carbon for all parameters except TSC.
[d] From Kitazawa (1977).
[e] MacFadyen (1970). Means of calculations by authors assumes NEP=0.
[f] From Satchell (1973).
[g] From Medwecka-Kornas, Lomnicki & Bandola-Ciolczyk (1974). (Values in parentheses are in kcal.)

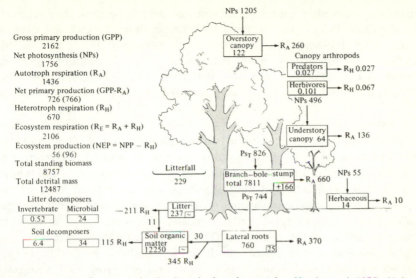

Fig. 9.3. Annual carbon cycle in the *Liriodendron* forest (after Harris *et al.* 1975). Units of measure are g C m^{-2} yr^{-1} and g C m^{-2}. Mean annual standing crops are given in the center of each box and net annual increment in the right-hand corner. Values of NPP and NEP in parenthesis are based on harvest/allometric methods; other values are based on gas exchange data.

exchange studies during the past several years as a part of the US–IBP. It is a second growth forest established on deep, alluvial, silt loam soil (Emory series). Mean annual temperature is 13.3°C and annual precipitation averages 126.5 cm. The forest is dominated by yellow poplar (*Liriodendron tulipifera* L.) interspersed with oaks (*Quercus velutina, Q. alba, Q. rubra*). The age of the oldest trees at the time of the study was about 50 yr. Redbud (*Cercis canadensis*) and flowering dogwood (*Cornus florida*) are the predominant understory species. Herbaceous groundcover is primarily Virginia creeper (*Parthenocissus quinquefolia*), woody hydrangea (*Hydrangea arborescens*), and Christmas fern (*Polystichum acrostichoides*).

The carbon budget for the yellow poplar forest is summarized in Fig. 9.3. Except for carbon exchanges between biota and atmosphere, the budget is based on harvested biomass which was analyzed for carbon content by flame photometric detection, a method discussed by Horton, Shultz & Meyer, (1971). Above-ground autotrophic carbon pools and increments were based on allometric relations of tree component weights and diameter breast height (dbh) (Sollins & Anderson, 1971; Harris *et al.*, 1975) and periodic inventories of tree diameters (Sollins, Reichle & Olson, 1973). The standing pool of lateral root biomass was determined from a series of soil core data collected through the year. Increment of lateral roots was determined from net seasonal differences in the lateral root biomass pool.

Litter was harvested monthly for carbon determinations. Soil carbon, to a depth of 75 cm, was assumed to average about 58% of soil organic matter (after Jackson, 1958). Canopy arthropods and litter and soil invertebrate carbon pools were determined from population analyses (Reichle & Crossley, 1967; Moulder & Reichle, 1972; McBrayer & Reichle, 1971). Soil microflora carbon pool was based on ATP analyses (Ausmus, 1973).

Transfers of carbon among ecosystem components and between certain components and the atmosphere involved measurements of photosynthesis, autotrophic respiration above-ground and below-ground, heterotrophic respiration from the forest floor, annual litterfall, tree mortality, and foliage consumption by insects. Net photosynthesis was determined from several hundred hours of *in situ* carbon dioxide measurements under natural temperature and light conditions by means of gas exchange analysis in controlled environment chambers (Dinger, 1972a). Above-ground autotrophic respiration was based on data of Woodwell & Botkin (1970). Preliminary analysis of woody shoot respiration of yellow poplar suggests good agreement with their data. Below-ground autotrophic respiration was estimated from *in vivo* lateral root respiration measurements and monthly root biomass (Reichle *et al.*, 1973b). Total respiration from the forest floor was measured for 24-hr intervals throughout the year by carbon dioxide analysis, using chambers inverted over the forest floor and infrared gas analysis (Edwards & Sollins, 1973). Total forest floor heterotrophic respiration was calculated by subtracting root respiration from total forest floor respiration. Transfer of carbon by litterfall was determined from litterfall traps (Edwards & Harris, 1977), and tree mortality was determined from stand inventory over an eight-year period (1962–70) (Sollins *et al.*, 1973). Consumption of leaves by insects was determined from measurements of hole expansion in leaves (Reichle *et al.*, 1973b).

Oak–pine forest

The structure and metabolism of a xeric oak–pine forest at Brookhaven National Laboratory, Long Island, New York, has been intensively studied over the past several years using harvest techniques, stand dimension measurements and gas exchange analyses (Whittaker & Woodwell, 1968, 1969 Woodwell & Botkin, 1970). The forest is dominated by small oak trees (12–20 cm dbh and 8–13 m tall) interspersed with pitch pine (*Pinus rigida*), often a little taller than the oaks, forming a canopy which is open and irregular (Whittaker & Woodwell, 1969). This openness permits a relatively good cover of shrubs. The principal shrubs are low Vaccinaceae such as *Gaylussacia baccata* (black huckleberry) and *Vaccinium vacillans* Torr. (low bush blueberry). Soils are sandy, well drained and podzolic. Mean annual temperature is 9.8°C and annual precipitation averages 124 cm (Whittaker & Woodwell, 1969).

Pine plantation

This plantation has been in agricultural production prior to planting of loblolly pine (*Pinus taeda* L.) in 1958. Trees were planted on a 2 × 2 m grid. Harvest methods and gas exchange analysis have been used to study the carbon cycle of this forest since canopy closure (Kinerson, Ralston & Wells, 1977). The forest is located in the Piedmont of North Carolina. The soil is sandy loam, varying from 20–33 cm in thickness, over sandy clay to clay subsoils. Mean annual air temperature is 15.6°C and average annual precipitation is 115 cm (Kinerson *et al.*, 1977).

Forest carbon cycle comparisons

The three forests discussed above have been studied in sufficient detail to permit reasonably accurate descriptions of their annual carbon cycles. Data obtained through measurements of various components of an ecosystem can be pooled and extrapolated so that the metabolism of one entire forest ecosystem can be compared with that of another (Fig. 9.4). The total standing crop (TSC) of the yellow poplar forest is considerably greater (8760 g C m^{-2}) than that of the pine forest (7062 g C m^{-2}) which is greater than the carbon pool of the oak–pine forest (5096 g C m^{-2}). However, the pine forest has a higher GPP (4124 g C m^{-2}yr^{-1}) than the yellow poplar forest (2162 g C m^{-2}yr^{-1}); probably a reflection of the differences in age of the two stands. The younger pine stand is still in the linear portion and to the left of the inflection point of the sigmoid growth curve used by population biologists, while calculated relative production (NPP/TSC) ratios for the yellow poplar and oak–pine forests place these systems on the upper (flatter) portion of the growth curve (Kinerson *et al.*, 1977). The NPP/TSC ratios for the yellow poplar and oak–pine forest are 0.08 and 0.11, respectively, while that of the pine forest is 0.29.

Gross primary production (GPP), sometimes referred to as gross photosynthesis, is also greatest in the pine forest. Since respiration of a plant increases with increased growth rates, we would expect that the younger, faster growing forest would also have the greatest autotrophic respiration (R_A). This is substantiated by the fact that the R_A of the pine forest is considerably greater than that of the yellow poplar forest (1068 g C m^{-2}yr^{-1} against 1436 g C m^{-2}yr^{-1}), even though both forests have very similar annual temperature and moisture regimes. Even though the pine forest loses more of its carbon by R_A than does the yellow poplar forest, the GPP of the younger pine forest is great enough to offset these losses

507

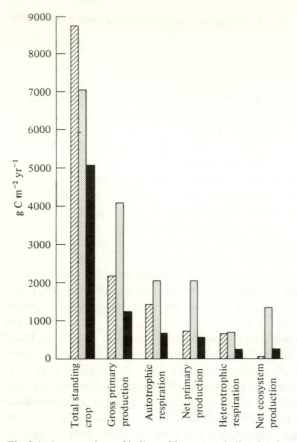

Fig. 9.4. A comparison of indices of forest metabolism for three Eastern US forests (Tennessee, yellow poplar; N. Carolina, loblolly pine; oak–pine, New York). Sources of data are: yellow poplar (▨) from Harris *et al.* 1975; loblolly pine (▨) from Kinerson *et al.* 1977; oak–pine (■) from Woodwell & Botkin, 1970. Data from the oak–pine forest was originally reported as biomass.

resulting in a net primary production ($NPP = GPP - R_A$) considerably greater than the yellow poplar forest (2056 g C $m^{-2}yr^{-1}$ against 726 g C $m^{-2}yr^{-1}$). The NPP of the oak–pine forest (592 g C $m^{-2}yr^{-1}$) is slightly less than that of the yellow poplar forest as are other parameters (TSC, GPP, R_A, and R_H), no doubt a reflection of the drier, well drained sandy soil and cooler environment of the oak–pine forest.

Heterotrophic respiration (R_H) appears to be affected almost exclusively by regional climates as evidenced by the similar R_H rates in the yellow poplar and pine forests (670 g C $m^{-2}yr^{-1}$ and 694 g C $m^{-2}yr^{-1}$, respectively), which have similar climatic regimes but very different substrates involved in decay. The R_H of the cooler, drier oak–pine forest is so small relative to the other two forests that net ecosystem production ($NEP = NPP - R_H$) is greater for the oak–pine forest than for the yellow poplar forest. NPP of the yellow poplar forest is 184 g C $m^{-2}yr^{-1}$ greater than NPP of the oak–pine forest. Thus, it appears that for these three forests at least, age of the stand and regional climate (primarily its effects on R_H) are two of the major factors controlling net ecosystem production.

Photosynthesis and photosynthate allocation

A thorough understanding of carbon dynamics in the biosphere is necessitated by the role of carbon in the chemistry of living systems, the coupling of biological systems by carbon through common atmospheric and detritus pools and the interaction of the biosphere and atmosphere to regulate global carbon balance. Forests play a large role in this cycle. Olson (1975) reports total terrestrial global production (gross primary production less plant respiration) of 75.4×10^9 metric tons annually. Of this amount, forests account for 59.4×10^9 metric tons (80%). The dominance of forests is also striking in terms of their contribution to the global terrestrial 'live carbon pool'. Forests contain 74% of the total terrestrial living carbon pool (Olson, 1975).

Despite the large net annual flux of carbon to forests, relatively few studies exist that describe the photosynthetic processes of individual forest canopies. The high cost of equipment, availability of suitable sites (e.g. proximity to electrical power, etc.) impose severe limitations to the extent of detailed physiological studies. Because of the limitations to location of detailed physiological studies, it is inappropriate to consider these limited studies generally representative of the forest type in which they happen to be located. The detailed investigations that have been undertaken do provide valuable insights into the factors influencing forest canopy photosynthesis and its subsequent utilization.

Photosynthesis

The conversion of photochemical energy into the biochemical energy of organic molecules by photosynthetic processes is the basis of initiation of metabolic cycles in terrestrial ecosystems. Thus, the study of the rates and controlling variables in these processes is a prerequisite for developing budgets of energy flow through both autotrophic and associated heterotrophic communities. The efficiency of conversion of radiant energy into photosynthate, while low for terrestrial ecosystems, is higher for forests (2.0–3.5%) than for perennial herbaceous communities or annual herb and crop communities, 1–2% and <1.5%, respectively (Kira, 1975). The relatively higher gross photosynthetic efficiency may be attributed in large

509

part to the maintenance of a large foliar surface area within the relatively deep forest canopy. Support of this geometrically complex photosynthesis surface has its costs, however, and maintenance of a large above-ground and below-ground structural support and supply system and its associated metabolic demands draws heavily on the GPP of forest systems as previously noted in the comparative metabolism of terrestrial ecosystems (Table 2, Reichle, 1975). Kira (1975) compared GPP, NPP, and NPP:GPP for world forests and found NPP:GPP ratios to range between 0.26 and 0.68, values generally less than or equal to other terrestrial communities. Realization that between 32% and 74% of gross photosynthate is used in the respiratory metabolism supporting annual production of biomass in forest systems emphasizes the tremendous production costs of forest trees and stresses the importance of both maintenance of high photosynthetic rates and of subsequent processes linking photosynthesis with net primary production.

The structural complexity of forest canopies and the associated spatial and temporal variability in biotic and abiotic variables has been a continuing challenge in efforts to accurately quantify gross photosynthetic production of forest stands. Kira (1975) lists the relatively few publications on the photosynthetic activity of forest canopies relative to structure and solar energy utilization. The interplay of solar energy and stand structure provides spatial gradients in a variety of variables influencing tree photosynthesis. These include: light (quality and quantity), temperature, relative humidity, carbon dioxide, water supply and demand (see reviews by Saeki, 1975; Shaedle, 1975; Hinckley, Lassoie & Running, 1978), nutrient flux, and assimilate supply and demand (see reviews by Wareing & Patrick, 1975; Saeki, 1975; Shaedle, 1975; Hinckley *et al.*, 1978). Associated with light-induced physical gradients are physiological gradients resulting from anatomical and morphological variations in foliage within individual canopies and between different individuals and species in the various canopy layers (see review by Boardman, 1977). Hozumi & Kirita (1970), for example, showed a 40% increase in simulated canopy photosynthesis when adjustments for higher photosynthetic efficiencies of shade grown foliage in the inner canopy of a Japanese evergreen oak forest were incorporated into the canopy photosynthesis model of Monsi & Saeki (1953). Thus, the ultimate photosynthetic production of a forest stand is the result of an integration of the physiological characteristics of the component species and spatially complex variations in environmental stresses imposed by interactions of abiotic variables and stand structure. These variables are integrated further over diurnal and seasonal cycles to determine ultimate levels of carbon fixation by the forest canopy.

Quantification of photosynthetic processes for the compositionally and structurally complex forests of the eastern United States has been approached with a variety of techniques at three forest sites (Brookhaven, Oak Ridge, and Research Triangle). Basic to all these approaches has been characterization of *in situ* gas exchange rates of a dominant overstory species over a range of environmental conditions. Measurement techniques have involved the use of in-canopy cuvettes in which carbon dioxide exchange rates of branchlets with attached leaves have been monitored by infrared gas analysis. Description of detailed methodologies can be found for measurements on scarlet oak at Brookhaven (Woodwell & Botkin, 1970); loblolly pine at Research Triangle (Kinerson, 1975; Higgenbotham, 1974), and yellow poplar, red maple, red oak, dogwood, and chestnut oak at Oak Ridge (Dinger, 1972*b*, 1973). Gas exchange characteristics of species measured at the Oak Ridge and Triangle sites have been summarized by Strain & Higgenbotham (1976) as shown in Table 9.3.

Their conclusions concerning the major variables controlling photosynthetic processes in these two systems are summarized below:

1. During the period winter 1971 to fall 1973, ambient air temperatures and incident solar radiation accounted for most of the variability in net photosynthesis. At no time did tissue water potential become low enough to decrease net photosynthesis.

2. For *Liriodendron* and *Pinus taeda*, irradiance is the primary factor influencing stomatal behavior. Stomatal opening was initiated below 0.04 cal cm^{-2} min^{-1} and minimal resistances were observed by 0.4 cal in *Liriodendron* and by 0.5 cal in *Pinus* with further increases in radiation having no direct effect on stomatal resistance.

3. *Liriodendron* appears to maintain canopy dominance by comparatively high photosynthesis whereas *Pinus taeda* takes advantage of an evergreen growth and consequent year-round photosynthesis.

4. Canopy position, season and leaf age all affect the photosynthetic potential of *Pinus taeda*. Seasonal potential must be determined and applied in simulation or predictive modeling of photosynthetic production.

One of the variables known to affect rates of photosynthesis is moisture availability. However, quantification of this effect is sometimes difficult in field experiments. The absence of a direct control of moisture supply on photosynthesis in the forests at both the Oak Ridge and Triangle sites during the period of study is interesting and in striking contrast to results from northeastern, midwestern, and western forests (see review by Hinckley *et al.*, 1978). These results, however, apparently reflect a fortuitous combination of relatively mesic site conditions and abundant rainfall during the study interval. Periodic growth analysis of increment cores from permanent plots at the Oak Ridge site has, in fact, shown reduced growth particularly at more xeric ridge top sites during years of low moisture supply as shown in Fig. 9.5. It should be noted that absence of limitations on photosynthesis

Table 9.3. *Gas exchange characteristics of some species of the deciduous forest biome. These annual means should be considered for general comparative purposes only. The species vary in acclimation potential and static values should not be used for predictive purposes*

Species	PhAr (equiv. m^{-2} sec^{-1}) at net photosynthetic compensation	PhAr (equiv. m^{-2} sec^{-1}) at net photosynthetic saturation	Optimum temperature (°C) for photosynthesis	Maximum photosynthesis (mg carbon dioxide dm^{-2}h^{-1})	Average photosynthesis	Stomatal resistance (sec/cm) at 46 μequiv m^{-2} sec^{-1} PhAr	Stomatal resistance (sec/cm) at 920 μequiv m^{-2} sec^{-1} PhAr	Range in xylem pressure potential (−bar)
Liriodendron tulipifera	69	1610	28	14	6	6	1.4	4–19
Quercus alba	46	920	26	8	4	—	—	—
Acer rubrum	92	460	26	12	6	—	—	—
Quercus rubra	115	690	31	6	3	—	—	—
Cornus florida	69	920	28	6	5	—	—	—
Pinus taedia	69	1380	20[a]	7	4	14	2.0	7–14

[a] *Pinus taedia* had extreme seasonal adjustment of temperature response with winter optimum of 10°C and summer of 25°C.

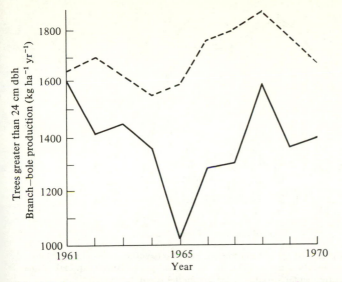

Fig. 9.5. Comparison of growth of trees from two classes of slope (—, 15%, ridge tops); ---, 30%) on the Walker Branch Watershed. Note differential effect of 1965 drought.

through hydroactive stomatal closure does not negate significant impacts on growth through differential effects on a variety of more sensitive biochemical processes associated with growth (Wardlaw, 1968a; Hsiao, 1973; Slatyer, 1973; Slavik, 1975). McConathy, McLaughlin, Reichle & Dinger (1976) measured *in situ* xylem potentials on yellow poplar foliage as low as -22 bars, a potential near that determined by Richardson, Dinger & Harris (1972) to be necessary for stomatal closure of yellow poplar seedlings and well within the range for effects on a variety of growth-related biochemical processes (Hsiao, 1973).

The study of limiting factors in photosynthetic production of forest trees is intriguing and may provide valuable information with implications for differential sensitivity of species to environmental stress and a physiological basis for growth, competition, and survival. Richardson *et al.* (1972) used multiple regression techniques to examine the influence of variations in radiation, soil moisture, soil nitrogen, and time of treatment, leaf pigment content, stomatal diffusion resistance, leaf water potential and tissue nitrogen on photosynthesis of greenhouse-grown yellow poplar seedlings. The large number of variables was included to develop a physiological basis for predicting photosynthetic potential. The most precise regression developed was for fully mature sun leaves collected from a single harvest and accounted for 99% of the total variability. Parameters included in this regression were leaf chlorophyll content ($a+b$), absorbance ratio (460/663),

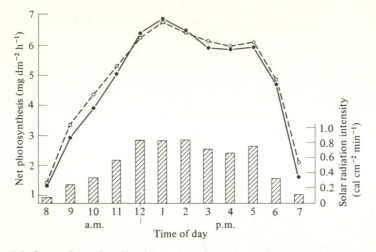

Fig. 9.6. Comparison of predicted (- - - -) to observed net photosynthesis (——) in relation to solar radiation intensity, for 19 August 1971 in the yellow poplar forest at Oak Ridge, Tennessee. Observed environmental variables used for model input.

soil water potential, leaf water potential, and percent total leaf nitrogen. The final regression which considered all sources of variability (430 data points) accounted for 61% of the variation in net carbon dioxide exchange and took the form:

$$P = 6.82 + 1.07 \ln C + 0.41 \ln A + 0.63 \ln R - 0.28 \ln S, \qquad (9.1)$$

where P = photosynthetic uptake, A = absorbance ratio 460/663, R = radiation, and S = soil water potential.

Evaluation of factors influencing photosynthetic production in the field has been approached by Reed, Hammerly & Dinger (1976) who used cuvette data of Dinger (1971) for yellow poplar to develop a simulation model calculating uptake as a function of four variables, light, air temperature, stomatal diffusion resistance, and ambient carbon dioxide levels. Developed around a random selection of 279 hourly observations, it was subsequently validated against 45 additional observations. Analysis of variance of simulated and observed photosynthetic rates indicated that 89% of the variability in the original data set and 78% in the test set was accounted for. A plot of predicted against observed values of photosynthesis for a single day is shown in Fig. 9.6 and indicates good agreement for observed and predicted photosynthetic uptake. A review of additional considerations and approaches in modeling photosynthetic production in forest canopies can be found in Murphy & Knoerr, 1972; Miller, 1972; Sestak & Catsky, 1976; Sinclair, Murphy & Knoerr, 1976.

While consideration of variables influencing photosynthetic production of

514

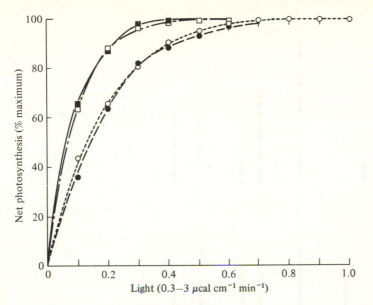

Fig. 9.7. Relative photosynthetic efficiency of four deciduous forest species in the yellow poplar forest as estimated from light saturation curves for leaf temperatures 26–30°C. O- - -O, *Liriodendron tulipifera*; ●——●, *Quercus alba*; □----□, *Cornus florida*; ■—■, *Acer rubrum*.

individual leaves or groups of leaves of a single species may be sufficient for estimating stand level productivity of forests of uniform composition, the evaluation of comparative photosynthetic efficiencies, percent composition, and canopy position of additional species becomes important in the typically complex forests of the Eastern United States. Fig. 9.7 illustrates responses of forest tree species representative of a temperate deciduous forest to light within a moderate range of temperatures (25–30°C) typical of the growing season canopy microclimate. The light-saturated photosynthetic capacity varies by a factor of three among the few species examined. Although measured light compensation points do not differ detectably between species, the more shade-tolerant red maple and dogwood trees exhibit higher photosynthetic capacities at the lower light intensities at which these trees function beneath the closed forest canopy.

Knowledge of individual species photosynthetic response can be used to estimate photosynthesis of the entire canopy if sufficient information is available on leaf area and radiant energy distribution within the canopy. As an approximation of stand-level carbon productivity of the yellow poplar dominated stand at Oak Ridge, Dinger (1971) used the relative photosynthetic rates of yellow poplar, red maple, and white oak, considered to be

515

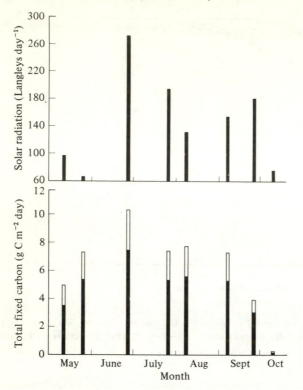

Fig. 9.8. Relationship between daily radiation inputs and total carbon fixed in a yellow poplar forest in Tennessee. ■, *Liriodendron tulipifera*; □, all other species.

the predominant species, and their relative contributions to the total leaf area of the stand (66% yellow poplar, 34% other) and seasonal radiation regimes to estimate daily productivity at eight intervals from mid-May to mid-October (Fig. 9.8). The trend suggested by this analysis indicates that daily productivity reaches a maximum in the middle of the growing season. The pattern is highly correlated with the course of leaf area development (Fig. 9.9). The absolute values obtained from such a reconstruction of forest canopy photosynthesis must be interpreted with caution because of the measurement uncertainties involved. A major uncertainty is the manner in which solar radiation averages within the forest are obtained. A number of investigators have noted that space or time averaging of solar radiation values are less than satisfactory because of the non-normality of the flux density frequency distributions induced by the presence of beam radiation (Ramann, 1911; Aleksew, 1963; Gay, Knoerr & Braaten, 1971). The mean of a non-normal radiation flux density distribution is an especially poor characterization of radiation climate for comparison with such processes as

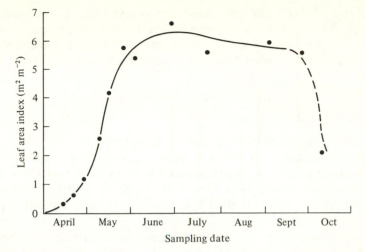

Fig. 9.9. Seasonal changes in leaf area index of a yellow poplar forest in Tennessee.

photosynthesis or evaporation which vary nonlinearly with radiation amounts. Thus, the curvilinear response of photosynthesis may lead to overestimation of average photosynthetic rates, particularly where a significant fraction of the total incoming radiation exceeds the light saturation level for photosynthesis.

In a detailed analysis of the radiation climate of a *Liriodendron* forest, Hutchison & Matt (1976) found little high flux density radiation within the forest, but most of the radiation reaching the forest floor was received at flux densities exceeding 40 nly min^{-1}, the photosynthetic compensation point for yellow poplar (Burgess & O'Neill, 1975). While the total quantity of radiation penetrating the forest may be sufficient, the quality of radiation might further limit photosynthesis because of the spectral selectivity of canopy transmission.

The preceding discussion has emphasized the complexity of some aspects of physical and physiological considerations involved in realistically describing photosynthetic processes of forest canopies. While physiological analysis of photosynthesis of entire forest canopies is presently limited by our lack of complete understanding of the complex temporal and spatial variables, the use of currently available gas exchange data can play an important part in estimates of gross primary production for construction of a budget of carbon flow through forest systems as well as developing an understanding of the physiological limitations of photosynthate production. Harris et al. (1975) used in-canopy photosynthesis measurements on yellow poplar to estimate annual gross primary production at the Oak Ridge site at 2162 g C m^{-2}. Estimates of NPP (GPP − RA = NPP) based on dimen-

517

sional analysis of NPP showed good agreement with physiological measurements (760 g C m^{-2} yr^{-1} against 720 g C m^{-2} yr^{-1}, respectively. While the confidence limits for such estimates remain large, the convergence of physiological and dimensional estimates based on harvest techniques is encouraging.

Photosynthate allocation

While the fixation of energy in the products of photosynthesis represents a starting point for carbon metabolism in terrestrial ecosystems, neither the rate nor the extent of net production necessarily bears a close relationship to gross production (Evans, 1975). The wide range of NPP:GPP ratios (0.25–0.62) for world forests (Table 9.2) is ample evidence of the extent of variations in the efficiency of photosynthate allocation to growth processes. Thus, energetic costs of the processes of translocation storage, remobilization, and biosynthetic conversion in support of basal metabolism and growth vary widely for forest systems. While considerable progress has been made in quantifying the net energetic costs of forest metabolism through gas exchange analysis of autotrophic losses (Yoda *et al.*, 1965; Woodwell & Botkin, 1970; Kinerson, 1973; Harris *et al.*, 1975; Kinerson *et al.*, 1977), relatively little is known about the allocation processes of whole trees. An obvious reason for this is their physical and physiological complexity.

Annual photosynthate production and allocation is strongly influenced by environmental stresses during both diurnal and seasonal cycles. Impacts of these cycles on allocation are mediated primarily through variations in temperature, light, water, nutrition (Wardlaw, 1968*b*) and hormone levels (Waring & Patrick, 1975). Diurnal cycles are strongly controlled by tissue moisture status which regulates rates of photosynthesis and translocation as well as the relative activity of the many competing sinks for that photosynthate within the tree (Wardlaw, 1968*b*; Slavik, 1975; Waring & Patrick, 1975). There is ample evidence that under some conditions photosynthate production may exceed plant capacity to translocate and use resulting assimilates (Wardlaw, 1968*a*; Slavik, 1975). Seasonal cycles are more strongly influenced by temperature and the occurrence of a 'dormant' season in non-tropical latitudes. Based on fluctuations in carbohydrate economy, Kimura (1969) recognized three phases in the annual cycle of a young *Abies* stand in Japan.

In the first phase (May to early June) photosynthesis was low and resulted in storage in older organs or respiratory consumption. The second phase (mid-June to late August) was characterized by rapid growth of new organs at the expense of current photosynthate and reserve carbohydrates in older organs. The final phase (September to November) constituted a period of storage for initiation of the subsequent annual cycle. For temperate deciduous forests of the US, seasonal phenology would suggest four distinct phases: March through June – mobilization of stored reserves and generation of the new canopy (a period of depletion); June through early August – full photosynthetic production with allocation primarily to new diameter growth; mid-August through October – completion of woody growth and restoration of storage reserves; and finally, November through March – reduced physiological activity and maintenance of basal metabolism.

Meeting the seasonally adjusted physiological demands of temperate climates results in well-defined patterns of storage and mobilization of food resources, principally starch and fats in deciduous trees (see reviews by Kramer & Kozlowski, 1960; Kozlowski & Keller, 1966). Seasonal patterns are less well defined in coniferous species (Kramer & Kozlowski, 1960). In general, trees may be classified as fat accumulators (predominantly diffuse porous species) or starch accumulators (predominantly ring porous species) (Kramer & Kozlowski, 1960; Ziegler, 1964). Although a considerable number of studies have been conducted on seasonal variations of starch, fat, or sugar content of various tree parts, few have considered the simultaneous allocation of carbon to a variety of constituents for whole forest trees.

A radiochemical approach which has been used during IBP to evaluate rates of utilization of photosynthate has provided additional insights into whole-tree carbon utilization rates for three forest tree species. Using inoculation with ^{14}C-sucrose, Edwards, Harris & Shugart, (1973) examined seasonal metabolism of *Liriodendron tulipifera*, *Pinus echinata*, and *Quercus rubra*. All specimens were subcanopy trees with a dbh range of 11.3–16 cm. Trees were inoculated in September and periodic sampling of ^{14}C in tree parts (stem, branches, roots, and foliage) as well as ^{14}C efflux from the soil above the roots was done to characterize turnover times and distribution patterns of endogenous carbon reserves.

Distribution of carbohydrates, as inferred from ^{14}C activity approximately nine months after inoculation, indicated that the bole and stump are major storage sinks with lesser and seasonably variable amounts of carbohydrate distributed to branches and roots <0.5 cm diameter (Table 9.4). The average ^{14}C residence time based on measured activity losses from initial (one week after inoculation) levels varied from 6.5 months for *Q. rubra* to 11.0 months for *P. echinata*, although variations in distribution patterns and residence times were probably related to phenologically based differences in physiological activity at the time of inoculation and during the dormant season. The relatively longer mean residence times for pine is interesting in that it indicates greater stability of carbon resources, a condition that would be expected from its evergreen life form. It should be noted, however, that turnover times of available carbon would actually be

Table 9.4. *Distribution of ^{14}C nine months after inoculation. Average ^{14}C residence time in each tree assumed an initial tag equal to that measured one week after inoculation.*

Species	^{14}C Distribution (%)			Average residence time (months)
	Branch	Bole	Roots	
Liriodendron tulipifera	2.7	95.8	1.5	9.0
Pinus echinata	8.2	91.0	0.8	11.0
Quercus rubra	88.0	8.0	4.0	6.5

faster than rates calculated in Table 9.4, since a significant but undetermined nonlabile fraction resulting from structural utilization of supplied label was not considered. The generally high turnover times of carbon in all three species provide preliminary indication that storage reserves of temperate zone trees may be barely sufficient to maintain basal metabolism during the winter season and initiate new canopy generation the following spring.

An interesting offshoot of this work was an evaluation of below-ground utilization of the supplied carbon using $^{14}CO_2$-trapping techniques (Edwards *et al.*, 1973). Rates of $^{14}CO_2$ efflux from the soil surface followed at approximately monthly intervals over a 10-month period indicated a rapid initial turnover with a second peak period occurring in June. The peak period of $^{14}CO_2$ efflux from yellow poplar at this time may reflect an increase in decay of sloughed roots <0.5 cm diameter. This is in good agreement with the seasonal pattern of small root production and turnover found by Harris & Todd, (1972) through periodic harvest techniques (Fig. 9.10). Furthermore, integration of $^{14}CO_2$ efflux rates over time and distance around the trees suggests losses of approximately 20% of the supplied sucrose through below-ground respiration. A follow-up study with white oak (McLaughlin, Edwards & Beauchamp, 1977) found similar loss rates. In addition this investigation indicated rather slow rates of ^{14}C-substrate movement through the root system ($\simeq 0.5$ m day^{-1}) and a relatively homogeneous pattern of radial release of $^{14}CO_2$ indicating uniform occupancy of the site by the root system of the sample tree.

A pervasive theme at many of the IBP intensive study sites has been merger of field data and physiologically based theory into mathematical simulations with which higher resolution processes could be explored and new questions asked. This convergence of approaches has been particularly useful in the area of photosynthate allocation.

The efforts of Shugart, Harris, Edwards & Ausmus (1977) to combine data from the tagging study discussed above with those from other process

520

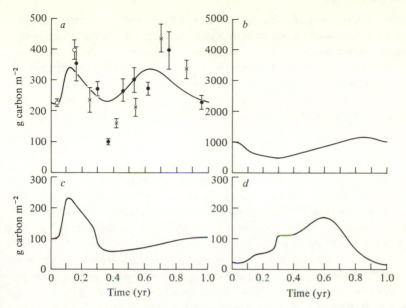

Fig. 9.10. Simulation of the dynamics of the 'measured small roots' and components of a yellow poplar forest in Tennessee (from Shugart *et al.*, 1977). Measured small roots (*a*) is assumed to equal live small roots (*c*) plus dead small roots (*d*) plus 10% of the storage (*b*). Time 0 is 1 January. *a*, measured roots less than 0.5 cm diameter; X, 1971; ●, 1972; ○, 1973; bar indicates 1 S.E. (Q4 + Q5 + 0.1Q1): *b*, stored photosynthate (Q1): *c*, live roots less than 0.5 cm diameter (Q4): *d*, dead roots less than 0.5 cm diameter (Q5).

level studies is a good example of the direction of this thrust toward below-ground allocation strategies. A model was developed to simulate the dynamics of root production and turnover using the data of Harris, Kinerson & Edwards (1977), a photosynthate input curve developed by Sollins *et al.* (1973), and a carbohydrate transport model developed from ^{14}C-sucrose experiments (J. R. Runkle & H. H. Shugart, unpublished data). Non-linear differential equations were used to describe changes in five state variables, photosynthate storage, living and dead large roots, and living and dead small roots. Small root (≤0.5 cm) dynamics described by this model are shown in Fig. 9.10. Comparison of the sum of simulated components (live roots, dead roots, plus 10% of stored photosynthate) with the seasonal lateral root dynamics (Harris *et al.*, 1977) shows good agreement. Quantitatively, the model overestimates lateral root biomass pools in the spring and underestimates measured values in the fall. These discrepancies are thought to be the result of the assumption that a constant 10% of stored photosynthate was present in lateral roots during the whole growing season. In actuality, reserve carbohydrates in roots are likely to be low in the spring as a result of new root growth and high in the fall following

521

resupply of storage pools by late season photosynthesis. The simulated seasonal inventory of stored photosynthate clearly shows these changes.

As with many modeling exercises, this simulation served not only as a focus for synthesis of the results of related process level studies but also as a stimulus for additional research. The need for obtaining more data on carbohydrate allocation of forest trees has been recognized and is being pursued largely as a result of synthetic models such as this.

Respiration

Measurement of total respiration of an ecosystem is a measure of the physiological activity of the system, and thus is the most conspicuous objective in studying ecosystem metabolism (Woodwell & Botkin, 1970). However, because of the structural and biological complexity of a forest ecosystem, respiration is perhaps the most difficult parameter to quantify. Specific problems encountered in measurement of ecosystem respiration are discussed later. Perhaps the most difficult problem is separating autotrophic (R_A) from heterotrophic (R_H) respiration, especially below-ground. Woodwell & Botkin (1970) measured R_E using infrared analysis of carbon dioxide concentrations in the Brookhaven forest during temperature inversions. Use of this method is limited to areas which have frequent temperature inversions and other atmospheric and topographic characteristics which result in a kind of large 'chamber effect' over the entire forest. A measurement of this type permits only the calculation of R_E and is useful in calculating NEP if GPP is known. It also is useful for comparisons with calculated R_E values from separate small chamber measurements of R_A and R_H. Quantification of both R_A and R_H is necessary for development of a comprehensive ecosystem carbon cycle.

Respiration measurements in the field almost always involve measurements of carbon dioxide, thus providing a very direct method for calculating carbon efflux. Usually, carbon dioxide is measured by absorption in alkali followed by acid titration or by infrared gas analysis. The alkali absorption method requires no expensive equipment and may be used in remote areas since no electrical or auxillary power source is needed. Infrared gas analysis requires expensive equipment and a power supply, but is much more sensitive than the alkali absorption method. Another advantage of the infrared gas analysis technique is that it can be used to obtain measurements over discrete short term time intervals. Oxygen uptake is often used in the laboratory as an indicator of respiration rates but is seldom used in the field. If oxygen measurements are to be used to estimate carbon efflux, it is necessary to determine the respiratory quotient (RQ), which requires periodic measurements of carbon dioxide as well as oxygen, or to know the amounts and kinds of substrates used in the respiratory process. This is impractical if not impossible in the field. Also, atmospheric oxygen concentrations are so high that small changes due to uptake during respiration require very controlled conditions and precise measurements to be statistically meaningful.

These conditions can be met in the laboratory using a Warburg apparatus or a differential respirometer. Carbon dioxide efflux from small field collected samples may also be measured under controlled laboratory conditions. If we can accept the assumptions necessary for extrapolating laboratory measurements of respiration rates in the field, then it is possible to separate the total *in situ* respiration rates of various component contributors to that total. For developing general forest carbon budgets, it is necessary to separate only R_A and R_H. However, a further breakdown to at least specific functional groups of autotroph organs and heterotrophic organisms permits a much more comprehensive picture of ecosystem metabolism.

Autotrophic respiration

Quantification of total forest ecosystem R_A requires separate respiration measurements of various autotrophic organs. Leaves, stems, roots, and reproductive parts of the same plant respire at different rates. These rates change not only with changes in environmental conditions, but also with progressive changes in the plant growth and development. Even changes in a plant's respiration rates with time of day become important when measuring total ecosystem R_A. Edwards & McLaughlin (1978) found night-time respiration rates of white oak and yellow poplar tree boles to be higher than day-time rates, in spite of higher day-time temperatures. Edwards & Sollins (1973) had found similar diel patterns of forest floor carbon dioxide efflux in yellow poplar dominated forest, probably a reflection of diel patterns of tree root respiration. Thus, R_A measurements must be taken over 24-hr periods to be accurate. As a general rule younger, faster growing organs respire faster per unit weight than older parts. Also the respiration rates vary with species. Thus, an ideal quantification of R_A for a complex forest ecosystem with measurements of all species and their separate organs is almost impossible and certainly impractical. The practical approach is to measure respiration rates of a few representative dominant species. Woodwell & Botkin (1970) measured total ecosystem respiration by measuring the rate of carbon dioxide accumulation during temperature inversions and compared results with that obtained by summing the data from three types of small chambers. From their small chamber measurements they were able to calculate R_A for the oak–pine forest. Their use of small chambers to measure ecosystem respiration is representative of the state-of-the-art and is discussed in some detail below.

523

One type of chamber used at Brookhaven in the oak–pine forest was a leaf chamber constructed of polyvinyl chloride and aluminum framing. One chamber contained a twig and its leaves. Temperature was kept near outside ambient air temperatures in the chambers by passing the air through at a very fast rate (10 l/min). Stem chambers were operated on a similar principle but were painted with an opaque reflective paint to control temperature. A third type of chamber covered 0.25 m^2 of ground surface and stood about 1 m high, which was sufficiently large to enclose both the groundcover and low shrubs of the forest. These ground chambers were temperature-controlled with a thermostat and a cooling compressor. By measuring differences in carbon dioxide concentrations of the air entering the chambers and that exhausting from the chambers the net gas exchange within the chambers was determined. The annual course of respiration was calculated from periodic 24-hr measurements over four years. Using dimension analysis of different species in earlier studies (Whittaker & Woodwell, 1967, 9) total bark surface for the oak–pine forest was calculated and used in conjunction with their carbon dioxide data to arrive at total respiration of stems, branches, groundcover and soil. Their estimate of R_E by small chambers is biased toward a lower value because respiration of leaves during photosynthesis could not be measured. Leaf respiration during the day is often assumed to equal the night-time rates (Reichle, O'Neill & Harris, 1975; Kinerson, 1975). However, Richardson *et al.* (1972) found that in yellow poplar seedlings, light respiration may account for four times as much carbon loss as dark respiration.

The ground chamber technique used in the Brookhaven forest did not permit the measurement of below-ground R_A directly due to the confounding of root respiration and soil organic matter respiration. However, using data from a previous study (Woodwell & Marples, 1968), which suggested a decay rate of 621 g m^{-2} yr^{-1} in the Brookhaven forest, they were able to calculate R_A. Soil and vegetation respiration within the 1-m high chambers totaled 1045 gdw m^{-2} yr^{-1}. Thus, autotrophic respiration within the chambers could be calculated ($1045 - 621 = 424$ g m^{-2} yr^{-1}). The mean estimate of bark and branch respiration (1096) plus 424 gives an estimate of 1520 gdw m^{-2} yr^{-1} for total ecosystem R_A. This converts to 690 g carbon m^{-2} yr^{-1}, compared to 1436 for the yellow poplar forest (Harris *et al.*, 1975) and 2068 for the pine forest in North Carolina (Kinerson *et al.*, 1977).

Techniques in determining R_A in the pine forest and the yellow poplar forest were similar to those used in the oak–pine forest at Brookhaven, except for the below-ground component. Root respiration was measured on various size classes of excised roots *in vivo* at both of these sites. At the oak–pine forest roots were placed in glass jars and air was passed through the jars to an infrared gas analyzer for measurement of carbon dioxide efflux (Kinerson, 1975). At the yellow poplar forest roots were placed in flasks and oxygen uptake was measured with a differential respirometer (Edwards & Harris, 1977). Thus in these two forests below-ground R_A was measured directly on excised root tissue as respiration rate per unit weight of root tissue. Root biomass was determined for the yellow poplar forest by periodic core sampling (Edwards & Harris, 1977) and by excavation of several entire root systems in the pine plantation (Kinerson *et al.*, 1977).

Heterotrophic respiration

The simplified techniques discussed above provide only estimates of R_A in complex systems such as forests. The heterotrophic community of a forest is even more complex with a myriad of organisms, most of them microscopic, feeding on the products of the autotrophs. Most of the heterotrophic activity within the forest floor by monitoring carbon dioxide efflux from its Just the physical separation of various functional groups is virtually impossible. Since most of the heterotrophic activity is in the forest floor, a number of studies have been designed to measure the total metabolic activity within the forest floor by monitoring carbon dioxide efflux from its surface. This measure of forest floor respiration includes root respiration, which must be measured (as discussed above) and subtracted from the total to give an estimate of forest floor R_H.

The contribution of dead root decay to total R_H can be calculated from root turnover estimates from periodic root excavations (Edwards & Harris, 1977; Kinerson et al., 1977). Their root turnover estimates were based on differences between seasonal maximum and minimum values for fine lateral root biomass. The contribution of the litter horizon to total R_H can be calculated from *in vivo* measurements of litter respiration (Edwards & Harris, 1977) or from weight loss measurements of litter confined in nylon or fiberglass mesh bags and placed back onto the mineral soil surface in the field (Fogel & Cromack, 1977; Kinerson *et al.*, 1977). Estimates of carbon losses by root decay and surface litter decay can then be subtracted from *in vitro* measurements of total forest floor R_H for an estimate of carbon losses from the slowly decomposing organic matter within the mineral soil profiles. In the yellow poplar forest at Oak Ridge a total of 1065 g C m^{-2} yr^{-1} was lost as carbon dioxide from the forest floor (Edwards & Harris, 1977). 35% of this total was accounted for by respiration of live roots, 42.2% from dead roots, 20.6% from litter and 2.2% from soil organic matter. Thus total forest floor R_H for the yellow poplar forest amounted to 692 g C m^{-2} yr^{-1}. Respiration rates of canopy feeding insects in the yellow poplar forest was calculated from body size metabolism regressions (Reichle, 1971) and mean body size for each age-class for the various insect species. The contribution to ecosystem R_H by the canopy insects was less than 0.1 g C m^{-2} yr^{-1}. A further delineation of the organisms contributing to R_H is discussed below for the yellow poplar forest. This approach may be considered academic because of the difficulty in separating, measuring, and reconstructing the metabolic pathways of such a complex and dynamic system. However, by doing such exercises we do gain some insight into the functioning of the total ecosystem.

Microflora respiration. Total ecosystem microfloral respiration is difficult to quantify because it is virtually impossible to separate fungi and bacteria from their substrates and from soil invertebrates without excessive substrate disturbance which affects respiration rates. We do know that the standing pool of carbon in microflora is small relative to most other ecosystem components, but large relative to invertebrate carbon pools. Harris *et al.* (1975) reported an annual mean standing carbon pool of 58 g m^{-2} in the Oak Ridge *Liriodendron* forest floor microflora as compared to 7 g m^{-2} in invertebrates. The total carbon in microflora was only 0.7% of the total ecosystem carbon pool (excluding soil organic detritus) in that forest and only 2.7% of gross primary production. These estimates were based on ATP analysis of litter and soil, and assumed a constant ratio of ATP to carbon (McElroy, 1947). Using this technique, Ausmus, Edwards & Witkamp (1976) claimed that at certain times of the year as much as 37% of the 01 litter carbon, 25% of the 02 litter carbon, and 73.8 g C m^{-2} in mineral soil horizons was immobilized by bacteria and fungi. These estimates should be considered only preliminary at best because ATP would be found only in living cells, and carbon found in dead fungi and bacteria was not considered. Also, the problem of heterogeneity of such small organisms in vast ecosystems is compounded by uncertainty of techniques for determining microflora biomass. Parkinson (1970) has reviewed the problems associated with various techniques for studying microfloral populations.

The problems with techniques for determining respiration rates of microflora are therefore primarily problems of separation of this component from other respiring components of the ecosystem. In fact, since disturbances by separation do affect respiration, measurement (without disturbance) of total carbon dioxide efflux from litter and soils as an estimate of an upper limit of total forest floor respiration followed by respiration measurements of separated substrates is perhaps the best approach. Using this technique, Edwards & Harris (1977) attributed 65% of total forest floor respiration to decomposers in a *Liriodendron* forest. However, no attempt was made to separate invertebrate decomposers from microfloral decomposers. Based on biomass estimates of these two groups, we can assume that most of the decomposer carbon turnover in the forest floor is by the microflora. If we divide the carbon dioxide efflux between the two groups according to the relative biomass estimates of the two groups, we calculate that over 600 g C m^{-2} cycled through the microflora annually and was lost as carbon dioxide to the atmosphere. Such a large turnover of carbon by such a small component of the ecosystem implies an extremely metabolically active group of organisms. This activity is influenced by many biotic and abiotic factors.

Factors which have been shown to affect microfloral activity include temperature and moisture (Witkamp, 1966; Reiners, 1968; Edwards, 1975), pH (Katznelson & Stephenson, 1956), substrate quality (Witkamp, 1966; Fogel & Cromack, 1977) and inhibitors (MacFadyen, 1973). Edwards (1975) found that 94% of the variability observed in *Liriodendron* forest floor carbon dioxide evolution rates was accounted for by the square of litter temperature. He observed moisture effects (Table 9.5) but no statistical correlation was found between forest floor respiration and moisture in this forest where moisture was seldom limiting. In other drier ecosystems, moisture has a more pronounced effect. Reiners (1968) and Froment (1972) reported direct correlations between soil moisture and soil respiration in oak forests while DeSanto, Alfani & Sapio (1976) found an inverse correlation between carbon dioxide evolution rates and soil moisture in a beech forest in Italy. Soil moisture in the beech forest was seldom lower than 70% of the soil dry weight.

Substrate quality effects on microfloral respiration in ecosystems has not been as extensively studied as abiotic variables. However, the C:N ratio of a substrate has long been known to affect decomposition rates of that substrate, i.e. the lower the C:N ratio the faster the decomposition rate. Fogel & Cromack (1977) found decomposition rates of leaf litter to be even more closely correlated with lignin content of the litter than with C:N ratio. Lignin decomposes very slowly and products of its decomposition are even more resistant to decay than lignin. Because of differences in chemical structure of chemical compounds found in forest litter and the ability of different micro-organisms to produce enzymes which will digest these compounds, some components of litter decompose very rapidly (sugar, starches, proteins, etc.) while others decompose very slowly (lignin, cellulose, etc.). Thus the slower decomposing materials tend to accumulate. The SCEP Report (1970) gives a mean carbon residence time of 3.1 yr for the 'fast' component in forests and a mean residence time of the 'slow' component of 59 yr.

Thus, the study of decomposition in forest soils can be very complex, especially since a relatively small component (the microflora) of the ecosystem is primarily responsible for this decomposition. Even though not much carbon is bound in microflora at one point in time, their high metabolic activity cycles large amounts of carbon rapidly back to the atmosphere. This 'loss' of carbon back to the atmosphere is the price the autotrophs pay for remineralization of nutrient elements for root uptake and perhaps more importantly the temporary immobilization of nutrients by the micro-organisms, thus slowing nutrient losses from the ecosystem.

Table 9.5. *Daily means of carbon dioxide evolution, temperature, and moisture are shown for the floor of a Liriodendron-dominated mixed deciduous forest in eastern Tennessee. Predicted means ($\pm 95\%$ CL) of carbon dioxide evolution are shown (from Edwards, 1975)*

| Date | Carbon dioxide evolution (g CO_2 m^{-2} day^{-1}) | | | | Temperature (°C) | Moisture (% of dry wt) | |
| | Total (litter + soil) | | Soil | | Litter | Litter | Soil |
	Measured[a]	Predicted	Measured[a]	Predicted	Litter	01	02
10 Jan 1972	2.68	5.37±0.66	0.58	4.39±0.84	11.0	219	229
25 Jan 1972	2.25	1.70±0.21	0.68	1.39±0.27	6.2	100	141
4 Feb 1972	0.91	0.01±0.001	0.23	0.01±0.003	-0.5	140	185
22 Feb 1972	1.37	0.54±0.07			3.5	146	176
8 March 1972	0.94	0.94±0.12	0.0	0.77±0.15	4.6	132	149
23 March 1971	1.20	2.43±0.30			7.4	26	125
12 April 1971	5.31	5.57±0.69	3.91	4.55±0.87	11.2	19	93
3 May 1971	4.66	5.57±0.69	4.31	4.55±0.87	11.2	26	52
17 May 1971	16.95	12.09±1.49	16.39	9.88±1.89	16.5	83	207
8 June 1971	18.72	15.20±1.88			18.5	199	107
21 June 1971	17.61	18.49±2.29	14.33	15.11±2.90	20.4	145	114
7 July 1971	13.87	20.34±2.51	11.48	16.62±3.19	21.4	52	78
26 July 1971	20.86	20.73±2.57	17.32	16.94±3.25	21.6	276	204
18 Aug 1971	23.16	28.22±3.40	20.82	23.06±4.43	25.2	60	53
7 Sept 1971	25.54	17.95±2.23	24.51	14.67±2.82	20.1	56	51
11 Oct 1971	14.90	11.37±1.41	6.13	9.29±1.78	16.0	95	183
25 Oct 1971	16.06	12.24±1.52			16.6	211	136
15 Nov 1971	6.19	7.05±0.87	3.77	5.76±1.11	12.6	48	85
6 Dec 1971	3.24	6.08±0.75			11.7	335	269
21 Dec 1971	1.53	2.05±0.25	0.09	1.67±0.32	6.8	91	169
Annual totals[b] (g^{-2})	3924	3840	3086	3095			

[a] $N = 72$/day (3 replicates measured hourly through one day).
[b] Annual total carbon dioxide evolution rates were calculated from monthly mean values which were based on daily total carbon dioxide (evolution rates

Table 9.6. *Ecological efficiency quotients for invertebrate trophic levels (after Reichle, 1977)*

	Assimilation efficiency $(A_n/I_n)^a$	Ecological growth efficiency $(NP_n/I_n)^a$	Tissue growth efficiency $(NP_n/A_n)^a$	Respiratory coefficient $(R_n/NP_n)^a$	Ecological efficiency $(I_n/I_{n-1})^a$	Trophic level production efficiency $(A_n/NP_{n-1})^a$
Saprovore	0.10–0.40	0.05–0.08	0.17–0.40	3.7–4.6	0.09–0.18	0.11–0.17
Herbivore	0.36–0.78	0.08–0.27	0.20–0.40	2.16–3.06	~ 0.10	0.02–0.07
Carnivore	0.47–0.92	~ 0.34	0.10–0.37	1.70–4.18	~ 0.07	~ 0.02

[a] Subscripts denote trophic levels. A, assimilation; I, ingestion; NP, net production; R, respiration.

Invertebrate respiration. The importance of invertebrates in the decomposition of detritus and subsequent mineralization of nutrients has long been recognized in a diversity of ecosystems. Odum (1971) summarizes studies which show that up to 90% of the primary productivity of ecosystems may be processed by detritus based food chains. Decomposers dominate the heterotrophic metabolism of forest ecosystems. Under normal conditions, herbivorous invertebrates may be responsible for only 3–7% (Bray, 1961; Reichle *et al.*, 1973*a*) of carbon flow by heterotrophs in forest ecosystems. Reichle *et al.* (1973*a*) have demonstrated that over 95% of heterotrophic metabolism in forests is due to the metabolic activities of decomposers. Microflora dominate the metabolism ($\sim 90\%$) of forest decomposers, but much of the substrate metabolized by microbes is eventually fragmented or processed digestively by the metabolism ($\sim 10\%$) of invertebrate decomposers.

Approximately 60% of the total soil invertebrate biomass is composed of fungivores that feed directly upon fungal mycelia and fruiting bodies (McBrayer & Reichle, 1971). Nearly 25% of the biomass was saprophagous, feeding directly upon organic detritus. The remaining biomass ($\sim 15\%$) is predaceous. Sufficient information on decomposer energetics now exists (Reichle, 1977) so that the efficiencies of energy utilization and respiration by decomposer invertebrates can be estimated (Table 9.6). Fungivores have energetic efficiencies comparable to the general herbivore classification used in the table. Saprovores utilize an energy-poor substrate (low assimilation efficiency) and, consequently, have a low ecological growth efficiency. Saprovores tend to be high metabolic to low productivity groups (high respiratory coefficient), but at the same time their trophic level production efficiency shows them to be among the most efficient trophic level convertors of assimilated energy. This relatively small biomass of decomposer invertebrates with high metabolic rates and efficiencies of energy utilization plays an important role in the overall metabolism of the forest ecosystem.

More than 50% of the carbon in a forest ecosystem occurs as dead organic matter (detritus) on the ground surface or incorported into the soil. Reichle *et al.* (1973*b*) estimated that 50% (12 300 g C m^{-2}) of the carbon biomass in the mesic yellow poplar forest ecosystem occurred as soil detritus. The corresponding biomass of invertebrate decomposers for this same forest does not exceed 21 g m^{-2} (Reichle, 1977). Earthworms, primarily, and pulmonates and nematodes accounted for approximately 50% of invertebrate decomposer biomass. 42% ($\sim$8.7 g m^{-2}) of heterotrophic invertebrate biomass was arthropods. These data seem fairly typical for forest ecosystems. McBrayer *et al.* (1977), in a comparison of soil and litter arthropod population densities across six forest types examined in the US–IBP, found densities to range between 4.3×10^4 m^{-2} and 13.7×10^4 m^{-2} (the average was 9.1×10^4 m^{-2}; *Liriodendron* forest $= 9.7 \times 10^4$ m^{-2}). These populations are segregated into a diverse number of trophic guilds – each specialized to exploit the diverse nature of the detritus substrate.

Heterotrophic respiration typically amounts to about 10% annually of the standing crop of forests. This value is higher in grasslands and may reach 100% of biomass which is decomposed annually (Reichle, 1975). Heterotrophic respiration has been shown to range between 280 and 670 g m^{-2} yr^{-1}, respectively, for mesic and xeric forest ecosystems. Heterotrophic respiration (R_H) represents between 0.41 and 0.47 (R_H/R_A) of autotrophic respiration (R_A) for the xeric and mesic forests, respectively (Reichle, 1977). Forests have higher absolute rates of heterotrophic respiration than do prairie and tundra ecosystems (Reichle, 1975). Although data on heterotrophic metabolism of different ecosystems are sparse, it appears that the high values of decomposer metabolism in forests are directly related to the high net primary production of these ecosystems.

Summary

The study of ecosystem carbon metabolism is especially difficult in forests because of measurement problems associated with vegetation size, heterogeneity of the biota, and variability of time constants associated with fluxes among components. However, because of the large land masses occupied by forests, they dominate all other terrestrial ecosystems in terms of carbon storage and turnover. Therefore, quantification of forest pools and fluxes is essential not only for understanding forest metabolism *per se*, but also for assessing the roles of the terrestrial vegetation in regulating global carbon cycles. In this chapter we have discussed ways that some of the measurement difficulties have been approached, while emphasizing similarities and differences between carbon metabolism of three forest types in the US and, in a more general way, other forests and non-forested ecosystems of the world.

Forest metabolism, in general, differs considerably from that of other terrestrial ecosystems. Forests, for example, have a much higher gross primary production than other ecosystems. Yet because of very large respiratory costs in maintaining their large standing crop, temperate forests have a net production only slightly higher than that of agricultural systems. Tundra and prairie ecosystems have considerably lower net production than forests or agricultural systems. These comparisons are quite general due to the high degree of variability even with a single ecosystem type, especially in forest ecosystems.

Biotic measurements necessary to arrive at carbon budgets and cycles of total ecosystems can be categorized into gravimetric analysis, chemical analysis of carbon concentrations in tissues, photosynthetic rates, carbon allocation within the plants, and carbon losses through heterotrophic and autotrophic activity. Techniques within each category were discussed and used in arriving at the ecosystem parameters compared in this chapter.

A comparison of several parameters of eight different forests in various locations of the world leads to a number of general observations. For example, production efficiency (R_A/GPP) is generally higher in forests with higher total standing carbon (TSC). This is expected since as forests grow the ratio of photosynthetic surface to non-photosynthetic surface decreases. The respiration allocation (R_H/R_A) ratio decreases generally with increased TSC. Heterotrophic respiration (R_H), on the other hand, does not appear to be related to TSC but is affected more by regional climate differences, and may have a greater influence on differences in net ecosystem production (NEP) than any other single biotic variable.

The importance of heterotrophic respiration as a regulator of NEP is exemplified by a comparison of three US forests of different types, ages, and geographic locations. A 50-yr-old yellow poplar forest in Tennessee, an oak–pine forest in New York (older trees $\simeq 45$ yr old) and a 20-yr-old pine plantation in North Carolina contain 8760, 7062, and 5096 g C m^{-2}, respectively. Because of age differences, NPP was much higher in the younger pine forest than in either of the other two forests. Heterotrophic respiration was about the same for the Tennessee forest and the North Carolina forest because of similar climates, and despite differences in soil and litter substrates. However, the cooler and drier forest floor of the New York forest lost less than half as much carbon by heterotrophic respiration as did the other two forests. Thus, despite having a lower NPP, the oak–pine forest in New York had nearly five times higher NEP than did the yellow poplar forest in Tennessee. The younger pine forest of North Carolina fixed carbon at such high rates that respiratory losses were offset, resulting in a NEP five times higher than the oak–pine forest and 25 times higher than the yellow poplar forest.

We emphasize that these comparisons are only as accurate as the measurement and extrapolation techniques necessary to arrive at parameters for total ecosystems. The parameters that are compared were determined from only periodic measurements of very small parts of highly variable ecosystems.

References

Aleksew, V. A. (1963). Light measurements under a forest canopy. *Soviet Plant Phys.* **10**, 109–201.

Andrews, R., Coleman, D. C., Ellis, J. E. & Singh, J. S. (1974). Energy flow relationships on a shortgrass prairie ecosystem, pp. 22–28. In *Proc. 1st Internat, Cong. Ecol.*, The Hague.

Ausmus, B. S. (1973). The use of the ATP assay in terrestrial decomposition studies. *Bull. Ecol. Res. Comm.* (Stockholm) **17**, 223–34.

Ausmus, B. S., Edwards, N. T. & Witkamp, M. (1976). Microbial immobilization of carbon, nitrogen, phosphorus, and potassium: implications for forest ecosystem processes. In *The role of terrestrial and aquatic organisms in decomposition processes. 17th symposium of the British ecological society 15–18 April 1975*, ed. J. M. Anderson & A. MacFadyen. Blackwell Scientific Publications, Oxford, London, Edinburgh & Melbourne.

Baes, C. F., Jr., Goeller, H. E., Olson, J. S. & Rotty, R. M. (1977). Carbon Dioxide and climate: The uncontrolled experiment. Amer. *Sci.* **65**, 310–20.

Boardman, N. K. (1977). Comparative photosynthesis of sun and shade plants. *Annu. Rev. Plant. Physiol.* **28**, 355–77.

Bray, J. R. (1961). Measurements of leaf utilization as an index of minimum level of primary consumption. *Oikos*, **12**, 70–4.

Burgess, R. L. & O'Neill, R. V. (1975). *Eastern deciduous forest biome report, September 1973– August 1974. EDFB/IBP–75.* Oak Ridge National Laboratory, Oak Ridge, Tennessee.

DeSanto, A. V., Alfani, A. & Sapio, S. (1976). Soil metabolism in beech forests of Monte Taburno (Campania Apennines). *Oikos*, **27**, 144–52.

Dinger, B. E. (1971). *Net photosynthesis and production of* Liriodendron tulipifera *as estimated from carbon dioxide exchange.* EDFB Report 71–75. Oak Ridge National Laboratory.

Dinger, B. E. (1972a). Gaseous exchange, forest canopy and meteorology, pp. 45–6. In *Ecological Sciences Division Annual Progress Report for period ending 30 Sept. 1971.* ORNL–4757. Oak Ridge National Laboratory, Oak Ridge, Tenn.

Dinger, B. E. (1972b). *Comparative photosynthetic efficiency of four deciduous forest species in relation to canopy environment.* EDFB Report 72–151. Oak Ridge National Laboratory.

Dinger, B. E. (1973). Equipment design and analysis of carbon dioxide exchange in yellow poplar (*Liriodendron tulipifera* L.). In *Terrestrial primary production*, ed. B. E. Dinger & W. F. Harris, pp. 10–24. EDFB–IBP Report 73–6. Oak Ridge National Laboratory, Oak Ridge, Tennessee.

Edwards, N. T. (1975). Effects of temperature and moisture on carbon dioxide evolution in a mixed deciduous forest floor. *Proc. Soil Sci. Soc. America*, **39**, 361–5.

Edwards, N. T., Harris, W. F. & Shugart, H. H. (1973). *Carbon Cycling in deciduous forest.* EDFB–IBP Report 73–80. Oak Ridge National Laboratory.

Edwards, N. T. & Harris, W. F. (1977). Carbon cycling in a mixed deciduous forest floor. *Ecology*, **58**, 431–7.

Edwards, N. T., Harris, W. F. & Shugart, H. H. (1977). Carbon cycling in deciduous forest. In: *The belowground ecosystem: a synthesis of plant-associated processes*, ed. J. K. Marshall, pp. 153–7. Range Science Dept. Science Series No. 26. Colorado State Univ., Fort Collins, Colo.

Edwards, N. T. & McLaughlin, S. B. (1978). Temperature-independent diurnal variations of respiration rates in *Quercus alba* L. and *Liriodendron tulipifera* L. *Oikos*, **31**, 200–6.

Edwards, N. T. & Sollins, P. (1973). Continuous measurement of carbon dioxide evolution from partitioned forest floor components. *Ecology*, **54**, 407–12.

Evans, L. T. (1975). Beyond photosynthesis – the role of respiration, translocation and growth potential in determining productivity, pp. 501–7. In *Photosynthesis and productivity in different environments*. Cambridge University Press, New York.

Fogel, R. & Cromack, K. Jr. (1977). Effect of habitat and substrate quality on Douglas fir litter decomposition in Western Oregon. *Can. J. Bot.* **55**, 1632–40.

Froment, A. (1972). Soil respiration in a mixed oak forest. *Oikos*, **23**, 273–7.

Gay, L. W., Knoerr, K. R. & Braaten, M. O. (1971). Solar radiation variability on the floor of a pine plantation. *Agric. Metcar.* **3**, 39–50.

Harris, W. F., Kinerson, R. S. Jr. & Edwards, N. T. (1977). Comparison of belowground biomass of natural deciduous forest and loblolly pine plantations. *Pedobiologia*, **17**, 369–81.

Harris, W. F., Sollins, P., Edwards, N. T., Dinger, B. E. & Shugart, H. H. Jr. (1975). Analysis of carbon flow and productivity in a temperate deciduous forest ecosystem. In *Productivity of world ecosystems*, ed. D. E. Reichle, J. F. Franklin & D. W. Goodall. Nat. Acad. of Sci. Wash., DC.

Harris, W. F. & Todd, D. E. (1972). *Forest root biomass production and turnover*. EDFB–Memo Report 72–156., Oak Ridge National Laboratory. Oak Ridge, Tennessee.

Harris, W. F. & Witherspoon, J. P. (1973). Effects of ionizing radiation in processes influencing tolerance of tree seedlings. In *Proceedings of Third National Symposium on Radioecology*, ed. D. J. Nelson, pp. 961–71. AEC–CONF–710501–P2. National Technical Information Service, Springfield, Virginia.

Higgenbotham, K. O. (1974). The influence of canopy position and the age of leaf tissue on growth and photosynthesis in loblolly pine. Ph.D. Dissertation, Duke University Durham, NC.

Hinckley, T. M., Lassoie, J. P. & Running, S. W. (1978). *Temporal and spatial variations in selected biological parameters indicative of water status in forest trees. For. Sci. Monograph.*

Horton, A. D., Shultz, W. D. & Meyer, A. S. (1971). Determination of nitrogen, sulfur, phosphorus, and carbon in solid ecological materials via hydrogenation and element-selective detection. *Analytical Letters*, **4**, 613–21.

Hozumi, K. & Kirita, H. (1970). Estimation of total photosynthesis in forest canopies. *Bot. Mag. Tokyo*, **17**, 238–305.

Hsiao, T. C. (1973). Plant response to water stress. *Annu. Rev. Plant Physiol.* **24**, 519–70.

Hutchison, B. A. Matt, D. R. (1976). Beam enrichment of diffuse radiation in a deciduous forest. *Agric. Meteor.* **17**, 93–110.

Jackson, M. L. (1958). *Soil chemical analysis*. Prentice-Hall, Inc. Englewood Cliffs, N J.

Katznelson, H. & Stevenson, I. L. (1956). Observations on the metabolic activity of the soil microflora. *Can. J. Microbiol.* **2**, 611–22.

Kimura, M. (1969). Ecological and physiological studies on the vegetation of Mt

Shimagare. VII. Analysis of production processes of young *Abies* stand based on carbohydrate economy. *Bot. Mag. Tokyo*, **82**, 6–18.

Kinerson, R. S. (1975). Relationships between plant surface area and respiration in loblolly pine. *J. appl. Ecol.* **12**, 965–71.

Kinerson, R. S., Ralston, C. H. & Wells, C. G. (1977). Carbon cycling in a loblolly pine plantation. *Oecologia*, **29**, 1–10.

Kira, T. (1975). Primary production of forests. In *Photosynthesis and productivity in different environments.* ed. J. P. Cooper, pp. 5–41. Cambridge University Press, Cambridge, London, New York.

Kitazawa, Y. (1977). Ecosystem metabolism of the Subalpine Coniferous Forest of the Shigayama IBP area. In *Ecosystem analysis of the subalpine coniferous forest of the shigayama IBP area*, Central Japan, ed. Y. Kitazawa, JIBP Synthesis **15**, pp. 181–99. University of Tokyo Press.

Kozlowski, T. T. & Keller, T. (1966). Food relations of woody plants. *Bot. Rev.* **32**, 293–382.

Kramer, P. J. & Kozlowski, T. T. (1960). *The physiology of trees.* McGraw Hill, New York.

Lindeman, R. L. (1942). The trophic–dynamic aspect of ecology. *Ecology*, **23**, 399–418.

McBrayer, J. F., Ferris, J. M., Metz, L. J., Gist, C. S., Cornaby, B. W., Kitazawa, Y., Kitazawa, T., Wernz, J. G., Krantz, G. W. & Jensen, H. (1977). Decomposer Invertebrate Populations in US Forest Biomass. *Pedobiologia*, **17**, 89–96.

McBrayer, J. F. & Reichle, D. E. (1971). Trophic structure and feeding rates of forest soil invertebrate populations. *Oikos*, **22**, 381–8.

McConathy, R. K., McLaughlin, S. B., Reichle, D. E. & Dinger, B. E. (1976). Leaf energy balance and transpirational relationships of tulip poplar (*Liriodendron tulipifera*). EDFB/IBP–76/6. Oak Ridge National Laboratory, Oak Ridge, Tennessee.

McElroy, W. D. (1947). The energy source for bioluminescence in an isolated system. *Proc. natl. Acad. Sci.* **33**, 342–5.

MacFadyen, A. (1970). Soil metabolism in relation to ecosystem energy flow and primary and secondary production. In *Methods of study in soil ecology*, pp. 167–72. UNESCO, Paris.

MacFadyen, A. (1973). Inhibitory effects of carbon dioxide on microbial activity in soil. *Pedobiologia*, **13**, 140–9.

McLaughlin, S. B., Edwards, N. T. & Beauchamp, J. J. (1977). Temporal and spatial variations in transport and respiratory allocation of ^{14}C-sucrose by White Oak (*Quercus alba l*) Roots. *Can. J. Bot.* **55**, 2971–80.

Medwecka-Kornas, A., Lomnicki, A. & Bandola-Ciolczyk, E. (1974). Energy flow in the Oak–Hornbeam Forest (IBP Project Ispina). *Bulletin de L'academie Polonaise Des Sciences. Serie des sciences biologiques cl. II.* **22**, 563–7.

Miller, P. C. (1972). Leaf temperature and primary production in red mangrove canopies of south Florida. *Ecology*, **53**, 22–45.

Monsi, M. & Saeki, T. (1953). uber den Lichtfaktor in den Pflanzengesellschaften und seine Bedeutung fur die Stoffproktion. *Jap. J. Bot.* **14**, 22–52.

Moulder, B. C. & Reichle, D. E. (1972). Significance of spider predation in the energy dynamics of forest floor arthropod communities. *Ecol. Monogr.* **42**, 473–98.

Murphy, C. E. Jr. & Knoerr, K. (1972). Modeling the energy balance processes of natural ecosystems. EDFB–IBP 72–10. Oak Ridge National Laboratory, Oak Ridge, Tennessee.

Olson, E. P. (1971). *Fundamentals of ecology.* W. B. Saunders, Philadelphia.

Olson, J. S. (1975). Productivity of world ecosystems, pp. 33–43. In *Productivity of world ecosystems*, ed. D. E. Reichle, J. F. Franklin & D. W. Goodall, pp. 33–43. National Academy of Sciences, Washington, DC.

Parkinson, D. (1970). Methods for the quantitative study of heterotrophic soil micro-organisms. In *Methods of study in soil ecology*, ed. J. Phillipson. UNESCO, Paris.

Ramann, E. (1911). Lichtmessungen in Fichtenbestanden. *Allg. Forst-u Jazdztz* **87**, 401–6.

Reed, K. L., Hammerly, E. R. & Dinger, B. E. (1976). An analytical model for gas exchange studies on photosynthesis. *J. appl. Ecol.* **13**, 915–24.

Reichle, D. E. (1971). Energy and nutrient metabolism of soil and litter invertebrates, pp. 465–477. In *Productivity of forest ecosystems*, ed. P. Duvigneaud, pp. 465–77. UNESCO, Paris.

Reichle, D. E. (1975). Advances in ecosystem analysis. *Bioscience*, **25**, 257–64.

Reichle, D. E. (1977). The Role of Soil Invertebrates in Nutrient Cycling. In *Soil organisms as components of ecosystems*, pp. 145–56. *Ecol. Bull.* (Stockholm) No. 25.

Reichle, D. E. & Crossley, D. A. Jr. (1967). Investigation of heterotrophic productivity in forest insect communities. In *Secondary productivity of terrestrial ecosystems*, ed. K. Petrusewicz, pp. 563–87. Polish Academy of Sciences, Warsaw.

Reichle, D. E., Dinger, B. E., Edwards, N. T., Harris, W. F. & Sollins, P. (1973*b*). Carbon Flow and Storage in a Forest Ecosystem. In *Carbon and the Biosphere*, ed. G. M. Woodwell & E. V. Pecan, pp. 345–65. CONF–720510. US–DOE Technical Information Center. Springfield, Virginia.

Reichle, D. E., Goldstein, R. A., Van Hook, R. I. & Dodson, G. J. (1973*a*). Analysis of Insect Consumption in a Forest Canopy. *Ecology*, **54**, 1076–84.

Reichle, D. E., McBrayer, J. F. & Ausmus, B. S. (1975). Ecological Energetics of Decomposer Invertebrates in a Deciduous Forest and Total Respiration Budget. In *Progress in Soil Zoology*, pp. 283–92. Academia, Prague.

Reichle, D. E., O'Neill, R. V. & Harris, W. F. (1975). Principles of energy and material exchange in ecosystems. In *Proc. First international congress of ecology*, pp. 27–43. Dr W. Junk, b.v., Publishers, The Hague.

Reiners, W. A. (1968). Carbon dioxide evolution from the floor of three Minnesota forests. *Ecology*, **49**, 471–83.

Richardson, C. J., Dinger, B. E. & Harris, W. F. (1972). The use of stomatal resistance, photopigments nitrogen, water potential, and radiation to estimate net photosynthesis in *Liriodendron tulipifera* L–a physiological index. EDFB–IBP–72–13. Oak Ridge National Laboratory, Oak Ridge, Tennessee.

Saeki, T. (1975). Distribution of radiant energy and CO_2 in terrestrial communities pp. 297–322. In *Photosynthesis and productivity in different environments*, ed. J. P. Cooper, pp. 297–322. Cambridge University Press, Cambridge, London, New York.

Satchall, J. (1973). Biomass model of a mixed oak forest, United Kingdom. In *Modeling forest ecosystems*, ed. D. E. Reichle, R. V. O'Neill & J. S. Olson. Oak Ridge National Laboratory. EDFB–IBP–72–7.

SCEP (1970). *Man's impact on the global environment: assessment and recommendations for action*. MIT Press, Cambridge & London.

Sestak, Z. & Catsky, J. (1976). Bibliography of reviews and methods of photosynthesis–35. *Photosynthetica*, **10**, 93–105.

Shaedle, M. (1975). Tree photosynthesis. *Annu. Rev. Plant Physiol.* **26**, 101–15.

Shugart, H. H. Harris, W. F., Edwards, N. T. & Ausmus, B. S. (1977). Modeling of

535

belowground processes associated with roots in deciduous forest ecosystems. *Pedobiologia*, **17**, 382–8.

Sinclair, T. R., Murphy, C. E. & Knoerr, K. R. (1976). Development and evaluation of simplified models for simulating canopy photosynthesis and transpiration. *J. appl. Ecol.* **13**, 813–30.

Slatyer, R. O. (1973). *The effect of internal water status on plant growth, development, and yield*, pp. 177–91. Plant response to climatic factors. Proc. Uppsala Symp. 1970 (Ecology and Conservation, 5). UNESCO, Paris.

Slavik, B. (1975). Water stress, photosynthesis and the use of photosynthates. In *Photosynthesis and productivity in different environments*, ed. J. P. Cooper, pp. 511–36. Cambridge University Press, Cambridge, London, New York.

Sollins, P. & Anderson, R. M. (1971). Dry weight and other data for trees and woody shrubs of the southeastern United States. ORNL–IBP–71–6. Oak Ridge National Laboratory, Oak Ridge, Tennessee.

Sollins, P., Reichle, D. E. & Olson, J. S. (1973). Organic matter budget and model for a southern Appalachian *Liriodendron* forest. EDFB–IBP–73–2. Oak Ridge National Laboratory, Oak Ridge, Tennessee.

Strain, B. R. & Higgenbotham, K. O. (1976). *A summary of gas exchange studies on forest trees*, pp. 99–102. Proceedings XVI IVFRO World Congress, Norway.

Tansley, A. G. (1935). The use and abuse of vegetational concepts and terms. *Ecology*, **16**, 2284–307.

Wardlaw, I. F. (1968a). The control and pattern of carbohydrate movement in plants. *Bot. Rev.* **34**, 79–105.

Wardlaw, I. F. (1968b). The effect of water stress on translocation in relation to photosynthesis and growth II. Effect driving leaf development in *Lolium tremulotum* L. *Aust. J. Biol. Sci.* **22**, 1–16.

Waring, T. F. & Patrick, J. (1975). Sowcca-Sink relations and the partition of assimilates in the plant. In *Photosynthesis and productivity in different environments*, ed. J. P. Cooper. Cambridge University Press, Cambridge, London, New York.

Whittaker, R. H. & Woodwell, G. M. (1967). Surface area relations of woody plants and forest communities. *Amer. J. Bot.* **54**, 931–9.

Whittaker, R. H. & Woodwell, G. M. (1968). Dimension and production relations of trees and shrubs in the Brookhaven Forest, New York. *J. Ecol.* **56**, 1–25.

Whittaker, R. H. & Woodwell, G. M. (1969). Structure, production and diversity of the oak-pine forest at Brookhaven, New York. *J. Ecol.* **57**, 155–74.

Witkamp, M. (1966). Rates of carbon dioxide evolution from the forest floor. *Ecology*, **47**, 492–4.

Woodwell, G. M. (1978). The carbon dioxide question. *Sci. Amer.* **238**, 34–43.

Woodwell, G. M. & Botkin, D. B. (1970). Metabolism of terrestrial ecosystems by gas exchange techniques. In *Analysis of temperate forest ecosystems*, ed. D. E. Reichle, pp. 73–85. Springer-Verlag, Berlin, Heidelberg, New York.

Woodwell, G. M. & Marples, T. G. (1968). Production and decay of litter and humus in an oak-pine forest and the influence of chronic gamma irradiation. *Ecology*, **49**, 456–65.

Yoda, K., Shinozaki, K., Ogawa, H., Hozumi, K. & Kira, T. (1965). Estimation of the total amount of respiration in woody organs of trees and forest communities. *J. Biol., Osaka City Univ.* **16**, 15–26.

Ziegler, H. (1964). Storage mobilization and distribution of reserve material in trees. In *Formation of wood in forest trees*, ed. M. L. Zimmerman, pp. 303–20. Academic Press, Chicago.

10. Analysis of forest growth and water balance using complex ecosystem models

P. SOLLINS, R. A. GOLDSTEIN, J. B. MANKIN, C. E. MURPHY & G. L. SWARTZMAN

Contents

In this chapter we examine the behavior and applicability of several complex models developed under the auspices of IBP. The three models chosen are of very different resolution in both time and space and are of very different mathematical structure. The models treat (1) stand energy balance over a period of several days considering both time and depth in the canopy as independent variables, (2) energy and water flow through a mixed deciduous forest, and (3) carbon and water flow through a mature coniferous forest. As a central theme we have chosen response of forests to change in leaf area such as might occur following insect outbreaks or herbicide treatment. Because leaf area is an important coupling between water and carbon flow, changes in leaf area should have profound effects on both sets of processes. For example, leaf area affects both photosynthesis and foliar respiration, transpiration, interception of water, and the energy balance of a forest stand.

Our procedure was to examine the response of each model to the same perturbations including both acute (instantaneous) decrease in foliage area and chronic defoliation (occurring continuously over an entire growing season or year). In the coniferous stand model we divided foliage into two age classes (current year and older), but in none of the models have we considered the actual distribution of branches or foliage either vertically or horizontally through the canopy. The models were used to examine the potential effects of a uniform removal of foliage throughout the canopy on carbon, water and energy fluxes through a forest ecosystem. A brief description of the structure and assumptions of each model and the results

537

of the simulations are presented below. Analyses are performed to determine whether similarity in model behavior of different forests is due to similarity of model assumptions and structure, or whether it can be attributed to accurate representation of actual behavior. We also consider problems of whether general purpose models are, in fact, applicable to a specific question such as effects of defoliation.

Description of the models

The Stand Energy Model

During the last two decades there has been a great deal of progress in understanding the micrometeorological processes which interact with forest vegetation to produce the forest microclimate (see Chapter 3). At the same time more has been learned about the dynamics of the growth processes in the forest and how they are affected by the forest microclimate. The model described below is an attempt to synthesize the information available from efforts in these two fields of research for the purpose of producing a quantitative description of how the mesoscale and synoptic scale of meteorological events act through the micrometeorological scale to affect tree growth through the photosynthetic process. The model was developed by C. E. Murphy, Jr, and K. R. Knoerr of Duke University under the auspices of the US–IBP Eastern Deciduous Forest Biome (Murphy & Knoerr, 1970, 2, 5; Murphy, Sinclair & Knoerr, 1974).

The model has four major components: (1) turbulent diffusion through the boundary layer above the stand, (2) turbulent diffusion through the trunk space, (3) foliage surface temperature and the related sink strength for sensible and latent heat, and (4) foliage source strength for carbon dioxide and the related net photosynthesis. The modeling of these components can be simplified greatly if it is assumed that all turbulent transfer of carbon dioxide, sensible heat, latent heat, and radiant transfer of heat are in the vertical direction. This assumption limits the application of the model to stands which are large with respect to the area of edge through which horizontal transport might take place. Another important simplification can be made by assuming that the forest is at steady state with respect to the exchanges of carbon dioxide, sensible heat and latent heat with the atmosphere. This assumption can be made when the environment above the stand changes slowly as compared to the time constant of the forest which is about 15–20 minutes on the basis of the physical transport.

Under the preceding conditions, a set of mathematical equations can be developed to describe the processes contributing to the microclimate and photosynthesis of the stand. The relationships used in this model are given in detail elsewhere (Murphy & Knoerr, 1970, 2, 5; Murphy *et al.*, 1974). The variables in the equations discussed below are those which have a significant role in determining the microclimate and growth.

Procedure

Turbulent diffusion in the boundary layer above the stand and in the stand air space can be described by

$$\frac{\partial}{\partial z}\left(K\frac{\partial x}{\partial z}\right) + Sx = 0, \tag{10.1}$$

where the variable x denotes the concentration of the material being transported, Sx is the source strength for that material, and z is the height above the forest floor. The source strengths are equal to zero in the boundary layer above the canopy and are equal to the flux divergence per unit height within the canopy. The diffusivities can be expressed approximately as

$$K = U_* k(z-d)/\phi, \; z > H, \; \text{and} \tag{10.2}$$

$$K = K_H \exp\left[a(z-H)\right], \; z < H, \tag{10.3}$$

where U_* is the friction velocity which is related to the shearing stress of the wind over the forest, k is Von Karman's constant ($\simeq 0.4$), d is the aerodynamic displacement height caused by the forest, ϕ is a correction for the effect of bouyancy, a is an empirical constant, and H is the height of the stand.

The source strengths for the various fluxes are determined at the foliage surfaces. There is interaction between these fluxes which can be described through the foliage energy balance equation

$$S + L - R - C - \lambda E - \Gamma P_n = 0, \tag{10.4}$$

which can be expanded to

$$\alpha S + L - \sigma T^4 - \frac{\rho C_p}{r_h}(T_a - T_l) - \frac{\rho\lambda}{(r_w + r_s)}(q_a - q_l) - \frac{\Gamma}{r_c + r_{sc} + r_{cw} + r_p}(C_a - C_l) = 0. \tag{10.5}$$

This equation demonstrates the relationships between the different fluxes as they are part of the energy exchange of the foliage. The radiant energy absorbed in the short wave (αS) and long wave (L) spectral bands is partitioned to long wave reradiation (R), sensible heat flux (C), latent heat flux (λE), and photochemical heat flux (ΓP_n). The proportioning is determined by the molecular diffusion resistances (1) through the boundary layer above the foliage element for heat (r_h), water vapor (r_w) and carbon

539

dioxide (r_c), (2) through the leaf stoma for water vapor (r_s) and carbon dioxide (r_{sc}), and (3) by diffusion through liquid water in the cell walls (r_{cw}) and the protoplasm (r_p) for carbon dioxide transport. Diffusion across these resistances is a function of the gradients between the leaf (l) and the adjacent air (a) of (1) temperature (T) for sensible heat flux, (2) specific humidity (q) for latent heat flux and (3) carbon dioxide concentration (c) for the photochemical flux. The last four terms of eqn 10.5 are functions of the leaf temperature, terms three (R) and four (C) explicitly and terms five (λE) and six (ΓP_n) implicitly through the effect of temperature on leaf surface specific humidity and the respiratory carbon dioxide flux.

The amount of radiation present in any layer of the stand is a function of the penetration of radiation in that spectral band through the canopy to the level at which the leaf is located. This is described by the equation

$$\alpha S_Z = \alpha S_o \exp(-\gamma A_Z), \tag{10.6}$$

where γ is a constant which can be determined empirically or, in some cases, theoretically.

Implementation

The six equations given above form the basis of the simulation model of microclimate and photosynthesis. The objective of a simulation of a steady state system is the solution of all the equations simultaneously, i.e. finding a set of values for the system variables that are constant for all of the equations. The method used to solve the set of equations in this simulation model is a type of successive approximation where reasonable first estimates of the variables are used to solve eqns 10.1, 10.2, 10.3, 10.4 and 10.6 for the profiles of absorbed radiation, air temperature, and air specific humidity. The values of these parameters at a number of heights in the canopy are then used to solve eqn 10.5 for leaf temperature and the source strengths at the foliage surfaces. This will give new values of the variables needed in eqns 10.1, 10.2, 10.3, 10.4 and 10.6. Each loop through this sequence produces values of the parameters which are closer to the final steady state solution for the system. The iterations are stopped when two successive estimates of convection differ by less than 0.0001 cal/cm² min.

The simulation program was written in FORTRAN IV. The simulations presented in this section were done at the Duke University Computation Center on an IBM 370/165 computer.

A simulation was carried out to investigate the effect of leaf area on the net photosynthesis of a loblolly pine (*Pinus taeda* L) plantation. The model was run for a standard day. The inputs of air temperature and solar radiation were varied sinusoidally (Fig. 10.1). The wind speed was held constant at 200 cm s⁻¹, the density of water vapor in the air above the

540

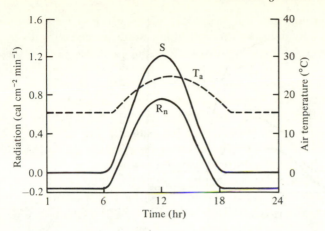

Fig. 10.1. Input values of solar radiation, (S), and air temperature (T$_a$), and simulated value of net radiation (R$_n$) for the standard day.

stand was held constant at 18 g m^{-3} and the longwave sky temperature was held constant at 240°K.

Results

The net radiation calculated from these inputs is also shown in Fig. 10.1, for the case of a leaf area index (LAI) of 7. Fig. 10.2 shows the trends of net photosynthesis, latent heat exchange and sensible heat exchange for the standard day and a LAI of 7. It is worth noting the asymmetry of these fluxes, a situation which is often seen in the field. In the simulation the asymmetry is caused by the interaction between the solar radiation changes which are symmetrical around noon and the air temperature change which has a lag of one hour behind the radiation.

Fig. 10.3 shows the effect of having canopies of different LAIs for the standard day. Notice that the model predicts a maximum photosynthesis rate. This maximum is found somewhat above the LAIs found in this stand through a normal growing season (Kinerson, Higgenbotham & Chapman, 1974). This is because the foliage added past a LAI value of approximately 14.5 contributes more to respiration than photosynthesis because of mutual shading of the leaves. However, the maximum value is not sharply peaked and through the range of LAIs found in this stand during the growing season the net photosynthesis is almost on a plateau.

The evaporation from the stand does not show the same peak rate in the range of leaf areas considered in the simulations. However, the evaporation from the stand does approach an upper limit for the environmental conditions hypothesized. This limit is related to the energy available from

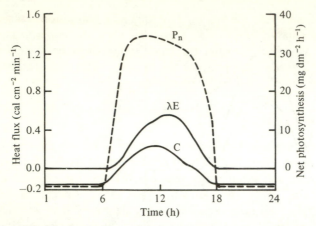

Fig. 10.2. Simulated values of net photosynthesis (P_n), latent heat flux (λE) and sensible heat flux (C) for the standard day.

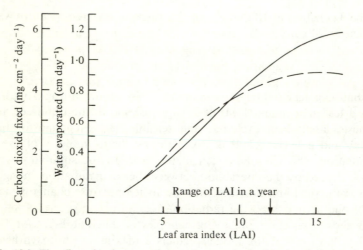

Fig. 10.3. The effect of leaf area on the net photosynthesis of (- - -) and evaporation from (———) a loblolly pine stand for the standard day. The values of net photosynthesis are the average for 10 hr centered around noon (mg cm^{-2} day^{-1}).

the radiation incident on the stand. It is interesting to note that the highest water use efficiency for photosynthesis is at a LAI of 7 which is within the range naturally found in this stand. However, the change in water use efficiency within the natural range of leaf areas is not large.

This simulation was run to show the type of results that can be obtained from models synthesizing environment physics with plant physiology. The results agree very well with experiments done at the US–IBP Eastern

Deciduous Forest Biome Triangle Site on a loblolly pine plantation. This in itself does not fully validate the model. However, when the experimental results are coupled with the fact that the model relies on a basic under-standing of the processes which produce the results, we are led to have added confidence in the validity of the simulation.

The TEEM-PROSPER model

The second model which we will consider is a Terrestrial Ecosystem Energy Model, TEEM, (Shugart, Goldstein, O'Neill & Mankin, 1974) coupled with a model of atmosphere–soil–plant water relations, PROSPER, (Goldstein, Mankin & Luxmoore, 1974). In contrast to the Stand Energy model, TEEM does not attempt to take detailed canopy geometry into account. TEEM is designed to simulate changes in organic components of a de-ciduous forest over an annual time scale, and it is divided into submodels to simulate primary production (PTEEM), consumption (CTEEM), and de-composition (DTEEM).

We will consider PTEEM, which simulates on a day-to-day basis the production, growth, and development of a close canopy *Liriodendron* forest stand assumed to have minimal heterogeneity. PTEEM is based on the assumption that the vegetation maximizes its production. PTEEM cal-culates the net production and allocates it in a manner that maximizes the production subject to survival constraints (e.g. the maintenance structural strength and sufficient reserves to carry the vegetation through the winter).

The functional relationship of the photosynthetic rate per unit surface area of light saturated leaf tissue, P_m, is

$$P_m = f(\psi_x, T, t_p),$$

where ψ_x is the plant water potential which is calculated in PROSPER, T is temperature, and t_p is physiological time. Gross photosynthesis is obtained by the relation

$$(1 - l_r)P_g = P_m \frac{S}{S + a_p}, \tag{10.7}$$

where P_g is the gross photosynthesis, l_r is the fraction of gross photo-synthesis utilized in carrying out photosynthesis, and the relationship in parenthesis on the right hand side is an empirical function (Rabinowitch, 1951; Monsi & Saeki, 1953; Monteith, 1965b; deWit, 1965; Duncan, Loomis, Williams & Hannon, 1967;) which expresses the relationship between light intensity, S, and photosynthesis. The net photosynthetic rate per unit surface area is determined by removing the maintenance respiration from gross production. Assuming that the solar radiation extinction through the canopy is an exponential function of the LAI, the total net photosynthesis

543

per unit ground area is obtained by integrating the net photosynthesis per unit surface area across the canopy yielding

$$P_N(F) = \frac{P_m}{\alpha_s} \ln\left[\frac{S_o + a_p}{S_o \exp(-\alpha_s F) + a_p}\right] - R_F F \tag{10.8}$$

where S_o is the solar radiation incident upon the canopy, α_s is the extinction coefficient, and R_F is the maintenance respiration per unit surface area.

Growth and development are described by nonlinear, ordinary differential equations. The four compartments are leaves, bole, roots, and storage (labile carbohydrates). The net photosynthesis is placed into storage as it is produced, and the respirational and growth energy demands are then removed from storage as they are required. This yields the equation

$$\frac{dx_s}{dt} = P_n - r_s x_s - r_B x_B - r_R x_R - G, \tag{10.9}$$

where x_s, x_B, and x_R are the standing crop of storage, bole, and roots, respectively; r_s, r_B, and r_R are the maintenance respirations per unit standing crop of storage, bole, and roots, respectively; and G is the growth requirement.

The optimal LAI (e.g., that LAI which maximizes production) may be determined from the equations for photosynthesis production, and foliage is grown as long as the actual LAI is less than optimal. It is hypothesized that it is to the tree's advantage to grow leaves as long as growth does not cause a net negative energy balance or dangerously deplete storage. Bud burst during the spring is controlled by physiological time, and the leaves are sloughed as litter when the energy required for respiration is greater than the energy produced by photosynthesis.

The equation for foliage growth is

$$\frac{dx_L}{dt} = a_L(\psi_x) x_L \left(1 - \frac{x_L}{a_F F_o}\right)\left(\frac{x_s}{x_s + S_1}\right) u(a_F F_o - x_L) + x_L^* \, \delta(t_p - b), \tag{10.10}$$

where x_L is the standing crop of leaves, $a_L(\psi_x)$ is the growth rate as a function of plant water potential, F_o is the optimum LAI, a_F is the conversion factor from standing crop to LAI, S_1 is the standing crop of storage below which storage starts to become dangerously depleted, x_L^* is the energy required for bad burst, t_p is physiological time, b is the point in physiological time at which bud burst occurs, u is the unit step function, and δ is the Dirac delta function.

The growth rate of bole material is a modified logistic form that decreases maximum potential growth rate as standing crop increases. Bole, in this context, includes all above-ground standing crop that is neither leaves nor labile storage. Bole growth is modified by storage, physiological time and plant water potential. The equation describing bole growth rate is

$$\frac{dx_B}{dt} = a_B(b, \ \psi_x)x_B\left(\frac{x_s - S_o}{S_1 + (x_s - S_o)}\right)u(x_s - S_o), \tag{10.11}$$

where a_B is a rate parameter and $S_o(S_o > S_1)$ is level of storage below which the vegetation will cease growing new bole material in order to conserve storage.

The functions of the roots are both support and water conduction. It is hypothesized that fine roots are grown in the spring in order to satisfy the increased water demand and that there is a net death of roots in the fall when transpiration demand declines. This is based on the assumption that the plant utilizes less stored energy in growing new fine roots in the spring than it does in maintaining fine roots through the winter. It is also assumed that roots are produced in the spring for transpirational purposes in proportion to the amount of foliage produced, and that roots are produced throughout the growing season in proportion to the amount of growth. A relationship which is consistent with these assumptions is

$$\frac{dx_R}{dt} = C_L\frac{dx_L}{dt} + C_B\frac{dx_B}{dt}, \tag{10.12}$$

where C_L and C_B are proportionality constants. The first term on the right accounts for changes in root standing crop due to changes in transpirational demand as a result of changes in leaf material. The second term accounts for changes in the amount of roots due to changes in above-ground material to support.

The model PROSPER is a phenomenological model which simulates atmosphere–soil–plant water relations on a forest stand (Goldstein *et al.*, 1974). Since it was originally developed to provide estimates of water stress for PTEEM, the initial emphasis was on closed canopy, deciduous vegetation. PROSPER was modified for applications to coniferous vegetation and cases of drastic vegetation alternation, such as forest clearing (Swift *et al.*, 1975).

PROSPER applies a water balance to a forest stand growing on a well developed soil profile. A hypothetical evapotranspiration (*ET*) surface which homogenizes the plant, soil surface, and litter characteristics is defined. This choice is based on the assumption that a model which simulates water dynamics on a day-to-day basis and uses environmental inputs which are daily means or totals does not require a detailed description of canopy geometry. *ET* is conceptualized as taking place from a single surface (surface 1 in Fig. 10.4) which is a combination of the ground and all canopy surfaces.

545

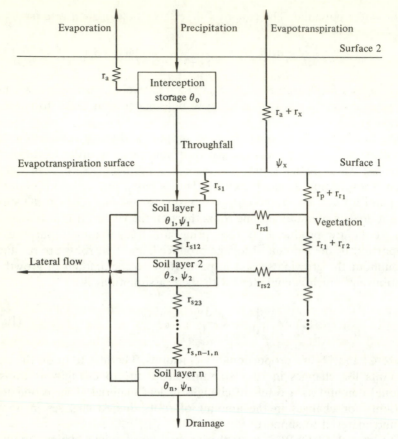

Fig. 10.4. Schematic representation of PROSPER. From Goldstein *et al.* (1974).

The equation for vapor flow

$$F_v = \left[\frac{(R_N - G)\Delta}{\sigma C_p \rho} + \frac{\rho_2^* - \rho_2}{r_a}\right] \bigg/ \left[\frac{r_a + r_x}{r_a} + \frac{L_v \Delta}{\sigma C_p \rho}\right] \qquad (10.13)$$

is derived by the combined energy balance–aerodynamic method (Monteith, 1965a). Here F_v is the vapor flow from the ET surface, R_N is the net radiation absorbed by the surface per day, G is the daily amount of heat energy transferred from the surface to the soil, σ is the ratio of convection area to ET area, C_p is the specific heat of air at constant pressure, ρ is the density of air, ρ_2^* is the saturation vapor density of surface 2 (Fig. 10.4), ρ_2 is the vapor density of surface 2, r_a is the resistance to transfer of water vapor between levels 1 and 2, r_x is the resistance of the evapotranspiration surface to the release of vapor, L_v is the latent heat of vaporization for water, and

546

$$\Delta = \frac{\rho_1^* - \rho_2^*}{T_1 - T_2}, \tag{10.14}$$

where ρ_1^* and ρ_2^* are the saturation vapor densities of levels 1 and 2, respectively, and T_1 and T_2 are the temperatures of levels 1 and 2, respectively.

The equation for liquid flow to the ET surface is derived by applying the law of conservation of mass and Darcy's Law (Gardner, 1960) to the vegetation which is viewed as a network of resistances. By applying standard techniques for solving network problems (Seshu & Balabanian, 1963), we can solve for F_L, the amount of liquid flowing to the ET surface, which is a function of ψ_x,

$$F_L = f(\psi_x). \tag{10.15}$$

The liquid and vapor flows are equated, and this resulting nonlinear equation is solved iteratively for ET. Cowan (1965) developed a similar description and solved some cases with an approximate analytical solution.

The energy balance equation is solved in units of energy per unit of effective ET surface area. The leaf area index is computed in PTEEM. Since all of the canopy is not equally active in the ET process, total leaf area is not used. The effective ET surface area is determined from an experiment described by Swank & Helvey (1970).

The values of interception, litter evaporation, transpiration and drainage are computed in PROSPER. Interception includes all water which is stored as surface moisture on leaves, stems, etc. PROSPER also computes the redistribution of moisture within and below the rooting zone.

Procedure and results

The combined PTEEM–PROSPER model was used to consider hypothetical insect infestations in a mesic *Liriodendron* stand. The insect infestations were assumed to take the form of an acute defoliation (a fixed percentage of the foliage removed on a specified day) or a chronic defoliation (a fixed rate of removal throughout the growing season). For the acute defoliation, we ran 15 cases: 10%, 25%, 50%, 75%, and 90% defoliation occurring at day 120, 180, and 240 of the calendar year. These results are shown in Table 10.1. For the case of chronic defoliation, we ran four cases: defoliation rates of 0.01, 0.05, 0.1 and 0.5 per day each initiated at day 120 so the canopy would be reasonably well developed before defoliation began. These results are shown in Table 10.2.

Acute defoliation at day 120 (early in the growing season) affected both growth of woody tissue (boles and roots) and total growth, and it had

547

Table 10.1. *Effects of acute defoliation on productivity and water balance of a deciduous forest. All values except evapotranspiration are kcal × 10^8/ha*

Percent defoliation	Year 1					Year 2				
	Net annual photosynthesis	Labile carbohydrate pool	Annual total tissue growth	Annual woody tissue growth	Annual evapotranspiration (cm)	Net annual photosynthesis	Labile carbohydrate pool	Annual total tissue growth	Annual woody tissue growth	Annual evapotranspiration (cm)
Day 120										
0	1.07	0.680	0.202	0.173	68.9	1.05	0.640	0.149	0.189	61.9
10	1.07	0.677	0.196	0.170	68.9	1.05	0.638	0.147	0.186	61.9
25	1.07	0.670	0.187	0.168	68.8	1.05	0.635	0.143	0.178	61.8
50	1.06	0.655	0.170	0.166	68.6	1.04	0.630	0.133	0.158	61.7
75	1.05	0.631	0.150	0.160	68.2	1.03	0.620	0.120	0.131	61.4
90	1.03	0.580	0.126	0.152	67.4	1.02	0.610	0.110	0.130	60.7
Day 180										
0	1.07	0.680	0.202	0.173	68.9	1.05	0.640	0.149	0.189	61.9
10	1.07	0.632	0.155	0.173	68.9	1.05	0.599	0.156	0.184	61.9
25	1.07	0.550	0.076	0.173	68.8	1.05	0.557	0.169	0.162	61.7
50	1.07	0.409	−0.068	0.173	68.6	1.04	0.430	0.180	0.159	61.3
75	1.05	0.261	−0.216	0.173	68.4	0.99	0.272	0.159	0.148	59.3
90	1.02	0.155	−0.314	0.173	67.4	0.91	0.175	0.074	0.054	53.6
Day 240										
0	1.07	0.680	0.202	0.173	68.9	1.05	0.640	0.149	0.189	61.9
10	1.07	0.669	0.192	0.173	68.8	1.05	0.632	0.152	0.189	61.8
25	1.07	0.629	0.152	0.173	68.6	1.05	0.612	0.160	0.177	61.7
50	1.07	0.535	0.083	0.173	68.4	1.04	0.609	0.165	0.091	61.4
75	1.05	0.511	0.034	0.173	67.4	1.02	0.604	0.182	0.072	60.9
90	1.00	0.514	0.037	0.173	63.8	0.97	0.607	0.181	0.076	59.1

Table 10.2. *Effects of chronic defoliations on productivity and water balance of a deciduous forest. All values except evapotranspiration are kcal × 10⁸/ha*

Defoliation rate constant (yr^{-1})	Year 1					Year 2				
	Net annual photo-synthesis	Labile carbo-hydrate pool	Annual total tissue growth	Annual woody tissue growth	Annual evapo-transpiration (cm)	Net annual photo-synthesis	Labile carbo-hydrate pool	Annual total tissue growth	Annual woody tissue growth	Annual evapo-transpiration (cm)
0	1.07	0.680	0.202	0.173	68.9	1.05	0.640	0.149	0.189	61.9
0.01	1.07	0.678	0.198	0.170	68.9	1.05	0.640	0.150	0.188	61.9
0.05	1.07	0.673	0.180	0.157	68.9	1.05	0.639	0.152	0.186	61.9
0.1	1.07	0.666	0.158	0.141	68.9	1.05	0.638	0.154	0.182	61.9
0.5	1.06	0.396	−0.208	0.047	68.4	1.04	0.530	0.195	0.060	61.6

roughly an equal effect during the second year. Acute defoliation on day 180 had negligible effect on the growth of woody tissue (there was some change in the fourth significant figure). Total growth was greatly decreased, and it became negative for higher defoliation. This negative growth was caused by a drastic reduction in storage which resulted in a very definite reduction in both woody tissue and total growth in the second year. Acute defoliation at day 240 had no effect on the growth of woody tissue. However, it did reduce storage which adversely affected growth in the second year. In general, it appeared that the visible effects of defoliation early in the growing season were seen in both years, while they were seen in the second year for mid and late season defoliation. There was a tendency toward increased total growth the second year for mid and late season defoliations. This was because the plants put most of their energy into building up reserves and less into putting on woody tissue. This resulted in somewhat reduced respiratory losses from woody tissue and a drastic reduction in growth respiration. Our simulations indicate that mid-season is the worst time that defoliation can occur. Storage reserves are low, and foliage has a more difficult time recovering (relative to an early season defoliation). Foliage has a very difficult time recovering from a late season defoliation. Another point is that a 75% defoliation at day 240 is more damaging than a 90% defoliation the same day. This is because the canopy cannot recover in either case, but little energy is wasted in the attempt in the case of 90% defoliation.

In most of the simulations the effect of an acute defoliation on transpiration was slight because the defoliation actually caused little change in leaf area. Fig. 10.5 shows canopy development throughout the year for no defoliation and 50% defoliation at day 120, 180, and 240. The canopy recovers quite rapidly from defoliation, particularly early in the season (see also Heichel & Turner, 1976). This pattern is a result of the basic hypotheses of the TEEM model: as long as there is a net energy advantage, the first priority of energy allocation is foliage development. The ability to recover diminishes throughout the growing season due to aging. For day 120 there was a decrease in the annual total transpiration of 0.3 cm; for day 180, 0.2 cm; and day 240, 0.5 cm. Table 10.3 shows total evapotranspiration separated into evaporation from the canopy, evaporation from the litter, and transpiration for the control case and 10%, 50% and 90% defoliation at days 120. 180, and 240. This shows slight decreases in water loss from the canopy, and, at least in one case, there is a slight increase in evaporation from the litter.

Fig. 10.6 shows the amount of energy in food reserves throughout the year. A large fraction of storage is put into growth, particularly of foliage, early in the growing season, and the later part of the growing season is devoted to replenishing these reserves in preparation for the next growing season. If one hypothesizes that the plant's basic function is to survive, this seems a logical course of action.

550

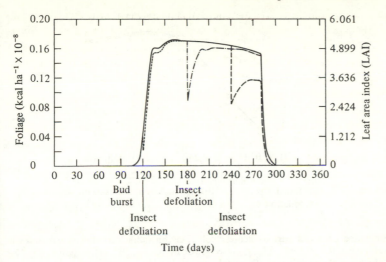

Fig. 10.5. Simulated effects of various acute defoliations on foliage levels in a deciduous forest in 1969.———, no defoliation; ····, 50% defoliation at day 120, –·–·–, 50% defoliation at day 180: ----, 50% defoliation at day 240.

Table 10.3. *Effect of various degrees of defoliation on evaporation and transpiration from a deciduous forest. All values are cm/yr*

% Defoliation	Evaporation from canopy	Evaporation from litter	Transpiration	Total evapotrans-piration
No defoliation	17.7	8.5	42.7	68.9
10% at day 120	17.7	8.5	42.7	68.9
10% at day 180	17.7	8.5	42.7	68.9
10% at day 240	17.6	8.6	42.6	68.8
50% at day 120	17.6	8.5	42.5	68.6
50% at day 180	17.5	8.5	42.6	68.6
50% at day 240	17.4	8.6	42.4	68.4
90% at day 120	17.4	8.7	41.3	67.4
90% at day 180	17.1	8.7	41.6	67.4
90% at day 240	16.9	9.1	37.8	63.8

The chronic defoliation results, shown in Table 10.2, are not surprising in view of the acute defoliation results. In the TEEM model, a chronic defoliation will not substantially reduce the amount of foliage. However, increasing amounts of energy will be expended in producing the foliage at the expense of the growth of woody tissue and storage reserves.

551

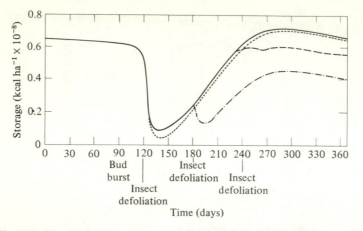

Fig. 10.6. Simulated effects of various acute defoliations on labile carbohydrate reserves in trees of a deciduous forest in 1969. ———, no insect defoliation; ----, 50% defoliation at day 120; —·—·—, 50% defoliation at day 180; ---- 50% defoliation at day 240.

Description of CONIFER

CONIFER is a model of carbon and water flow through a coniferous forest designed to simulate response of the system to such perturbations as defoliation, climatic change, irrigation, and thinning. Typical model outputs include production of foliage, wood and roots, evaporation, transpiration, and transfer of water from subsoil to groundwater. The model was developed by the US–IBP Coniferous Forest Biome and has been implemented for a 450-year-old Douglas fir stand located in the Cascade Mountains near Blue River, Oregon (Grier & Logan, 1977; Sollins *et al.*, in press).

The model consists of about 20 coupled difference equations with a time step of one day for water and energy equations, and one week for carbon equations. The model considers a unit area of land surface (1 ha) and assumes that all storages are distributed uniformly across the area. A portion of the area is assumed to be without vegetative cover, although this portion is also assumed to be distributed uniformly across the land area.

Important driving variables (Zs) include daily precipitation, daily short wave radiation (corrected for slope and aspect), mean 24 hr air temperature, mean day-time air temperature, mean 24 hr dewpoint temperature, all of which were measured, and daily incident long wave radiation which was calculated from other Zs. The model contains some 140 intermediate variables (Gs) calculated from Zs, state variables (Xs), and other Gs. It is this process coupling, the choice of Gs, Xs, and Zs used to calculate any particular flux, which characterizes the model far more than the form of the equations. An overview of this structure is given in Table 10.4. For a more

Table 10.4. *Interactions among modules of CONIFER.* (*From CFB Modeling Group*, 1977)

Effect of carbon variables on water and energy flows	Comments
A. Foliage biomass affects: 1. Transpiration.	—
2. Fraction of rain incident to canopy that strikes foliage (and therefore also fraction striking nonfoliage).	2. This and following two affect drip, litter, and soil moisture dynamics. There are also indirect effects through percent cover.
3. Water retention capacity of canopy.	—
4. Distribution of retention capacity between foliage and nonfoliage.	—
5. Fraction of rainfall passing directly to forest floor.	5. Through percent cover.
6. Net long wave radiation input to canopy.	6. Through percent cover, which affects input and loss.
B. Stem biomass affects: 1. Percent cover (and therefore numbers 2–6 above)	—
C. Fine, leaf and woody litter mass affects: 1. Water retention capacity of litter.	—

Effect of water variables on carbon and energy flows

A. Soil moisture affects: 1. New and old foliage photosynthesis.	1. Via stomatal resistance.
2. Fine root death.	2. Via plant moisture stress.
3. Dead root + soil organic matter decomposition processes.	—
B. Litter moisture affects: 1. Litter decomposition processes.	—
C. Snowpack ice affects: 1. Litter temperature.	—
D. Snowfall affects: 1. Heat loss from snowpack due to snowfall.	—
2. Albedo of snowpack.	—
E. Drip plus direct rainfall affect: 1. Litter and soil temperature.	—

Effect of energy variables on carbon and water flows

A. Heat input to canopy affects: 1. Potential evaporation from canopy.	—
2. Transpiration.	—
B. Litter temperature affects: 1. Litter decomposition processes.	—
2. Potential evaporation from litter.	—

Table 10.4. *continued*

Effect of carbon variables on water and energy flows	Comments
C. Soil temperature affects: 1. Large and fine root respiration and growth.	—

D. Net heat input to snowpack and heat deficit of snowpack affect:
  1. Net transfer between free water and ice
    in snowpack.
	—

complete description, the documentation should be consulted (Coniferous Forest Biome Modeling Group, 1977).

The carbon part of the model is similar to an earlier model of a *Liriodendron* stand in Tennesse (Sollins, Reichle & Olson, 1973, Sollins, Harris & Edwards, 1976). In both models carbohydrate (CH_2O) reserves throughout the trees were assumed to constitute a single well mixed pool providing substrate for tissue growth and respiration. Growth and net assimilation were considered as distinct processes. Net assimilation was assumed to be regulated by conditions in the canopy while growth processes were assumed to be regulated by conditions in cambia and other meristematic regions. For example, wood production in CONIFER was assumed to be weakly dependent on the value of the carbohydrate pool and strongly dependent on air temperature (in the case of stems and branches) and soil temperature (in the case of roots). Net assimilation thus affected wood production only indirectly in that it was one of the processes regulating the level of the carbohydrate pool.

New features in the carbon model include (1) dependence of photosynthesis on stomatal resistance which is in turn controlled, at least part of the time, by soil moisture; (2) a priority scheme for allocating carbohydrate among various growth and respiratory processes; and (3) a bud compartment to allow for regulation of foliage development during a current year by bud-set the previous year.

The water part of CONIFER is similar to ones constructed by Goldstein *et al.* (1974) and Rogers (Chapter 4). Evaporation from the canopy is calculated with a Penman-type equation using 24 hr mean air temperature and an aerodynamic resistance calculated from wind speed measured in a nearby clearcut. The transpiration equation is similar (Monteith, 1965*a*) but uses average day-time air and dewpoint temperature and a canopy resistance equal to stomatal resistance divided by LAI. Stomatal resistance is regulated either by day-time vapor pressure deficit or by soil moisture, whichever results in the greater resistance. Transpiration thus increases with increasing LAI.

554

Interception, storage and drip from the canopy are all affected by leaf and stem biomass. Equations for energy flow through the canopy, snowpack and litter also include dependence on foliage area and stem biomass. These effects of foliage area were included in the model specifically to allow simulation of response to defoliation and thinning and have not, to our knowledge, been included in other hydrology models.

Procedure and results

As with TEEM–PROSPER simulations, we considered effects of both chronic and acute defoliations. For the acute case we considered eight combinations of 10%, 25%, 50% and 90% removal of old, current year or both old and current year foliage on days 135, 182, 240 and 350 of the first year (Table 10.5). For the case of chronic defoliation, we considered defoliation rates of 0.005, 0.01, and 0.02 wk^{-1} which resulted in year-end foliage values which were 25%, 50% and 75% of those occurring without defoliation (Table 10.6). In all cases the defoliation was assumed to occur uniformly throughout the old-growth canopy.

Effects of both chronic and acute defoliation on the stand water balance were quite striking. Removal of foliage substantially reduced interception and subsequent evaporative loss (up to 20% for 90% defoliation on day 182). Transpiration was also reduced (up to 50%) as was the length of time snow was on the ground (up to 47%). The model predicted that acute defoliation during the summer causes a larger decrease in transpiration and evaporation than does an equivalent removal of foliage chronically over an entire year. This is reasonable because the period of greatest evaporative demand occurs during the summer and the chronic defoliation allows higher foliage levels during summer than does the acute defoliation.

With respect to production processes, effects of acute defoliation on end-of-year CH_2O levels and subsequent year's new foliage production were substantial (Figs. 10.7, 10.8). Removal of old foliage in general stimulated net assimilation by new foliage (Table 10.5, 10.6). However, the general effect was to decrease end-of-year levels of CH_2O reserves because the new foliage accounts for only a small part of total net assimilation. A 90% defoliation on day 182 had a disastrous effect and even a 50% loss on day 182 resulted in a decrease of 50% in the subsequent year's foliage production. The worst time for defoliation appeared to be day 182. Acute defoliations on days 135 and 240 caused slightly smaller decreases in end-of-year CH_2O levels and all three caused approximately equal reductions in new foliage production in the subsequent year.

Effects of chronic defoliation were even more striking. Again, almost any degree of defoliation increased net assimilation by current year foliage, but this was more than offset by decreases in net assimilation by older foliage.

555

Table 10.5. *Effects of acute defoliation upon water dynamics and primary production*

Criterion	Year	No defoliation	Defoliation/perturbation							
			10% new foliage day 182	25% new foliage 10% old foliage day 182	25% new foliage 25% old foliage day 182	25% new foliage 25% old foliage day 135	25% new foliage 25% old foliage day 240	50% new foliage 50% old foliage day 182	90% new foliage 90% old foliage day 182	50% old foliage day 350
Evaporation (m^3 ha^{-1} yr^{-1})	1	3200	3190	3130	3070	3060	3080	2900	2330	3030
	2	2570	2560	2490	2490	2490	2490	2370	1850	2370
Transpiration (m^3 ha^{-1} yr^{-1})	1	3340	3320	3290	3210	3220	3210	2980	1660	3030
	2	2385	2365	2365	2340	2340	2340	2265	1760	2265
% change in fraction of time snow is on ground	1	—	0	−2.4	−4.4	−4.4	−4.4	−5.6	−47.4	−6.8
	2	—	0	−0.4	−1.0	−1.0	−1.0	−3.7	−20.7	−3.9
New carbon assim. by new foliage (ton ha^{-1} yr^{-1})	1	1.53	1.55	1.70	1.97	2.10	1.67	2.22	1.70	1.53
	2	1.19	1.18	1.31	1.63	1.63	1.64	2.40	1.30	2.40
Net carbon assim. by old foliage (ton ha^{-1} yr^{-1})	1	32.2	32.4	32.8	32.7	32.7	31.7	31.1	16.0	27.9
	2	26.5	26.3	26.1	25.6	25.7	25.7	24.1	13.9	24.2
Change in old foliage carbon[a] (ton ha^{-1})	1	0	−0.12	−0.63	−1.28	−1.14	−1.14	−2.28	−4.10	−2.28
	2	0.03	−0.06	−0.11	0.17	0.06	0.09	0.02	0.25	0.02
Change in CH$_2$O pool carbon[a] (ton ha^{-1})	1	1.06	1.08	1.16	0.69	0.49	0.23	−1.36	−13.11	−2.82
	2	−0.11	−0.42	−1.43	−1.76	−1.69	−1.48	−3.21	−3.10	−1.97
Max. new foliage carbon mass (ton ha^{-1})	1	1.40	1.40	1.40	1.40	1.11	1.44	1.40	1.44	1.44
	2	1.50	1.33	1.11	1.11	1.11	1.11	0.74	0.32	0.74

[a] Good ... of year value less beginning-of-year value.

Table 10.6. *Effects of chronic defoliation on water dynamics and primary production. Upper values show day 131 1972–day 130 1973. Lower values are for the following year. Defoliation continued through both years.*

	Percent reduction in foliage at end of year			
	0%	25%	50%	75%
Evaporation (m^3 ha^{-1} yr^{-1})	3140	3100	2970	2735
	2500	2450	2290	1890
Transpiration (m^3 ha^{-1} yr^{-1})	3290	3230	3070	2630
	2320	2230	2030	1875
Percent change in fraction of time snow on ground	—	−1.2	−2.0	−12.2
		−1.7	−5.8	−21.7
Net carbon assimilation by new foliage (ton ha^{-1} yr^{-1})	1.64	1.74	1.91	1.90
	1.50	1.72	2.43	4.19
Net carbon assimilation by old foliage (ton ha^{-1} yr^{-1})	32.5	32.2	31.0	27.2
	25.0	23.6	19.6	15.6
Old foliage carbon (ton)	3.86	3.43	2.47	1.33
	3.41	2.76	1.51	0.49
CH_2O pool carbon (ton)	16.5	16.0	14.4	10.2
	14.4	12.6	7.7	2.3
Maximum new foliage carbon mass (ton ha^{-1})	1.33	1.24	0.94	0.60
	1.25	1.10	0.73	0.36

In general, effects on end-of-year CH_2O levels were greater after two years of defoliation.

Under low levels of acute defoliation on day 182, end-of-year CH_2O levels actually increased, due almost entirely to an increase in net assimilation by both old and new foliage. Increased net assimilation was due partly to a decrease in foliar respiration (a direct result of decreased leaf biomass) accompanied by little change in photosynthesis. Increased light penetration essentially offset the decreased leaf area resulting in little change in photosynthesis over the range of LAIs. A second cause of increased net assimilation was the decrease in moisture stress experienced by the remaining foliage. Defoliation resulted in decreased evaporation and transpiration losses which in turn resulted in higher soil moisture levels and lower stomatal resistances. Reduced stomatal resistance in turn increased photosynthesis but had no effect on respiration. In fact, the increase in end-of-year CH_2O level would be greater but for an increase in root respiration caused by increase in soil moisture. Root death increases, in CONIFER, with increasing moisture stress; defoliation caused less root death and thus greater root respiration.

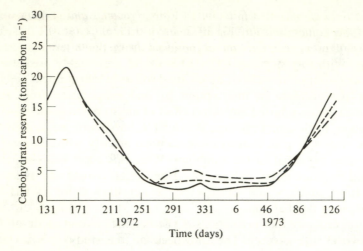

Fig. 10.7. Simulated seasonal pattern of carbohydrate reserves under varying intensity defoliation of an old growth coniferous forest. ——, no defoliation; ----, 25% defoliation on day 182; — —, 50% defoliation on day 182.

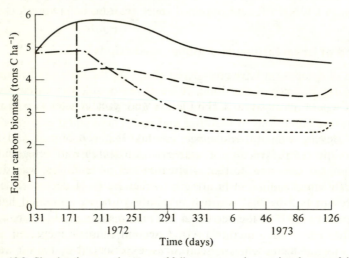

Fig. 10.8. Simulated seasonal pattern of foliage mass under varying degrees of defoliation of an old growth coniferous forest. Symbols same as in Figure 10.7.

Discussion

CONIFER was specifically designed to examine potential interactions between vegetation and the hydrologic cycle under a wide range of perturbations. Although many interactions between water flow, energy flow

558

and production were included in CONIFER, many were omitted. Inclusion of some of these could alter our conclusions. For example, CONIFER lacks any effect of light penetration on evaporation from the litter and litter temperature. With this included, CONIFER would presumably predict an increase in evaporation from litter following defoliation which might somewhat offset the decrease in interception by and evaporation from the canopy. We have also omitted any direct effect of moisture stress on growth and respiration. Little direct evidence is available, but Whitmore & Zahner (1967) have shown in pine a decrease in cambial activity with decreasing (negative) water potential in the cambium. Were this effect included, the increases soil moisture might well increase rates of cell growth and division and respiration which would cause even greater draw-down of CH_2O reserves following defoliation.

Finally, we recognize that predictions of effects of defoliation on duration of the snowpack are suspect. The snowpack on the study sites in fact persists for at most a few weeks on several occasions each winter. Clear-felling increases both snow accumulation and the rate of snowmelt (Rothacher, 1965) and we would expect defoliation to have a similar effect. Since performing the simulations reported here, we have modified the assumptions in CONIFER and obtained more realistic behavior.

Comparison of observations and model predictions

Few studies of defoliation have considered effects on the water and nutrient cycles or energy balance of the stand. It seems reasonable to us that severe defoliations would affect snow accumulation and melt as well as moisture and temperature conditions in the litter and soil. These could in turn affect growth by altering stomatal resistance, cambial water potential, soil and possibly canopy temperatures, and nutrient availability and uptake. The general consensus seems to be that a given reduction in foliage causes an approximately equal reduction in growth (review by Kulman, 1971) which may or may not be true, but which is certainly a gross oversimplification. Mattson & Addy (1975) concluded that mild defoliation of less than 40 to 50% has little effect on growth, but that growth reduction may be proportional to foliage losses when defoliation exceeds 40%. Defoliations which recur during several growing seasons have in general more pronounced effects than even a complete single defoliation of a deciduous tree (Kulman, 1971). However, Heichel & Turner (1976) found that even severe defoliation of *Quercus rubra* and *Acer rubrum* caused no mortality and only mild reduction in growth because of the ability of these hard-wood species to replace foliage quickly following defoliation..

Results from all three models agree with these generalizations in that mild defoliations have little effect on net assimilation, growth or CH_2O reserves while, as defoliations increase beyond about 50%, effects become proportionately greater. Two of the models actually predict slight increases in net assimilation with moderate leaf area reduction at high LAI values.

TEEM–PROSPER and CONIFER predicted substantially different CH_2O pool patterns both with and without defoliation (Fig. 10.6). Minimum pool values were predicted for the winter in the coniferous stand and for the summer in the deciduous. In CONIFER defoliation resulted in CH_2O values which were greater in the winter but lower the following spring, while with TEEM, CH_2O values after defoliation were always lower.

There are some major differences between the forests and their respective models that explain to some extent this radically different behavior. In the *Liriodendron* forest foliage grows indeterminately and recovers from defoliation, while in the coniferous forest, foliage growth is not only determinate but also ceases after the first year (Fig. 10.7). In CONIFER we further assume that early season defoliation removes not only existing foliage, but also reduces the potential foliar development during the remainder of the growing season (see Reichle, Goldstein, Van Hook & Dodson, 1973).

Since in CONIFER there is no increase in foliar growth after defoliation, the immediate impact on the carbohydrate pool is not as drastic as in TEEM. However, even in CONIFER reduced photosynthesis does eventually take its toll on the carbohydrate pool and, during the year following defoliation, carbohydrate reserves, though depleted less, rise more slowly than in the control run.

Whether either the CONIFER or TEEM–PROSPER prediction is correct is at this stage a moot point, since to our knowledge no one has measured carbohydrate levels following defoliation. In fact, we know of only three recently published studies of food reserve dynamics in woody parts of trees (Hepting, 1945; Kimura, 1969; Mooney & Hays, 1973). In an *Abies Veitchii* community, Kimura (1969) found carbohydrate reserve patterns similar to those predicted by CONIFER. A maximum occurred in early June but unlike the CONIFER output, the data showed a distinct minimum in August following which reserves returned to June levels. Unfortunately, the climates of the two sites are sufficiently different that comparison is probably meaningless.

Even without data for comparison we can assume that our predictions of food reserve dynamics could be improved. Although predicted patterns of photosynthesis, respiration and mortality may be fairly accurate, lack of data has precluded realistic modeling of factors regulating timing and rates of tissue growth. For example, direct effects of moisture stress on tissue growth are missing from CONIFER, while effects of defoliation on stem and soil temperatures are missing from both, as are any nutritional regulatory mechanisms. Nor do any of the models include possible effects of defoliation on persistence of the remaining foliage and timing of bud break the subsequent year.

With respect to stand water and energy balance, all three models predicted reduced transpiration and evaporation with decreased LAI, which agrees with measurements of outflow from gauged watersheds following clearcut (Hewlett & Helvey, 1970; Hornbeck, Pierce & Federer, 1970; Hornbeck, 1973; Rothacher, 1970; Cheng *et al.*, 1975; Harr, 1976). Stephens, Turner & DeRoo (1972) found an increase in soil moisture in a mixed deciduous forest following severe defoliation despite abundant rainfall. They found an increase in xylem water potential from about − 20 bars at low levels of defoliation to nearly − 4 bars at complete defoliation. These results contradicted those of Nichols (1968) who found that internal pressure potential decreased with increasing defoliation. Our predictions of effects of defoliation on the hydrologic cycle are speculative but strongly suggest that even mild defoliations might cause significant changes. J. D. Helvey and A. R. Tiedemann of the US Forest Service, Wenatchee, Washington, have initiated a study of effect of defoliation of mixed conifer stands on rainfall interception loss and snow accumulation and melt which should provide data with which to compare our simulation results.

Conclusions

One goal of IBP was to understand and contribute to increasing the world's food, fiber and timber producing capability. IBP sought basic knowledge on the functioning of natural ecosystems so that we might better manage our biological resources in the face of increasing demands and increasing frequency of disruption. Through IBP we have now developed complex models of ecosystem behavior. Do these models help? What do they tell us that we did not already know? Specifically, can they predict effects of, say, insect outbreaks on forest productivity?

The answers to the questions respectively are: 'Yes, much, and we don't know'. The data are as yet insufficient to judge the accuracy of the models. Because controlled defoliation experiments have never been conducted in conjunction with an ecosystem analysis program, we lack basic data on effects of defoliation on canopy interception and storage capacity, canopy energy balance and net assimilation. But it is clear that the model development, and particularly the attempt at application to a specific problem described herein, has at least identified some specific pieces of information which are still needed.

The first set of missing 'pieces' would connect production ecology and plant physiology. Existing models of stand growth (e.g., Botkin, Janak & Wallis, 1972; Stage, 1973) do not consider the physiological mechanisms through which changes in net assimilation rates affect wood production. The two long term models described here (TEEM and CONIFER) make more use of the results of plant physiological research but are still inadequate. Agricultural crop models have been developed which more realistically describe the growth processes (Fick, Williams & Loomis, 1973; Fick, Loomis & Williams, 1975). In one of the most promising approaches, Hunt & Loomis (1976) grew tobacco callus on media of differing sucrose concentration and used the data to describe effects of storage reserves on growth in their sugar-beet model. In contrast to the agricultural situation, it appears that after decades of forestry research we still have virtually no information on factors regulating cell elongation and division in cambial regions of commercially important trees. We can assume that stem temperature, water status, and levels of carbohydrates, nutrients and various growth hormones regulate cambial activity, but without actual data physiologically realistic models of tree growth are impossible.

The second set of needed data involves the interface between micrometeorology, production and decomposition ecology and hydrology. Although much excellent work has been done we still need data on the relation between foliage mass and the intercepting and storage capacity of the canopy and on factors regulating the energy balance of the litter layer and snowpack under canopies of differing LAI. Particularly for deciduous hardwoods, a much better understanding of stomatal control mechanisms is still needed. Micrometeorological models are vital but by themselves they cannot answer questions posed by land managers.

The third area of data deficiency, one not even addressed by any of the models described, here, concerns forest nutrient cycles. Defoliation may affect growth through any of several possible nutritional mechanisms. Nutrient levels in leaf and root meristems may affect food reserve and water status of the entire tree and thus cambial activity.

Internal nutrient levels in the trees can be affected by water flow rates through the system which we already know are affected by defoliation. Nutrient uptake rates are affected by nutrient mineralization rates. An increased input of readily decomposable, nutrient-rich frass occurs following defoliation by insects (Kimmins, 1972), and this input could certainly affect mineralization rates. Several of the modeling groups participating in this study are, by developing annual budgets and whole system models of nutrient flow in forest ecosystems, attempting first to determine which processes are important in regulating growth and next to develop models which include these processes. The accuracy and realism of our predictions will probably increase as these mechanisms are incorporated into the models, but again, without whole system data for validation, the models will become even more speculative than they are already.

It seems that in order to answer a specific question such as might be posed by a land-manager, 'What are the effects of defoliation on wood production?' we must integrate the knowledge and talents of the meteorologist, hydrologist, plant physiologist, entomologist, decomposition ecologist and perhaps even soil chemist. Mechanistic, whole-system models seem to offer the only hope for integrating this knowledge, but it is also clear from the exercise described here that the models work better when designed from the outset with the particular question in mind. Nonetheless, posing such a question after the models have been developed is still useful. It causes re-examination of the assumptions and objectives of the modeling and, perhaps belatedly, brings the disciplines together. In retrospect it seems that, regardless of whether IBP ecosystems analysis programs and attendant models have had a significant impact on forest management practices, these projects have caused a profound re-organization of ecological research and perhaps even a redefinition of ecology.

References

Botkin, D. B., Janak, J. F. & Wallis, J. R. (1972). Some ecological consequences of a computer model of forest growth. *J. Ecol.* **60**, 849–72.

Cheng, J. D., Black, T. A., de Vries J., Willington, R. P. & Goodell, B. C. (1975). The evaluation of initial changes in peak streamflow following logging of a watershed on the west coast of Canada. *Assoc. Int. Sci. Hydrol. Publ.* **117**, 475–86.

Coniferous Forest Biome Modeling Group. (1977). *CONIFER: A model of carbon and water flow through a coniferous forest.* Coniferous Forest Biome. Bull. No. 8, Univ. Wash., Seattle, Wash.

Cowan, I. R. (1965). Transport of water in the soil–plant–atmosphere system. *J. appl. Ecol.* **2**, 221–39.

Duncan, W. G., Loomis, R. S., Williams, W. A. & Hannon, R. (1967). A Model for Simulating Photosynthesis in Plant Communities. *Hilgardia,* **38**, 181–205.

Fick, G. W., Loomis, R. S. & Williams, W. A. (1975). Sugar beet. In *Crop physiology: some case studies,* ed. L. T. Evans, pp. 259–95. Cambridge University Press, Cambridge.

Fick, G. W., Williams, W. A. & Loomis, R. S. (1973). Computer simulation of dry matter distribution during sugar beet growth. *Crop Sci.* **13**, 413–7.

Gardner, W. R. (1960). Dynamic aspects of water availability to plants. *Soil Sci.* **89**, 63–73.

Goldstein, R. A., Mankin, J. B. & Luxmoore, R. J. (1974). *Documentation of PROSPER: A model of atmosphere–soil–plant water flow.* USAEC Rep. No. EDFB-IBP-73-9, Oak Ridge National Laboratory, Oak Ridge, Tennessee.

Grier, C. C. & Logan, R. S. (1977). Organic matter distribution and net production in plant communities of a 450-year-old Douglas fir ecosystem. *Ecol. Monogr.* **47**, 373–400.

Harr, R. D. (1976). *Forest practices and streamflow in western Oregon.* USDA For. Serv. Gen. Tech. Rep. PNW-49, Pac. Northwest For. Range Exp. Sta., Portland, Oregon.

Heichel, G. H. & Turner, N. C. (1976). Phenology and leaf growth of defoliated hardwood trees. In *Perspectives in forest entomology,* ed. J. F. Anderson & H. K. Kaya, pp. 31–40. Academic Press, New York.

Hepting, G. H. (1945). Reserve food storage in shortleaf pine in relation to little-leaf disease. *Phytopathology*, **35**, 106–19.

Hewlett, J. D. & Helvey, J. D. (1970). Effects of forest clear-felling on the storm hydrograph. *Water Resour. Res.* **6**, 768–82.

Hornbeck, J. M. (1973). Stormflow from hardwood-forested and cleared watersheds in New Hampshire. *Water Resour. Res.* **9**, 346–54.

Hornbeck, J. M., Pierce, R. S. & Federer, C. A. (1970). Streamflow changes after forest clearing in New England. *Water Resour. Res.* **6**, 1124–32.

Hunt, W. F. & Loomis, R. S. (1976). Carbohydrate limited growth kinetics of tobacco (*Nicotiana rustica* L.) callus. *Plant Physiol.* **57**, 802–5.

Kimmins, J. P. (1972). Relative contributions of leaching, litter-fall and defoliation by *Neodiprion sertifer* (Hymenoptera) to the removal of cesium-134 from red pine. *Oikos*, **23**, 226–34.

Kimura, M. (1969). Ecological and physiological studies on the vegetation of Mt Shimagare. VII. Analysis of production processes of young *Abies* stand based on the carbohydrate economy. *Bot. Mag.* (*Tokyo*), **82**, 6–19.

Kinerson, R. S., Higgenbotham, K. O. & Chapman, R. C. (1974). The dynamics of foliage distribution within a forest canopy. *J. appl. Ecol.* **11**, 347–53.

Kulman, H. M. (1971). Effects of insect defoliation on growth and mortality of trees. *Annu. Rev. Entomol.* **16**, 289–324.

Mattson, W. J. & Addy, N. D. (1975). Phytophagous insects as regulators of forest primary production. *Science*, Washington, **190**, 515–22.

Monsi, M. & Saeki, T. (1953). Über den Lichtfaktor in den Pfanzengesellschaften und zeine Bedentung für die Stoffprodurodktion. *Jap. J. Bot.* **14**, 22–52.

Monteith, J. L. (1965a). Evaporation and environment. In *The state and movement of water in living organisms*, ed. G. E. Fogg, pp. 205–34. Academic Press, New York.

Monteith, J. L. (1965). Light distribution and photosynthesis in field crops. *Ann. Bot. N.S.* **19**, 17–37.

Mooney, H. A. & Hays, R. I. (1973). Carbohydrate storage cycles in two California Mediterranean-climate trees. *Flora*, **162**, 295–304.

Murphy, C. E. Jr., & Knoerr, K. R. (1970). A general model for the energy exchange and microclimate of plant communities. In *Proceedings 1970 Summer Computer Simulation Conference*, pp. 786–97. Simulations Council Inc., La Jolla, California.

Murphy, C. E. Jr., & Knoerr, K. R. (1972). *Modeling the energy balance processes of natural ecosystems.* USAEC Rep. No. EDFB-IBP-72-10. Oak Ridge National Laboratory, Oak Ridge, Tennessee. 135 p.

Murphy, C. E. Jr., & Knoerr, K. R. (1975). The evaporation of intercepted rainfall from a forest stand: An analysis by simulation. *Water Resour. Res.* **11**, 273–80.

Murphy, C. E., Sinclair, T. S. & Knoerr, K. R. (1974). Modeling the photosynthesis of plant stands. In *Vegetation and environment. Part VI. Handbook of vegetation science*, ed. B. R. Strain & W. D. Billings, pp. 125–47. Junk, Publ. Co., The Hague, Netherlands.

Nichols, J. O. (1968). Oak mortality in Pennsylvania–a ten year study. *J. Forest.* **66**, 681–94.

Rabinowitch, E. T. (1951). *Photosynthesis and related process. Vol. II, Part I.* Interscience, New York.

Reichle, D. E., Goldstein, R. A., Van Hook, R. I. & Dodson, G. J. (1973). Analysis of insect consumption in a forest canopy. *Ecology*, **54**, 1076–84.

Rothacher, J. (1965). Snow accumulation and melt in strip cuttings on the west slope

564

of the Oregon Cascades. USDA For. Serv. Res. Note PNW–23, Pac. Northwest For. Range Exp. Sta., Portland, Oregon.

Rothacher, J. (1970). Increases in water yield following clear-cut logging in the Pacific Northwest. *Water Resour. Res.* **6**, 653–8.

Seshu, S. & Balabanian, N. (1963). *Linear network analysis.* John Wiley and Sons, London.

Shugart, H. H., Goldstein, R. A., O'Neill, R. V. & Mankin, J. B. (1974). TEEM: A Terrestrial Ecosystem Energy Model for forests. *Oecolog. Plannar.* **9**, 231–64.

Sollins, P., Grier, C. C., McCorison, F. M., Cromack, K., Fogel, R., and Fredriksen, R. L. The internal element cycles of an old-growth Douglas-fir stand in western Oregon. *Ecol. Monogr.* (in press).

Sollins, P., Harris, W. F. & Edwards, N. T. (1976). Simulating the physiology of a temperate deciduous forest. In Systems Analysis and Simulation in Ecology, Vol. 4, ed. B. C. Patten, pp. 173–218. Academic Press, New York.

Sollins, P., Reichle, D. E. & Olson, J. S. (1973). Organic matter budget and model for a southern Appalachian *Liriodendron* forest. USAEC Rep. EDFB-IBP-73-2. Oak Ridge National Laboratory, Oak Ridge, Tennessee.

Sollins, P. & Swartzman, G. L. A model of carbon and water flow through a coniferous forest: Structure, assumptions and behavior (in prep).

Stage, A. R. (1973). *Prognosis model for stand development.* USDA Forest Service Research Paper INT-137, Intermountain For. Range Exp. Sta., Ogden, Utah.

Stephens, G. R., Turner, N. C. & DeRoo, H. C. (1972). Some effects of defoliation by gypsy moth (*Porthetria dispar* L.) and elm spanworm (*Ennomos subsignarius* Hbn.) on water balance and growth of deciduous forest trees. *For. Sci.* **18**, 326–30.

Swank, W. T. & Helvey, J. D. (1970). Reduction of streamflow increases following regrowth of clearcut hardwood forests. Symposium on the results of research on representative and experimental basins. *Wellington. Assoc. Int. Sci. Hydrol. Publ.* **96**, 346–60.

Swift, L. W., Swank, W. T., Mankin, J. B., Luxmoore, R. J. & Goldstein, R. A. (1975). Simulation of evapotranspiration and drainage from mature and clear-cut deciduous forests and young pine plantation. *Water Resour. Res.* **11**, 667–73.

Whitmore, F. W. & Zahner, R. (1967). Evidence for a direct effect of water stress on tracheid cell wall metabolism in pine. *For. Sci.* **13**, 397–400.

Wit, C. T. de. (1965). *Photosynthesis of leaf canopies.* Agr. Res. Report 663. Inst. Biol. Chem. Res. on Field Crops and Herbage.

11. Productivity of forest ecosystems studied during the IBP: the woodlands data set*

D. L. DeANGELIS, R. H. GARDNER & H. H. SHUGART

Contents

The IBP Woodlands Synthesis consists of contributions from 116 international forest research sites, all but a few of which were associated with projects committed to the International Biological Programme. The data from these sites have been collected into a Woodlands Data Set, the purpose of which is to facilitate comparisons involving a large number of diverse woodland ecosystems. Analysis based on this data set is done in various chapters of this volume. The projects submitting data to the Woodlands Synthesis are diverse both internationally and in the scope and emphasis of their research. Because of this, the data did not always conform easily to the uniform format in which it is presented here. Repeated communications with members of various projects were often employed before deciding on appropriate values. We have tried to follow faithfully the wishes of the projects in this regard.

Following the IBP Woodlands Workshop at Oak Ridge in 1972, data forms were drawn up. Data were submitted on these forms from 55 sites and appear in the report on the IBP Woodlands Workshop at Göttingen in 1973 (Ulrich, Mayer & Heller, 1974). Subsequently, at a meeting in Jädraas, Sweäden, new data forms were prepared. Additional data from some of the original sites as well as data from 61 new sites were collected on these and several other types of forms. The remaining review and organ-

* Research supported by the Eastern Deciduous Forest Biome, US–IBP, funded by the National Science Foundation under Interagency Agreement AG–199, DEB76–00761 with the US Dept. of Energy Publication No. 1511, Environmental Sciences Division, Oak Ridge National Laboratory. Operated by Union Carbide Corporation for the U.S. Department of Energy. Contribution No. 342 from the Eastern Deciduous Forest Biome, US–IBP.

567

ization of the data have been performed at the Oak Ridge National Laboratory, where the data are stored on computer tapes for easy access and updating.

Content of the data sets

Representatives of almost every kind of forest ecosystem are present in the data set. Included are ecosystems of the following types with the number of each in parentheses (see classification in Chapter 1): mediterranean broad-leaved evergreen (3), tropical broad-leaved evergreen (4), tropical broad-leaved deciduous (4), tropical broad-leaved deciduous plantation (12), temperate broad-leaved evergreen (1), temperate needle-leaved deciduous plantation (1), temperate needle-leaved evergreen (7), temperate needle-leaved evergreen plantation (4), temperate broad-leaved deciduous (42), boreal needle-leaved evergreen (24), and boreal needle-leaved evergreen plantation (5).

The data for each site are divided into two general parts. The first part consists of 'general site description' data, including edaphic characteristics, average meteorological conditions, and basic descriptive and quantitative information on the vegetation. In the second, or 'biomass compartment tabulation' section, the organic constituents of the site are divided into 38 compartments. Tables of the measured amounts and increments in these compartments, as well as the fluxes between them, are presented.

Some useful definitions for interpreting the information in this Chapter are the following:

Compartment. A division of the ecosystem for which we may measure organic matter in terms of dry weight.

Increment (Net annual). The change in the size of a compartment during a year. For example, leaf mass in July 1970 = 812 g^{-2} and leaf mass in July 1971 = 826 g^{-2}. The increment is therefore 14 g^{-2} yr^{-1}. Note that increment in this case does not include leaf litterfall. Similar definitions are used for branch and bole increments, with branch and bole litterfall not included in the increment.

Flux (Net annual). The annual amount of organic matter being transferred (e.g. by litterfall) from one compartment to another.

Productivity (Net annual). The sum of litterfall plus all biomass increments plus consumption not accounted for in productivity.

Overstory. The trees forming the main canopy.

Understory (or Substory). The sume of sub-canopy and shrub layers.

Field Layer. Herbaceous plants, mosses and lichens. Woody plants less than 50 cm in height are included here.

568

Stocking Density. Number of stems per hectare.

Basal Area Increment. The amount of woody tissue added to the stand basal area per year through growth.

Global Radiation. The sum of irradiance on a horizontal surface caused by direct solar radiation and diffuse short-wave radiation due to clouds, dust and molecular scattering from all parts of the sky. (Definition from Van Wijk & Scholte Ubing, 1966.)

Radiation Balance. The difference between short-wave radiation from the sky and the sum of short-wave reflected radiation and long-wave terrestrial radiation.

Standing Dead. Dead plants or parts of plants still standing in the forest.

Frass. Includes insect excreta and may include other unidentifiable material.

Soil Top Organic. Includes recently fallen litter (L layer) as well as decomposing organic layers on top of the mineral soil (F and H layers).

The meteorological and climatological measurements in the 'general site descriptions' in some cases represent values averaged over many years, and in other cases are values only for the years during which biomass measurements were made. In some cases the meteorological measurements were made directly at the site, while in other cases they were made at nearby stations. Lack of precise and generally agreed upon definitions is evident with respect to some quantities. Because of the variety of data forms submitted, with different compartmentalizations of the soil layers, values of the various soil compartments are not always consistent from site to site.

Occasionally discrepancies occur between values which are listed as 'sums' of two or more compartments and the actual values in those compartments. The same holds true for net annual productivity; the value given for this is not always equal to the sum of fluxes and increments (see definitions above). Often, the reason for this is that the researchers have used their own insight to take into consideration other factors such as consumption, which could not be directly measured and, therefore, were not given in the biomass data.

Rigorous comparisons among sites are complicated because of the above data constraints and because for each site there may be additional highly pertinent facts which could not be included in the data summaries. The chapters of this volume treat such comparisons and discuss the interpretations in detail. The reader wishing to pursue similar analyses may find it useful to consult the literature publications relevant to each site, which are cited in the general site descriptions.

While the above caveats are necessary, we hope they will not inhibit the

Table 11.1. *Abbreviations relevant to general site description data*

Abbreviation	Meaning	Units
AS	Age of stand	Years
LAI	Leaf area index	Square meters/square meter
BA	Basal area	Square meters/hectare
BAI	Basal area increment	Square meters/hectare
SH	Stand height	Meters
SD	Stocking density	Stems/hectare
SCA	Standing crop above-ground	Grams/square meter
SCB	Standing crop below ground	Grams/square meter
PA	Productivity above-ground	Grams/square meter/year
PB	Productivity below-ground	Grams/square meter/year
LAT	Latitude	Degrees and minutes
LNG	Longitude	Degrees and minutes
ALT	Altitude	Meters
MAT	Mean annual temperature	Degrees Centigrade
MAP	Mean annual precipitation	Millimeters
MAR	Mean annual radiation (Global)	Calories/square centimeter/year
RBA	Radiation balance	Calories/square centimeter/year
LGS	Length of growing season	Days
TGS	Temperature during growing season	Degrees Centrigrade
PGS	Mean precipitation in growing season	Millimeters
RGS	Mean radiation in growing season (Global)	Calories/square centimeter/year
RBG	Radiation balance in growing season	Calories/square centimeter/year
SPH	Soil pH	
DRZ	Depth of rooting zone	Centimeters

use of the Woodlands Data Set in stimulating query, posing scientific hypotheses, and comparative analyses among the various forest ecosystems.

Definition of terms

Abreviations are used so that the data can be presented in a compact form. These are listed in Tables 11.1 and 11.2.

Several symbols are used in the biomass compartments tabulation:

1. The number 0.0 means that a value was measured and found to be insignificantly small.
2. The symbol IN. OV. is used to indicate that a particular understory compartment is lumped with the corresponding overstory compartment.
3. The symbol ** indicates a lumping of compartments. When this symbol appears it means that the particular compartment in question is lumped together with the next compartment below it in the column which has a numerical value attached.

Table 11.2. *Abbreviations relevant to compartment biomass data*

Abbreviation	Meaning	Units
OL	Overstory leaves	Grams/square meter
OFF	Overstory fruits, flowers	Grams/square meter
OBR	Overstory branches–sum	Grams/square meter
OBRB	Overstory branches–bark	Grams/square meter
OBRW	Overstory branches–wood	Grams/square meter
OBO	Overstory bole sum	Grams/square meter
OBOB	Overstory bole bark	Grams/square meter
OBOW	Overstory bole wood	Grams/square meter
OSTD	Overstory standing dead	Grams/square meter
OSUM	Sum overstory	Grams/square meter
UL	Understory leaves	Grams/square meter
UFF	Understory fruits, flowers	Grams/square meter
UBR	Understory branches–sum	Grams/square meter
UBRB	Understory branches–bark	Grams/square meter
UBRW	Understory branches–wood	Grams/square meter
UBO	Understory bole–sum	Grams/square meter
UBOB	Understory bole–bark	Grams/square meter
UBOW	Understory bole–wood	Grams/square meter
USTD	Understory standing dead	Grams/square meter
USUM	Sum understory	Grams/square meter
HERB	Herbaceous field	Grams/square meter
EPIP	Epiphytes total	Grams/square meter
PSUM	Total plants	Grams/square meter
RL	Living roots sum	Grams/square meter
RLL	Living roots >5 mm	Grams/square meter
RLS	Living roots <5 mm	Grams/square meter
RD	Dead roots	Grams/square meter
STO	Soil top organic	Grams/square meter
RZ	Rooting zone	Grams/square meter
IRZ	Intensive rooted	Grams/square meter
ERZ	Extensive rooted	Grams/square meter
SUBS	Subsoil	Grams/square meter
HETR	Heterotrophs sum	Grams/square meter
AGC	Above-ground consumers	Grams/square meter
DCFA	Decomposer fauna	Grams/square meter
DCFL	Decomposer flora	Grams/square meter
CS	Consumption total	Grams/square meter
CSF	Consumption foliage total	Grams/square meter
CSO	Consumption overstory	Grams/square meter
CSOW	Consumption wood–overstory	Grams/square meter/year
CSUW	Consumption wood–understory	Grams/square meter/year
CSR	Consumption roots–total	Grams/square meter/year
CSU	Consumption–understory	Grams/square meter/year
CSHB	Consumption–herbaceous	Grams/square meter/year
CSW	Consumption wood total	Grams/square meter/year
LF	Litterfall total	Grams/square meter/year
LFL	Litterfall leaf	Grams/square meter/year
LFFF	Litterfall flower, fruit	Grams/square meter/year
LFBR	Litterfall branch	Grams/square meter/year
LFBO	Litterfall bole	Grams/square meter/year
LFFR	Litterfall frass	Grams/square meter/year

571

Table 11.2 (*cont.*)

Abbreviation	Meaning	Units
LFEP	Litterfall epiphytes	Grams/square meter/year
ATIN	Atmospheric input	Grams/square meter/year
PRE	Precipitation	Grams/square meter/year
DRP	Dry particulates (dust)	Grams/square meter/year
GAF	Gaseous fixation	Grams/square meter/year
LEAC	Leaching	Grams/square meter/year
LWAS	Leaf wash	Grams/square meter/year
STMF	Stem flow	Grams/square meter/year
LSOL	Leaching soil layers	Grams/square meter/year
OUTP	Output	Grams/square meter/year
OPWI	Output wind erosion	Grams/square meter/year
OPWA	Output water erosion	Grams/square meter/year
OPPR	Output percolation	Grams/square meter/year

Acknowledgments

The task of organizing the Woodlands Data Set could not have been accomplished without the wholehearted co-operation of the contributing projects. In addition, we have relied on the special assistance of many people along the way. Professor Bernhard Ulrich of the Institute for Soil Science of Forest Fertilization hosted the Göttingen Workshop and did his utmost to facilitate the initial phase of this work. Professor Heinz Ellenberg, Dr Hans Heller, Dr Robert Mayer and Professor Bernhard Ulrich drew up the initial data forms. Dr Peter Attiwill, Dr Folke Andersson and others drew up subsequent data forms. Dr. Robert Mayer and Dr Partap Khanna, as well as the technical staff of the Institute, contributed much effort to reviewing the data. Dr Folke Andersson served as host to the working meeting in Jädraas. Dr Alan Gordon of the Forestry Biology Laboratory in Saulte Ste Marie, Canada and Prof. Helmut Lieth of the Department of Botany of the University of North Carolina have taken the time to provide useful comments. At Oak Ridge, we have been ably assisted in the computer programming aspects by David Taylor, J. A. Watts, and M. L. Tharp, Dr W. C. Johnson has been our principal contact with the Russian projects and is personally responsible for the inclusion of many sites from the USSR. Dr David E. Reichle encouraged and facilitated the data organization and Dr Robert L. Burgess, Dr Robert ·V. O'Neill, Dr Jerry S. Olson, Dr W, Frank Harris, and Dr Gray S. Henderson reviewed the manuscript and contributed useful suggestions.

References

Ulrich, B., Mayer, R. & Heller, H. (1974). *Göttinger Bodenkundliche Berichte* 30, University of Goettingen, West Germany.
Van Wijk, W. R. & Scholte Ubing, D. W. (1966). Physics of Plant Environment, ed. W. R. Van Wijk, Northern-Holland Publishing, Amsterdam.

Listing of the Data Sets*

(in alphabetical order by country)

MT DISAPPOINTMENT, AUSTRALIA

INVESTIGATOR: P. Attiwill   ADDRESS: Botany School, University of Melbourne, Parkville, Victoria 3052, Australia
SELECTED CITATIONS: Attiwill, P. (1962). *Forest Science*, **8**, 132–41; Attiwill, P. (1966). *Ecology*, **47**, 795–804; Attiwill, P. (1968). *Ecology*, **49**, 142–145 (1968); Attiwill, P. (1966). *Plant and Soil*, **24**, 390–406; Attiwill, P. (1972). *Aust. Forest-Tree Nutrition Conf.* 39–46, 125–34.
YEAR(S): 1955–72   FOREST TYPE: Broad-leaved Evergreen, Eucalyptus
SOIL TYPE: Krasnozem (Red Friable Porous Earth)   GEOLOGY: Metamorphosed Silurian Mudstones, Sandstones and Shales   SOIL DRAINAGE: Excellent
PRINCIPAL PLANT TYPES: *Eucalyptus obliqua*

AS	51	LAI	4.1	BA	54.1–63.3	BAI	0.55
SH	21.7–29.1	SD	704–976	SCA	31236	SCB	7534
PA	852	PB	148	LAT	S 37° 25′	LNG	E 145° 10′
ALT	545	MAT	11.15	MAP	982	LGS	365
TGS	11.15	PGS	982	SPH	5.21–5.88	DRZ	100

	Amount	Increment	Flux	
OL	688.00	12.40	LFL	189.00
OBR	2200.00	60.30	LFBR	101.00
OBO	27250.00	383.20	LFBO	47.00
OBOB	4401.00	33.60	LFEP	19.00
OBOW	22850.00	349.60		
OSTD	1100.00	40.30		
OSUM	31240.00	496.20		
USUM	374.00			
RL	7534.00			
STO	1824.00			

---

*See appendix to Chapter 6 for data sets on elemental concentrations, pools and fluxes for many of the forest ecosystems at these IBP sites.

VIRELLES, BELGIUM

INVESTIGATORS: P. Duvigneaud, A. Galoux   ADDRESS: Université De Bruxelles, 28 Av. Heger, 1050, Bruxelles, Belgium

OTHER INVESTIGATORS: A. Galoux, S. Denaeyer, M. Tanghe, P. Bourdeau

SELECTED CITATIONS: Galoux, A., Schnock, G. & Grulois, J. (1967). *Trav. Sta. Rech. Groenendael, Ser. A*, **12**; Duvigneaud, P. (1968). *Bull. Soc. Bot. Belg.* **101**, 129–39; Duvigneaud, P., Denaeyer-De Smet, S., Ambros, P. & Timperman, J. (1971). *Mem. Inst. Sci. Nat. Belg.*, No. 164, 101

YEAR(S): 1964–9   FOREST TYPE: Mixed Oak Forest

SOIL TYPE: Calcareous brown soil   GEOLOGY: Devon chalk   SOIL DRAINAGE: 454.2 mm

PRINCIPAL PLANT TYPES: Overstory: *Quercus robur, Carpinus betulus, Fagus sylvatica, Acer campestris, Prunus avium.* Substory: *Carpinus betulus, Crataegus sp., Euonymus europaeus, Corylus avellana.* Field Layer: *Hedera helix, Mercurialis perennis, Lamium galeobdolon.* Other: *Scilla bifolia, Narcissus pseudonarcissus, Cardamine pratensis*

AS	80	LAI	6.8	BA	21.2	BAI	0.6
SH	16–17	SD	1250	SCA	12100	SCB	3500
PA	1224	PB	233	LAT	N 50° 4′	LNG	E 4° 21′
ALT	245	MAR	93131	RBA	41149.5	MAT	8.5
MAP	951.6	LGS	155	TGS	13.8	PGS	450
RGS	57198	RBG	27875	DRZ	30–50		

MANAUS, BRAZIL

INVESTIGATOR: H. Klinge ADDRESS: Max Planck Institute fuer Limnologie, Abteilung Tropenoekologie, D 232 Ploen, Postfach 165, Federal Republic of Germany
OTHER INVESTIGATORS: W. A. Rodrigues
SELECTED CITATIONS: Klinge, H. & Rodriguez, W. A. (1968). *Amazonia*, **1**, 287–302, 303–10; Fittkau, E. J. & Klinge, H. (1973). *Biotropica*, **5**, 2–14; Klinge, H., Rodriguez, W. A., Brunig, E. & Fittakau, E. J. (1975). In *Tropical Ecological System*, ed. Golley & Medina, Springer Verlag
YEAR(S): 1970 FOREST TYPE: Tropical rain forest
SOIL TYPE: Pale yellow latosol GEOLOGY: Plio-pleistocene SOIL DRAINAGE: Good
PRINCIPAL PLANT TYPES: Leguminoseae, Sapotaceae, Lauraceae, Rosaceae, Rubiaceae, Burseraceae, Palmae, Annonaceae, Lecythidaceae, Moraceae, Violaceae

AS	Mature	BA	30.7	SH	38.1	SD	93780
SCA	40600	SCB	6720	LAT	4°	LNG	60°
ALT	90	MAT	27.2	MAP	1771	LGS	365
TGS	27.2	PGS	1771	SPH	3.1–4.4	DRZ	110

	Amount		*Flux*	
OL	930.00		LFL	560.00
OBR	9770.00		LFFF	35.00
OBO	29890.00		LFBR	135.00
OSTD	760.00			
OSUM	41350.00			
UL	IN. OV.			
UBR	IN. OV.			
UBO	IN. OV.			
HERB	IN. OV.			
PSUM	41350.00			
RLL	2740.00			
RLS	3980.00			
STO	11100.00			
RZ	36000.00			
SUBS	1259000.00			

ONTARIO SITE REGION 5, SITE 1 (DRY), CANADA

INVESTIGATOR: Alan G. Gordon ADDRESS: Ontario Ministry of Natural Resources, Forest Research Branch, P. O. Box 490, Sault Ste Marie, Ontario, Canada
SELECTED CITATIONS : Gordon, A. G. (1976). *Can. J. Bot.* **54**, 781–813. Gordon, A. G., (1975). In '*Energy Flow – Its Biological Dimensions*' Royal Soc. Canada, Ottawa.
YEAR(S): 1961–75 FOREST TYPE: Spruce
SOIL TYPE: Lithic Ferro-Humic Podzol GEOLOGY: Precambrian Granite Gneiss
SOIL DRAINAGE: Well drained
PRINCIPAL PLANT TYPES: Overstory: *Picea rubens, P. mariana, P. glauca, Pinus strobus, Abies balsamea, Tsuga canadensis, Thuja occidentalis, Betula papyrifera, Acer rubrum, Fagus grandifolia.* Substory: *Picea rubens, P. glauca, Tsuga canadensis, Abies balsamea, Pinus strobus, Acer rubrum, Betula papyrifera, Amelanchier arborea.* Field Layer: *Vaccinium angustifolium, Pteridium aquilinum, Abies balsamea, Tsuga canadensis, Cladonia rangiferina.*

AS	84	LAI	11.6	BA	32.31	BAI	0.31
SH	14.9	SD	3311	SCA	12150.7	SCB	2799.6
PA	451.18	LAT	N 45° 14'	LNG	W 78° 23'	ALT	465
MAT	4.0	MAP	1242.5	LGS	167	TGS	14.0
PGS	468.6	SPH	4.3	DRZ	13.0		

	Amount	*Increment*	*Flux*	
OL	1585.00	19.11	LFL	120.25
OFF	21.04	0.27	LFFF	24.84
OBR	2105.00	29.71	LFBR	22.96
OBO	8431.00	212.10	LFBO	12.04
OBOB	828.90	20.32	LFFR	9.88
OBOW	7602.00	191.80		
OSTD	583.70	11.45		
UL	0.00	0.00		
UFF	**	**		
UBR	**	**		
UBO	0.03	0.00		
HERB	8.50			
RLL	2705.00	69.60		
RLS	94.63	2.43		
STO	**			
RZ	3095.60			

ONTARIO SITE REGION 5, SITE 2 (FRESH), CANADA

INVESTIGATOR: Alan G. Gordon ADDRESS: Ontario Ministry of Natural Resources, Forest Research Branch, P. O. Box 409, Sault Ste Marie, Ontario, Canada
SELECTED CITATIONS: Gordon, A. G. (1976). *Can J. Bot.* **54**, 781–813. Gordon, A. G., (1975). In '*Energy Flow-Its Biological Dimensions*'. Royal Soc. Canada, Ottawa.
YEAR(S): 1961–75 FOREST TYPE: Spruce
SOIL TYPE: Orthic Ferro-Humic Podzol GEOLOGY: Precambrian Granite Gneiss
SOIL DRAINAGE: Well drained
PRINCIPAL PLANT TYPES: Overstory: *Picea rubens, Abies balsamea, Tsuga canadensis, Acer rubrum.* Substory: *Picea rubens, Abies balsamea, Thuja occidentalis, Fagus grandifolia, Acer rubrum.* Field Layer: *Vaccinium brittonii, V. myrtilloides, Gaultheria hispidula, Coptis groenlandica, Picea rubens, Abies balsamea, Tsuga canadensis*

AS	246	LAI	15.0	BA	46.75	BAI	0.25
SH	18.3	SD	3879	SCA	26394.5	SCB	6196.0
PA	870.4	LAT	N 45° 32'	LNG	W 78° 49'	ALT	495
MAT	4.0	MAP	1242.5	LGS	167	TGS	14.0
PGS	468.6	SPH	4.1	DRZ	30		

	Amount	Increment	Flux	
OL	2048.00	35.63	LFL	219.10
OFF	14.57	0.29	LFFF	38.24
OBR	3640.00	70.73	LFBR	68.12
OBO	20690.00	409.10	LFBO	10.45
OBOB	1763.00	40.91	LFFR	18.75
OBOW	18920.00	368.10		
OSTD	853.80	16.59		
UL	0.01	0.00		
UFF	**	**		
UBR	**	**		
UBO	0.03	0.00		
HERB	5.52			
RLL	5987.00	102.50		
RLS	209.40	3.59		
STO	**			
RZ	18729.00			

Dynamic properties of forest ecosystems

ONTARIO SITE REGION 5, SITE 3 (MOIST), CANADA

INVESTIGATOR: Alan G. Gordon   ADDRESS: Ontario Ministry of Natural Resources, Forest Research Branch, P. O. Box 490, Sault Ste Marie, Ontario, Canada

SELECTED CITATIONS: Gordon, A. G., *Can. J. Bot.* (1976). **54**, 781–813. Gordon, A. G., (1975). In *'Energy Flow – Its Biological Dimensions'* Royal Soc. Canada, Ottawa.

YEAR(S): 1961–75   FOREST TYPE: Spruce

SOIL TYPE: Orthic Ferro-Humic Podzol   GEOLOGY: Precambrian Granite Gneiss

SOIL DRAINAGE: Well drained

PRINCIPAL PLANT TYPES: Overstory: *Picea rubens, Tsuga canadensis, Abies balsamea, Thuja occidentalis, Betula lutea, Fagus grandifolia.* Substory: *Acer pennsylvanicum, A. spicatum, A. saccharum, A. rubrum, Fagus grandifolia, Corylus cornuta.* Field Layer: *Oxalis montana, Dryopteris intermedia, Aralia nudicaulis, Lycopodium lucidulum, Abies balsamea, Picea rubens, Acer pennsylvanicum*

AS	212	LAI	16.9	BA	59.44	BAI	0.10
SH	25.6	SD	3065	SCA	46222.8	SCB	10456.03
PA	989.53	LAT	N 45° 17'	LNG	W 78° 17'	ALT	480
MAT	4.0	MAP	1242.5	LGS	167	TGS	14.0
PGS	468.6	SPH	4.5	DRZ	25		

	Amount	Increment	Flux	
OL	2122.00	34.56	LFL	225.55
OFF	11.65	0.20	LFFF	10.51
OBR	6449.00	100.70	LFBR	27.13
OBO	37640.00	578.30	LFBO	5.71
OBOB	3163.00	52.48	LFFR	6.88
OBOW	34470.00	525.80		
OSTD	1817.00	28.74		
UL	0.03	0.00		
UFF	**	**		
UBR	**	**		
UBO	0.20	0.01		
HERB	3.03			
RLL	10100.00	150.40		
RLS	353.40	5.26		
STO	**			
RZ	16326.00			

ONTARIO SIZE REGION 5, SITE 4 (WET), CANADA

INVESTIGATOR: Alan G. Gordon   ADDRESS: Ontario Ministry of Natural Resources, Forest Research Branch, P. O. Box 490, Sault Ste Marie, Ontario, Canada
SELECTED CITATIONS: Gordon, A. G. (1976). *Can. J. Bot.* **54**, 781–813. Gordon, A. G., (1975). In '*Energy Flow – Its Biological Dimensions*' Royal Soc. Canada, Ottawa.
YEAR(S): 1961–75   FOREST TYPE: Spruce
SOIL TYPE: Hydric Humisol   GEOLOGY: Precambrian Granite Gneiss   SOIL DRAINAGE: Poorly drained
PRINCIPAL PLANT TYPES: Overstory: *Picea rubens, P. mariana.* Substory: *Picea rubens, Abies balsamea.* Field Layer: *Sphagnum capillaceum, Sphagnum palustre, Carex trisperma, Abies balsamea, Picea rubens*

AS	130	LAI	7.136	BA	45.66	BAI	0.78
SH	20.7	SD	9058	SCA	16923.9	SCB	4748.03
PA	383.21	LAT	N 45° 17'	LNG	W 78° 16'	ALT	503
MAT	4.0	MAP	1242.5	LGS	167	TGS	14.0
PGS	468.6	SPH	3.74	DRZ	18		

	Amount	Increment	Flux	
OL	1001.64	26.62	LFL	142.28
OFF	15.79	0.10	LFFF	5.46
OBR	2046.09	34.70	LFBR	39.54
OBO	13844.25	114.24	LFBO	14.51
OBOB	1277.72	12.07	LFFR	5.76
OBOW	12566.53	102.17		
OSTD	834.90	10.73		
UL	0.01	0.00		
UFF	**	**		
UBR	**	**		
UBO	0.08	0.00		
HERB	16.05			
RLL	4569.97	62.31		
RLS	178.06	2.42		
STO	**			
RZ	12007.8			

Dynamic properties of forest ecosystems

BAB, CZECHOSLOVAKIA

INVESTIGATOR: V. Biskupsky ADDRESS: Institute of Botany, Czechoslovakian Academy of Sciences, Dukelska 145, 37982 Trevon, Czechoslovakia
OTHER INVESTIGATORS: P. Pelisek, E. Bublinec, J. Kristek, J. Bernat, M. Penka, F. Smolen, R. Intribus, J. P. Ondok, J. Kolek
SELECTED CITATIONS: Research Project BAB, IBP Progress Report 1 (1970). Research Project BAB, IBP Progress Report 2 (1975).
YEAR(S): 1972 FOREST TYPE: Fageto–Quercetum, Primulae Verris Carpinetum
SOIL TYPE: Parabraunerde GEOLOGY: Loess SOIL DRAINAGE: Very good
PRINCIPAL PLANT TYPES: *Carpinus betulus, Acer campestris, Quercus cerris, Quercus petraea Cornus mas, Crataegus oxyacantha, Ulmus carpinifolia, Rosa canina*

AS	50–70	LAI	5.19	BA	25.59	SH	19
SD	733	SCA	16096	SCB	7605.06	PA	1070
LAT	N 48° 11′	LNG	E 17° 54′	ALT	208–210	MAT	10.0
MAP	569.8	LGS	245	TGS	13.68	PGS	450.4
SPH	5.8–6.1	DRZ	150				

	Amount	Increment
OL	342.40	
OFF	169.00	
OBR	4284.00	401.20
OBRB	930.00	71.54
OBRW	3354.00	329.70
OBO	11470.00	330.00
OBOB	1434.00	41.27
OBOW	10040.00	
OSUM	16260.00	1243.00
UL	11.89	
UBO	35.12	1.92
UBOB	9.95	0.54
UBOW	25.17	1.38
USUM	82.13	13.81
HERB	35.00	
PSUM	16380.00	
RL	7505.00	
RLL	5638.00	
RLS	1919.00	
RD	453.00	

HESTEHAVEN, DENMARK

INVESTIGATOR: H. M. Thamdrup   ADDRESS: Naturhistorisk Museum, Universitetsparken, 8000 Aarhus C, Denmark,
OTHER INVESTIGATORS: E. Holm, T. F. Jensen, H. K. Hughes, V. Jensen, M. S. Luxton B. Overgaard Nielsen, H. Petersen, T. Secher Jensen, G. W. Yeates and others
SELECTED CITATIONS: Holm E. & Jensen, V. (1972). *Oikos*, **23**, 248–60; Overgaard Nielsen, B. (1974). *Vid. Medd.* **137**, 95–124; Petersen H. (1971). *Ann. Zool. Ecol. An.* (*Hors Serie*), 235–54; Yeates, G. W. (1973). *Oikos*, **24**, 179–85.
YEAR(S): 1970   FOREST TYPE: Temperate Deciduous–Beech
SOIL TYPE: Parabraunerde (Grey Brown Podzolic) mollic hapludalf   GEOLOGY: Wurm Moraine   SOIL DRAINAGE: Moderately well drained
PRINCIPAL PLANT TYPES: *Fagus sylvatica, Anemone nemorosa, Melica uniflora, Asperula odorata, Carex sylvatica*

AS	85–90	LAI	5.0	BA	28.46	BAI	0.48
SH	28.6	SD	370	SCA	22126	SCB	4322
PA	1499	PB	375	LAT	N 56° 18'	LNG	E 10° 29'
ALT	11–28	MAR	88362	MAT	7.1	MAP	660
LGS	145–152	TGS	14.1	PGS	277	RGS	65520
SPH	4.5–5.8	DRZ	60				

	Amount	Increment	Flux	
OL	210.00	0.0	CS	5.50
OBR	4320.00	445.00	CSF	5.50
OBO	17040.00	517.00	LFL	269.00
OBOB	740.00	26.00	LFFF	23.00
OBOW	16300.00	491.00	LFBR	90.00
OSUM	21570.00	936.00		
USUM	540.00	79.00		
HERB	16.00	0.0		
PSUM	22130.00	1015.00		
RL	4322.00	0.0		
RLL	3000.00			
RLS	1322.00			
STO	750.00			
RZ	14140.00			

Dynamic properties of forest ecosystems

OULU, FINLAND

INVESTIGATOR: P. Havas ADDRESS: Botanical Institute, University of Oulu, Finland
OTHER INVESTIGATORS: S. Sulkava, F. Virano
SELECTED CITATIONS: Havas, P. (1973). In *Modeling Forest Ecosystems*, EDFB–IBP 73–7, Oak
Ridge, USA; Havas, P. (1972). *Aquilo, Ser. Botanica*, **11**.
YEAR(S): 1967–71 FOREST TYPE: Hylocomium–Myrtillus type
SOIL TYPE: Podsol GEOLOGY: Quartzite (Dolomite) SOIL DRAINAGE: Poor (soil almost wet)
PRINCIPAL PLANT TYPES: *Picea excelsa, Vaccinium myrtillus, Vaccinium vitis-idaea, Hylocomium
splendens, Pleurozium schreberi*

AS	260	LAI	4.9	BA	22.2	BAI	0.0
SH	16.2	SD	550	SCA	10194.3	SCB	3755.7
PA	420.8	PB	20.0	LAT	66° 22'	LNG	29°
ALT	270	MAR	75139	RBA	15630	MAT	0.0
MAP	500	LGS	120	TGS	11.1	PGS	250
RGS	46800	RBG	18300	SPH	4.7	DRZ	15

	Amount	Photosynthesis	Flux	
OL	661.80		LF	250.90
OFF	20.00		LFL	124.70
OBR	1712.00		LFFF	20.00
OBO	6732.00		LFBR	**
OBOB	1074.00		LFBO	10.00
OBOW	5659.00		LFEP	3.30
OSTD	417.70			
OSUM	9543.00			
UL	3.30			
UBR	3.30			
UBO	6.80			
USUM	13.40			
HERB	550.10	450.00		
EPIP	86.70			
PSUM	10190.00			
RL	3753.00	450.00		
RLL	2365.00			
RLS	1316.00			
STO	6760.00			
AGC	0.60			
DCFA	3.00			

FONTAINEBLEAU, FRANCE

INVESTIGATOR: G. Lemee ADDRESS: Laboratoire d'Ecologie Végétale, Université de Paris-Sud, 91506-Orsay, Paris, France
OTHER INVESTIGATORS: N. Bichaut, J. Bouchon, A. Faille, A. M. Robin, A. Schmitt
SELECTED CITATIONS: Lemee, G. (1966). *Bull. Soc. Bot. Fr.* **113**, 305–23; Lemee, G. & Bichaut, N. (1971). *Oecol. Plant*, **6**, 133–49; Lemee, G. & Bichaut, N. (1973). *Oecol. Plant*, **8**, 153–74; Lemee, G. (1974). *Oecol. Plant*. **9**, 187–200.
YEAR(S): 1968–74 FOREST TYPE: Temperate Deciduous, Beech
SOIL TYPE: Sol Lessive GEOLOGY: Sable Eolien Quaternaire Sur Calcaire Oligocene
SOIL DRAINAGE: Good
PRINCIPAL PLANT TYPES: *Fagus sylvatica, Brachypodium sylvaticum, Melica uniflora, Festuca heterophylla, Ruscus aculeatus*

AS	Mature	LAI	6.9	BA	27.5	BAI	0.41
SH	34	SD	190	SCA	22200	PA	440
LAT	N 48° 25′	LNG	E 02° 38′	ALT	135	MAR	99780
MAT	10.15	MAP	674	LGS	160	TGS	16.1
PGS	270	RGS	63660	SPH	3.5–5.0	DRZ	150

	Amount	Increment	Flux	
OL	345.00		LFL	345.00
OFF	48.00		LFFF	48.00
OBR	5800.00	**	LFBR	106.00
OBO	23200.00	476.00		
HERB	120.00			
RL	5000.00	60.00		
STO	1400.00			
RZ	14600.00			

583

MADELEINE, FRANCE

INVESTIGATOR: P. Lossaint   ADDRESS: Centre D'Etudes Phytosociologiques et Ecologiques, B.P. 1018, Route de Mende, Montpellier, France
OTHER INVESTIGATORS: M. Rapp
SELECTED CITATIONS: Rapp, M. (1971). In *Ecologie du Sol, Recherche Cooperative sur Programme N. 40, C.N.R.S., Paris*, **2**, 19–188; Lossaint, P. (1967). *Oecol. Plant.* **2**, 341–66; Rapp, M. (1969). *Oecol. Plant.* **4**, 377–410.
SOIL TYPE: Rendzina   GEOLOGY: Hard limestone
PRINCIPAL PLANT TYPES: *Quercus ilex*

BA	34.11	SH	15	SD	527	LAT	N 48° 36'
LNG	1° 65'	ALT	10	MAT	14.1	MAP	754
SPH	7.8	DRZ	100				

ROUQUET, FRANCE

INVESTIGATOR: P. Lossaint ADDRESS: Centre d'Etudes Phytosociologiques et Ecologiques,
B. P. 1018, Route de Mende, Montpellier, France
OTHER INVESTIGATORS: M. Rapp
SELECTED CITATIONS: Rapp, M. (1971). In *Ecologie du Sol, Recherche Cooperative sur Programme N. 40, C.N.R.S., Paris*, **2**, 19–188; Lossaint, P. (1967). *Oecol. Plant.* **2**, 341–66; Rapp, M. (1969), *Oecol. Plant.* **4**, 377–410
YEAR(S): 1966–70 FOREST TYPE: Mediterranean Evergreen Oak Forest
SOIL TYPE: Brunified Mediterranean Red Soil GEOLOGY: Portlandian–Kimmeridgian Limestone SOIL DRAINAGE: Good
PRINCIPAL PLANT TYPES: *Quercus ilex* (100 *per cent*), *Viburnum tinus, Arbutus unedo, Phillyria media, Juniperus oxycedrus*

AS	150	LAI	4.5	BA	31.2	SH	11
SD	1440	SCA	26200.	SCB	5000	PA	650
LAT	N 43° 42′	LNG	E 3° 46′	ALT	180	MAR	124000
MAT	12.4	MAP	987	LGS	365	TGS	13.4
PGS	987	RGS	124000	SPH	7.5–7.8	DRZ	50–1000

	Amount	Increment		Flux
OL	700.00		LF	384.20
OFF	59.60		LFL	244.60
OBR	2700.00	110.00	LFFF	59.60
OBO	23500.00	150.00	LFBR	80.00
OSTD	0.00			
OSUM	26900.00			
UL	4.90			
UBRB	**			
UBRW	29.10			
UBO	0.00			
USTD	0.00			
USUM	34.00			
PSUM	26930.00			
RL	450.00			
RLL	450.00			
STO	1140.00			

Dynamic properties of forest ecosystems

SIKFOKUT, HUNGARY

INVESTIGATOR: P. Jakucs ADDRESS: Botanical Department, L. Kossuth University, H–4010
Debrecen, Hungary
OTHER INVESTIGATORS: Z. Varga, K. Viragh, Cs. Aradi, J. A. Toth, M. Szabo, L. B.-Papp, M.
Papp, L. Hargitai, M. Kovacs, M. Kecskes, T. Simon, L. Nagy, I. Karasz and others
SELECTED CITATIONS: Jakucs P. (1974). *MTA Biol. Oszt. Koezl.* **16**, 11–25, (In Hungarian);
Jakucs P. & Papp, M. (1974). *Acta Bot Acad. Sci. Hung.* **20**, 295–308; B-Papp, L. (1974). *Acta
Bot Acad. Sci. Hung.* **20**, 333–9.
YEAR(S): 1972–3 FOREST TYPE: Oak-Wood/Quercetum Petraeae-Cerris
SOIL TYPE: Brown forest soil GEOLOGY: Miocene Clay–Gravel SOIL DRAINAGE: Good
PRINCIPAL PLANT TYPES: Overstory: *Quercus petraea, Q. cerris.* Substory: *Cornus mas, C.
sanguinea, Acer campestre, A. tataricum, Crataegus monogyna, Euonymus verrucosus, Ligustrum.*
Field Layer: *Poa nemoralis, Carex montana, C. michelii, Melica uniflora.* Other:
*Chrysanthemum corymbosum, Lathyrus niger, L. vernus, Fragaria vesca, Dactylis polygama,
Galium schulthesii*

AS	65–68	LAI	8.14	BA	15.11	SH	17.4
SD	816	SCA	22191.0	SCB	3559.98	PA	715.31
LAT	N 47° 54′	LNG	E 20° 28′	ALT	250–280	MAR	89391
MAT	9.9	MAP	582	LGS	180–200	TGS	17
PGS	316	RGS	70235	SPH	4.04–6.16	DRZ	40–100

	Amount	Increment	Flux	
OL	337.40		LFL	371.20
OBR	5907.00	**	LFBR	45.78
OBO	14040.00	300.00		
UL	33.88			
UFF	**			
UBR	**	30.00		
UBO	379.20			
HERB	14.08			
RL	3560.00			

CHAKIA FOREST, VARANASI, INDIA

INVESTIGATORS: V. K. Sharma   ADDRESS: Department of Botany, Banaras Hindu University, Varanasi, Uttar Pradesh, India
YEAR(S): 1971
FOREST TYPE: Tropical Dry Deciduous Forest
SOIL TYPE: Metamorphic sandstone and Dhaural quartz of Vindhyan Range   GEOLOGY: Two facies of deposits: Upper = estaurine and lower = marine and calcareous   SOIL DRAINAGE: Good
PRINCIPAL PLANT TYPES: Overstory: *Shorea robusta, Buchanania lanzan.* Substory: *Zizyphus xylophyrus.* Field Layer: *Cassia tora (only in rainy season)*

AS	38	BA	11.36	SH	3.0–12.7	SD	729
SCA	2820	SCB	670	PA	301	LAT	N 25°
LNG	E 83° 2′	ALT	300	SPH	6.5–6.8		

	Amount		Flux	
OL	238.00		LFL	151.00
OBR	581.00		LFBR	58.00
OBO	2045.00			
RL	669.00			

CHAKIA FOREST, VARANASI, INDIA

INVESTIGATOR: K. P. Singh   ADDRESS: Department of Botany, Banaras Hindu University, Varanasi, Uttar Pradesh, India
YEAR(S): 1974   FOREST TYPE: Tropical dry deciduous forest
SOIL TYPE: Vindhyan System   GEOLOGY: Two facies of deposits: Upper-estuarine and lower-marine and calcareous   SOIL DRAINAGE: Land is cut by several water courses
PRINCIPAL PLANT TYPES: Overstory: *Shorea robusta, Madhuca indica, Buchanania lanzan, Diospyros melanoxylon, Terminalia tomentosa, Semicappus anacardium.* Substory: *Carrisa opaca, Zizyphus glaberima.* Field Layer: *Cassia tora*

AS	9.7–120.2	BA	12.8	SH	2.0–15.0	SD	1019
SCA	4628	SCB	739	PA	537.4	LAT	N 25°
LNG	E 83°	ALT	350	MAT	30.1	MAP	741.6
SPH	5.7–6.1						

	Amount	Increment
OL	363.40	
OBR	1156.00	39.98
OBO	2914.00	101.10
RL	954.60	33.03
STO	787.00	

CHAKIA, INDIA

INVESTIGATOR: D. Bandhu   ADDRESS: Department of Botany, Swami Shraddhanand College, University of Delhi, Alipur, Delhi–36, India

OTHER INVESTIGATORS: R. Misra, K. C. Misra, S. S. Ramam, R. S. Ambasht, H. R. Sant, K. P. Singh, B. Gopal, Q. Foruqi, A. K. Singh, R. P. Singh, P. C. Misra, D. Satyanarayana, V. K. Sharma, P. K. Agarwal

SELECTED CITATIONS: *Progress report of The Indian IBP–PT* (1967–71); Bandhu, D. (1973). In *Modeling Forest Ecosystems*, EDFB–IBP 73–7, Oak Ridge, USA.

YEAR(S): 1971   FOREST TYPE: Tropical-Dry, Deciduous

SOIL TYPE: Red, Sandy to Clay Loam   GEOLOGY: Red sandstone and Dhaurau Quartzite

PRINCIPAL PLANT TYPES: *Shorea robusta, Buchanania lanzan, Anogeissus latifolia, Terminalia tomentosa*

AS	60	LAI	6	BA	30.62	SH	5–18
SD	664	SCA	20540.6	SCB	3427.3	PA	793.8
PB	136.4	LAT	N 25° 20′	LNG	E 83°		
ALT	350	MAP	844	LGS	270	SPH	6.65–6.8
DRZ	Up to 150						

	Amount	*Increment*	*Flux*	
OL	580.90		CSF	0.00
OFF	5.38		CSO	0.00
OBR	5429.00	233.20	CSOW	0.00
OBO	13390.00	576.30	LF	678.10
OSUM	19410.00		LFL	566.00
UL	9.40		LFFF	**
UBR	83.69	4.80	LFBR	**
UBO	105.60	5.70	LFBO	112.10
USUM	198.70			
PSUM	19610.00			
RL	326.80	29.20		
RLL		**		
IRZ	7877.00			
ERZ	5075.00			

SAL PLANTATION (5 YR OLD), GORAKHPUR FOREST DIVISION, INDIA

INVESTIGATOR: Q. Foruqi   ADDRESS: Department of Botany, Banaras Hindu University, Varanasi, Uttar Pradesh, India
YEAR(S): 1972   FOREST TYPE: Tropical Deciduous Plantations
PRINCIPAL PLANT TYPES: *Shorea robusta*

AS	5	LAI	2.1	BA	7.62	SH	4.66
SD	3150	SCA	1550	SCB	440	PA	310
PB	88	LAT	N 27°	LNG	E 83° 53′	ALT	81

	Amount
OL	250.00
OBR	140.00
OBO	1160.00
RL	440.00
STO	260.00

SAL PLANTATION (8 YR OLD), GORAKHPUR FOREST DIVISION, INDIA

INVESTIGATOR: Q. Foruqi   ADDRESS: Department of Botany, Banaras Hindu University, Varanasi, Uttar Pradesh, India
YEAR(S): 1972   FOREST TYPE: Tropical Deciduous Plantations
PRINCIPAL PLANT TYPES: *Shorea robusta*

AS	8	LAI	5.8	BA	18.69	SH	7.35
SD	2568	SCA	5940	SCB	1590	PA	1463
PB	380	LAT	N 27°	LNG	E 83° 53′	ALT	81

	Amount
OL	680.00
OBR	740.00
OBO	4520.00
RL	1590.00
STO	670.00

589

.SAL PLANTATION (14 YR OLD), GORAKHPUR FOREST DIVISION, INDIA

INVESTIGATOR: Q. Foruqi   ADDRESS: Department of Botany, Banaras Hindu University, Varanasi, Uttar Pradesh, India
YEAR(S): 1972   FOREST TYPE: Tropical Deciduous Plantations
PRINCIPAL PLANT TYPES: *Shorea robusta*

AS	14	LAI	6.3	BA	20.36	SH	11.12
SD	1660	SCA	8990	SCB	1690	PA	442
PB	330	LAT	N 27°	LNG	E 83° 53'	ALT	81

	Amount
OL	620.00
OBR	870.00
OBO	7310.00
RL	1690.00
STO	682.00

.SAL PLANTATION (26 YR OLD), GORAKHPUR FOREST DIVISION, INDIA

INVESTIGATOR: Q. Foruqi   ADDRESS: Department of Botany, Banaras Hindu University, Varanasi, Uttar Pradesh, India
YEAR(S): 1972   FOREST TYPE: Tropical Deciduous Plantations
PRINCIPAL PLANT TYPES: *Shorea robusta*

AS	26	LAI	6.9	BA	46.25	SH	15.5
SD	1620	SCA	21660	SCB	5510	PA	1072
PB	318	LAT	N 27°	LNG	E 83° 53'	ALT	81

	Amount
OL	810.00
OBR	1760.00
OBO	19090.00
RL	5510.00
STO	891.00

SAL PLANTATION (30 YR OLD), GORAKHPUR FOREST DIVISION, INDIA

INVESTIGATOR: Q. Foruqi ADDRESS: Department of Botany, Banaras Hindu University, Varanasi, Uttar Pradesh, India
YEAR(S): 1972 FOREST TYPE: Tropical Deciduous Plantations
PRINCIPAL PLANT TYPES: *Shorea robusta*

AS	30	LAI	11.4	BA	58.62	SH	17.15
SD	1496	SCA	33590	SCB	6330	PA	2982
PB	205	LAT	N 27°	LNG	E 83° 53'	ALT	81

	Amount
OL	1350.00
OBR	3010.00
OBO	29230.00
RL	6330.00
STO	995.00

SAL PLANTATION (40 YR OLD), GORAKHPUR FOREST DIVISION, INDIA

INVESTIGATOR: Q. Foruqi ADDRESS: Department of Botany, Banaras Hindu University, Varanasi, Uttar Pradesh, India
YEAR(S): 1972 FOREST TYPE: Tropical Deciduous Plantations
PRINCIPAL PLANT TYPES: *Shorea robusta*

AS	40	LAI	13.4	BA	66.55	SH	21.2
SD	1134	SCA	54090	SCB	10090	PA	2150
PB	376	LAT	N 27°	LNG	E 83° 53'	ALT	81

	Amount
OL	1580.00
OBR	4220.00
OBO	48220.00
RL	10090.00
STO	1140.00

TEAK PLANTATION (5 YR OLD), GORAKHPUR FOREST DIVISION, INDIA

INVESTIGATOR: Q. Foruqi   ADDRESS: Department of Botany, Banaras Hindu University, Varanasi, Uttar Pradesh, India
YEAR(S): 1972   FOREST TYPE: Tropical Deciduous Plantations
PRINCIPAL PLANT TYPES: *Tectona grandis*

AS	5	LAI	8.6	BA	15.32	SH	9.2
SD	2068	SCA	4960	SCB	1150	PA	990
PB	230	LAT	N 27°	LNG	E 83° 53′	ALT	81

	Amount
OL	1000.00
OBR	660.00
OBO	3296.00
RL	1150.00
STO	680.00

TEAK PLANTATION (8 YR OLD), GORAKHPUR FOREST DIVISION, INDIA

INVESTIGATOR: Q. Foruqi   ADDRESS: Department of Botany, Banaras Hindu University, Varanasi, Uttar Pradesh, India
YEAR(S): 1972   FOREST TYPE: Tropical Deciduous Plantations
PRINCIPAL PLANT TYPES: *Tectona grandis*

AS	8	LAI	14.6	BA	25.25	SH	9.9
SD	1943	SCA	10080	SCB	2760	PA	1706
PB	536	LAT	N 27°	LNG	E 83° 53′	ALT	81

	Amount
OL	1680.00
OBR	2098.00
OBO	6310.00
RL	2760.00
STO	1190.00

592

TEAK PLANTATION (14 YR OLD), GORAKHPUR FOREST DIVISION, INDIA

INVESTIGATOR: Q. Foruqi ADDRESS: Department of Botany, Banaras Hindu University, Varanasi, Uttar Pradesh, India
YEAR(S): 1972 FOREST TYPE: Tropical Deciduous Plantations
PRINCIPAL PLANT TYPES: *Tectona grandis*

AS	14	LAI	10.6	BA	32.59	SH	13.2
SD	1022	SCA	15890	SCB	3120	PA	968
PB	60	LAT	N 27°	LNG	E 83° 53'	ALT	81

	Amount
OL	1230.00
OBR	2670.00
OBO	11990.00
RL	3120.00
STO	1201.00

TEAK PLANTATION (26 YR OLD), GORAKHPUR FOREST DIVISION, INDIA

INVESTIGATOR: Q. Foruqi ADDRESS: Department of Botany, Banaras Hindu University, Varanasi, Uttar Pradesh, India
YEAR(S): 1972 FOREST TYPE: Tropical Deciduous Plantations
PRINCIPAL PLANT TYPES: *Tectona grandis*

AS	26	LAI	11.5	BA	40.64	SH	17.1
SD	791	SCA	26560	SCB	6470	PA	889
PB	279	LAT	N 27°	LNG	E 83° 53'	ALT	81

	Amount
OL	1330.00
OBR	5220.00
OBO	20000.00
RL	6470.00
STO	1240.00

TEAK PLANTATION(30 YR OLD), GORAKHPUR FOREST DIVISION, INDIA

INVESTIGATOR: Q. Foruqi ADDRESS: Department of Botany, Banaras Hindu University, Varanasi, Uttar Pradesh, India
YEAR(S): 1972 FOREST TYPE: Tropical Deciduous Plantations
PRINCIPAL PLANT TYPES: *Tectona grandis*

AS	30	LAI	17.4	BA	54.93	SH	17.8
SD	682	SCA	30190	SCB	7497	PA	907
PB	257	LAT	N 27°	LNG	E 83° 53'	ALT	81

	Amount
OL	2030.00
OBR	4380.00
OBO	23780.00
RL	7497.00
STO	1514.00

TEAK PLANTATION (40 YR OLD), GORAKHPUR FOREST DIVISION, INDIA

INVESTIGATOR: Q. Foruqi ADDRESS: Department of Botany, Banaras Hindu University, Varanasi, Uttar Pradesh, India
YEAR(S): 1972 FOREST TYPES: Tropical Deciduous Plantations
PRINCIPAL PLANT TYPES: *Tectona grandis*

AS	40	LAI	17.2	BA	97.85	SH	19.2
SD	545	SCA	60550	SCB	13789	PA	2036
PB	629	LAT	N 27°	LNG	E 83° 53'	ALT	81

	Amount
OL	2000.00
OBR	13370.00
OBO	45170.00
RL	13790.00
STO	1560.00

SAL PLANTATION (10 YR OLD) LACCHMIPUR RANGE, GORAKHPUR FOREST, INDIA

INVESTIGATORS: S. S. Ramam   ADDRESS: Department of Botany, Banaras Hindu University.
Varanasi, Uttar Pradesh, India
OTHER INVESTIGATORS: D. Satyanarayana
YEAR(S): 1972   FOREST TYPE: North Tropical Dry Deciduous Forest (Plantations)
PRINCIPAL PLANT TYPES: *Shorea robusta*

AS	10	LAI	6.44	BA	15.17	SH	8.6
SD	2229	SCA	3850	SCB	1221	PA	507
ALT	81	MAR	172879	LAT	N 27°	LNG	E 83° 30'
				MAT	27.5	MAP	1158.1

	Amount		Flux	
OL	363.00		LFL	29.00
OBR	318.00			
OBO	3169.00			
RL	1221.00			
STO	330.00			

SAL PLANTATION (16 YR OLD) LACCHMIPUR RANGE, GORAKHPUR FOREST, INDIA

INVESTIGATOR: S. S. Ramam   ADDRESS: Department of Botany, Banaras Hindu University,
Varanasi, Uttar Pradesh, India
OTHER INVESTIGATORS: D. Satyanarayana
YEAR(S): 1972
FOREST TYPE: North Tropical Deciduous Forest (Plantations)
PRINCIPAL PLANT TYPES: *Shorea robusta*

AS	16	LAI	6.16	BA	14.01	SH	12.3
SD	1461	SCA	4621.	SCB	1372	PA	370
LAT	N 27°	LNG	E 83° 30'	ALT	81	MAR	172879
MAT	27.5	MAP	1158.1				

	Amount		Flux	
OL	350.00		LFL	260.00
OBR	446.00			
OBO	3785.00			
RL	1372.00			
STO	310.00			

SAL PLANTATION (22 YR OLD) LACCHMIPUR RANGE, GORAKHPUR FOREST, INDIA

INVESTIGATOR: S. S. Ramam   ADDRESS: Department of Botany, Banaras Hindu University, Varanasi, Uttar Pradesh, India
OTHER INVESTIGATORS: D. Satyanarayana
YEAR(S): 1972   FOREST TYPE: North Tropical Deciduous Forest (Plantations)
PRINCIPAL PLANT TYPES: *Shorea robusta*

AS	22	LAI	5.76	BA	19.86	SH	14.2
SD	1687	SCA	10434	SCB	2359	PA	580
LAT	N 27°	LNG	E 83° 30′	ALT	81	MAR	172879
MAT	27.5	MAP	1158.1				

	Amount		Flux	
OL	373.00		LFL	280.00
OBR	367.00			
OBO	9693.00			
RL	2359.00			
STO	330.00			

SAL PLANTATION (28 YR OLD) LACCHMIPUR RANGE, GORAKHPUR FOREST, INDIA

INVESTIGATOR: S. S. Ramam   ADDRESS: Department of Botany, Banaras Hindu University, Varanasi, Uttar Pradesh, India
OTHER INVESTIGATORS: D. Satyanarayana
YEAR(S): 1972   FOREST TYPE: North Tropical Deciduous Forest (Plantations)
PRINCIPAL PLANT TYPES: *Shorea robusta*

AS	28	LAI	11.66	BA	31.15	SH	17.1
SD	1594	SCA	16743	SCB	3881	PA	730
LAT	N 27°	LNG	E 83° 30′	ALT	81	MAR	172879
MAT	27.5	MAP	1158.1				

	Amount		Flux	
OL	551.00		LFL	380.00
OBR	1638.00			
OBO	14550.00			
RL	3881.00			
STO	460.00			

596

SAL PLANTATION (35 YR OLD) LACCHMIPUR RANGE, GORAKHPUR FOREST, INDIA

INVESTIGATOR: S. S. Ramam ADDRESS: Department of Botany, Banaras Hindu University, Varanasi, Uttar Pradesh, India
OTHER INVESTIGATORS: D. Satyanarayana
YEAR(S): 1972 FOREST TYPE: North Tropical Deciduous Forest (Plantations)
PRINCIPAL PLANT TYPES: *Shorea robusta*

AS	35	LAI	14.02	BA	41.58	SH	17.6
SD	1741	SCA	24837	SCB	6220	PA	880
LAT	N 27°	LNG	E 83° 30′	ALT	81	MAR	172879
MAT	27.5	MAP	1158.1				

	Amount		Flux	
OL	696.00		LFL	510.00
OBR	1448.00			
OBO	22690.00			
RL	6220.00			
STO	600.00			

SAL PLANTATION (38 YR OLD) LACCHMIPUR RANGE, GORAKHPUR FOREST, INDIA

INVESTIGATOR: S. S. Ramam ADDRESS: Department of Botany, Banaras Hindu University, Varanasi, Uttar Pradesh, India
OTHER INVESTIGATORS: D. Satyanarayana
YEAR(S): 1972 FOREST TYPE: North Tropical Deciduous Forest (Plantations)
PRINCIPAL PLANT TYPE: *Shorea robusta*

AS	38	LAI	8.51	BA	31.64	SH	20.9
SD	742	SCA	24646	SCB	4752	PA	770
LAT	N 27°	LNG	E 83° 30′	ALT	81	MAR	172879
MAT	27.5	MAP	1158.1				

	Amount		Flux	
OL	639.00		LFL	490.00
OBR	3304.00			
OBO	20700.00			
RL	4752.00			
STO	490.00			

597

BANCO (PLATEAU), IVORY COAST

INVESTIGATOR: G. Lemee   ADDRESS: Laboratoire d'Ecologie Vegetale, Universite de Paris–Sud, 91506–Orsay, Paris, France

OTHER INVESTIGATORS: F. Bernhard-Reversat, Ch. Huttel, D. Alexandre, G. Delaunay

SELECTED CITATIONS: Bernhard, F. (1970). *Oecol. Plant.* **5**, 247–66. Bernhard-Reversat, F. (1972) *Oecol. Plant.* **7** 279–300. Bernhard-Reversat, F., Huttel, C. & Lemee, G. (1972). In *Tropical Ecology Symp. New Delhi*, 1971, 217–34.

YEAR(S): 1966–74   FOREST TYPE: Equatorial

SOIL TYPE: Ferrallitique fortement desature   GEOLOGY: Sables tertiaires   SOIL DRAINAGE: Well drained

PRINCIPAL PLANT TYPES: *Turraeanthus africana, Dacryodes klaineana, Strombosia glaucescens, Berlinia confusa, Coula edulis, Chrysophyllum sp., Combretodendron africanum, Allanblackia floribunda*

AS	Mature	LAI	8–10	BA	30	BAI	0.26
SD	265	SCA	51000.	SCB	4900	PA	1630
LAT	N 5° 23'	LNG	W 4° 2'	ALT	50	MAR	135960
MAT	26.2	MAP	2095	LGS	365	TGS	26.2
PGS	2095	RGS	135960	SPH	4.1–5.1	DRZ	100

	Amount		*Flux*	
OL	900.00	LF	1186.00
PSUM	43000.00	LFL	819.00
RL	6000.00	LFFF	109.00
STO	400.00	LFBR	258.00
RZ	7800.00		

YAPO, IVORY COAST

INVESTIGATOR: G. Lemee   ADDRESS: Laboratoire d'Ecologie Végétale, Université de Paris-Sud, 91506–Orsay, Paris, France
OTHER INVESTIGATORS: F. Bernhard-Reversat, Ch. Huttel, D. Alexandre, G. Delaunay
SELECTED CITATIONS: Bernhard, F., (1970) *Oecol. Plant.* **5**, 247–66. Bernhard-Reversat, F., (1972) *Oecol. Plant.* **7**. Bernhard-Reversat, F., Huttel C. & Lemee, G. (1972). In *Tropical Ecology, Symp. New Delhi*, 1971, 217–34.
YEAR(S): 1966–1974   FOREST TYPE: Equatorial
SOIL TYPE: Ferrallitique fortement desature, gravillonnaire   GEOLOGY: Schistes arkosiques   SOIL DRAINAGE: Poor
PRINCIPAL PLANT TYPES: *Dacryodes klaineana, Allanblackia floribunda, Coula edulis, Strombosia glaucescens, Scottelia chevalieri, Scytopetalum thieghemii, Piptadeniastrum africanum, Tarrietia utilis*

AS	Mature	LAI	8–10	BA	31	BAI	0.25
SD	427	SCA	45000	PA	1480	LAT	N 5° 42'
LNG	W 4° 6'	ALT	70	MAP	1739	LGS	365
PGS	1739	SPH	4.3–4.8	DRZ	100		

	Amount		*Flux*	
PSUM	36000.00		LF	961.00
RL	6000.00		LFL	712.00
STO	320.00		LFFF	105.00
			LFBR	144.00

599

JPTF–66–KOIWAI, KOIWAI, JAPAN

INVESTIGATOR: T. Satoo ADDRESS: Department of Forestry, University of Tokyo, Tokyo 113, Japan
FOREST TYPE: Larch Plantation
SOIL TYPE: Black soil of volcanic ash origin
PRINCIPAL PLANT TYPES: Overstory: *Larix leptolepis.* Understory: *Morus bombycis, Prunus grayana, Stachyurus praecox, Viburnum opulus (var. galvescens).* Field Layer: *Calyptranthe petiolu (var. ovalifolia).*

AS	39	LAI	6.68	BA	37.29	SH	19.4
SD	1155	SCA	16943	SCB	3794	PA	1449
PB	264	LAT	N 39° 45′	LNG	E 141°	ALT	360
MAR	120085	MAT	10.2	MAP	1806	LGS	150
RGS	57525	DRZ	155				

	Amount	Increment	Flux	
OL	359.00		LFL	359.00
OBR	1550.00			
OBO	14530.00	580.00		
UL	42.00			
UBR	**			
UBO	**			
USTD	361.00			
HERB	96.00			
RLL	3756.00			
RLS	38.00			
STO	1390.00			

SCHOOL FOREST, ASHU, KYOTO, JAPAN

INVESTIGATOR: T. Shidei   ADDRESS: Department of Forestry, Faculty of Agriculture, Kyoto University, Kyoto, Japan
OTHER INVESTIGATORS: T. Kira, T. Satoo, Y. Kitazawa, M. Morisita, T. Hosokawa
SELECTED CITATIONS: Progress Report 1966–71 of IBP–JIBP Special Committee in Japanese Science Congress.
YEAR(S): 1966   FOREST TYPE: Natural beech stand
SOIL TYPE: Brown forest soil   GEOLOGY: Palaeozoic   SOIL DRAINAGE: Good
PRINCIPAL PLANT TYPES: Overstory: *Fagus crenata, Carpinus laxiflora, C. tschonoskii, Quercus mongolica var. grosserrata.* Substory: *Acer japonicum, A.sieboldiana, Magnolia salicifolia, Prunus grayana.* Field Layer: *Viola vaginata, Oxalis griffithii, Sasa sp.* Other: *Shizophragma hydrangioides, Rhus ambiqua*

AS	150	LAI	4.5	BA	43.32	SH	14.3
SD	785	SCA	29240	SCB	4520	PA	1010
PB	150	LAT	N 35° 20′	LNG	E 135° 45′	ALT	680
MAT	11.3	MAP	2787.6	LGS	244	TGS	16.4
PGS	1822	DRZ	100				

	Amount			Flux	
OL	300.00			LFL	349.00
OBR	9500.00			LFBR	58.20
OBO	19430.00			LFBO	**
STO	11.60			LFFR	14.40
RZ	182.50				

OKINAWA, JAPAN

INVESTIGATOR: Japan IBP–PF–F
YEAR(S): 1973   FOREST TYPE: Evergreen Broad-leaved
SOIL TYPE: Clay Loam
PRINCIPAL PLANT TYPES: Overstory: *Castanopsis cuspidata var. sieboldii.* Substory: *Schefflera octophylla, Neolitsea sericea, Distylium racemosum, Styrax japonica.* Field Layer: *Cyathea spp.*

AS	55	LAI	5.99	BA	47.93	SH	12
SD	2900	SCA	19326	PA	1368	LAT	N 26° 47′
LNG	E 128° 05′	ALT	320	MAT	21.5	MAP	2630

	Amount	Increment	Flux	
OL	770.00		LFL	385.00
OBR	4888.00	259.00		
OBO	13670.00	724.00		
STO	809.00			

601

JPTF–OKITA, OKITA, JAPAN

INVESTIGATOR: T. Satoo   ADDRESS: Department of Forestry, University of Tokyo, Tokyo 113, Japan
FOREST TYPE: Pine
SOIL TYPE: Moderately Moist Brown Forest Soil
PRINCIPAL PLANT TYPES: *Pinus densiflora.* Field Layer: *Quercus serrata* (Sapling), *Carex lanceolata, Pteridium aquilinum var. latiusculum*

AS	20	BA	32.28	SH	10.2	SD	6600
SCA	9329	SCB	2287	PA	1638	LAT	N 39° 02'
LNG	E 141° 21'	ALT	300	MAT	11.3	MAP	1467
LGS	180	SPH	5.4				

	Amount	Increment	Flux	
OL	676.00	125.00	LFL	343.50
OFF	1.00		LFFF	**
OBR	1560.00	**	LFBR	**
OBO	7092.00	1133.00	LFBO	35.20
RLL	2277.00		LFFR	11.50
RLS	10.00			

SHIGAYAMA, JAPAN

INVESTIGATOR: Y. Kitazawa   ADDRESS: Biology Department, Tokyo Metropolitan Univ.,
Fukazawa 2-1-1, Setagaya-Ku, Tokyo 158, Japan
OTHER INVESTIGATORS: M. Mitsudera, T. Kurotori, Y. Oshima, Y. Takai, T. Yoshida, J. Aoki,
K. Haneda, Y. Imaizumi
SELECTED CITATIONS: Kimura, M. (1963) *Japan. J. Bot.* **18**, 255–87.
YEAR(S): 1968–72   FOREST TYPE: Coniferous–Subalpine
SOIL TYPE: Wet (Humid) Podsolic Soil   GEOLOGY: Volcanic Lava flow of Pleistocene
PRINCIPAL PLANT TYPES: *Tsuga diversifolia, Abies mariesii, Betula ermani*

AS	290	LAI	6.8	BA	53.1	BAI	0.04
SH	18	SD	1199	SCA	19250	SCB	6410
PA	610	PB	140	LAT	36° 40′	LNG	138° 30′
ALT	1790	RBA	117377	MAT	4.2	MAP	1455
LGS	180	TGS	11.7	PGS	922	RBG	69984
SPH	3.6–4.2						

	Amount	Increment	Respiration	Flux	
OL	990.00	0.00		LF	342.10
OFF	0.10	0.00		LFL	211.90
OBR	5170.00	178.70		LFFF	6.15
OBO	13090.00	180.10		LFBR	92.58
OSUM	19250.00	0.00		LFBO	1.74
RL			1870.00	LFFR	12.70
STO	26840.00		4960.00	LFEP	10.06
HETR	411.40		3100.00		
AGC	0.26		9.00		
DCFA	5.13		195.00		
DCFL	406.00		2890.00		

Dynamic properties of forest ecosystems

JPTF–70 YUSUHARA KUBOTANIYAMA, JAPAN

INVESTIGATOR: Takashi Ando ADDRESS: Shikoku Branch, Government Forest Experiment Station, 915 Tei Asakura, Kochi 780, Japan
OTHER INVESTIGATORS: T. Shidei, T. Satoo, K. Negishi, K. Hozumi, K. Chiba, T. Nishimura, T. Tanimoto
SELECTED CITATIONS: Progress Report 1966–71 of IBP–JIBP Special Committee in Japanese Science Congress
YEAR(S): 1970 FOREST TYPE: Natural Forest
SOIL TYPE: Brown Forest Soil (moderately dry) GEOLOGY: Cretaceous–Sandstone
PRINCIPAL PLANT TYPES: Overstory: *Tsuga sieboldii, Chamaecyparis obtusa, Pinus densiflora*. Substory: *Cleyera japonica, Eurya japonica, Illicium religiosum, Quercus acuta, Q. salicina, Clethra barbinervis*

AS	120–443	LAI	7.4	BA	87.997	SH	21.2
SD	1948	SCA	56691.9	SCB	16556.5	PA	803.7
PB	121.4	LAT	N 33° 20′	LNG	E 133°	ALT	720
MAT	13.6	MAP	2748	SPH	4.44–4.87	DRZ	70

	Amount	Increment	Flux	
OL	784.00	4.30	LF	688.70
OFF	0.00		LFL	341.30
OBR	9234.00	51.50	LFBR	254.20
OBO	34840.00	114.60		
UL	234.90	0.90		
UBR	**	**		
UBO	11590.00	105.50		
HERB	0.00			
RL	16560.00	84.50		
STO	5078.00			
SUBS	0.00			

JPTF–71 YUSUHARA TAKATORIYAMA, JAPAN

INVESTIGATOR: Takashi Ando   ADDRESS: Shikoku Branch, Government Forest, Experiment Station, 915 Tei Asakura, Kochi 780, Japan
OTHER INVESTIGATORS: T. Shidei, T. Satoo, K. Yoda, K. Hatiya, R. Kato, K. Chiba, T. Nishimura, T. Tanimoto
SELECTED CITATIONS: Progress Report 1966–71 of IBP–JIBP Special Committee in Japanese Science Congress
YEAR(S): 1971   FOREST TYPE: Natural Forest
SOIL TYPE: Brown Forest Soil (moderately wet)   GEOLOGY: Cretaceous/Mudstone
PRINCIPAL PLANT TYPES: Overstory: *Abies firma.* Substory: *Actinodaphne lancifolia, Castanopsis cuspidata, Cleyera japonica, Illicium religiosum, Quercus salicina, Magnolia obvata, Sorbus japonica*

AS	97–145	LAI	9.2	BA	81.712	SH	26.3
SD	2077	SCA	50151.9	SCB	14564.7	PA	1304.3
PB	179.5	LAT	N 33° 20′	LNG	E 133°	ALT	420
MAT	13.6	MAP	2748	SPH	4.49–5.07	DRZ	70

	Amount	Increment	Flux	
OL	1513.00	12.60	LF	760.50
OBR	5776.00	66.00	LFL	424.20
OBO	30670.00	264.00	LFBR	255.10
UL	269.90	2.90		
UBR	**	**		
UBO	11930.00	197.70		
RL	14560.00	179.50		
STO	427.90			
SUBS	0.00			

PASOH, WEST MALAYSIA

INVESTIGATOR: J. A. Bullock   ADDRESS: Department of Zoology, University of Leicester, University Road, Leicester LE1 7RH, England
SELECTED CITATIONS: Bullock, J. A. & Khoo, B. K. (1969). *Malay. Nat. J.* **22**, 136–43. Guha, M. M. (1970) *Malay. Forester*, **32**, 423–33. Bullock, J., (1971). *Proc. Symp. Planned Util. of Trop. Lowland Forests, Bogar.*
YEAR(S):1971–2   FOREST TYPE: Equatorial–Lowland Dipterocarp
SOIL TYPE: Sandy clay loam–clay   GEOLOGY: Shale–Granite   SOIL DRAINAGE: Poor–Medium
PRINCIPAL PLANT TYPES: Families: Dipterocarpaceae, Fagaceae, Burseraceae, Leguminosae, Euphorbiaceae, Myrtaceae

AS	Mature	LAI	8–10	BA	27.34	SH	35–40
SD	590.7	SCA	35360.8	SCB	2652.0	LAT	N 2° 59′
LNG	102° 18′	ALT	100	MAR	130140	MAT	26
MAP	1800	LGS	365	TGS	26	PGS	1800
SPH	4.3–4.8	DRZ	100				

	Amount	Increment	Respiration	Photo synthesis	Flux	
OL	475.10	−4.35	3262.00	10840.00	CS	37.50
OFF	3.88		30.00	76.50	LF	1729.00
OBR	5943.00	−28.29	1798.00	5972.00	LFL	748.80
OBO	28710.00	−158.60	1188.00	3608.00	LFFF	46.50
OSTD	1387.00				LFBR	309.50
OSUM	36520.00		6278.00		LFBO	596.60
UL	107.10		825.90		LFFR	27.08
UBR	196.20		208.00		PRE	**
UBO	1037.00		330.30		DRP	4.92
USUM	1340.00		1364.00		LWAS	**
HERB	2959.00				STMF	13.04
PSUM	37990.00		7442.00			
RL	2959.00	−17.45	563.60	1631.00		
RD	10300.00					
STO	1257.00		737.10			
RZ	8989.00					
IRZ			2022.00			
AGC	6.93		10.39			

MEERDINK, NETHERLANDS

INVESTIGATOR: J. Van Der Drift ADDRESS: Research Institute for Nature Management
Kemperbergerweg 11, Arnhem, Netherlands
OTHER INVESTIGATORS: (Mrs. Dr.) H. M. De Boois, (Mrs. Dr.) J. C. Went, Ir. J. H. De Gunst
SELECTED CITATIONS: Final Report 1966–71 of The Netherlands contribution to the IBP,
North Holland Publ. Co., Amsterdam, 26–32 (1974).
YEAR(S): 1967–70 FOREST TYPE: Oak
SOIL TYPE: Mor Layer of about 10 cm thickness on humus infiltrated sands GEOLOGY:
Glacial driftsand on boulder clay and oligocenous clay SOIL DRAINAGE: Poorly drained
PRINCIPAL PLANT TYPES: *Quercus petraea, Fagus sylvatica, Sorbus aucuparia, Frangula alnus,*
Field Layer: *Pteridium aquilinum, Rubus spp., Hedera helix*

AS	140	LAI	5.2	BA	33.8	BAI	0.59
SH	27.2	SD	300	SCA	27300	SCB	4180
PA	1120	LAT	N 51° 55′	LNG	E 6° 42′	ALT	45
MAT	8.6	MAP	780	LGS	180		

	Amount	Increment	Flux	
OL	333.00		LFL	438.00
OFF	79.00		LFFF	79.00
OBR	2850.00	80.00	LFBR	107.00
OBO	23650.00	350.00	LFFR	24.00
UL	34.00			
UBR	90.00	15.00		
UBO		55.00		
UBOW	780.00			
PSUM	60.00			
RL	4180.00	70.00		
STO	20550.00			
RZ	15940.00			
DCFA	5.10			
DCFL	7.10			

GEOBOTANICAL STATION, BIALOWIEZA, POLAND

INVESTIGATOR: J. B. Falinski   ADDRESS: Stacja Geobotaniczna U. W., Bialowieza Woj., Bialystok, Poland
YEAR(S): 1966–70   FOREST TYPE: Querco–Carpinetem
SOIL TYPE: Lessive soil type   GEOLOGY: Ground Moraine   SOIL DRAINAGE: Good to 1.8 m
PRINCIPAL PLANT TYPES: *Carpinus betulus, Tilia cordata, Picea excelsa, Acer platanoides, Quercus robur, Ulmus campestris, Fraxinus excelsior*

AS	120–210	LAI	8.29	BA	33.23	BAI	0.28
SD	367	LAT	52° 45′	LNG	23° 52′	ALT	157.5
MAR	97144	MAT	5.8–5.3	MAP	715–649	LGS	197–190
TGS	13.8–13.6	PGS	430–418	RGS	79107	SPH	5.2
DRZ	20						

	Amount
UFF	15.74
HERB	22.10

ISPINA, NIEPOLOMICE NEAR KRAKOW, POLAND

INVESTIGATOR(S): E. Medwecka-Kornas A. Bandolo-Ciolczyk ADDRESS: Nature Conservation Research Center, Polish Academy of Sciences, Lubicz 46, 31–512 Cracow, Poland
OTHER INVESTIGATORS: J. Banasik, M. Czarnowski, J. Dziewolski, M. Karkanis, J. Klein
SELECTED CITATIONS: Bandolo-Ciolczyk, E. (1974) *Stud. Natur.* **A9**, 29–91. Dziewolski, J. (1974) *Stud. Natur.* **A9**, 7–28. Medwecka-Kornas, A., Lomnicki, A. & Bandola-Ciolczyk, E. (174) *Bull. Pol. Acad. Sci., Ser. Biol.* **22**, 563–7.
YEAR(S): 1967–9   FOREST TYPE: Oak–Hornbeam Forest (Tilio–Carpinetum)
SOIL TYPE: Leached Brown   GEOLOGY: Alluvia   SOIL DRAINAGE: Well Drained
PRINCIPAL PLANT TYPES: Overstory: *Quercus robur, Carpinus betulus, Tilia cordata.* Substory: *Carpinus betulus, Tilia cordata.* Herbaceous: *Aegopodium podagraria, Anemone nemorosa, Impatiens noli-tangere*

AS	100	LAI	4.72	BA	23.57	BAI	0.203
SH	25	SD	297	SCA	24330	SCB	5003
PA	1006.1	PB	124.3	LAT	N 50° 6'	LNG	E 20° 22'
ALT	180–185	MAR	56700	RBA	48300	MAT	7.8
MAP	729.4	LGS	227	TGS	13.8	PGS	669
RGS	31800	RBG	27000	SPH	5.2	DRZ	90

	Amount	Increment	Respiration	Photo-synthesis	Flux	
OL	200.90		764.70	2025.00	CSF	91.00
OFF	7.64				LFL	292.10
OBR	1750.00	30.00			LFFF	7.64
OBO	20640.00	427.00			LFFR	33.48
OSTD	1.40				DRP	37.22
UL	90.60		231.10	850.90		
UBR	1508.00	33.50				
UBO	1860.00	37.00				
HERB	100.20		336.10	583.00		
RL	5003.00	128.90				
STO	214.00					

KAMPINOS NATIONAL PARK, POLAND

INVESTIGATOR: T. Traczyk ADDRESS: Institute of Ecology, Polish Academy of Sciences, Djiekanow Lesny K. Warszawy, 05–150 Lomnianki, Poland
OTHER INVESTIGATORS: A. Stachurski, J. R. Zimka, L. Gruem, H. Traczyk
SELECTED CITATIONS: Moszynska, B. (1970). *Ekol. Pol.* **18**, 779–803. Traczyk, H. & Traczyk, T. *Ekol. Pol.* (1967). **15**, 823–35. Traczyk, T. (1967). *Ekol. Pol.* **15**, 837–67. Traczyk, T., Traczyk, H. & Moszynska, B. (1973). *Ekol.* **21**, 37–55.
YEAR(S): 1971–5 FOREST TYPE: Pine–Blueberry
SOIL TYPE: Podzol GEOLOGY: Glacial Till SOIL DRAINAGE: Well drained
PRINCIPAL PLANT TYPES: Overstory: *Pinus sylvestris, Quercus robur, Betula verrucosa.* Substory: *Vaccinium myrtillus, Calluna vulgaris, Cornus sanguinea.* Field Layer: *Cladonia rangiferina, Anemone nemorosa, Oxalis acetosella, Majanthemum bifolium*

AS	85	SH	25	SD	1020	SCA	26538.25
SCB	7701.3	LAT	N 52° 20′	LNG	E 20° 50′	ALT	105
MAP	547.6	LGS	207	PGS	485.0	SPH	3.7–4.7
DRZ	100						

	Amount		Flux	
OL	225.00		LF	442.20
OBR	1728.00		LFL	183.90
OBO	11460.00		LFFF	13.60
HERB	118.90		LFFR	2.20
RL	17360.00		LFBR	22.90

610

BABADAG, SITE 1, RUMANIA

INVESTIGATORS: N. Donita, C. Bindiu, V. Mocanu   ADDRESS: Institutal de Cercetari si Amenajari Silvice, Bucuresti 505, Pipera 46, Sector 2, Rumania
OTHER INVESTIGATORS: I. Popescu-Zeletin
SELECTED CITATIONS: Dihoru, G. & Donita, N. (1970). *Ed. Academiei RSR*, Bucuresti.
Popescu-Zeletin, I. *et al.* (1971) *Ed. Academiei RSR*, Bucuresti.
YEAR(S): 1960–5   FOREST TYPE: Submediterranean Deciduous Forest
SOIL TYPE: Rendzina   GEOLOGY: Cretacic Calcar   SOIL DRAINAGE: Well drained
PRINCIPAL PLANT TYPES: *Quercus pubescens, Cotinus coggygria, Galium dasypodium, Paeonia peregrina*

AS	37	LAI	2.54	BA	22.74	BAI	0.52
SH	6.8	SD	2730	SCA	6019.5	PA	437.9
LAT	44° 54′	LNG	28° 43′	ALT	170	MAT	10.6
MAP	480	LGS	178	TGS	17.9	PGS	181
SPH	7.0	DRZ	30				

	Amount	Increment	Flux	
OL	237.60		LF	287.60
OBR	773.90	20.50	LFL	237.60
OBRB	113.90	3.10		
OBRW	660.00	17.40		
OBO	4958.00	129.70		
OBOB	1518.00	39.90		
OBOW	3440.00	89.80		
OSUM	5969.00			
HERB	50.00			

BABADAG, SITE 2, RUMANIA

INVESTIGATORS: N. Donita, C. Bindiu, V. Mocanu ADDRESS: Institutal de Cercetari si Amenajari Silvice, Bucuresti 505, Pipera 46, Sector 2, Rumania
OTHER INVESTIGATORS: L. Popescu-Zeletin
SELECTED CITATIONS: Dihoru, G. & Donita, N. (1970) *Ed. Academiei RSR*, Bucuresti. Popescu-Zeletin, I. *et al.* (1971). Ed. *Academiei RSR*, Bucuresti.
YEAR(S): 1960–5 FOREST TYPE: Submediterranean–Continental Deciduous
SOIL TYPE: Leached Chernozem GEOLOGY: Loess SOIL DRAINAGE: Well drained
PRINCIPAL PLANT TYPES: *Quercus pedunculiflora, Acer tataricum, Brachypodium pinnatum, Centaurea stenolepis*

AS	37	LAI	3.43	BA	19.06	BAI	0.75
SH	8.6	SD	1970	SCA	5305.6	PA	626.8
LAT	44° 54′	LNG	28° 43′	ALT	178	MAT	10.2
MAP	500	LGS	163	TGS	17.2	PGS	189
SPH	6.4	DRZ	80				

	Amount	Increment	Flux	
OL	326.40		LF	422.40
OBR	334.50	28.40	LFL	326.40
OBRB	50.70	4.40		
OBRW	283.80	24.00		
OBO	4548.00	176.00		
OBOB	1948.00	75.20		
OBOW	2600.00	100.80		
OSUM	5209.00			
HERB	96.00			

SINAIA, SITE 1, RUMANIA

INVESTIGATORS: I. Popescu-Zeletin, C. Bindiu, V. Mocanu ADDRESS: Institutul de Cercetari si Amenajari Silvice, Bucuresti 505, Pipera 46, Sector 2, Rumania
SELECTED CITATIONS: Bindiu C. (1973). Doctoral Thesis. Academia de Stunte Agricole si Silvice.
YEAR(S): 1967–70 FOREST TYPE: Beech–Fir Forest
SOIL TYPE: Brown Forest Soil Eubasic GEOLOGY: Conglomerates covered by
Loam SOIL DRAINAGE: Well drained
PRINCIPAL PLANT TYPES: *Fagus sylvatica* (65%), *Abies alba* (35%). Field Layer: *Pulmonaria rubra, Dentaria glandulosa, Symphytum cordatum, Asperula odorata, Oxalis acetosella*

AS	1–450	LAI	9.74	BA	41.96	BAI	1.08
SH	38.5–42.5	SD	842	SCA	29494.7	PA	925.0
LAT	45° 23′	LNG	23° 15′	ALT	950	MAR	117
MAT	5.7	MAP	895	LGS	155–175	TGS	10.7
PGS	645	RGS	83.5	SPH	6.1	DRZ	120

	Amount	Increment	Flux	
OL	1410.00		LF	395.00
OBR	1433.00	75.00	LFL	375.00
OBRB	182.00	6.00		
OBRW	1251.00	69.00		
OBO	26630.00	455.00		
OBOB	2316.00	37.00		
OBOW	24310.00	418.00		
OSUM	29470.00			
HERB	20.00			

613

SINAIA, SITE 2, RUMANIA

INVESTIGATOR: C. Bindiu   ADDRESS: Institutul de Cercetari si Amenajari Silvice, Bucuresti 505, Pipera 46, Sector 2, Rumania
SELECTED CITATIONS: Bindiu C. (1973). Doctoral Thesis. Academia de Stunte Agricole si Silvice.
YEAR(S): 1968–70   FOREST TYPE: Fir Forest
SOIL TYPE: Brown Forest Soil Eubasic   GEOLOGY: Flysch   SOIL DRAINAGE: Well drained
PRINCIPAL PLANT TYPES: *Abies alba.* Field Layer: *Oxalis acetosella, Pleurozium schreberi*

AS	110	LAI	12.0	BA	76.62	BAI	1.48
SH	36.0	SD	485	SCA	47009.0	PA	1130.0
LAT	45° 23′	LNG	23° 15′	ALT	960–1010	MAR	117
MAT	5.1	MAP	1025	LGS	175	TGS	10.4
PGS	718	TGS	83.5	SPH	6.8–6.2	DRZ	80

	Amount	Increment	Respiration	Photo-synthesis	Flux	
OL	2670.00		4880.00	6430.00	LFL	334.50
OBR	3784.00	190.00				
OBRB	490.00					
OBRW	3294.00					
OBO	40550.00	605.50				
OBOB	4055.00	60.00				
OBOW	36500.00	545.50				

SAN JUAN, SPAIN

INVESTIGATOR: J. Puigdefabregas   ADDRESS: Centro Pirenaico de Biologia Experimental, AP. 64 Jaca (Huesca), Spain
OTHER INVESTIGATORS: B. Alvera, J.-R. Vericad, C. Pedrocchi
SELECTED CITATIONS: Alvera, B. (1973). *Pirineos*, **109**, 17–29. Pedrocchi, C. (1973). *Pirineos*, **109**, 73–7.
YEAR(S): 1972   FOREST TYPE: Pine–Holly Forest
GEOLOGY: Oligocene Calcareous Conglomerate   SOIL DRAINAGE: Well drained
PRINCIPAL PLANT TYPES: Overstory: *Pinus sylvestris*. Substory: *Ilex aquifolium*. Field Layer: *Brachypodium pinnatum, Pseudoscleropodium purum*. Other: *Fagus sylvatica, Vaccinium myrtillus*

AS	40–140	LAI	11.87	BA	52.32	BAI	0.624
SH	15.31	SD	1916	SCA	20415	SCB	5354
PA	1756	LAT	N 42° 30′	LNG	W 0° 39′	ALT	1230
MAT	8.0	MAP	802	LGS	202	TGS	11.6
PGS	510	SPH	4.5–6.0	DRZ	50–60		

	Amount	Increment	Flux	
OL	1148.00	213.70	LF	488.20
OFF	7.00	0.00	LFL	245.40
OBR	3504.00	483.00	LFFF	105.00
OBO	14610.00	425.80	LFBR	133.60
OBOB	1574.00		LFBO	4.20
OBOW	13040.00		LFFR	0.00
OSTD	1990.00			
UL	176.00	35.90		
UFF	**	0.00		
UBR	**	**		
UBRB	**	**		
UBRW	**	**		
UBO	**	**		
UBOB	**	**		
UBOW	946.00	109.40		
USTD	20.00			
HERB	22.00			
RLL	5354.00			

ANDERSBY ANGSBACKAR III, SWEDEN

INVESTIGATOR: H. Hytteborn   ADDRESS: Institute of Ecological Botany, P. O. Box 559, S–751 22, Uppsala, Sweden
OTHER INVESTIGATORS: H. Persson, B. M. P. Larsson, B. Axelsson, U. Lohm, T. Persson, O. Tenow
SELECTED CITATIONS: Hytteborn, H. (1975). *Acta Phytogeographica Suecica*, **61**. Persson, H. (1975). *Acta Phyto. Sueca*, **62**. Larsson, B. M. P. (1971) *Fauna Flora*, Stockholm. Axelsson, B., Lohm, U., Persson, H. & Tenow, O. (1974). *Zoon*, **2** (Uppsala).
YEAR(S): 1968–73   FOREST TYPE: Deciduous, *Quercus-Betula-Corylus*
SOIL TYPE: Brown earth with mull and moder   GEOLOGY: Till and clay
PRINCIPAL PLANT TYPES: Overstory: *Quercus robur, Betula pubescens and B. verrucosa.*
Substory: *Corylus avellana, Lonicera xylosteum, Ribes alpinum.* Field Layer: *Fragaria vesca, Anemone nemorosa, A. hepatica, Convallaria majalis, Geranium sylvaticum, Lathyrus montanus, Rubus saxatilis, Trifolium medium, Calamagrostis arundinaca, Deschampsia flexuosa, Luzula pilosa*

AS	40–200	LAI	4.0–5.4	BA	12.3	BAI	0.24
SH	23	SD	452	SCA	8330	PA	723.2
LAT	60°	LNG	17°	ALT	30	MAR	81000
MAT	5.5	MAP	566	LGS	138	TGS	13.5
PGS	266	SPH	5.2–7.0				

	Amount	Increment		Flux	
OL	161.40			LFL	170.70
OFF	20.40			LFFF	20.70
OBR	2030.00	132.20		LFBR	132.10
OBRB	530.00				
OBRW	1500.00				
OBO	5550.00	91.20			
OSTD	130.00				
OSUM	7892.00				
UL	53.80				
UBR	**				
UBO	650.60	55.20			
USUM	742.60				
HERB	68.60				
STO	3890.00				
RZ	8515.00				
HETR	0.00				
DCFA	8.86				

616

KONGALUND BEECH SITE, SWEDEN

INVESTIGATOR: B. Nihlgard   ADDRESS: Ekologihuset, Helgonavaegen 5, S–223 62 Lund, Sweden

SELECTED CITATIONS: Nihlgard, B. (1070) *Oikos*, **21**, 208–17. Nihlgard, B. (1971). *Oikos*, **22**, 301–13. Nihlgard, B. (1972). *Oikos*, **23**, 69–81. Nihlgard, B. (1969) *Bot. Notiser*, **122**, 333–52.

YEAR(S): 1966–1969   FOREST TYPE: Beech forest, Stellaria–Lamium-type

SOIL TYPE: Brown forest soil (acid)   GEOLOGY: Cambrian shales and sandstone, stony-sandy moraine   SOIL DRAINAGE: Good

PRINCIPAL PLANT TYPES: *Fagus sylvatica, Stellaria nemorum, Lamium galeobdolon, Oxalis acetosella, Anemone nemorosa*

AS	45–130	LAI	3.4	BA	31.4	SH	25
SD	240	SCA	32400	SCB	5100	PA	1540
PB	240	LAT	55° 59′	LNG	13° 10′	ALT	120
MAR	90000	MAT	7	MAP	800	LGS	230
TGS	12.9	PGS	350	RGS	70000	SPH	4.0–4.5
DRZ	70						

	Amount	Increment	Flux	
OL	**	0.00	LF	569.00
OFF	390.00	0.00	LFL	357.00
OBR	9900.00	**	LFFF	103.00
OBRB		**	LFBR	109.00
OBRW		**		
OBO	22100.00	**		
OBOB	900.00	**		
OBOW	21200.00	1010.00		
OSUM		0.00		
USUM		0.00		
HERB	10.00	0.00		
PSUM	32400.00	0.00		
RL	5100.00			
RLL	4500.00			
RLS	600.00			
STO	520.00			
RZ	29800.00			
IRZ	12700.00			
ERZ	17100.00			

Dynamic properties of forest ecosystems

KONGALUND SPRUCE SITE, SWEDEN

INVESTIGATOR: B. Nihlgard   ADDRESS: Ekologihuset, Helgonavaegen 5, S-223 62 Lund, Sweden
SELECTED CITATIONS: Nihlgard, B. (1970). *Oikos*, **21**, 208–17. Nihlgard, B. (1971). *Oikos*, **22**, 301–13. Nihlgard, B. (1972). *Oikos*, **23**, 69–81. Nihlgard, B. (1969). *Bot. Notiser*, **122**, 333–52.
YEAR(S): 1966–9   FOREST TYPE: Planted spruce, *Oxalis* type
SOIL TYPE: Brown Forest soil (acid)   GEOLOGY: Cambrian shales and sandstone, stony-sandy moraine   SOIL DRAINAGE: Good
PRINCIPAL PLANT TYPES: *Picea abies, Oxalis acetosella, Rubus idaeus*

AS	60	LAI	11.5	BA	55.6	SH	25
SD	880	SCA	30800	SCB	5900	PA	1370
PB	260	LAT	55° 59′	LNG	13° 10′	ALT	120
MAR	90000	MAT	7	MAP	800	LGS	230
TGS	12.9	PGS	350	RGS	70000	SPH	4.0
DRZ	70						

	Amount	Increment	Flux	
OL	**		LF	571.00
OFF	330.00		LFL	436.00
OBR	2500.00	**	LFFF	16.00
OBRB		**	LFBR	119.00
OBRW		**		
OBO	26200.00	**		
OBOB	2200.00	**		
OBOW	24000.00	990.00		
USUM	0.00	0.00		
HERB	1.00	1.00		
PSUM	30800.00			
RL	5900.00			
RLL	5700.00			
RLS	200.00			
STO	1850.00			
RZ	28700.00			
IRZ	11300.00			
ERZ	17400.00			

618

LANGAROD, SWEDEN

INVESTIGATORS: L. Lindgren, B. Nihlgard   ADDRESS: Ekologihuset, Helgonavaegen 5, S–223 62 Lund, Sweden
SELECTED CITATIONS: Nihlgard, B. & Lindgren, L. (1977). *Oikos*, **28**, 95–104. Lindgren, L. (1970). *Bot. Notiser*, **123**, 401–24. Lindgren, L. (1975). In *Vegetation and Substrate*, ed. R. Tuxen.
YEAR(S): 1966–9   FOREST TYPE: Beech, *Deschampsia*-type
SOIL TYPE: Podsol   GEOLOGY: Cambrian sandstone, overlaid by primary rock moraine, dominated by sand   SOIL DRAINAGE: Good (dry)
PRINCIPAL PLANT TYPES: *Fagus sylvatica.* Field Layer: *Deschampsia flexuosa*

AS	100	LAI	3.2	BA	29.7	SH	22
SD	320	SCA	22600	SCB	3660	PA	1060
PB	170	LAT	55° 45′	LNG	13° 55′	ALT	150
MAR	85–90000	MAT	6	MAP	900	LGS	220
TGS	12	PGS	400	RGS	70000	SPH	4.0–4.5
DRZ	70						

	Amount	Increment	Flux	
OL	**	0.00	LF	382.50
OFF	330.00	0.00	LFL	240.00
OBR	5600.00	**	LFFF	35.00
OBRB		**	LFBR	107.50
OBRW		**		
OBO	16600.00	**		
OBOB	800.00	**		
OBOW	15800.00	630.00		
OSUM		630.00		
USUM	0.00	0.00		
HERB	1.00	0.00		
PSUM	22600.00			
RL	3660.00			
STO	820.00			
RZ	16500.00			

LINNEBJER, SWEDEN

INVESTIGATOR: F. Andersson   ADDRESS: Swedish Coniferous Forest Project, P. O. Box 7008, S–750 07, Sweden
SELECTED CITATIONS: Andersson, F. (1970). *Opera Botanica No.* 27, 190 p. Andersson, F. (1970). *Bot. Not.*, **123**, 8–51. Andersson, F. (1973). In *Modeling Forest Ecosystems*, EDFB–IBP 73–7, Oak Ridge, USA.
YEAR(S): 1964–8   FOREST TYPE: Deciduous Oak–Hazel
SOIL TYPE: Brown Forest soil of gley type   GEOLOGY: Silurian slate   SOIL DRAINAGE: Good (zonal)
PRINCIPAL PLANT TYPES: *Quercus robur, Tilia cordata, Sorbus aucuparia, Corylus avellana Anemone nemorosa, Oxalis acetosella, Convallaria majalis*

AS	125–190	LAI	5.4	BA	21.6	SH	23
SD	4725	SCA	20100	SCB	3900	PA	1290
PB	230	LAT	N 55° 44′	LNG	E 13° 18′	ALT	60
MAR	90480	MAT	7–8	MAP	644	LGS	230
TGS	13	PGS	437	RGS	76300	SPH	4.4–5.0
DRZ	60						

	Amount	Increment	Flux	
OL	300.00	0.00	CS	40.00
OBR	4100.00	−130.00	LF	476.80
OBO	11100.00	200.00	LFL	326.00
OBOB	1100.00		LFFF	51.00
OBOW	10000.00		LFBR	107.00
OSUM	15500.00			
UL	210.00			
UBR	1200.00	**		
UBO	31000.00	300.00		
UBOB	5000.00			
UBOW	26000.00			
USUM	4510.00			
HERB	20.00	0.00		
PSUM	20010.00	0.00		
RL	3920.00	1940.00		
RLL	3320.00			
RLS	600.00			
STO	610.00			
RZ	28800.00			

OVED, SWEDEN

INVESTIGATOR: L. Lindgren, B. Nihlgard   ADDRESS: Ekologihuset, Helgonavaegen 5, S–223 62 Lund, Sweden

OTHER INVESTIGATIONS: L. Lindgren

SELECTED CITATIONS: Nihlgard, B. & Lindgren, L. (1977). *Oikos*, **28**, 95–104. Lindgren, L. (1970). *Bot. Notiser*, **123**, 401–24. Lindgren, L. (1975) In *Vegetation and Substrate*, ed. R. Tuxen.

YEAR(S): 1966–9   FOREST TYPE: Beech, Mercurialis-type

SOIL TYPE: Brown forest soil with gley horizon   GEOLOGY: Silurian slate overlaid by clay and slate and primary rock moraine   SOIL DRAINAGE: Poor

PRINCIPAL PLANT TYPES: *Fagus sylvatica, Mercurialis perennis, Allium ursinum, Anemone nemorosa, A. ranunculoides*

AS	80–100	LAI	4.3	BA	31.1	SH	26–32
SD	180	SCA	31500	SCB	4340	PA	1670
PB	230	LAT	55° 42′	LNG	13° 38′	ALT	60
MAR	85–90000	MAT	7	MAP	650	LGS	230
TGS	13	PGS	350	RGS	70000	SPH	6.5–7.5
DRZ	70						

	Amount	Increment	Flux	
OL	**		LF	421.60
OFF	480.00		LFL	275.00
OBR	6400.00	**	LFFF	70.60
OBO	24500.00	**	LFBR	76.00
OBOB	1100.00	**		
OBOW	23400.00	1000.00		
USUM	0.00	0.00		
HERB	90.00			
PSUM	31500.00			
RL	4340.00			
STO	340.00			

KOINAS, ARKANGELSK REGION, USSR

INVESTIGATORS: Rudneva, Tonkonogov, Dorochova
YEAR(S): 1965   FOREST TYPE: North Taiga, Picetum Hylocomiosum
SOIL TYPE: Gley podsol loamy   GEOLOGY: Sand on the moraine with stony clay
PRINCIPAL PLANT TYPES: *Picea abies, Juniperus communis, Vaccinium myrtillus, V. vitis-idaea.*
Field Layer: *Hylocomium proliferum*

AS	125	SH	15	SD	1050	SCA	13284
SCB	4070	PA	537	PB	60	LAT	64° 40′
LNG	47° 30′	RBA	27000	MAT	−1.2	MAP	499
LGS	76	TGS	13.5	PGS	171	SPH	3.4–4.6

	Amount
OL	728.00
OFF	72.00
OBR	2126.00
OBO	9780.00
UL	32.00
UBR	70.00
UBO	142.00
PSUM	13280.00
HERB	236.00
EPIP	77.00
RL	4070.00

CAUCASUS BIRCH SITE 1, AZERBAIJAN, USSR

INVESTIGATOR: O. G. Merzoev   ADDRESS: Botanical Institute, Academy of Sciences of Azerbaijan SSR, Baku, USSR
SELECTED: Merzoev, O. G. (1975). *Variability of the Reserve of Litter in Birch Forests* (Betula pubescens) *of Different Ages', Izvestia,* **1.**
YEAR(S): 1974   FOREST TYPE: Birch
PRINCIPAL PLANT TYPES: Overstory: *Betula pendula.* Field Layer: *Rosa* spp.

AS	20	SH	4.5	SD	9480	SCA	2615
PA	410	LAT	N 41°	LNG	E 48°	ALT	2000
MAP	450	LGS	210	TGS	7.8		

	Amount		*Flux*	
OL	234.00		LF	234.00
OBR	529.00			
OBO	1796.00			
OSTD	29.00			
HERB	27.00			
STO	154.00			

CAUCASUS BIRCH SITE 2, AZERBAIJAN, USSR

INVESTIGATOR: O. G. Merzoev   ADDRESS: Botanical Institute, Academy of Sciences of Azerbaijan SSR, Baku, USSR
SELECTED CITATIONS: Merzoev, 0. G. (1975). *Variability of the Reserve of Litter in Birch Forests* (Betula pubescens) *of Different Ages, Izvestia,* **1**.
YEAR(S): 1974   FOREST TYPE: Birch
PRINCIPAL PLANT TYPES: Overstory: *Betula pendula.* Field Layer: *Rosa* spp.

AS	30	SH	6.5	SD	3560	SCA	2262
PA	250	LAT	N 41°	LNG	E 48°	ALT	2000
MAP	450	LGS	210	TGS	7.8		

	Amount			Flux	
OL	127.00			LF	127.00
OBR	413.00			LFBR	7.00
OBO	1686.00				
HERB	28.00				
STO	134.00				

TALLISH IRONWOOD (SUBTROPICAL) SITE 1, AZERBAIJAN, USSR

INVESTIGATOR: A. A. Esmekanova   ADDRESS: Botanical Institute, Academy of Sciences of Azerbaijan SSR, Baku, USSR
YEAR(S): 1973–4   FOREST TYPE: *Parrotia persica*
SOIL TYPE: Yellow podzolic
PRINCIPAL PLANT TYPES: Overstory: *Parrotia persica.* Field Layer: *Rosa, Ruscus*

AS	31	SH	11	SD	950	SCA	3649
LAT	N 38° 50′	LNG	E 48° 30′	ALT	550	MAT	10
MAP	1200	LGS	235				

TALLISH IRONWOOD (SUBTROPICAL) SITE 2, AZERBAIJAN, USSR

INVESTIGATOR: A. A. Esmekanova   ADDRESS: Botanical Institute, Academy of Sciences of Azerbaijan SSR, Baku, USSR
YEAR(S): 1973–4   FOREST TYPE: *Parrotia persica*
SOIL TYPE: Yellow podzolic
PRINCIPAL PLANT TYPES: Overstory: *Parrotia persica*

AS	21	SH	7	SD	4850	SCA	4840
LAT	N 38° 50′	LNG	E 48° 30′	ALT	550	MAT	10
MAP	1200	LGS	235				

TALLISH IRONWOOD (SUBTROPICAL) SITE 3, AZERBAIJAN, USSR

INVESTIGATOR: A. A. Esmekanova   ADDRESS: Botanical Institute, Academy of Sciences of
Azerbaijan SSR, Baku, USSR
YEAR(S): 1973–4   FOREST TYPE: *Parrotia persica*
SOIL TYPE: Yellow podzolic
PRINCIPAL PLANT TYPES: Overstory: *Parrotia persica, Quercus castaneifolia, Carpinus caucasica.*
Field Layer: *Rosa, Ruscus*

SH	16	SD	912	SCA	4716	LAT	N 38° 50′
LNG	E 48° 30′	ALT	550	MAT	10	MAP	1200
LGS	235						

TALLISH IRONWOOD (SUBTROPICAL) SITE 4, AZERBAIJAN, USSR

INVESTIGATOR: A. A. Esmekanova   ADDRESS: Botanical Institute, Academy of Sciences of
Azerbaijan SSR, Baku, USSR
YEAR(S): 1973–4   FOREST TYPE: *Parrotia persica*
SOIL TYPE: Yellow podzolic
PRINCIPAL PLANT TYPES: Overstory: *Parrotia persica, Quercus castaneifolia, Carpinus caucasica.*
Field Layer: *Rosa, Ruscus*

AS	40	SH	22	SCA	9314	LAT	N 38° 50′
LNG	E 48° 30′	ALT	550	MAT	10	MAP	1200
LGS	235						

624

TALLISH OAK (SUBTROPICAL) SITE 1, AZERBAIJAN, USSR

INVESTIGATORS: K. G. Djhalilov, I. S. Safarov   ADDRESS: Botanical Institute, Academy of
Sciences of Azerbaijan SSR, Baku, USSR
SELECTED CITATIONS: Safarov, I. S. & Djhalilov, K. G. (1973). *Lesovedeniya*, **3**, 40–6.
FOREST TYPE: Sub-tropical Oak (Quercetum nudum)
SOIL TYPE: Subtropical yellow soils
PRINCIPAL PLANT TYPES: Overstory: *Quercus castaneifolia, Zelkova carpinifolia, Parrotia persica*

AS	75–80	BA	30.9	SH	27.5	SD	490
SCA	35000	PA	858	LAT	N 38° 50′	LNG	E 48° 31′
ALT	450	MAT	10	MAP	700	LGS	230–250

	Amount		Flux	
OL	723.00		LFL	365.00
OBR	4970.00		LFFF	32.00
OBO	29280.00		LFBR	91.00
HERB	20.00		LFBO	8.00
EPIP	6.00			
STO	1658.00			

TALLISH OAK (SUBTROPICAL) SITE 2, AZERBAIJAN, USSR

INVESTIGATORS: K. G. Djhalilov, I. S. Safarov   ADDRESS: Botanical Institute, Academy of
Sciences of Azerbaijan SSR, Baku, USSR
SELECTED CITATIONS: Safarov, I. S. & Djhalilov, K. G. (1973). *Lesovedeniya*, **3**, 40–6.
YEAR(S): 1968–72   FOREST TYPE: Sub-Tropical Oak (Quercetum nudum)
SOIL TYPE: Subtropical yellow soils
PRINCIPAL PLANT TYPES: Overstory: *Quercus castaneifolia* (50%), *Alnus barbata* (30%), *Ulmus
carpinifolia* (10%), *Parrotia persica* (5%), *Fraxinus excelsion* (5%)

AS	50–60	SH	22.6	SD	420	SCA	34246
PA	Bole–1229	LAT	N 38° 40′	LNG	E 48° 30′	ALT	−22
MAT	15	MAP	1350	LGS	230–250		

	Amount		Flux	
OL	820.00		LFL	567.00
OBR	7900.00		LFFF	15.00
OBO	25000.00		LFBR	78.00
HERB	324.00		LFBO	7.00
STO	1312.00			
EPIP	2.00			

LES NA VORSKLE, PLOT 7, BELGOROD REGION, USSR

INVESTIGATOR: T. K. Goryshina   ADDRESS: Leningrad State University, Leningrad, USSR,
OTHER INVESTIGATORS: U. N. Neshatov, M. B. Metena, S. E. Samelyak, L. S. Chastnaya, E. A.
Tepeshenkova, V. G. Plavnikov, O. V. Petrov
SELECTED CITATIONS: Goryshina, T. K. (Ed.), (1974). *Biological Productivity and Its Factors In
The Oaks Of The Forest Steppe.* Scientific Notes, Series of Biological Science, No. 367, **2.**
Leningrad University Press, 212 p.
YEAR(S): 1966–71   FOREST TYPE: Quercetum Pilosi–Caricosum
SOIL TYPE: Grey and dark grey soils   GEOLOGY: Quaternary loess intermingled with loam
and sandy loam soils   SOIL DRAINAGE: Well drained
PRINCIPAL PLANT TYPES: Overstory: *Quercus robur, Tilia cordata, Acer platanoides, Ulmus sca-
bra.* Substory: *Malus sylvestris, Euonymus verrucosa, E. europaea, Crataegus curvisepala.* Field
Layer: *Carex pilosa, Aegopodium podagraria, Scilla sibirica, Anemone ranunculoides, Corydalis
halleri, Stellaria holostea*

AS	80	BA	25	SH	25	SD	446
SCA	25080.0	SCB	6440.0	PA	975.0	LAT	N 50° 38'
LNG	E 35° 58'	ALT	195	MAR	100000	RBA	35000
MAT	6.0	MAP	537	LGS	120	TGS	12.5
PGS	200–250	RGS	77400	RBG	27000	SPH	4.9–6.4
DRZ	100						

	Amount	Increment		Flux	
OL	267.00		LF	1176.00	
OBR	5527.00	164.00	LFL	356.00	
OBRB	1324.00		LFBR	140.00	
OBRW	4203.00				
OBO	18300.00	252.00			
OBOB	2174.00				
OBOW	16120.00				
OSTD		43.00			
OSUM	24190.00				
UL	8.00				
UBR	300.00	99.00			
UBRB	82.00	31.00			
UBRW	218.00	68.00			
UBO	495.00				
UBOB	154.00				
UBOW	341.00				
USUM	804.00				
HERB	90.00				
RL	4589.00				

LES NA VORSKLE, PLOT 8, BELGOROD REGION, USSR

INVESTIGATOR: T. K. Goryshina   ADDRESS: Leningrad State University, Leningrad, USSR,
OTHER INVESTIGATORS: U. N. Neshatov, M. B. Metena, S. E. Samelyak, L. S. Chastnaya, E. A.
Tepeshenkova, V. G. Plavnikov, O. V. Petrov
SELECTED CITATIONS: Goryshina, T. K. (Ed.) (1974), *Biological Productivity and Its Factors In
The Oaks of The Forest Steppe*. Scientific Notes, Series of Biological Science, No. 367, **2**.
Leningrad University Press, 212 p.
YEAR(S): 1966–71   FOREST TYPE: Tilieto–Quercetum Aegopodiosum
SOIL TYPE: Grey and dark grey soils   GEOLOGY: Quaternary loess intermingled with loam and
sandy loam soils   SOIL DRAINAGE: Well drained
PRINCIPAL PLANT TYPES: Overstory: *Quercus robur, Tilia cordata, Acer platanoides, Ulmus sca-
bra*. Substory: *Acer campestre, Euonymus verrucosa, E. europaea, Crataegus curvisepala*. Field
Layer: *Aegopodium podagraria, Scilla sibirica, Gagea lutea, Carex pilosa, Stellaria holostea*

AS	250	BA	37	SH	31	SD	557
SCA	30741.0	SCB	12493.0	PA	964.0	LAT	N 50° 38′
LNG	E 35° 58′	ALT	200	MAR	100000	RBA	35000
MAT	6.0	MAP	537	LGS	120	TGS	12.5
PGS	200–250	RGS	77400	RBG	27000	SPH	5.7–6.4
DRZ	100						

	Amount	Increment	Flux	
OL	360.00		LF	1316.00
OBR	6793.00	127.00	LFL	376.00
OBRB	1838.00		LFB	230.00
OBRW	4955.00			
OBO	22800.00	237.00		
OBOB	3694.00			
OBOW	19110.00			
OSTD		220.00		
OSUM	30050.00			
UL	10.00			
UBR	322.00	58.00		
UBRB	90.00	35.00		
UBRW	232.00	23.00		
UBO	283.00			
UBOB	33.00			
UBOW	250.00			
USUM	617.00			
HERB	70.00			
RL	9169.00			

CENTRAL FOREST RESERVE, USSR

INVESTIGATOR: V. G. Karpov ADDRESS: Laboratory of Experimental Phytocoenology, Geobotany Department, Komarov Botanical Institute, Leningrad, USSR
SELECTED CITATIONS: Karpov, V. G. (Ed.), (1973). *Structure and Productivity of Spruce Forests of the Southern Taiga.* Nauka, Leningrad Branch, Academy of Sciences, 303 p.
YEAR(S): 1967–73 FOREST TYPE: Taiga, Piceetum–Myrtilloso–Oxalidosum
SOIL TYPE: Clayed weak podzol GEOLOGY: Moraine debris, clay silty loam SOIL DRAINAGE: Adequate
PRINCIPAL PLANT TYPES: *Picea abies, Vaccinium myrtillus.* Field Layer: *Linnaea borealis, Oxalis acetosella, Majanthemum bifolium, Trientalis europaea, Ramischia secunda, Rubus saxatilis, Luzula pilosa, Dryopteris austriaca, Thelypteris phegopteris, Stellaria holostea*

AS	110	LAI	9.7	SH	26.5	SD	678
SCA	20550	SCB	7420	PA	530	PB	115
LAT	56° 30′	LNG	32° 40′	ALT	200	RBA	35000
MAT	3.4	MAP	640	LGS	128	TGS	14.4
PGS	353	SPH	4.0–4.5	DRZ	20		

	Amount
OL	1250.00
OBR	2350.00
OBO	16850.00
OBOB	1260.00
OBOW	15590.00
OSUM	20450.00
HERB	146.00
RL	6810.00

SOUTHERN KARELIAN SPRUCE, SITE 1, KARELIA, USSR

INVESTIGATORS: N. I. Kazimirov, R. M. Morozova   ADDRESS: Karelian Branch of the Russian
SSR Academy of Sciences, Petrozovodsk, USSR
SELECTED CITATIONS: Kazimirov, N. I. & Morozova, R. M. (1973). *Biological Cycling of
Matter In Spruce Forests of Karelia.* Nauka, Leningrad Branch, Academy of Sciences, 168 p.
YEAR(S): 1965–9   FOREST TYPE: Cladinosum Saxatile
SOIL TYPE: Eluvium debris   GEOLOGY: Crystalline Base
PRINCIPAL PLANT TYPES: *Picea abies*

AS	37	LAI	1.8	BA	13.3	BAI	1.1
SH	4.2	SD	9010	SCA	2828	SCB	661
PA	287	PB	43	LAT	62°	LNG	34°
ALT	200	MAR	72100	RBA	31500	MAT	2.2
MAP	650	LGS	150	TGS	11.9	PGS	380
RGS	51400	RBG	30000	SPH	3.3	DRZ	7–10

	Amount	Increment	Flux	
OL	570.00	8.00	CS	355.00
OBR	609.00	12.00	CSF	150.00
OBO	1649.00	94.00	CSR	161.00
OBOB	198.00	11.00	CSW	44.00
OBOW	1451.00	83.00	LF	195.00
OSTD	195.00	4.00	LFL	140.00
OSUM	2983.00	118.00	LFBR	16.00
USUM	0.00	0.00	LFBO	39.00
HERB	301.00	3.00		
PSUM	3284.00	121.00		
RL	661.00	27.00		
RLL	425.00	21.50		
RLS	60.00	2.40		
RD	80.00	1.50		
STO	1870.00	7.00		
IRZ	980.00			

SOUTHERN KARELIAN SPRUCE, SITE 2, KARELIA, USSR

INVESTIGATORS: N. I. Kazimirov, R. M. Morozova   ADDRESS: Karelian Branch of the Russian SSR Academy of Sciences, Petrozovodsk, USSR
SELECTED CITATIONS: Kazimirov, N. I. & Morozova, R. M. (1973). *Biological Cycling of Matter in Spruce Forests of Karelia.* Nauka, Leningrad Branch, Academy of Sciences, 168 p.
YEARS: 1965–9   FOREST TYPE: Vacciniosum
SOIL TYPE: Humus iron podsol   GEOLOGY: Sand moraine
PRINCIPAL PLANT TYPES: *Picea abies*

AS	45	LAI	2.6	BA	17.9	BAI	1.3
SH	6.9	SD	9620	SCA	4592	SCB	1014
PA	478	PB	70	LAT	62°	LNG	34°
ALT	170	MAR	72100	RBA	31500	MAT	2.2
MAP	650	LGS	150	TGS	11.5	PGS	380
RGS	51400	RBG	30000	SPH	3.8	DRZ	40–45

	Amount	Increment	Flux	
OL	821.00	13.00	CS	581.00
OBR	868.00	17.00	CSF	239.00
OBO	2903.00	151.00	CSR	271.00
OBOB	319.00	17.00	CSW	71.00
OBOW	2584.00	134.00	LF	323.00
OSTD	370.00	7.40	LFL	220.00
OSUM	4962.00	188.40	LFBR	29.00
USUM	0.00	0.00	LFBO	74.00
HERB	183.00	5.00		
PSUM	5145.00	193.40		
RL	1014.00	36.00		
RLL	850.00	29.00		
RLS	60.00	2.50		
RD	129.00	2.60		
STO	1930.00	11.00		
IRZ	5140.00			

SOUTHERN KARELIAN SPRUCE   SITE 3, KARELIA, USSR

INVESTIGATORS: N. I. Kazimirov, R. M. Morozova   ADDRESS: Karelian Branch of the Russian
SSR, Academy of Sciences, Petrozovodsk, USSR
SELECTED CITATIONS: Kazimirov, N. I. & Morozova, R. M. (1973). *Biological Cycling of
Matter in Spruce Forests of Karelia.* Nauka, Leningrad Branch, Academy of Sciences, 168 p.
YEAR(S): 1965–69   FOREST TYPE: Myrtillosum
SOIL TYPE: Humus iron podsol   PRINCIPAL PLANT TYPES: *Picea abies*

AS	39	LAI	3.4	BA	20.6	BAI	1.4
SH	7.8	SD	9980	SCA	6966	SCB	1461
PA	629	PB	93	LAT	62°	LNG	34°
ALT	130	MAR	72100	RBA	31500	MAT	2.2
MAP	650	LGS	150	TGS	11.9	PGS	380
RGS	51400	RBG	30000	SPH	4.2	DRZ	50–55

	Amount	*Increment*	*Flux*	
OL	1021.00	17.00	CS	755.00
OBR	1123.00	22.00	CSF	281.00
OBO	4822.00	205.00	CSR	371.00
OBOB	482.00	21.00	CSW	103.00
OBOW	4340.00	184.00	LF	405.00
OSTD	530.00	10.60	LFL	261.00
OSUM	7496.00	254.60	LFBR	38.00
USUM	0.00	0.00	LFBO	106.00
HERB	147.00	4.00		
PSUM	7643.00	258.60		
RL	1461.00	48.00		
RLL	1259.00	41.00		
RLS	70.00	2.70		
RD	161.00	3.20		
STO	1980.00	13.00		
IRZ	5640.00			

SOUTHERN KARELIAN SPRUCE SITE 4, KARELIA, USSR

INVESTIGATORS: N. I. Kazimirov, R. M. Morozova ADDRESS: Karelian Branch of the Russian
SSR, Academy of Sciences, Petrozovodsk, USSR
SELECTED CITATIONS: Kazimirov, N. I. & Morozova, R. M. (1973). *Biological Cycling of
Matter in Spruce Forests of Karelia.* Nauka, Leningrad Branch, Academy of Sciences, 168 p.
YEAR(S): 1965–9 FOREST TYPE: Oxalidoso–Myrtillosum
SOIL TYPE: Sand podsol GEOLOGY: Sand moraine SOIL DRAINAGE: Good
PRINCIPAL PLANT TYPES: *Picea abies*

AS	43	LAI	4.3	BA	23.2	BAI	1.4
SH	9.8	SD	6310	SCA	7938	SCB	1675
PA	716	PB	104	LAT	62°	LNG	34°
ALT	150	MAR	72100	RBA	31500	MAT	2.2
MAP	650	LGS	150	TGS	11.9	PGS	380
RGS	51400	RBG	30000	SPH	4.4	DRZ	60–65

	Amount	Increment	Flux	
OL	950.00	15.00	CS	845.00
OBR	1246.00	25.00	CSF	311.00
OBO	5842.00	234.00	CSR	423.00
OBOB	527.00	23.00	CSW	111.00
OBOW	5315.00	211.00	LF	449.00
OSTD	620.00	12.40	LFL	285.00
OSUM	8558.00	286.40	LFBR	40.00
USUM	0.00	0.00	LFBO	114.00
HERB	104.00	2.00	LSOL	124.00
PSUM	8662.00	288.40		
RL	1675.00	53.00		
RLL	1459.00	46.20		
RLS	67.00	2.10		
RD	178.00	3.60		
STO	2060.00	12.00		
IRZ	7240.00			

SOUTHERN KARELIAN SPRUCE, SITE 5, KARELIA, USSR

INVESTIGATORS: N. I. Kazimirov, R. M. Morozova ADDRESS: Karelian Branch of the Russian
SSR, Academy of Sciences, Petrozobodsk, USSR
SELECTED CITATIONS: Kazimirov, N. I. & Morozova, R. M. (1973). *Biological Cycling of
Matter in Spruce Forests of Karelia.* Nauka, Leningrad Branch, Academy of Sciences, 168
YEAR(S): 1965–9 FOREST TYPE: Oxalidosum
SOIL TYPE: Sand podsol GEOLOGY: Moraine SOIL DRAINAGE: Good
PRINCIPAL PLANT TYPES: *Picea abies*

AS	38	LAI	4.4	BA	25.4	BAI	1.5
SH	12.2	SD	4480	SCA	8748	SCB	1827
PA	779	PB	117	LAT	62°	LNG	34°
ALT	160	MAR	72100	RBA	31500	MAT	2.2
MAP	650	LGS	150	TGS	11.9	PGS	380
MGS	51400	RBG	30000	SPH	4.6	DRZ	70–75

	Amount	Increment	Flux	
OL	991.00	16.00	CS	915.00
OBR	1223.00	27.00	CSF	330.00
OBO	6534.00	250.00	CSR	459.00
OBOB	523.00	23.00	CSW	126.00
OBOW	6011.00	227.00	LF	494.00
OSTD	685.00	13.70	LFL	312.00
OSUM	9433.00	306.70	LFBR	45.00
USUM	0.00	0.00	LFBO	137.00
HERB	78.00	1.00		
PSUM	9511.00	307.70		
RL	1827.00	61.00		
RLL	1598.00	53.40		
RLS	75.00	2.50		
RD	192.00	3.80		
STO	2000.00	10.00		
IRZ	11740.00			

SOUTHERN KARELIAN SPRUCE, SITE 6, KARELIA, USSR

INVESTIGATORS: N. I. Kazimirov, R. M. Morozova   ADDRESS: Karelian Branch of the Russian SSR Academy of Sciences, Petrozovodsk, USSR
SELECTED CITATIONS: Kazimirov, N. I. & Morozova, R. M. (1973). *Biological Cycling of Matter in Spruce Forests of Karelia.* Nauka, Leningrad Branch, Academy of Sciences, 168 p. (1973).
YEAR(S): 1965–9   FOREST TYPE: Politrichosum
SOIL TYPE: Peat   GEOLOGY: Sand moraine
PRINCIPAL PLANT TYPES: *Picea abies*

AS	42	LAI	2.0	BA	14.8	BAI	1.2
SH	5.8	SD	9410	SCA	3450	SCB	798
PA	368	PB	52	LAT	62°	LNG	34°
ALT	100	MAR	72100	RBA	31500	MAT	2.2
MAP	650	LGS	150	TGS	11.9	PGS	380
RGS	51400	RBG	30000	SPH	3.6	DRZ	40–50

	Amount	Increment	Flux	
OL	653.00	10.00	CS	482.00
OBR	700.00	14.00	CSF	220.00
OBO	2097.00	120.00	CSR	209.00
OBOB	230.00	13.00	CSW	53.00
OBOW	1867.00	107.00	LF	226.00
OSTD	270.00	5.40	LFL	151.00
OSUM	3270.00	149.40	LFBR	21.00
USUM	0.00	0.00	LFBO	54.00
HERB	350.00	7.00		
PSUM	4070.00	156.40		
RL	798.00	29.00		
RLL	640.00	23.30		
RLS	70.00	2.50		
RD	103.00	2.00		
STO	3380.00	14.00		
IRZ	12000.00			

SOUTHERN KARELIAN SPRUCE, SITE 7, KARELIA, USSR

INVESTIGATORS: N. I. Kazimirov, R. M. Morozova   ADDRESS: Karelian Branch of the Russian
SSR, Academy of Sciences, Petrozovodsk, USSR
SELECTED CITATIONS: Kazimirov, N. I. & Morozova, R. M. (1973). *Biological Cycling of
Matter is Spruce Forests of Karelia.* Nauka, Leningrad Branch, Academy of Sciences, 168 p.
YEAR(S): 1965–9   FOREST TYPE: Uliginio–Herbosum
SOIL TYPE: Peat   GEOLOGY: Peat on clay moraine
PRINCIPAL PLANT TYPES: *Picea abies*

AS	41	LAI	2.4	BA	17.7	BAI	1.3
SH	6.7	SD	9930	SCA	4176	SCB	947
PA	432	PB	63	LAT	62°	LNG	34°
ALT	80	MAR	72100	RBA	31500	MAT	2.2
MAP	650	LGS	150	TGS	11.9	PGS	380
RGS	51400	RBG	30000	SPH	6.0	DRZ	40–50

	Amount	Increment	Flux	
OL	753.00	11.00	CS	573.00
OBR	819.00	16.00	CSF	225.00
OBO	2604.00	137.00	CSR	247.00
OBOB	260.00	15.00	CSW	71.00
OBOW	2344.00	122.00	LF	275.00
OSTD	340.00	6.80	LFL	181.00
OSUM	4516.00	170.80	LFBR	26.00
USUM	0.00	0.00	LFBO	68.00
HERB	246.00	3.00		
PSUM	4762.00	173.80		
RL	947.00	33.00		
RLL	787.00	26.40		
RLS	60.00	2.10		
RD	124.00	2.50		
STO	6100.00	20.00		
IRZ	24900.00			

SOUTHERN KARELIAN SPRUCE, SITE 8, KARELIA, USSR

INVESTIGATORS: N. I. Kazimirov, R. M. Morozova ADDRESS: Karelian Branch of the Russian SSR Academy of Sciences, Petrozovodsk, USSR
SELECTED CITATIONS: Kazimirov, N. I. Morozova, R. M. (1973). *Biological Cycling of Matter in Spruce Forests of Karelia*. Nauka, Leningrad Branch, Academy of Sciences, 168 p.
YEAR(S): 1965–9 FOREST TYPE: Myrtillosum
SOIL TYPE: Humus iron podsol GEOLOGY: Sand moraine
PRINCIPAL PLANT TYPES: *Picea abies*

As	22	LAI	1.8	BA	10.6	BAI	1.4
SH	2.6	SD	34800	SCA	2490	SCB	620
PA	350	PB	65	LAT	62°	LNG	34°
ALT	130	MAR	72100	RBA	31500	MAT	2.2
MAP	650	LGS	150	TGS	11.9	PGS	380
RGS	51400	RBG	30000	SPH	4.3	DRZ	50

	Amount	Increment	Flux	
OL	550.00	30.00	CS	445.00
OBR	650.00	29.00	CSF	174.00
OBO	1390.00	118.00	CSR	197.00
OBOB	153.00	13.00	CSW	74.00
OBOW	1237.00	105.00	LF	193.00
OSTD	104.00	16.00	LFL	143.00
OSUM	2694.00	193.00	LFBR	24.00
USUM	0.00	0.00	LFBO	26.00
HERB	130.00	1.00		
PSUM	2824.00	194.00		
RL	620.00	40.00		
RLL	478.00	30.80		
RLS	57.00	3.70		
RD	46.00	6.50		
STO	1740.00	11.00		

SOUTHERN KARELIAN SPRUCE, SITE 9, KARELIA, USSR

INVESTIGATORS: N. I. Kazimirov, R. M. Morozova ADDRESS: Karelian Branch of the Russian
SSR, Academy of Sciences, Petrozovodsk, USSR
SELECTED CITATIONS: Kazimirov, N. I. & Morozova, R. M. (1973). *Biological Cycling of
Matter in Spruce Forests of Karelia.* Nauka, Leningrad Branch, Academy of Sciences, 168 p.
YEAR(S): 1965–9 FOREST TYPE: Myrtillosum
SOIL TYPE: Humus iron podsol GEOLOGY: Sand Moraine
PRINCIPAL PLANT TYPES: *Picea abies*

AS	37	LAI	3.0	BA	21.9	BAI	1.3
SH	6.8	SD	13750	SCA	6200	SCB	1410
PA	539	PB	90	LAT	62°	LNG	34°
ALT	110	MAR	72100	RBA	31500	MAT	2.2
MAP	650	LGS	150	TGS	11.9	PGS	380
RGS	51400	RBG	30000	SPH	4.1	DRZ	60

	Amount	Increment	Flux	
OL	910.00	21.00	CS	665.00
OBR	1060.00	24.00	CSF	252.00
OBO	4230.00	185.00	CSR	312.00
OBOB	423.00	18.00	CSW	101.00
OBOW	3807.00	167.00	LF	334.00
OSTD	335.00	16.50	LFL	231.00
OSUM	6535.00	246.50	LFBR	36.00
USUM	0.00	0.00	LFBO	67.00
HERB	150.00	1.00		
PSUM	6685.00	247.50		
RL	1410.00	48.00		
RLL	1180.00	40.00		
RLS	87.00	3.00		
RD	112.00	5.50		
STO	1850.00	12.00		

SOUTHERN KARELIAN SPRUCE, SITE 10, KARELIA, USSR

INVESTIGATORS: N. I. Kazimirov, R. M. Morozova   ADDRESS: Karelian Branch of the Russian SSR Academy of Sciences, Petrozovodsk, USSR
SELECTED CITATIONS: Kazimirov, & Morozova, R. M. (1973). *Biological Cycling of Matter in Spruce Forests of Karelia*. Nauka, Leningrad Branch, Academy of Sciences, 168 p.
YEAR(S): 1965–9   FOREST TYPE: Myrtillosum
SOIL TYPE: Humus iron podsol   GEOLOGY: Sand moraine
PRINCIPAL PLANT TYPES: *Picea abies*

AS	45	LAI	3.2	BA	23.5	BAI	1.1
SH	8.8	SD	9240	SCA	7820	SCB	1580
PA	585	PB	104	LAT	62°	LNG	34°
ALT	140	MAR	72100	RBA	31500	MAT	2.2
MAP	650	LGS	150	TGS	11.9	PGS	380
RGS	51400	RBG	30000	SPH	4.2	DRZ	60

	Amount	Increment	Flux	
OL	980.00	14.00	CS	725.00
OBR	1210.00	19.00	CSF	266.00
OBO	5630.00	206.00	CSR	344.00
OBOB	507.00	18.00	CSW	115.00
OBOW	5123.00	188.00	LF	370.00
OSTD	418.00	14.00	LFL	251.00
OSUM	8238.00	253.00	LFBR	40.00
USUM	0.00	0.00	LFBO	79.00
HERB	160.00	2.00		
PSUM	8398.00	255.00		
RL	1580.00	52.00		
RLL	1213.00	43.20		
RLS	101.00	3.30		
RD	119.00	4.50		
STO	1920.00	14.00		

SOUTHERN KARELIAN SPRUCE, SITE 11, KARELIA, USSR

INVESTIGATORS: N. I. Kazimirov, R. M. Morozova   ADDRESS: Karelian Branch of the Russian
SSR, Academy of Sciences, Petrozovodsk, USSR
SELECTED CITATIONS: Kazimirov, N. I. & Morozova, R. M. (1973). *Biological Cycling of
Matter in Spruce Forests of Karelia*. Nauka, Leningrad Branch, Academy of Sciences, 168 p.
YEAR(S): 1965–9   FOREST TYPE: Myrtillosum
SOIL TYPE: Humus iron podsol   GEOLOGY: Sand moraine
PRINCIPAL PLANT TYPES: *Picea abies*

AS	54	LAI	3.6	BA	24.8	BAI	1.1
SH	11.1	SD	4820	SCA	9810	SCB	2160
PA	621	PB	110	LAT	62°	LNG	34°
ALT	130	MAR	72100	RBA	31500	MAT	2.2
MAP	650	LGS	150	TGS	11.9	PGS	380
RGS	51400	RBG	30000	SPH	4.4	DRZ	65–70

	Amount	*Increment*	*Flux*	
OL	1090.00	7.00	CS	774.00
OBR	1420.00	15.00	CSF	282.00
OBO	7300.00	215.00	CSR	369.00
OBOB	583.00	17.00	CSW	123.00
OBOW	6717.00	198.00	LF	412.00
OSTD	562.00	10.60	LFL	273.00
OSUM	10370.00	247.60	LFBR	42.00
USUM	0.00	0.00	LFBO	97.00
HERB	210.00	3.00		
PSUM	10580.00	250.60		
RL	2160.00	53.00		
RLL	1834.00	45.00		
RLS	124.00	3.00		
RD	162.00	3.00		
STO	2200.00	16.00		

SOUTHERN KARELIAN SPRUCE, SITE 12, KARELIA, USSR

INVESTIGATORS: N. I. Kazimirov, R. M. Morozova ADDRESS: Karelian Branch of the Russian SSR, Academy of Sciences, Petrozovodsk, USSR
SELECTED CITATIONS: Kazimirov, N. I. & Morozova R. M. (1973). *Biological Cycling of Matter in Spruce Forests of Karelia.* Nauka, Leningrad Branch, Academy of Sciences, 168 p.
YEAR(S): 1965–9 FOREST TYPE: Myrtillosum
SOIL TYPE: Humus iron podsol GEOLOGY: Sand Moraine
PRINCIPAL PLANT TYPES: *Picea abies*

AS	68	LAI	3.8	BA	29.9	BAI	0.9
SH	14.2	SD	2336	SCA	13260	SCB	2910
PA	617.0	PB	116.0	LAT	62°	LNG	34°
ALT	120	MAR	72100	RBA	31500	MAT	2.2
MAP	650	LGS	150	TGS	11.9	PGS	380
RGS	54100	RBG	30000	SPH	4.3	DRZ	80

	Amount	Increment	Flux	
OL	1150.00	0.0	CS	777.00
OFF	9.00	1.30	CSF	288.00
OBR	1510.00	9.00	CSR	360.00
OBO	10600.00	204.00	CSW	129.00
OBOB	742.00	15.00	LF	433.00
OBOW	9858.00	189.00	LFL	286.00
OSTD	688.00	6.00	LFBR	41.00
OSUM	13960.00	220.30	LFBO	106.00
USUM	0.00	0.00		
HERB	220.00	3.00		
PSUM	14180.00	223.30		
RL	2910.00	50.00		
RLL	2510.00	43.00		
RLS	152.00	2.60		
RD	189.00	1.60		
STO	2470.00	19.00		

SOUTHERN KARELIAN SPRUCE, SITE 13, KARELLA, USSR

INVESTIGATORS: N. I. Kazimirov, R. M. Morozova   ADDRESS: Karelian Branch of the Russian SSR, Academy of Sciences, Petrozovodsk, USSR
SELECTED CITATIONS: Kazimirov, N. I. & Morozova, R. M. (1973). *Biological Cycling of Matter in Spruce Forests of Karelia.* Nauka, Leningrad Branch, Academy of Sciences, 168 p.
YEAR(S): 1965–9   FOREST TYPE: Myrtillosum
SOIL TYPE: Humus iron podsol   GEOLOGY: Sand moraine
PRINCIPAL PLANT TYPES: *Picea abies*

AS	82	LAI	3.8	BA	32.3	BAI	0.7
SH	17.1	SD	1898	SCA	14420	SCB	3320
PA	570	PB	104	LAT	62°	LNG	34°
ALT	140	MAR	72100	RBA	31500	MAT	2.2
MAP	650	LGS	150	TGS	11.9	PGS	380
RGS	51400	RBG	30000	SPH	4.1	DRZ	80

	Amount	Increment	Flux	
OL	1140.00	−2.00	CS	743.00
OFF	29.00	0.29	CSF	288.00
OBR	1680.00	6.00	CSR	330.00
OBO	11600.00	174.00	CSW	125.00
OBOB	696.00	11.00	LF	438.00
OBOW	10900.00	163.00	LFL	288.00
OSTD	810.00	2.90	LFBR	36.00
OSUM	15260.00	181.80	LFBO	114.00
USUM	0.00	0.00		
HERB	320.00	3.00		
PSUM	15580.00	184.80		
RL	3320.00	42.00		
RLL	2842.00	36.00		
RLS	175.00	2.20		
RD	225.00	0.90		
STO	2600.00	20.00		

Dynamic properties of forest ecosystems

SOUTHERN KARELIAN SPRUCE, SITE 14, KARELIA, USSR

INVESTIGATORS: N. I. Kazimirov, R. M. Morozova ADDRESS: Karelian Branch of the Russian SSR, Academy of Sciences, Petrozovodsk, USSR
SELECTED CITATIONS: Kazimirov, N. I. & Morozova, R. M. (1973). *Biological Cycling of Matter in Spruce Forests of Karelia*. Nauka, Leningrad Branch, Academy of Sciencies, 168 p.
YEAR(S): 1965–9 FOREST TYPE: Myrtillosum
SOIL TYPE: Humus iron podsol GEOLOGY: Sand moraine
PRINCIPAL PLANT TYPES: *Picea abies*

AS	98	LAI	3.6	BA	33.1	BAI	0.5
SH	19.6	SD	1319	SCA	18530	SCB	4100
PA	481	PB	92	LAT	62°	LNG	34°
ALT	110	MAR	72100	RBA	31500	MAT	2.2
MAP	6500	LGS	150	TGS	11.9	PGS	380
RGS	51400	RBG	30000	SPH	4.0	DRZ	90

	Amount	Increment	Flux	
OL	1080.00	−4.00	CS	638.00
OFF	38.00	0.50	CSF	261.00
OBR	1650.00	4.00	CSR	265.00
OBO	15800.00	130.00	CSW	112.00
OBOB	885.00	7.00	LF	394.00
OBOW	14900.00	129.00	LFL	263.00
OSTD	798.00	0.50	LFBR	26.00
OSUM	19370.00	131.00	LFBO	105.00
USUM	0.00	0.00		
HERB	310.00	3.00		
PSUM	19680.00	134.00		
RL	4100.00	25.00		
RLL	3573.00	21.50		
RLS	194.00	1.20		
RD	207.00	0.10		
STO	2960.00	21.00		

SOUTHERN KARELIAN SPRUCE, SITE 15, KARELIA, USSR

INVESTIGATORS: N. I. Kazimirov, R. M. Morozova   ADDRESS: Karelian Branch of the Russian SSR, Academy of Sciences, Petrozovodsk, USSR
SELECTED CITATIONS: Kazimirov, N. I. & Morozova, R. M. (1973). *Biological Cycling of Matter in Spruce Forests of Karelia.* Nauka, Leningrad Branch, Academy of Sciences, 168 p.
YEAR(S): 1965–9   FOREST TYPE: Myrtillosum
SOIL TYPE: Humus iron podsol   GEOLOGY: Sand moraine
PRINCIPAL PLANT TYPES: *Picea abies*

AS	109	LAI	3.2	BA	38.9	BAI	0.5
SH	20.0	SD	1080	SCA	19230	SCB	4500
PA	409.0	PB	72.0	LAT	62°	LNG	34°
ALT	110	MAR	72100	RBA	31500	MAT	2.2
MAP	650	LGS	150	TGS	11.9	PGS	380
RGS	51400	RBG	30000	SPH	3.9	DRZ	80

	Amount	*Increment*	*Flux*	
OL	970.00	−5.00	CS	554.00
OFF	41.00	0.30	CSF	246.00
OBR	1760.00	2.00	CSR	214.00
OBO	16500.00	97.00	CSW	94.00
OBOB	858.00	5.00	LF	363.00
OBOW	15640.00	92.00	LFL	248.00
OSTD	711.00	−1.20	LFBR	25.00
OSUM	19980.00	93.10	LFBO	90.00
USUM	0.00	0.00		
HERB	360.00	4.00		
PSUM	20340.00	97.10		
RL	4500.00	15.00		
RLL	3968.00	13.00		
RLS	192.00	0.70		
RD	194.00	−0.30		
STO	3280.00	21.00		

Dynamic properties of forest ecosystems

SOUTHERN KARELIAN SPRUCE, SITE 16, KARELIA, USSR

INVESTIGATORS: N. I. Kazimorov, R. M. Morozova ADDRESS: Karelian Branch of the
Russian SSR, Academy of Sciences, Petrozovodsk, USSR
SELECTED CITATIONS: Kazimirov, N. I. & Morozova, R. M. (1973). *Biological Cycling of
Matter in Spruce Forests of Karelia.* Nauka, Leningrad Branch, Academy of Sciences, 168 p.
YEAR(S): 1965–9 FOREST TYPE: Myrtillosum
SOIL TYPE: Humus iron podsol GEOLOGY: Sand moraine
PRINCIPAL PLANT TYPES: *Picea abies*

AS	126	LAI	2.7	BA	40.5	BAI	0.3
SH	22.6	SD	856	SCA	20870	SCB	4600
PA	309.0	PB	56.0	LAT	62°	LNG	34°
ALT	120	MAR	72100	RBA	31500	MAT	2.2
MAP	650	LGS	150	TGS	11.9	PGS	390
RGS	51400	RBG	30000	SPH	3.8	DRZ	80.0

	Amount	Increment	Flux	
OL	810.00	− 6.00	CS	448.00
OFF	47.00	0.10	CSF	220.00
OBR	1660.00	1.00	CSR	145.00
OBO	18400.00	41.00	CSW	81.00
OBOB	902.00	2.50	LF	328.00
OBOW	17490.00	38.50	LFL	225.00
OSTD	712.00	− 1.90	LFBR	14.00
OSUM	21630.00	34.20	LFBO	89.00
USUM	0.00	0.00		
HERB	400.00	4.00		
PSUM	22030.00	38.20		
RL	4600.00	6.00		
RLL	4050.00	5.20		
RLS	202.00	0.30		
RD	178.00	− 0.40		
STO	3550.00	20.00		

SOUTHERN KARELIAN SPRUCE, SITE 17, KARELIA, USSR

INVESTIGATORS: N. I. Kazimirov, R. M. Morozova   ADDRESS: Karelian Branch of the Russian
SSR, Academy of Sciences, Petrozovodsk, USSR
SELECTED CITATIONS: Kazimirov, N. I. & Morozova, R. M. (1973). *Biological Cycling of
Matter in Spruce Forests of Karelia.* Nauka, Leningrad Branch, Academy of Sciences, 168 p.
YEAR(S): 1965–9   FOREST TYPE: Myrtillosum
SOIL TYPE: Humus iron podsol   GEOLOGY: Sand moraine
PRINCIPAL PLANT TYPES: *Picea abies*

AS	138	LAI	2.4	BA	38.0	BAI	0.2
SH	22.8	SD	1087	SCA	20050	SCB	4750
PA	255	PB	45	LAT	62°	LNG	34°
ALT	130	MAR	72100	RBA	31500	MAT	2.2
MAP	650	LGS	150	TGS	11.9	PGS	380
RGS	51400	RBG	30000	SPH	3.9	DRZ	90

	Amount	Increment	Flux	
OL	740.00	−8.00	CS	394.00
OFF	44.00	—	CSF	212.00
OBR	1710.00	1.00	CSR	109.00
OBO	17600.00	5.00	CSW	73.00
OBOB	810.00	0.30	LF	320.00
OBOW	16790.00	4.70	LFL	217.00
OSTD	744.00	−2.00	LFBR	10.00
OSUM	20840.00	−4.00	LFBO	93.00
USUM	0.00	0.00		
HERB	460.00	4.00		
PSUM	21300.00	0.00		
RL	4750.00	2.00		
RLL	4137.00	1.70		
RLS	201.00	0.10		
RD	201.00	−0.60		
STO	3900.00	20.00		

TIGROVAYA FLOODPLAIN, CENTRAL ASIA, TADJIKISTAN, USSR

INVESTIGATOR: Y. I. Molotovsky   ADDRESS: Botanical Institute, Academy of Sciences of Tadjik SSR, Dushanbe, USSR
OTHER INVESTIGATORS: R. C. Kabilov
SELECTED CITATIONS: Molotovsky, Y. I. & Kabilov, R. C. (1973). *Izvestia TSSR* **3**, (52), p. 3–14.
YEAR(S): 1963–9   FOREST TYPE: Tugai (Populus-Elaeganus)
SOIL TYPE: Alluvial-Meadow type, Tugai soils   GEOLOGY: Floodplain   SOIL DRAINAGE: Well drained
PRINCIPAL PLANT TYPES: Overstory: *Populus prunosa, Elaeagnus angustifolia, Imperata, Erianthus, Phragmites.* Substory: *Tamarix ramosissima.* Field Layer: *Glycyrrhiza glabra, Erianthus, Imperata, Phragmites*

AS	20–40	SH	7	SD	888	SCA	44600
SCB	68800	LAT	N 37° 20′	LNG	E 68° 30′	MAR	109900
MAT	17.3	MAP	185.6	LGS	240	PGS	111.4
SPH	7.4–8.4	DRZ	200–250				

	Amount		Flux	
OL	254.00		LFL	254.00
OBR	**			
OBO	42070.00			

MEATHOP WOOD, UNITED KINGDOM

INVESTIGATOR: J. E. Satchell   ADDRESS: The Nature Conservancy, Merlewood Research Sta,
Grange-Over-Sands, Lancashire, England
OTHER INVESTIGATORS: K. L. Bocock, A. H. F. Brown, R. G. H. Bunce, E. J. White, J. C.
Frankland, A. F. Harrison, O. W. Heal, P. J. A. Howard, A. D. Bailey
SELECTED CITATIONS: Bunce, R. G. H. (1968). *J. Ecology*, **56**, p. 759–75. Sykes, J. M. &
Oikos, **21**, Bunce, R. G. 326–9. Satchell, J. E. (1971). In *Productivity of the Forest Ecosystems
of the World*, ed. P. Duvigneaud. UNESCO, Paris.
YEAR(S): 1967–72   FOREST TYPE: Mixed deciduous
SOIL TYPE: Glacial drift and brown earths   GEOLOGY: Carboniferous
limestone   SOIL DRAINAGE: Free
PRINCIPAL PLANT TYPES: *Quercus petraea, Quercus robur, Betula pendula, B. pubescens,
Fraxinus excelsior, Corylus avellana*

AS	80	LAI	5.3	BA	25.3	BAI	0.5
SH	15	SD	759	SCA	12841.5	SCB	7521.3
PA	992.3	PB	268.8	LAT	54° 12.5'	LNG	2° 53.5'
ALT	45	MAR	72018	MAT	7.8	MAP	1115
LGS	244	TGS	10.0	PGS	797	RGS	62287
SPH	4.0–6.1	DRZ	20–30				

	Amount	Increment		Flux	
OL	268.80			LFL	324.10
OFF	18.20			LFFF	21.10
OBR	3364.00	94.80		LFBR	100.30
OBRB	439.80	16.40		LFBO	76.00
OBRW	2925.00	82.40		LFFR	39.00
OBO	7592.00	216.20		PRE	8.00
OBOB	1023.00	37.90		LWAS	41.60
OBOW	6570.00	186.90		STMF	5.60
OSTD	490.70	13.60			
UL	55.30				
UBR	456.20				
UBO	969.40				
USTD	54.90				
HERB	72.10	39.30			
RL	7534.00	197.80			
RLL	6462.00	138.50			
RLS	1072.00	43.50			
RD	747.60				
STO	671.80				
RZ	13800.00				
IRZ	9722.00				
ERZ	4083.00				
AGC	0.40	0.30			
DCFA	16.90	20.40			

647

BLACK SPRUCE–FEATHER MOSS SITE, ALASKA, USA

INVESTIGATOR: K. Van Cleve ADDRESS: Forest Soils Laboratory, University of Alaska, Fairbanks, Alaska 99701

OTHER INVESTIGATORS: Richard Barney

SELECTED CITATIONS: US Dept of Commerce. Local climatological data, annual summary with comparative data, Fairbanks, Alaska (1970).

YEAR(S): 1968–71 FOREST TYPE: Black Spruce, Feather Moss

SOIL TYPE: Pergelic cryaquept GEOLOGY: Bedrock: Birch Creek schist: basic in reaction

SOIL DRAINAGE: Moderately good

PRINCIPAL PLANT TYPES: Overstory: *Picea mariana* Substory: *Ledum groenlandicum, Vaccinium vitis-idaea, Geocaulon lividum, Alnus crispa, Equisetum spp.* Field Layer: *Hylocomium splendens, Pleurozium schreberi, Ptilium crista–castrensis*

AS	130	BA	34.7	BAI	2.4	SH	13.7
SD	5000	SCA	23098.7	SCB	5169.7	PA	159.0
PB	79.5	LAT	N 64°	LNG	W 148°	SPH	5.4–6.4
DRZ	20						

	Amount	Increment	Flux	
OL	892.60	7.4	LFL	22.74
OFF	527.00	7.3	LFFF	7.09
OBR	1295.00	25.9	LFBR	16.55
OBO	8605.00	118.4	LFBO	7.05
OSTD	3108.00			
UL	40.86			
UFF	**			
UBR	**			
UBO	473.30			
USTD	IN. OV.			
HERB	245.70			
RL	5170.00			
RZ	4749.00			

BLACK SPRUCE MUSKEG, SITE 1, ALASKA, USA

INVESTIGATOR: K. Van Cleve   ADDRESS: Forest Soils Laboratory, University of Alaska, Fairbanks, Alaska 99701
OTHER INVESTIGATORS: Richard Barney
SELECTED CITATIONS: US Dept of Commerce. Local Climatological Data, Annual Summary with Comparative Data, Fairbanks, Alaska (1970).
YEAR(S): 1968–71   FOREST TYPE: Black Spruce Muskeg
SOIL TYPE: Pergelic cryaquept   GEOLOGY: Parent material is loess: Alkaline in reaction
SOIL DRAINAGE: Poor
PRINCIPAL PLANT TYPES: Overstory: *Picea mariana* Substory: *Ledum groenlandicum, Vaccinium uliginosum, Vaccinium vitis-idaea, Rosa acicularis, Salix* spp. Field Layer: *Pleurozium schreberi, Hylocomium splendens, Polytrichum perinum, Sphagnum girgensohnii, Cladonia rangiferina, Cladonia* spp. *Peltigera apthosa, Peltigera scabrosa*

AS	51	BA	18.18	BAI	0.27	SH	2.9
SD	27335	SCA	11153.4	SCB	1245.7	PA	51.4
PB	41.1	LAT	N 64°	LNG	W 148°	ALT	166.7
MAT	−3.4	MAP	268.8	LGS	60–80	TGS	11.9
PGS	150.6	SPH	5.3	DRZ	20		

	Amount	Increment
OL	376.20	9.8
OFF	125.00	2.2
OBR	359.00	10.9
OBO	799.50	28.5
OSTD	430.80	
UL	44.70	
UFF	**	
UBR	**	
UBO	71.10	
USTD	IN. OV.	
HERB	748.10	
RL	1246.00	
STO	8172.00	
RZ	3951.00	

BLACK SPRUCE MUSKEG, SITE 2, ALASKA, USA

INVESTIGATOR: K. Van Cleve   ADDRESS: Forest Soils Laboratory, University of Alaska, Fairbanks, Alaska 99701
OTHER INVESTIGATORS: Richard Barney
SELECTED CITATIONS: US Dept of Commerce. Local Climatological Data, Annual Summary With Comparative Data, Fairbanks, Alaska (1970).
YEAR(S): 1968–71   FOREST TYPE: Black Spruce Muskeg
SOIL TYPE: Pergelic cryaquept   GEOLOGY: Bedrock: Birch Creek schist: Loessic soil
SOIL DRAINAGE: Good
PRINCIPAL PLANT TYPES: Overstory: *Picea mariana.* Substory: *Ledum groenlandicum, Vaccinium uliginosum.* Field Layer: *Pleurozium schreberi, Hylocomium splendens, Polytrichum perinium, Dicranum sp.* Other: *Peltigera apthosa, Cladonia sp., Nephroma articum*

AS	55	BA	22.00	BAI	0.70	SH	3.1
SD	14820	SCA	16370.6	SCB	1040.1	PA	111.4
PB	48.02	LAT	N 64°	LNG	W 148°	ALT	469.7
MAT	−3.4	MAP	286.8	LGS	60–80	TGS	13.0
PGS	164.3	SPH	4.6–5.6	DRZ	30–50		

	Amount	Increment
OL	459.20	17.2
OFF	71.80	2.2
OBR	469.10	22.1
OBO	1402.00	69.9
OSTD	262.40	
UL	53.70	
UFF	**	
UBR	**	
UBO	95.30	
USTD	IN. OV.	
HERB	479.30	
RL	1040.00	
STO	12890.00	
RZ	3535.00	

HUBBARD BROOK, NEW HAMPSHIRE, USA

INVESTIGATORS: R. H. Whittaker *et al.*   ADDRESS: Department of Ecology and Systematics, Cornell University, Ithaca, New York

OTHER INVESTIGATORS: F. H. Bormann, G. E. Likens, T. G. Siccama, J. R. Gosz

SELECTED CITATIONS: Bormann, F. H. *et al.* (1974). *Ecol. Monog.* **44**, 233–55. Siccama, T. G. *et al* (1970). *Ecol. Monog.* **40**, 389–402. Gosz, J. R. *et al.* (1972). *Ecology*, **53**, 769–84.

YEAR(S): 1961–5   FOREST TYPE: Northern Hardwoods

SOIL TYPE: Boulders, Glacial Till. Podzolic–Harplorthod   GEOLOGY: Littleton Gneiss

SOIL DRAINAGE: Downslope, none through bedrock

PRINCIPAL PLANT TYPES: Overstory: *Acer saccharum, Betula lutea, Fagus grandifolia.* Substory: *Acer spicatum, Viburnum alnifolium.* Field Layer: *Dryopteris spinulosa, Dennstaedtia punctilobula*

AS	110	LAI	6.0	BA	25	BAI	0.451
SH	16.8	SD	1290	SCA	13281	SCB	2846
PA	957	PB	190	LAT	N 44° 00′	LNG	W 71° 30′
ALT	550–710	MAP	1250	LGS	110	TGS	17.5

	Amount	Increment	Flux	
OL	315.00		CSF	21.00
OFF	16.00		LFL	342.00
OBR	3857.00	286.00	LFFF	27.00
OBO	9094.00	291.00	LFBR	**
OBOB	872.00	29.20	LFBO	217.00
OBOW	8222.00	261.40	LFFR	18.00
OSUM	10110.00	577.00		
UL	8.80			
UBR	**			
UBO	12.00			
USUM		22.00		
HERB	7.30			
RL	2826.00	181.30		
STO	4680.00	602.00		
RZ	17490.00			

BROOKHAVEN, NEW YORK, USA

INVESTIGATOR(S): R. H. Whittaker, G. M. Woodwell ADDRESS: Department of Ecology and Systematics, Cornell University, Ithaca, New York
SELECTED CITATIONS: Whittaker, R. H. & Woodwell, G. M. (1968). *J. Ecol.* **56,** 1–25.
Whittaker, R. H. & Woodwell, G. M. (1969). *J. Ecol.* **57,** 155–74.
YEAR(S): 1964 FOREST TYPE: Oak–Pine Coastal
SOIL TYPE: Sandy Podzolic GEOLOGY: Sandy glacial outwash plain
SOIL DRAINAGE: Rain percolation and nutrient loss to water table
PRINCIPAL PLANT TYPES: Overstory: *Quercus alba, Q. coccinea, Pinus rigida.* Substory: *Gaylussacia baccata, Vaccinium vacillans.* Field Layer: *Gaultheria procumbens* Other: *Quercus velutina*

AS	43 mean	LAI	3.8	BA	15.6	BAI	0.356
SH	8.9 mean	SD	185.4	SCA	6563	SCB	3635
PA	859	PB	337	LAT	N 40° 50′	LNG	W 72° 54′
ALT	0.0	MAT	9.8	MAP	1240	DRZ	50–200

	Amount	Increment	Flux	
OL	412.00		LFL	382.00
OFF	22.00		LFFF	22.00
OBR	1645.00	247.30		
OBO	4323.00	175.00		
OBOB	812.00	26.50		
OBOW	3511.00	148.90		
OSUM	6043.00	422.30		
UL	31.00			
UFF	2.02			
UBR	55.10	21.03		
UBO	70.70	6.64		
UBOB	16.40	1.76		
UBOW	54.30	4.88		
USUM	158.50	27.67		
HERB	2.20			
PSUM	6561.00	450.00		
RL	3631.00	332.70		

WATERSHED 1, COWEETA, NORTH CAROLINA, USA

INVESTIGATOR: W. T. Swank ADDRESS: Coweeta Hydrologic Laboratory, P.O. Box 601, Franklin, North Carolina 28734

OTHER INVESTIGATORS: H. T. Schreuder

SELECTED CITATIONS: Swank, W. T. & Schreuder, H. T. (1973). In *IUFRO Biomass Studies, Inter. Un. of Forest Research Organ., Univ. of Maine, Orono*, 171–82. Swank, W. T. & Schreuder, H. T. (1974). *Forest Science*, **20**, 91–100.

YEAR(S): 1967, 9, 72 FOREST TYPE: White Pine plantation

SOIL TYPE: Saluda stony loam (Topic hapludult, loamy, mixed, mesic, shallow family)

GEOLOGY: Granite, mica shists, and gneisses (bedrock) SOIL DRAINAGE: Well drained

PRINCIPAL PLANT TYPES: *Pinus strobus*

AS	15	LAI	17.8	BA	23.4	BAI	3.6
SH	12	SD	1760	SCA	6959.9	PA	1350
LAT	N 35° 04′	LNG	W 83° 26′	ALT	706–988	MAR	149830
MAT	13.6	MAP	1628	LGS	150	TGS	19.2
PGS	697	RGS	67412	SPH	5.6	DRZ	80

	Amount	Increment	Flux	
OL	466.40		LFL	363.20
OFF	5.40			
OBR	2283.00	305.20		
OBO	4211.00	684.40		
OSTD	8.60			
OSUM	0.00			
HERB	0.00			
STO	1422.00			
RZ	12410.00			

Dynamic properties of forest ecosystems

WATERSHED 18, COWEETA, NORTH CAROLINA, USA

INVESTIGATORS: C. D. Monk, F. P. Day   ADDRESS: Department of Botany, University of
Georgia, Athens, Georgia, 30601, USA
YEAR(S): 1970–3   FOREST TYPE: Oak–Hickory
SOIL TYPE: Saluda stony loam   GEOLOGY: Granite, Mica Shist and Gneisses (Bedrock)
SOIL DRAINAGE: Well drained
PRINCIPAL PLANT TYPES: Overstory: *Acer rubrum, Quercus prinus.* Substory: *Rhododendron maximum, Kalmia latifolia.* Field Layer: Ferns

AS	60–200	BA	25.6	SH	25	SD	3046
SCA	13990	SCB	3064.6	PA	875.4	LAT	N 35° 03′
LNG	W 83° 26′	ALT	726–993	MAR	111129	MAT	12.6
MAP	1945	LGS	150	TGS	18.4	PGS	810
RGS	63697	SPH	5.2	DRZ	80		

	Amount		*Flux*	
OL	347.40		LFL	319.70
OFF	41.90		LFFF	41.90
OBR	2205.00		LFBO	134.10
OBO	9706.00			
OSTD	958.10			
UL	214.10			
UBR	422.40			
UBO	1095.00			
USUM	1731.00			
HERB	5.60			
RL	3065.00			
STO	885.20			
RZ	13740.00			

DUKE FOREST, NORTH CAROLINA, USA

INVESTIGATOR(S): C. Wells, J. Jorgensen
YEAR(S): 1968 FOREST TYPE: Pine Plantation
SOIL TYPE: Granville coarse sandy loam GEOLOGY: Appalachian Piedmont SOIL DRAINAGE:
Moderately well drained
PRINCIPAL PLANT TYPES: *Pinus taeda*

AS	16	BA	49	BAI	2.18	SH	15.0
SD	2243	SCA	15600	SCB	3630	PA	1300
LAT	36°	LNG	79°	ALT	149	MAT	15.6
MAP	1150	LGS	231	TGS	20.3	PGS	749.3
SPH	4.5	DRZ	40				

	Amount	Increment	Flux	
OL	798.00	0.00	LF	774.90
OBR	1459.00	190.00		
OBO	12480.00	560.00		
OBOB	1523.00			
OBOW	10960.00			
OSTD	860.00			
OSUM	15600.00	750.00		
RL	3630.00	200.00		

TRIANGLE SITE, SAXAPAHAW, NORTH CAROLINA, USA

INVESTIGATOR(S): C. W. Ralston, R. S. Kinerson, C. Wells, C. E. Murphy
YEAR(S): 1972 FOREST TYPE: Pine Plantation
SOIL TYPE: Appling Sandy Loam GEOLOGY: Appalachian Piedmont SOIL DRAINAGE:
Somewhat restricted perched water table (30 cm) above B-horizon at times in winter.
PRINCIPAL PLANT TYPES: *Pinus taeda*

AS	15	LAI	3.9–6.6	BA	41.3	SH	11.58
SD	1470	SCA	9263	SCB	2184	PA	1063
PB	165	LAT	36°	LNG	79°	ALT	144
RBA	130620	MAT	15.6	MAP	1150	LGS	231
TGS	20.3	PGS	749.3	SPH	4.5	DRZ	40

	Amount	Increment	Respiration	Flux	
OL	601.00	30.00	47.90	LF	544.80
OBR	1179.00	79.00	705.00	LFL	507.40
OBO	7483.00	954.00	209.00	LFBR	37.40
OSUM	9263.00	1063.00	961.90		
PSUM	9263.00	1063.00	961.90		
RL	2184.00	165.00	107.20		
STO	600.00				

ANDREWS EXPERIMENTAL FOREST, WATERSHED 10, OREGON, USA

INVESTIGATOR: Charles C. Grier   ADDRESS: School of Forestry, Oregon State University, Corvallis, Oregon
OTHER INVESTIGATORS: R. H. Waring, P. Sollins
SELECTED CITATIONS: Franklin, J. F. & Dyrness, C. T. (1973). *USDA For. Ser. Res. Paper PNW–8*. Fujimori, T. (1971). *USDA For. Ser. Res. Paper PNW*–123, 1–11.
YEAR(S): 1972   FOREST TYPE *Pseudotsuga Menziesii*
SOIL TYPE: Inceptosol   GEOLOGY: Miocene tuffs and brecias   SOIL DRAINAGE: Good
PRINCIPAL PLANT TYPES: *Pseudotsuga menziesii, Tsuga heterophylla, Castanopsis chrysophylla, Thuja plicata, Pinus lambertiana, Acer macrophyllum*

AS	450	LAI	12.5	BA	62.7	BAI	0.25
SH	60	SD	290	SCA	77103.6	SCB	8249
PA	1171	PB	722	LAT	N 44° 15′	LNG	W 122° 20′
ALT	430–670	MAR	124100	MAT	8.5	MAP	2250
LGS	150	TGS	14	PGS	200	RGS	76500
SPH	6.2	DRZ	100				

	Amount	Increment	Respiration	Flux	
OL	1329.00	0.50		LF	446.00
OBR	4783.00	34.50		LFL	217.00
OBO	64590.00	234.40		LFFF	44.40
OBOB	7033.00	30.00		LFBR	65.80
OBOW	57560.00	204.40		LFBO	27.00
OSTD	5520.00	190.00		LFFR	40.00
OSUM	76230.00	459.40		LFEP	35.00
UL	359.50	75:40			
UBR		**			
UBO	510.40	108.60			
USUM	869.90	184.00			
HERB	6.50	6.00			
EPIP	110.00	0.00			
PSUM	77100.00				
RL	15770.00	57.20			
RLL	14800.00	57.20			
RLS	970.00				
RD	314.00				
STO	11400.00		366.00		
RZ	15800.00		616.00		
SUBS	7000.00				
DCFA	0.27	0.00	12.20		
DCFL	40.40	0.00			

LIRIODENDRON SITE OAK RIDGE, TENNESSEE, USA

INVESTIGATORS: D. E. Reichle, N. T. Edwards, W. F. Harris, P. Sollins   ADDRESS: Oak Ridge National Laboratory, P.O. Box X, Oak Ridge, Tennessee 37830
OTHER INVESTIGATORS: B. E. Dinger, R. K. McConathy, B. A. Hutchinson, D. R. Matt, D. E. Reichle, F. G. Taylor, H. H. Shugart
SELECTED CITATIONS: Harris, W. F., Sollins, P., Edwards, N. T., Dinger, B. E. & Shugart, H. H. (1975). In *Productivity of World Ecosystems*, ed. D. E. Reichle, J. F. Franklin and D. W. Goodall, NAS, Wash., D.C. Taylor, F. G., Jr. (1974). In *Phenology and Modeling of Seasonality*, Springer-Verlag.
YEARS: 1968–72   FOREST TYPE: Liriodendron–Mesic Deciduous
SOIL TYPE: Deep Alluvial Emory Silt Loam   GEOLOGY: Knox Dolomite Limestone   SOIL DRAINAGE: Well drained
PRINCIPAL PLANT TYPES: *Liriodendron tulipifera, Quercus spp., Carya tomentosa, Cornus florida, Cercis canadensis, Pinus echinata*

AS	50	LAI	5.1	BA	22.1	BAI	0.36
SH	30	SCA	13378	SCB	3600	PA	742
PB	710	LAT	N 35° 55′	LNG	W 80° 77′	ALT	225
MAR	116000	RBA	123500	MAT	13.3	MAP	1400
LGS	180	TGS	18.1	PGS	584	RGS	74500
RBG	81604	SPH	5.8	DRZ	70		

	Amounts	Increment	Autotrophic respiration[a]	Heterotrophic respiration[a]	Net photosynthesis
OL	323.00		649.00		3009.00
OFF	23.00				
OBR	2711.00	57.00			
OBO	9440.00	225.40			
OBOW					
OSTD	620.00				
OSUM	13117.00	282.90			
UL	50.00		340.00		1239.00
UBR	213.00	2.00			
UBO	589.00	6.00			
USUM	853.00	8.00			
HERB	28.20		25.00		137.00
PSUM			3586.00[b]		4385.00
RL	2726.00	100.00	931.00		
RLL	2046.00[c]	50.00[c]	224.00		
RLS	680.00	50.00	707.00		
STO	557.00			1124.00	173.00
RZ	22155.00			547.00	
DCFA	20.80			57.00	
DCFL	11.60				

[a] CO_2 fluxes converted to organic matter by multiplying by 0.681 (assumes metabolism of glucose).

[b] This value includes all autotrophic respiration (above- and below-ground).

[c] Includes stump with large lateral roots.

WALKER BRANCH SITE 1, OAK RIDGE, TENNESSEE, USA

INVESTIGATOR(S): W. F. Harris, G. S. Henderson   ADDRESS: Oak Ridge National Laboratory, P.O. Box X, Oak Ridge, Tennessee 37830
OTHER INVESTIGATORS: T. Grizzard, R. J. Luxmoore, S. B. McLaughlin, N. T. Edwards, D. S. Shriner, D. E. Todd, Jr.
SELECTED CITATIONS: Harris, W. F., Henderson, G. S. & Goldstein, R. A. (1973). *IUFRO Symp.*, U. Maine Press. Henderson, G. S. & Harris, W. F. (1975). In *Forest Soils and Forest Land Management*, pp. 179–93. Les Presses de L'Université Laval, Quebec. Grizzard, T., Henderson, G. S., Clebsch, E. E. C. & Reichle, D. E. (1976). ORNL–TM–5254
FOREST TYPE: Oak–Hickory
SOIL TYPE: Typic paleudults derived from Dolomitic residuum   GEOLOGY: Knox Dolomite   SOIL DRAINAGE: Well drained
PRINCIPAL PLANT TYPES: Overstory: *Carya spp., Quercus alba, Q. prinus, Q. velutina.* Substory: *Cornus florida, Nyssa sylvatica, Oxydendrum arboreum, Acer rubrum*

AS	30–80	BA	20	BAI	0.25	SH	25
SCA	12160	SCB	3330	PA	775	PB	200
LAT	N 35° 58′	LNG	W 84° 17′	ALT	265–360	MAR	116000
MAT	13.3	MAP	1400	LGS	180	TGS	18.1
PGS	584	RGS	74500	SPH	4.0–6.5	DRZ	75

	Amount	*Increment*	*Flux*	
OL	420.00		LF	480.00
OBR	2690.00		LFL	410.00
OBO	9050.00		LFBR	40.00
OSTD	118.00		LFBO	30.00
OSUM	483.00			
RL	3333.00	200.00		
RLL	860.00	40.00		
RLS	790.00	50.00		

WALKER BRANCH SITE 2, OAK RIDGE, TENNESSEE, USA

INVESTIGATORS: W. F. Harris, G. S. Henderson   ADDRESS: Oak Ridge National Laboratory,
P. O. Box X, Oak Ridge, Tennessee 37830
OTHER INVESTIGATORS: T. Grizzard, R. J. Luxmoore, S. B. McLaughlin, N. T. Edwards, D. S.
Shriner, D. E. Todd, Jr.
SELECTED CITATIONS: Harris, W. F., Henderson, G. S. & Goldstein, R. A. (1973). *IUFRO
Symp.*, U. Maine Press. Henderson, G. S. & Harris, W. F. (1975). In *Forest Soils and Forest
Land Management*, pp. 179–93. Les Presses de L'Université Laval, Quebec. Grizzard, T.,
Henderson, G. S., Clebsch, E. E. C. & Reichle, D. E. (1976). ORNL–TM–5254.
FOREST TYPE: Pine Forest
SOIL TYPE: Typic paleudults derived from Dolomitic residuum   GEOLOGY: Knox
Dolomite   SOIL DRAINAGE: Well drained
PRINCIPAL PLANT TYPES: Overstory: *Pinus echinata, Liriodendron tulipifera*. Substory:
Oxydendrum arboreum, Cornus florida

AS	30	BA	21.3	BAI	−0.29	SH	12–25
SCA	12140	SCB	3350	PA	676	PB	200
LAT	N 35° 58′	LNG	W 84° 17′	ALT	265–360	MAR	116000
MAT	13.3	MAP	1400	LGS	180	TGS	18.1
PGS	584	RGS	74500	SPH	4.0–6.5	DRZ	60

	Amount	Increment	Flux	
OL	460.00		LF	413.00
OBR	2760.00		LFL	340.00
OBO	8960.00		LFBR	20.00
OSTD		190.00	LFBO	53.00
OSUM		415.00		
RL	3350.00	200.00		
RLL	860.00	50.00		
RLS	790.00	50.00		
STO	4650.00			

Dynamic properties of forest ecosystems

WALKER BRANCH SITE 3, OAK RIDGE, TENNESSEE, USA

INVESTIGATORS: W. F. Harris, G. S. Henderson ADDRESS: Oak Ridge National Laboratory, P. O. Box X, Oak Ridge, Tennessee 37830
OTHER INVESTIGATORS: T. Grizzard, R. J. Luxmoore, S. B. McLaughlin; N. T. Edwards D. S. Shriner, D. E. Todd, Jr.
SELECTED CITATIONS: Harris, W. F. Henderson, G. S. & Goldstein, R. A. (1973). *IUFRO Symp.*, U. Maine Press. Henderson, G. S. & Harris, W. F. (1975). In *Forest Soils and Forest Land Management*, pp. 179–92. Les Presses de L'Universite Laval, Quebec. Grizzard, T., Henderson, G. S., Clebsch, E. E. C. & Reichle, D. E. (1973). ORNL–TM–5254.
FOREST TYPE: Chestnut Oak
SOIL TYPE: Typic Paleudults derived from Dolomitic residuum
GEOLOGY: Knox Dolomite SOIL DRAINAGE: Well drained
PRINCIPAL PLANT TYPES: Overstory: *Quercus prinus, Carya spp., Quercus alba, Q. rubra, Q. velutina.* Substory: *Cornus florida, Nyssa sylvatica, Oxydendrum arboreum*

AS	30–80	BA	25.8	BAI	0.59	SH	12–25
SCA	13790	SCB	3330	PA	952	PB	200
LAT	N 35° 58′	LNG	W 84° 17′	ALT	265–360	MAR	116000
MAT	13.3	MAP	1400	LGS	180	TGS	18.1
PGS	584	RGS	74500	SPH	4.0–6.5	DRZ	60

	Amount	Increment	Flux	
OL	470.00		LF	445.00
OBR	3030.00		LFL	390.00
OBO	10290.00		LFBR	40.00
OSTD		55.00	LFBO	15.00
OSUM		540.00		
RL	3330.00	200.00		
RLL	860.00	50.00		
RLS	790.00	50.00		
STO	4900.00			

660

WALKER BRANCH SITE 4, OAK RIDGE, TENNESSEE, USA

INVESTIGATORS: W. F. Harris, G. S. Henderson ADDRESS: Oak Ridge National Laboratory, P. O. Box X, Oak Ridge, Tennessee 37830

OTHER INVESTIGATORS: T. Grizzard, R. J. Luxmoore, S. B. McLaughlin, N. T. Edwards, D. S. Shriner, D. E. Todd, Jr.

SELECTED CITATIONS: Harris, W. F. Henderson, G. S. & Goldstein, R. A. (1973). *IUFRO Symp.*, U. Maine Press (1973). Henderson, G. S. & Harris, W. F. (1975). In *Forest Soils and Forest Land Management*, pp 179–93. Les Presses de L'Université Laval, Quebec.

FOREST TYPE: Yellow Poplar

SOIL TYPE: Typic paleudults derived from Dolomitic residuum

GEOLOGY: Knox Dolomite SOIL DRAINAGE: Well drained

PRINCIPAL PLANT TYPES: Overstory: *Liriodendron tulipifera, Carya spp., Quercus alba, Q. rubra.* Substory: *Acer rubrum, Cornus florida, Oxydendrum arboreum*

AS	30–80	BA	19.03	BAI	0.22	SH	12–25
SCA	10860	SCB	3330	PA	608	PB	200
LAT	N 35° 58'	LNG	W 84° 17'	ALT	265–360	MAR	116000
MAT	13.3	MAP	1400	LGS	180	TGS	18.1
PGS	584	RGS	74500	SPH	4.0–6.5	DRZ	60

	Amount	Increment	Flux	
OL	390.00		LF	433.00
OBR	2120.00		LFL	370.00
OBO	8350.00		LFBR	30.00
OSTD		118.00	LFBO	33.00
OSUM		355.00		
RL	3330.00	200.00		
RLL	860.00	50.00		
RLS	790.00	50.00		
STO	2340.00			

Dynamic properties of forest ecosystems

THOMPSON RESEARCH CENTER, SEATTLE, WASHINGTON, USA

INVESTIGATORS: S. P. Gessel, P. Sollins   ADDRESS: Coniferous Forest Biome, Biome Central Office AR–10, Seattle, Washington 98195
OTHER INVESTIGATORS: D. W. Cole, K. L. Reed, H. Riekerk, D. R. M. Scott, F. C. Ugolini, R. B. Walker
SELECTED CITATIONS: Cole, D. W., Gessel, S. P. & Dice, S. F. (1967). In *Proc. Symp. on Primary Prod. and Mineral Cycling in Natural Ecosystems'*, *AAAS Meeting, 1967*, pp. 191–232. University of Maine Press. Grier, C. C. & McColl, J. C. (1971). *Soil Sci. Soc. Am. Proc.* **35**, 988–91.
YEAR(S): 1961–Present   FOREST TYPE: Douglas Fir Plantation
SOIL TYPE: Typic haplorthod, Everett gravelly sandy loam   GEOLOGY: River Terraces
SOIL DRAINAGE: Well drained
PRINCIPAL PLANT TYPES: *Pseudotsuga menziesii, Tsuga heterophylla*

AS	36 (1967)	BA	3.77	SH	18	SD	2223
SCA	17430	SCB	3070	PA	1436	PB	363
LAT	N 47° 23'	LNG	W 121° 57'	ALT	210	MAR	99280
RBA	70557	MAT	9.8	MAP	1360	LGS	214
TGS	14	PGS	670	SPH	5.5	DRZ	60–100

	Amount	Increment	Flux	
OL	909.70	32.10	LF	183.00
OBR	1389.00	42.00	LWAS	3.00
OBOB	1873.00	69.10	STMF	0.02
OBOW	12170.00	604.90		
OSTD	814.50			
OSUM	17150.00			
HERB	101.00			
RL	3299.00	118.50		
STO	2277.00			
RZ	11160.00			

662

NOE WOODS (LAKE WINGRA), WISCONSIN, USA

INVESTIGATOR(S): O. L. Loucks, G. J. Lawson   ADDRESS: Department of Biology, University of Wisconsin, Madison, Wisconsin
SELECTED CITATIONS: Lawson, G. J. & Cottam, G. (1971). Eastern Deciduous Forest Biome Memo 71–51. Lawson, G. J., Cottam, G. & Loucks, O. L. (1972). Eastern Deciduous Forest Biome Memo 72–98.
YEAR(S): 1970–2   FOREST TYPE: Oak: Wisconsin (Prairie–Border)
SOIL TYPE: Miami Silt Loam (Dodge)   GEOLOGY: Loess, over Pleistocene outwash, over Ordovician Limestone + Dolomite   SOIL DRAINAGE: Deep
PRINCIPAL PLANT TYPES: *Quercus alba, Q. velutina, Prunus serotina*

AS	130	LAI	4.4	BA	33.2	BAI	0.32
SH	23.7	SD	422	SCA	26594.9	SCB	6603
PA	819	PB	662	LAT	N 43° 02′	LNG	W 89° 24′
ALT	273.6	RBA	40000	MAT	6.9	MAP	777
LGS	171	TGS	19.5	PGS	292	SPH	5.8
DRZ	150						

	Amounts	Increment	Respiration	Photo-synthesis	Flux	
OL	386.40		888.70	502.30	LFL	428.90
OFF	7.70		17.70	10.00	LFFF	7.70
OBR	5078.00	72.50	362.00	290.00	LFBR	90.10
OBO	20920.00	270.50	459.90	189.40	LFBO	149.00
OSTD	2058.00	349.20		9.90		
OSUM	26490.00		1729.00	1002.00		
UL	24.40		110.70	21.10		
UBRW	61.70					
UBO		8.80				
USTD	8.80	8.80				
USUM	86.10		110.70			
HERB	18.10	18.10	41.50	21.70		
PSUM	26590.00					
RL	6603.00	660.30	1176.00	517.80		
RD				540.80		
STO	470.80	609.30		578.40		
RZ	17580.00	149.90		150.10		

NAKOMA URBAN FOREST, WISCONSIN, USA

INVESTIGATORS: O. L. Loucks, G. J. Lawson   ADDRESS: Department of Biology, University of Wisconsin, Madison, Wisconsin

SELECTED CITATIONS: Lawson, G. J. & Cottam, G. (1971). Eastern Deciduous Forest Biome Memo 71–51. Lawson, G. J., Cottam, G. & Loucks, O. L. (1972). Eastern Deciduous Forest Biome Memo 72–98.

YEAR(S): 1970–2   FOREST TYPE: Urban oak remnant

PRINCIPAL PLANT TYPES: Overstory: *Quercus alba, Q. velutina.* Substory: *Lonicera spp., Syringa vulgaris.* Field Layer: *Poa pratensis*

AS	130	LAI	2.9	SH	27.4	LAT	N 43° 02'
LNG	W 89° 24'	ALT	273.6	LGS	171	PGS	292
DRZ	150						

	Amount	Increment	Flux	
OL	259.40		LFL	259.40
OBR	2600.00	59.00		
OBO	9902.00	222.90		
HERB	174.80			

FA. EGLHARTING, ABT. 27A, FEDERAL REPUBLIC OF GERMANY

INVESTIGATORS: A. Baumgartner, H. Hager   ADDRESS: D–8 Munchen 40, Amalienstrasse 52,
Meteorological Institute, Federal Republic of Germany
OTHER INVESTIGATORS: Koche, Droste, Klemmer, Tajchmann, Kerner, Strauss
SELECTED CITATIONS: Baumgartner, A. (1969). *Photosynthetica*, **3**, 127–49. Droste zu Huhlshoff,
B. (1969). Dissertation, Univ. Munich. Hager, H., (1975). Univ. Munich–Meteor. Inst.
Wissensch. Mitteilungen **26**, 183p. Strauss, R. (1971). Univ. Munich–Meteor. Inst. Wissensch.
Mitteilungen **22**, 66p.
YEAR(S): 1964–72   FOREST TYPE: Artificial Spruce (Natural Hardwoods)
SOIL TYPE: Parabraunerde   GEOLOGY: Old Wurmgravel   SOIL DRAINAGE: Average
PRINCIPAL PLANT TYPES: *Spruce*

AS	80	LAI	21.6	BA	57.42	BAI	1.03
SH	30.09	SD	800	SCA	32200	PA	1551.0
LAT	48°	LNG	12°	ALT	552	MAR	99000
RBA	54600	MAT	7.0	MAP	875	LGS	160–170
TGS	15	PGS	475	RGS	80850	SPH	4.6–6.5
DRZ	65			RBG	44550		

	Amount	Increment
OL	1585.00	0.00
OBR	2828.00	329.00
OBO	26800.00	588.00
OSTD	998.00	
OSUM	32210.00	

665

SOLLING PROJECT, SITE B 1, FEDERAL REPUBLIC OF GERMANY

INVESTIGATOR: H. Ellenberg ADDRESS: Systematisch–Geobotanisches Institut der Universitaet, 34 Goettingen, Untere Karspule 2, Federal Republic of Germany
OTHER INVESTIGATORS: H. Heller, S. Ulrich, W. Funke and others
SELECTED CITATIONS: Integrated Experimental Ecology, (1971). *Ecol. Studies*, **2**, Springer Verlag. Vegetationsuntersuchungen im Solling, (1970). *Schriftenr. Veg. Kunde*, **5**. Runge, M. (1973). *Energie Umsaetze in den Biozoenosen Terrestr. Oekosysteme, Scripta Geobotanica* **4**.
YEAR(S): 1969 FOREST TYPE: Luzulo–Fagetum
SOIL TYPE: Acid braunerde GEOLOGY: Buntsandstein SOIL DRAINAGE: Well drained
PRINCIPAL PLANT TYPES: *Fagus sylvatica*

AS	122	LAI	5.88	BA	28.29	BAI	0.56
SH	26.5	SD	243	SCA	27500.0	SCB	3710.0
PA	1122.8	PB	250.0	LAT	N 51° 49'	LNG	E 9° 35'
ALT	500	MAT	6.1	MAP	1063	LGS	144
TGS	12.9	PGS	530.7	RGS	60000	RBG	30000
SPH	3.7	DRZ	60				

	Amount	Increment	Flux	
OL	307.80		CSF	18.80
OFF	36.00		CSHB	1.30
OBR	3248.00	78.40	LFL	297.70
OBRB	650.00	16.00	LFFF	77.60
OBRW	2598.00	62.40		
OBO	23840.00	649.00		
OBOB	1550.00	42.00		
OBOW	22290.00	607.00		
OSTD	68.70			
OSUM	27500.00			
HERB	1.30			
RLL	3004.00	66.00		
RLS	707.00	0.00		
STO	2970.00			
IRZ	19000.00			

SOLLING PROJECT, SITE B 3, FEDERAL REPUBLIC OF GERMANY

INVESTIGATOR: H. Ellenberg ADDRESS: Systematisch–Geobotanisches Institut der Universitaet, 34 Goettingen, Untere Karspule 2, Federal Republic of Germany
OTHER INVESTIGATORS: H. Heller, B. Ulrich, W. Funke and others
SELECTED CITATIONS: Integrated Experimental Ecology, (1971). *Ecol. Studies*, **2**, Springer Verlag. Vegetationsuntersuchungen im Solling, (1970). *Schriftenr. Veg. Kunde*, **5**. Runge, M. (1973). *Energie Umsaetze in den Biozoenosen Terrestr. Oekosysteme, Scripta Geobotanica* **4**.
YEAR(S): 1969 FOREST TYPE: Luzulo–Fagetum
SOIL TYPE: Acid braunerde GEOLOGY: Buntsandstein SOIL DRAINAGE: Well drained
PRINCIPAL PLANT TYPES: *Fagus sylvatica*

AS	80	LAI	6.72	BA	25.19	BAI	0.76
SH	20.3	SD	1190	SCA	15908.4	PA	1007.8
LAT	N 51° 45′	LNG	E 9° 34′	ALT	470	MAT	6.1
MAP	1063	LGS	144	TGS	12.9	PGS	530.2
RGS	60000	RBG	30000	SPH	3.5	DRZ	60

	Amount	Increment	Flux	
OL	329.40		CSF	17.80
OFF	32.70		LFL	327.00
OBR	2587.00	44.70	LFFF	27.60
OBRB	517.40	8.90		
OBRW	2069.00	35.80		
OBO	12960.00	590.70		
OBOB	842.40	38.40		
OBOW	12120.00	552.30		
RLL	2208.00	63.30		
STO	3900.00			
IRZ	21200.00			

Dynamic properties of forest ecosystems

SOLLING PROJECT, SITE B 4, FEDERAL REPUBLIC OF GERMANY

INVESTIGATOR: H. Ellenberg ADDRESS: Systematisch-Geobotanisches Institut der Universitaet, 34 Goettingen, Untere Karspule 2, Federal Republic of Germany
OTHER INVESTIGATORS: H. Heller, B. Ulrich, W. Funke and others
SELECTED CITATIONS: Integrated Experimental Ecology, (1971). *Ecol. Studies*, **2**, Springer Verlag. Vegetationsuntersuchungen im Solling, (1970). *Schriftenr. Veg. Kunde*, **5**. Runge, M. (1973). *Energie Umsaetze in den Biozoenosen Terrestr. Oekosysteme, Scripta Geobotanica* **4**.
YEAR(S): 1969 FOREST TYPE: Luzulo-Fagetum
SOIL TYPE: Acid braunerde GEOLOGY: Buntsandstein SOIL DRAINAGE: Well drained
PRINCIPAL PLANT TYPES: *Fagus sylvatica*

AS	59	LAI	6.51	BA	30.44	BAI	1.36
SH	15.1	SD	3620	SCA	15493.2	PA	1224.4
LAT	N 51° 45'	LNG	E 9° 36'	ALT	430	MAT	6.3
MAP	1063	LGS	145	TGS	13	PGS	530.2
RGS	60000	RBG	30000	SPH	3.6	DRZ	60

	Amount	Increment	Flux	
OL	315.80		CSF	17.10
OFF	27.30		LFL	316.40
OBR	4150.00	98.90	LFFF	25.40
OBRB	829.90	19.80		
OBRW	3320.00	79.10		
OBO	11000.00	766.60		
OBOB	715.00	49.80		
OBOW	10290.00	716.80		
RLL	2395.00	126.00		
STO	2900.00			
IRZ	18600.00			

SOLLING PROJECT, SITE F 1, FEDERAL REPUBLIC OF GERMANY

INVESTIGATOR: H. Ellenberg   ADDRESS: Systematisch–Geobotanisches Institut der Universitaet, 34 Goettingen, Untere Karspule 2, Federal Republic of Germany
OTHER INVESTIGATORS: H. Heller, B. Ulrich, W. Funke and others
SELECTED CITATIONS: Integrated Experimental Ecology, (1971). *Ecol. Studies*, 2, Springer Verlag. Vegetationsuntersuchungen im Solling, (1970). *Schriftenr. Veg. Kunde* **5**. Runge, M. (1973). *Energie Umsaetze in den Biozoenosen Terestr. Oekosysteme, Scripta Geobotanica* **4**.
YEAR(S): 1969   FOREST TYPE: Spruce Forest (Plantation)
SOIL TYPE: Acid braunerde   GEOLOGY: Buntsandstein
PRINCIPAL PLANT TYPES: *Picea abies*

AS	87	BA	44.79	BAI	0.88	SH	24.9
SD	595	SCA	2446.5	PA	935.8	LAT	N 51° 49'
LNG	E 9° 35'	ALT	505	MAT	5.9	MAP	1063
LGS	132	TGS	12.4	PGS	530.7	RGS	60000
SPH	3.6	DRZ	60				

	Amount	*Increment*	*Flux*	
OL	1788.00		LFL	339.30
OBR	2821.00	60.30		
OBRB	423.20	9.00		
OBRW	2398.00	51.30		
OBO	19840.00	536.20		
OBOB	1587.00	42.90		
OBOW	18250.00	493.30		
OSTD	26.00			
OSUM	24500.00	1484.00		
RLL	7172.00			
STO	4900.00			
IRZ	19000.00			

669

Dynamic properties of forest ecosystems

SOLLING PROJECT, SITE F 2, FEDERAL REPUBLIC OF GERMANY

INVESTIGATOR: H. Ellenberg   ADDRESS: Systematisch-Geobotanisches Institut der Universitaet, 34 Goettingen, Untere Karspule 2, Federal Republic of Germany
OTHER INVESTIGATORS: H. Heller, B. Ulrich, W. Funke and others.
SELECTED CITATIONS: Integrated Experimental Ecology, (1971). *Ecol. Studies* **2**, Springer Verlag. Vegetationsuntersuchungen im Solling, (1970). *Schriftenr. Veg. Kunde* **5**. Runge, M. (1973). *Energie Umsaetze in den Biozoenosen Terrestr. Oekosysteme, Scripta Geobotanica* **4**.
YEAR(S): 1969   FOREST TYPE: Spruce Forest (Plantation)
SOIL TYPE: Acid braunerde   GEOLOGY: Buntsandstein   SOIL DRAINAGE: Well drained
PRINCIPAL PLANT TYPES: *Picea abies*

AS	115	BA	37.67	BAI	0.56	SH	31.3	
SD	300	SCA	23303.4	PA	747.0	LAT	N 51° 44′	
LNG	E 9° 34′	ALT	440	MAT	5.9	MAP	1063	
LGS	132	TGS	12.4	PGS	530.7	RGS	60000	
SPH	3.3	DRZ	50					

	Amount	Increment	Flux	
OL	1266.00		LFL	307.60
OBR	2463.00	39.00		
OBRB	369.40	5.90		
OBRW	2094.00	33.20		
OBO	19570.00	400.40		
OBOB	1566.00	32.00		
OBOW	18010.00	368.40		
OSUM		652.00		
USUM	0.00			
RLL	7493.00	85.30		
STO	11100.00			
IRZ	25100.00			

enter

SOLLING PROJECT, SITE F 3, FEDERAL REPUBLIC OF GERMANY

INVESTIGATOR: H. Ellenberg   ADDRESS: Systematisch-Geobotanisches Institut der Universitaet, 34 Goettingen, Untere Karspule 2, Federal Republic of Germany
OTHER INVESTIGATORS: H. Heller, B. Ulrich, W. Funke and others
SELECTED CITATIONS: Integrated Experimental Ecology, (1971). *Ecol. Studies*, **2**, Springer Verlag. Vegetationsuntersuchungen im Solling, (1970). *Schriftenr. Veg. Kunde* **5**. Runge, M. (1973). *Energie Umsaetze in den Biozoenosen Terrestr. Oekosysteme, Scripta Geobotanica* **4**.
YEAR(S): 1969   FOREST TYPE: Spruce Forest (Plantation)
SOIL TYPE: Acid Braunerde   GEOLOGY: Buntsandstein   SOIL DRAINAGE: Well drained
PRINCIPAL PLANT TYPES: *Picea abies*

AS	34	BA	35.47	BAI	1.47	SH	17.5
SD	1490	SCA	14277.9	PA	849.2	LAT	N 51°45'
LNG	E 9° 35'	ALT	390	MAT	5.9	MAP	1063
LGS	132	TGS	12.4	PGS	530.7	RGS	60000
SPH	3.4	DRZ	60				

	Amount	Increment	Flux	
OL	1887.00		LFL	292.40
OBR	1873.00	62.70		
OBRB	281.00	9.40		
OBRW	1592.00	53.30		
OBO	10510.00	489.50		
OBOB	841.00	39.20		
OBOW	9671.00	450.30		
OSUM		850.00		
RLL	3456.00	159.30		
STO	5200.00			
IRZ	19000.00			

Dynamic properties of forest ecosystems

LUBUMBASHI, ZAIRE

INVESTIGATOR: F. Malaisse   ADDRESS: Service de Botanique et Climatologie, Universite
Nationale Du Zaire, B. P. 1825, Lubumbashi, Zaire
OTHER INVESTIGATORS: R. Freson, G. Goffinet, M. Malaisse-Mousset
SELECTED CITATIONS: Malaisse, F., Alexadre, J., Freson, R., Goffinet, G. & Malaisse-Mousset,
M. (1972). In *Tropical Ecology, With an Emphasis on Organic Production*, pp. 363–405.
Athens, Georgia, USA.
YEAR(S): 1968–72   FOREST TYPE: Miombo-Woodland
SOIL TYPE: Latosol   SOIL DRAINAGE: Medium
PRINCIPAL PLANT TYPES: *Marquesia macroura*

AS	120	LAI	3.47	BA	21.96	BAI	0.14
SH	14–18	SD	446	SCA	16955	SCB	2547
PA	1110	PB	150	LAT	S 11° 29′	LNG	27° 36′
ALT	1208	MAR	168265	MAT	20.3	MAP	1273
LGS	118	TGS	22.2	PGS	1230	RGS	93509
SPH	5.0	DRZ	80				

	Amount		Flux	
OL	261.00		LF	600.10
OFF	53.70		LFL	425.50
OBR	7820.00		LFFF	53.70
OBRB	1603.00		LFBR	65.20
OBRW	6217.00		LFBO	52.90
OBO	6350.00		LFFR	2.70
OBOB	1073.00		LFEP	0.10
OBOW	5277.00			
OSTD	2470.00			
OSUM	16950.00			
HERB	310.60			
PSUM	17270.00			
RL	2547.00			
RLL	2115.00			
RLS	202.00			

INDEX

NB References in bold type refer to large passages or chapters.
References in italic type refer to figures & tables.

674

Oregon, USA, *5*; succession modelling in coniferous forest, 52–65
organic matter: and biomass production, 443–7, 485–6. *487*; and nitrogen accumulation, *353, 354–5*
Ostrya virginiana, 75
Oulu, Finland, *3*, 26, 29, *176, 582*
Oved, Sweden, *4*, 23, *621*
Oxalis acetosella, 18, 23, 28, 29
Oxalis griffithii, 21
Oxalis montana, 32
Oxalis spp., 28
Oxydendron arboreum, 24
oxygen absorption, by soil, 300

Paeonia peregrina, 23
Palmae, *3*, 9, *11, 575*
Panama tropical forests, *182*
Paris, France, *187*
Parrotia persica, 5, 21
Parthenocissus quinquefolia, 24, 505
Pasoh, Malaysia, *4*, 8, 10, *11, 175, 606*
Paltigera apthosa, 29
Paltigera scobrosa, 29
percolation, simulation of, 288–93
phosphorus: atmospheric input to soil, *296–7*, 301; and cycling of, 346, *355–9, 361–2, 365–6*; and litter input to soil, 304; and phosphate equilibria in soils, 309–11; site studies, *376–407*
photosynthesis: and carbon metabolism, 509–22; and moisture availability, 511–13; photosynthate allocation, 518–22; and Stand Energy model, 541, *542*; and TEEM-Prosper model, *548–9*
photosynthetically active radiation (PAR), flux of, 127–31
Phragmites spp., 21
Phryma leptostchya, 24
physical environment, modelling of, 111–20; atmosphere and plant, 111–16; plant-water-soil relationships, 116–19
physiognomy and phytosociology of woodlands, 1–35
phytosociology, 1–35
Picea abies, *4–6*, 26, *27*; biomass, *427*; carbon dioxide flux, 160; mineral cycling in, *344*, 348–51, 363, *373*; radiation flux, 129; water storage by canopy, 215; *see also* spruce forets
Picea engelmannii, 17
Picea excelsa, *3*, *4*, 22, 29
Picea glauca, *3*, *29*; and environmental factors, *69–70*; radiation flux, *123*
Picea mariana, *3*, 5, 29; and environmental factors, *69–70*; mineral cycling, *344, 376–8*

Picea pungens, 129–30
Picea rubens, *3*, *14*, 29
Picea sitchensis, *17*, *123*, *229*, *234*
Picea spp., 26
pine forests: carbon metabolism, *504*, 507; and chemical input to soil, 302; radiation flux, *149*; water balance, *183*, *192*; *see also Pinus* spp.
Pinus banksiana, 142
Pinus contorta, 142
Pinus densiflora, *4*, 16, 18
Pinus echinata, 5, 24, *25*, *26*; mineral cycling in, *344*, *388*; photosynthate allocation, 519, *520*
Pinus flexilis, 17
Pinus lambertiana, 5, 19
Pinus longifolia, *183*
Pinus ponderosa, 235, 247
Pinus radiata: biomass dynamics, 477; carbon dioxide flux, 154; heat flux, 142; net radiation balance, 135; Weibull density function of, 43
Pinus resinosa: and global radiation transmission to soil, 127; heat flux, 142; water storage, 215, *228*
Pinus rigida, 5, 23, 506
Pinus strobus, *3*, 5, 20, 29; biomass dynamics, 477; mineral cycling in, *344, 389, 397*; water storage, 215
Pinus sylvestris, *4*, 18, *27*, 302; heat flux, 142, 147; and mineral cycling, 348; root organic matter turnover, 320; transpiration, *230*; water balance, *192*, 215
Pinus taeda, 5, 19, 507; carbon dioxide balance, 154; global radiation transmission to soil, 127; heat flux, 147, *148–51*; interception, 124; and mineral cycling, 348; photosynthetic processes, 511, 540–3; radiation flux, *122*, 131, *132*, 135, 138; and root organic matter turnover, 320–1; and water-soil relationships, 117–19
Piptadeniastrum africanum, *4*, 10
Piptadeniastrum spp., 10
plant, water movement in, model of, 116–19
Pleurozium schreberi, 29
Poa pratensis, 24
Polystichum acrostichoides, 505
Polystichum munitum, 247
Polytrichum commune, 28
Polytrichum perinum, 29
Polytrichum spp., 28
Populus deltoides, 129, 348
Populus pruinosa, 5, 21
Populus tremula, 147
potassium: atmospheric input to soil, *296*,

Index

297; litter and input to soil, 304; and mineral cycling, 346, 349–53, *355–9*, *361–2, 365–6*; site studies, *376–407*

precipitation, *249*, 266–71, *546*; and chemical input to soil, 294–302, 299–300, 315; and productivity, 416–17; site studies, *376–407*; *see also* rain, snow

primary production, and ecosystem diversity, 70–4; *see also* productivity

production *see* productivity

productivity, **411–46**; and annual average temperature, 415–16; and available light, 422–4; and effects of defoliation using models, 547–52; and litter fall, 417–22; and mineral cycling, 364–8; and precipitation, 416–17; and stand age, 413–15; *see also* biomass

profile methods: of carbon dioxide flux determination, 104; instrumentation for, 106–8

PROSPER hydrology model, 242, 245–55, 545–7; *see also* TEEM-Prosper model

Protium spp., 10

Prunus avium, 3, 21, *574*

Prunus grayana, 16, 21

Prunus serotina, 5, 24

Pseudomonas spp., 317

Pseudoscleropodium purum, 18

Pseudotsuga menziesii, 5, *17*, 19, *56*, 246; mineral cycling in, *344*, 348–9, *350*, 363, *380–1*, *382*; and root organic matter turnover, 322; water potential gradient, 233–4; water storage, 215

Pteridium aquilinum, 18, 32

Ptilium crista-castrensis, 29

Pulmonaria rubra, 23

Quercus alba, 5, 20, 23, 24, 505; photosynthetic efficiency, *515*

Quercus acuta, 16

Quercus castaneifolia, 5, 21

Quercus cerris, 3, 21, 22

Quercus coccifera, 8

Quercus coccinea, 5, 21, 23

Quercus gambelli, 247

Quercus ilex, 3, 8, 130; mineral cycling in, *344, 407*; water balance, *186*, 195

Quercus macrocarpa, and environmental influences, 75, *75*

Quercus mongolica, 4, 21

Quercus pedunculiflora, 4, 21, 23

Quercus petraea, 3, 4, 5, 21, 22, 23

Quercus prinus, 5, 21, 23, *25–6*; mineral cycling in, *344*, 396

Quercus pubescens, 4, 21, 23

Quercus robur, 3–5, 18, 21, *22*, 23, *574*

Quercus rubra, 21, 24, 505, 519, *520*, 559

Quercus salicina, 16

Quercus serrata, 18

Quercus spp., *14*, 20, 74; and mineral cycling in, *344, 394–5, 399–400, 404*; *see also* oak forests

Quercus velutina, 5, 21, 23, 25, 75–6, 505

radiant energy transfer in forests, 140–57, 168–76; air temperature gradients, 142–3; sensible heat, 141–2, 147–52; and Stand Energy model, 540, *541*; in turbulent air, 142; vapor tension gradients, *145*; *see also* radiation flux

radiation: and the Earth, 89–91; and forest characteristics, 432–6

radiation flux, 87–98, 569; from biosphere to space, *91*, 93–4; energy transfer in forests, 140–57; entrophy flow and entroy production, 97–8; instrumentation, 98, *100–1*; long wave, 131–5; net radiation balance, 94–6, 135–40; short wave, 120–31, *see also* global radiation; from space to biosphere, *91*, 92–3; and Stand Energy model, 541; *542*; *see also* heat balance, radiant energy

rain, 270–1; instrumentation, 109–10

reproduction models, of tree growth, 50–1

respiration: autotrophic, 523–5, *see also* photosynthesis; heterotrophic, 508–9, 525–30; of invertebrates, 529–30; of microflora, 526–8

Rhizobium spp., 300

Rhododendron macrophyllum, 247

Rhododendron maximum, 23

Rhus ambigua, 21

Ribes alpinum, 23

Rocklin, California, USA; *184*, 185, *186*

roots, dynamics of: biomass, 485–6; organic accumulation, 323–6; sloughing, 327–8; and turnover root organic matter, 320–3; uptake of chemical elements, 316–17

root zone water, 227–8, 240

Rosa acicularis, 29

Rosaceae, 9, *11, 575*

Rosa sempervivens, 8

Rosa spp., 21

Rouquet, France, *3*, 7–8, *585*, mineral cycling in, *344, 407*

Rubiaceae, *3, 11, 575*

Rubus idaeus, 28

Rubus saxatilis, 28

Rubus spidigerus, 22

Ruscus aculeatus, 8, 22

Ruscus spp., 21

Russia (USSR): boreal forests, 26, 28; temperate forests, 15

Index